Bird Population Stu

Oxford Ornithology Series
Edited by C. M. Perrins

Bird Population Studies

RELEVANCE TO CONSERVATION AND MANAGEMENT

Edited by

C. M. PERRINS
Director, Edward Grey Institute of Field Ornithology
University of Oxford

J.-D. LEBRETON
Directeur de Recherche, Centre d'Ecologie Fonctionnelle et Evolutive
CNRS, Montpellier

G. J. M. HIRONS
Ecologist, Royal Society for the Protection of Birds

Oxford   New York   Tokyo
OXFORD UNIVERSITY PRESS

Oxford University Press, Walton Street, Oxford OX2 6DP

Oxford New York Toronto
Delhi Bombay Calcutta Madras Karachi
Kuala Lumpur Singapore Hong Kong Tokyo
Nairobi Dar es Salaam Cape Town
Melbourne Auckland Madrid

and associated companies in
Berlin Ibadan

Oxford is a trade mark of Oxford University Press

Published in the United States by
Oxford University Press Inc., New York

© Oxford University Press, 1991

First published 1991
First published in paperback 1993
Reprinted 1994

A catalogue record for this book is available from the British Library

Library of Congress Cataloging in Publication Data
Bird population studies: relevance to conservation and management/
edited by C. M. Perrins, J-D. Lebreton, G. J. M. Hirons.
p.   cm.—(Oxford ornithology series: 1)
Includes bibliographical references and index.
1. Bird populations.   2. Birds, Protection of.   I. Perrins,
Christopher M.   II. Lebreton, J. D. (Jean-Dominique).   III. Hirons,
G. J. M.   IV. Series.
QL677.4.B57   1991   639.9'78—dc20   90–41974
ISBN 0 19 854082 5 (Pbk)

Printed in Great Britain by Bookcraft (Bath) Ltd
Midsomer Norton, Avon

Foreword

FRANÇOIS BOURLIÈRE
UNIVERSITÉ RENÉ DESCARTES, PARIS

The successful conservation of any threatened species obviously requires an adequate knowledge of both its population biology and its ecological requirements.

It was therefore appropriate that the first International Symposium organized by the 'Station Biologique de la Tour du Valat'—an institution whose major objective is to promote the effective protection and management of threatened mediterranean biotas—be devoted to bird population studies and their relevance to conservation and management. Birds, together with a relatively small number of mammal and fish species of economic importance, are indeed the vertebrates whose population biology has been studied most carefully, thanks to close cooperation between professional and amateur ornithologists. The results of these studies, some of them long-term, are therefore of prime importance to conservation biology as a whole.

The initial decision to hold a symposium on this subject was taken by the board of the Fondation Tour du Valat late in 1987. During the following months, the Scientific Director, Dr Patrick Duncan, made a preliminary selection of potential contributors with three well-known specialists, Drs C. M. Perrins, J.-D. Lebreton, and G. J. M. Hirons, who were appointed as convenors of the 1988 Symposium. Invitations to speakers were accepted gladly, and work started immediately to allow comprehensive summaries of their presentations to be circulated to participants well in advance of the meeting.

The symposium took place from December 12 to 16, 1988, at the Station Biologique in the Camargue, and was attended by 69 participants from 9 different countries. Thirty papers were read, each of them followed by lively discussions. About half the papers were comprehensive reviews, while the others were case studies chosen to illustrate particular problems. The presentations were exemplary in the way their authors managed to combine theoretical and empirical approaches, and emphasize the conservation implications of their findings. The lessons from the eight sessions were ably set out by Ian Newton at the end of the meeting. It is unfortunate that the schedule of the sessions was too busy for the participants to enjoy to the full the wilds of the Camargue and its more spectacular birds . . .

The space available at the Station did not allow a large audience to profit from our debates. It was proposed, and immediately agreed, to publish most of the papers, once reviewed, in book form, for the benefit of a wider circle of ornithologists and conservation biologists. An early offer from Oxford University Press was therefore gratefully accepted, and the three conveners, now turned editors, spent months refereeing and editing the contributions—a task that was ably completed in a remarkably short time.

Now the fruit of their labours is in the reader's hands. I am quite sure that you will derive as much profit and pleasure reading these pages as the actual participants of the Symposium did during their week at la Tour du Valat.

Preface

Birds impinge on our lives in a wide variety of ways; hunters shoot them for food and/or sport; aviculturists keep them; bird-watchers watch them; farmers, fishermen, and gamekeepers curse some species and endeavour to reduce their numbers; ecologists and conservationists worry about species becoming extinct. All these groups are concerned, in one way or another, about the size of bird populations.

In 1988 the ICBP world checklist collated by N. J. Collar and P. Andrews described 11 per cent of the world's 9,000 or so species of birds as threatened with extinction. The plight of these species is but one aspect of a more general concern about the steady loss of biological diversity on our planet resulting from man's activities. While there is every likelihood that the situation will continue to deteriorate, there is one bright spot in this otherwise gloomy outlook: at long last, environmental matters have become sufficiently important to the 'man-in-the-street' that politicians are being forced to consider them. One undoubted consequence is that the number of questions asked of ecologists will increase and, therefore, so must their capacity to answer them.

At present our knowledge of the factors that affect the size of bird populations is far from perfect and a clearer understanding is urgently needed. For, whether one wishes to augment or reduce a population, it is important that one's actions should have the desired effect. Often, however, action is taken with little thought of trying to monitor its effectiveness. Schemes are implemented with a stated aim of reducing the population of a pest species, during which many birds are killed at great effort and expense but with scant regard to determining the effectiveness of the action. Similarly, there are examples of expensive attempts to augment populations of endangered species by the release into the wild of captive bred stock, where little or no attention has been paid to overcoming the reasons for the species becoming endangered in the first place, or even to monitoring the effectiveness of the release programme.

At a superficial level, effective management or conservation action can sometimes be taken with little prior knowledge and a reasonable chance of success; for example, scaring birds from a crop can lead to a reduction in crop damage. Likewise, many species are endangered because their habitat is threatened. So, protecting the habitat is an obvious first step in conserving some species, even in the absence of detailed ecological knowledge.

Unfortunately, it is becoming increasingly difficult to save a habitat in its pristine state: seas are fished, forests have become reduced and fragmented etc., and as a result new complications may arise which are difficult to predict, such as an invasion of predators from surrounding areas. So, in order to conserve a species it is becoming increasingly important to have a more complete understanding of its ecology. In other words, effective management requires knowledge. This is equally true whether one is trying to increase a species or control it.

Unfortunately again, and in spite of the amount of research carried out on birds, the lack of knowledge about many aspects of their population dynamics and ecology frequently means that we cannot give specific advice on management or conservation. In part this is because each problem is individual and its solution requires general knowledge to be adapted for a more local or specific focus. Often however, our understanding of basic principles is insufficient to make safe generalizations. As an example, take the setting of bag limits for wildfowl in North America. This is made complicated because we know little about the interaction between losses due to hunting and those arising from natural mortality. Yet this question is fundamental to any attempt to manage wildfowl populations. Two papers by Nichols and Boyd show that this key question lies at the heart of any strict management regime. Key experiments might help a great deal, but although theoretically possible, would be extremely unpopular and perhaps politically unacceptable. Yet if we cannot obtain this information for some of the most-studied birds in the wild, how can we be expected to provide answers about other species?

These proceedings stem from a meeting held at the Station Biologique de la Tour du Valat in the Camargue, where a group of people involved in various studies of bird populations—being carried out for a wide range of purposes—were invited to review the current state of knowledge and discuss the relevance of their work to management and conservation. The participants were chosen intentionally to reflect a wide diversity of approaches to studies of bird populations, ranging from the classic, long-term population studies of single species being undertaken for essentially academic reasons through purely theoretical approaches to applied studies by ornithologists who are required to give practical advice about bird populations. It soon became clear that, in practice, there are probably no truly 'pure' (as opposed to applied) studies, since all those present had important practical contributions to make concerning the management of populations. Equally, we were all agreed about how little critical knowledge there was for many aspects of birds' lives and population processes. In part this stems from the sheer complexity of avian population dynamics, where intrinsic changes interact with extrinsic ones. It was clear also how little is known about dispersal from and between populations, the scale of which varies greatly between species. The problem of the scale of such effects is one example of a major issue in ecology, where the same phenomena occur at different levels in different species. Another point clearly brought out during the conference was how critical has been the integration of many of the different approaches towards increasing our understanding of how bird populations are regulated.

We ourselves certainly came away from the meeting considerably wiser, and trust that the information that we individually provided at the conference may also help to broaden the reader's understanding of the problems to be faced when having to try to control or protect a species.

We met under the guidance of the Station Biologique's Scientific Director, Dr Patrick Duncan, and all the staff who fulfilled the role of an organizing committee to the conference, and at the invitation of its President, Dr Luc Hoffman, to whom we extend our gratitude for the idea, the stimulating atmosphere, and the hospitality.

Organization of the book

The papers documented here reflect, we hope, the range of the types of population studies of birds which are currently being undertaken; obviously not all the studies could be included in this presentation. In trying to arrange the papers into subjects we have almost certainly failed to reach our objective, since so many papers touched on several different aspects of the subject. Nevertheless, we felt it might be helpful to make some attempt at categorization. Accordingly, the ordering of the papers is chosen such that we begin with the more general and theoretical contributions, followed by studies on different species starting with the shorter-lived birds and ending with the longer-lived ones. There then follows a series of papers concentrating on topics rather than on species, followed by a series of contributions with important bearing on the management of species and its effectiveness. We conclude with two papers on endangered species and a synthesis of the meeting.

Oxford   C.M.P.
Montpellier   J.-D. L.
Arles   G.J.M.H.
March, 1990

Contents

Further Issues

Species Management

Synthesis

Contributors

NICHOLAS J. AEBISCHER, The Game Conservancy, Fordingbridge, Hants. SP6 1EF, UK.

PETER ARCESE, The Ecology Group, Department of Zoology, University of British Columbia, 6270 University Boulevard, Vancouver, B.C., Canada V6T 2A9.

FRANZ BAIRLEIN, Zoologisches Institut, Universität Köln, 5000 Köln 41, Weyertal 119, Germany.

PETER H. BECKER, Institut für Vogelforschung 'Vogelwarte Helgoland', An der Vogelwarte 21, D-2940 Wilhelmshaven 15, Germany.

CARLOS BERNSTEIN, Laboratoire de Biométrie, Université Claude Bernard, Lyon 1, 69622 Villeurbanne Cedex, France.

JEFFREY M. BLACK, Wildfowl and Wetlands Trust, Slimbridge, Gloucester GL2 7BT, UK.

JACQUES BLONDEL, Centre Louis Emberger, CNRS, B.P. 5051, 34033 Montpellier Cedex, France.

HUGH BOYD, Canadian Wildlife Service, Ottawa, Ontario, Canada K1A OH3.

JEAN CLOBERT, Laboratoire d'Ecologie, Ecole Normale Supérieure, 46 rue d'Ulm, 75230 Paris Cedex 05, France.

EVAN G. COOCH, Department of Biology, University of Pennsylvania, Philadelphia, PA 19104, USA.

FRED COOKE, Department of Biology, Queen's University, Kingston, Ontario, Canada K7L 3N6.

JOHN C. COULSON, Department of Biological Sciences, University of Durham, Science Laboratories, South Road, Durham DH1 3LE, UK.

JOHN P. CROXALL, British Antarctic Survey, NERC, High Cross, Madingley Road, Cambridge CB3 0ET, UK.

ANDREW P. DOBSON, Biology Department, Princeton University, Princeton, NJ 08544, USA.

PETER R. EVANS, Department of Biological Sciences, University of Durham, Science Laboratories, South Road, Durham DH1 3LE, UK.

CHRISTOPHER J. FEARE, Agricultural Development and Advisory Service, Central Science Laboratory, Tangley Place, Worplesdon, Guildford, Surrey GU3 3LQ, UK.

JOHN W. FITZPATRICK, Archbold Biological Station, P.O. Box 2057, Lake Placid, FL 33852, USA.

RHYS E. GREEN, Royal Society for the Protection of Birds, The Lodge, Sandy, Beds. SG19 2DL, UK.

MICHAEL P. HARRIS, Institute of Terrestrial Ecology, Hill of Brathens, Banchory, Kincardineshire AB31 4BY, Scotland.

GRAHAM J. M. HIRONS, Station Biologique de la Tour du Valat, Le Sambuc, 13200 Arles, France.

WESLEY HOCHACHKA, The Ecology Group, Department of Zoology, University of British Columbia, 6270 University Boulevard, Vancouver, B.C., Canada V6T 2A9.

PETER J. HUDSON, The Game Conservancy, Upland Research Group, Crubenmore Lodge, Newtonmore, Inverness-shire, PH20 1BE, Scotland.

ALAN R. JOHNSON, Station Biologique de la Tour du Valat, Le Sambuc, 13200 Arles, France.

PIERRE JOUVENTIN, C.E.B.A.S.—CNRS, 79360 Beauvoir/Niort, France.

ALEX KACELNIK, Edward Grey Institute, Department of Zoology, University of Oxford, South Parks Road, Oxford OX1 3PS, UK.

JOHN R. KREBS, AFRC Unit of Ecology and Behaviour, Department of Zoology, University of Oxford, South Parks Road, Oxford OX1 3PS, UK.

RUSSELL LANDE, Department of Ecology & Evolution, University of Chicago, Whitman Labs, 915 East 57th St., Chicago, Illinois 60637, USA.

JEAN-DOMINIQUE LEBRETON, CEPE/CNRS, Route de Mende, BP 5051, 34033 Montpellier Cedex, France.

ROBERT M. MAY, Department of Zoology, University of Oxford, South Parks Road, Oxford OX1 3PS, UK.

ROBIN H. McCLEERY, Edward Grey Institute, University of Oxford, Department of Zoology, South Parks Road, Oxford OX1 3PS, UK.

IAN NEWTON, Institute of Terrestrial Ecology, Monks Wood Experimental Station, Abbots Ripton, Huntingdon, Cambs. PE17 2LS, UK.

JAMES D. NICHOLS, U.S. Fish and Wildlife Service, Patuxent Wildlife Research Center, Laurel, MD 20708, USA.

MYRFYN OWEN, Wildfowl and Wetlands Trust, Slimbridge, Gloucester GL2 7BT, UK.

CHRISTOPHER M. PERRINS, Edward Grey Institute, University of Oxford, Department of Zoology, South Parks Road, Oxford OX1 3PS, UK.

G. RICHARD POTTS, The Game Conservancy, Fordingbridge, Hants. SP6 1EF, UK.

MICHAEL R. W. RANDS, International Council for Bird Preservation, 32 Cambridge Road, Girton, Cambridge CB3 0PJ, UK.

PHILIP ROTHERY, British Antarctic Survey, NERC, High Cross, Madingley Road, Cambridge CB3 0ET, UK.

IAN ROWLEY, Division of Wildlife & Ecology, CSIRO, Locked Bag No. 4, P.O. Midland, Western Australia 6056, Australia.

ELEANOR RUSSELL, Division of Wildlife & Ecology, CSIRO, Locked Bag No. 4, P.O. Midland, Western Australia 6056, Australia.

JAMES N. M. SMITH, The Ecology Group, Department of Zoology, University of British Columbia, 6270 University Boulevard, Vancouver, B.C., Canada V6T 2A9.

SARAH WANLESS, Institute of Terrestrial Ecology, Hill of Brathens, Banchory, Kincardineshire AB31 4BY, Scotland.

HENRI WEIMERSKIRCH, C.E.B.A.S.—CNRS, 79360 Beauvoir/Niort, France.

GLEN E. WOOLFENDEN, Department of Biology, University of South Florida, Tampa, FL 33852, USA.

The Comparative Approach

1 Population limitation in birds of prey: a comparative approach

IAN NEWTON

Introduction

In this paper I shall discuss the natural limitation of breeding density in raptors, updating Newton (1979). I shall also discuss the impact of various human activities on raptors and the various management procedures needed to restore depleted populations. The approach is comparative and examples are drawn from diurnal raptors, but many of the same points would apply to owls.

Although raptor populations over much of the world have been reduced by human activities—by habitat destruction, deliberate persecution, and pesticide use—there have been enough studies of intact populations to suggest how densities are naturally limited. I shall begin with the natural limitation of densities, and later discuss the role of chemicals and other human-related factors.

Evidence that densities are limited

For some raptor species, the idea that breeding density is limited, rather than fluctuating at random, is based on four main findings: (1) the stability of the breeding population in both size and distribution, over periods of many years; (2) the existence of surplus adults, physiologically capable of breeding, but attempting to do so only when a territory is made available through the death or removal of a previous occupant; (3) the re-establishment of populations, after their removal by man, to the same level as previously; and (4) in areas where nest sites are not restricted, a regular spacing of breeding pairs. In the sections below, each of these points is examined in turn.

Stability of breeding population

Raptors show some of the most stable breeding densities known among birds, as can be documented from long-term studies. Those involving at least ten pairs over at least 10 years are listed in Table 1.1. They include at least eleven studies of seven species, in none of which did breeding

Table 1.1 Stability in breeding populations of raptors that exploit fairly stable food sources. From studies involving at least ten pairs over at least ten years of populations free from serious human influence

Species	No. pairs	Period	Region	Authority
Buzzard	28–30	11 years, 1941–51	Germany	Wendland 1951–52
Buteo buteo	33–37	11 years, 1961–71	England	Tubbs 1974
Golden Eagle	10–13	26 years, 1944–69	Scotland	Watson 1970
Aquila chrysaetos	c. 12, 13, 16 in three areas	10 years, undated	Scotland	Brown & Watson 1964
Black Eagle	52–59	13 years, 1964–76	Rhodesia	Gargett 1977
Aquila verreauxi				
Sparrowhawk	29–39	15 years, 1972–86	Scotland	Newton 1988
Accipiter nisus				
Kestrel	10–13	11 years, 1941–51	Germany	Wendland 1952
Falco tinnunculus				
Hobby	10–12	11 years, 1941–51	Germany	Wendland 1952
Falco subbuteo				
Peregrine	10–13	13 years, 1935–47	Massachusetts	Hagar 1969
Falco peregrinus	31–36	20 years, 1952–71	Alaska	White & Cade 1971
	c. 12, 18, 25 in three areas	16 years, 1945–60	Britain	Ratcliffe 1962

Note. Evidence involving fewer pairs or fewer years is available for other populations of the above species, and also for at least 11 other species (Newton 1979).

numbers fluctuate by more than 15 per cent of the mean over the period concerned. Compared to findings on other birds (Lack 1954, 1966), and to what is theoretically possible, this represents a remarkable degree of stability. Other evidence, involving fewer pairs over fewer years, is available for other populations of the same species, and for at least eleven other species (Newton 1979).

Evidence for long-term stability also comes from studies of the total raptor fauna of particular areas. Near Berlin, Wendland (1952–3) found constancy in the number of individual species and hence of the raptor population as a whole over an 11-year period. Craighead and Craighead (1956) obtained similar results in 3 study years covering a 7-year period in Michigan.

Stability of breeding densities, as shown in the above studies, would be expected only in stable environments. It would not be expected in habitats that were changing rapidly through vegetation succession or human action. Nor would it be expected in those populations recovering from pesticide impacts, which have attracted so much attention in recent years.

Surplus birds

In some populations, non-breeding, non-territorial adults can be seen near nest sites, occasionally fighting with breeders (Gargett 1975; Monneret 1988). However, the main evidence for the existence of surplus birds, which breed only when a place becomes available, is that lost mates are often replaced in the same season by other individuals, which then breed themselves. Replacement sometimes occurs within a few days, and is evidently widespread among raptors. Specific instances, reported incidentally in the literature, involve at least twenty-six species, from small falcons to large vultures (Newton 1979). At some nest sites, more than one replacement was recorded (where shooting was continued), and at others individuals of both sexes were replaced. Most instances referred to females, perhaps because they were more easily shot than males, or because they were more numerous among surplus birds. Replacements included some birds in adult plumage and others in immature plumage.

Documented instances were mainly from the early literature and, as evidence for the existence of surplus birds, were often deficient. Thus if replacement occurred early in the season, the replacement bird might otherwise have moved on and bred elsewhere; if replacement occurred late, it might have been a bird which had a failed on another territory; and in no instance was it stated whether neighbouring territories were checked for the possibility of movements. Also, a loss followed by a replacement was more likely to have been documented than a loss followed by no replacement, so it was not possible from these early records to tell how often replacement occurred.

In recent years, however, properly controlled removal experiments, of the kind done on other birds, have been conducted on at least three species of raptors, namely the European Kestrel *Falco tinnunculus*, the American Kestrel, *F. sparverius* and the Sparrowhawk *Accipiter nisus* (Village 1983; Bowman and Bird 1986; Newton unpublished). In all these species, replacement of removed individuals occurred, sometimes within days, indicating the presence of surplus birds of both sexes, capable of breeding when a vacancy appeared. In each case, birds from neighbouring territories were marked, so it could be shown that the replacement individuals had not simply moved from other sites nearby. Some results from the European Kestrel are summarized in Table 1.2.

Re-establishment of populations

In Britain, instances are known of populations being removed or depleted by human action, and then recolonizing or recovering to about the same level as previously, with pairs in the same nesting places. This occurred

Table 1.2 Results of experimental removal of breeding European Kestrels. From Village (1983)

Stage of removal	Bird(s) removed	Bird(s) replaced (age of replacement)
Pre-lay	Pair	No
Pre-lay	Female	Yes (first-year)
Clutch incomplete	Female	No
Clutch complete	Male	Yes (first-year)
Clutch complete	Female	Yes (adult)

with Peregrines *Falco peregrinus* shot during the last war on the south coast of England, and depleted by pesticide poisoning during the 1960s in many other regions (Ratcliffe 1980). Some data for Peregrines in five regions of Britain are given in Table 1.3. In each case, not only did the newcomers use the same cliffs as their predecessors, but also the same nest ledges. Similar instances have been recorded for Merlins *F. columbarius*, Sparrowhawks and other species (Rowan 1921–2; Newton 1979), implying constancy in the carrying capacity of the environment over the years, with the same limitations on raptor breeding numbers. Again this would be expected only in landscapes that remained reasonably stable over the years, and not in those altered by human action.

Table 1.3 Evidence for a ceiling in British Peregrine breeding densities. Details from Ratcliffe (1980)

Region	Number pairs during earlier period	Number pairs during later period
Snowdonia	1950–55 16–17	1978–79 14
Lake District	1930–60 32–36	1975–79 30–43
Southern Uplands	1950–60 24–26	1970–79 23–26
Southern Highland Fringe	1960–62 28	1979 29
Central Highlands	1960–62 47	1977 47

Regular spacing

In continuously suitable nesting habitat, a regular spacing of breeding pairs has been documented in many kinds of raptors, and is evidently wide-spread in solitary nesting species. Such spacing is consistent with the idea of density limitation, especially when the same pattern holds over many years. It would not, of course, be expected where nest sites were sparse and irregular in distribution, constraining the distribution of breeding pairs.

These four arguments, taken together, provide circumstantial evidence: (a) that breeding density is limited, (b) that the limitation results from competition for breeding space (or 'territories'); and (c) that stability of breeding density is helped by the existence of surplus birds, encouraged to breed only when a territory becomes vacant. This does not necessarily mean that surplus birds are available at all times in all populations; nor does it imply that all non-breeders are capable of breeding that year if a place is made available to them.

In many studies of fairly stable populations, some previous territories remained vacant, even in the peak years. The usual occupancy of British Peregrine territories pre-1939 was 85 per cent (Ratcliffe 1972), and in some Buzzard *Buteo buteo* territories it was 77–83 per cent (Tubbs 1974). But territories vary in quality—that is in the opportunities they offer for existence and successful breeding—and possibly some poor territories are suitable for occupation only in certain years, or only by certain birds (Newton 1988*a*). In some removal experiments on Sparrowhawks, birds taken from known good territories were quickly replaced, while those taken from known poor territories nearby were not (Newton, unpublished).

Density limitation in relation to food supplies

In the limitation of breeding numbers, within the habitats that are occupied two resources seem of major importance, namely food-supply and nesting places. I shall consider first the relationship between breeding density and food-supply in areas where nest sites are freely available.

Regional variations in breeding density

In some species regional variations in breeding density are associated with regional variations in prey supplies. In Sparrowhawks, breeding densities in the woods of 12 different regions varied in relation to the local densities of prey birds (Fig. 1.1). The hawks nested closer together, at higher density, in areas where their prey were most numerous. The woods in all these areas were of roughly similar structure, and the differences in prey densities were associated with variation in elevation and soil type. Similarly, the densities of Kestrels in different areas and in different years were closely correlated with the densities of Field Voles *Microtus agrestis*, which formed the main prey (Village 1982).

For other species, the information is less quantitative, but is still consistent with the idea of a link between density and food supply. Thus Buzzard breeding densities in Britain are generally highest in areas where Rabbits *Oryctolagus cuniculus* are most abundant (Table 1.4). All the areas listed were fully occupied by territorial Buzzards, yet pair densities were up to 16 times greater in the best area than in the worst. Likewise, in

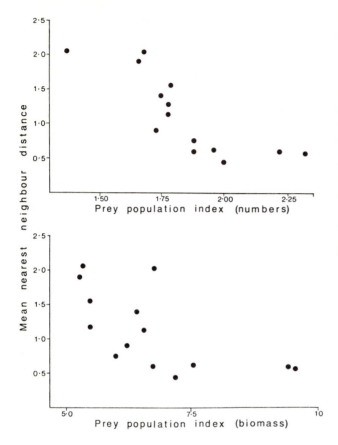

Fig. 1.1. Spacing of Sparrowhawk nesting places in the woodland of 14 districts of Britain, shown in relation to food-supply. Mean nearest neighbour distances decrease (so densities increase) with rise in prey numbers and biomass. Relationship between spacing and prey numbers: $r = 0.77$, $P \div 0.01$; between spacing and prey biomass, $r = 0.61$, $P \div 0.05$. From Newton *et al.* (1986).

British Peregrines, density is broadly related to land productivity and food supply, with highest breeding densities along coasts rich in sea birds, lower densities in hill areas of relatively high fertility or near to fertile valleys, and even lower densities in hill areas of low fertility or far from fertile valleys. Such variations in density would almost certainly correlate with prey abundance, though this was not measured (Ratcliffe 1972). Peregrine densities even higher than those in Britain were recorded on the Canadian Queen Charlotte Islands, again linked with massive concentrations of sea birds (Beebe 1960), and even lower ones in parts of inland Alaska (Cade 1960).

Unusually high densities of raptors are invariably associated with an unusual abundance of food. Examples of natural situations are the abundance of Peregrines around large sea-bird colonies, of Ospreys *Pandion haliaetus* around fisheries (Henny 1988), and of Black Eagles *Aquila verreauxi* around hyrax concentrations (Gargett 1975), but even greater concentrations of raptors occur in some African and Asian cities,

Table 1.4 Buzzard densities in relation to rabbit numbers in different parts of Britain

	Number of pairs	Pairs per 100 km²	Reference
Good rabbit areas			
South Wales	12	143	Moore (1957)
Devon	14–16	42–48	Dare (1961)
Central Wales	140	29	Newton *et al*. (1982)
Inverness	33	21	D. Weir, *in litt*.
Poor rabbit areas			
Hampshire	33–37	11–13	Tubbs (1974)
Yorkshire	5	11	Holdsworth (1972)
North Wales A	36–38	9	P. Dare, *in litt*.
North Wales B	61–70	8–9	P. Dare, *in litt*.

where human activity provides the food. The city of Delhi, in India, covers 150 km² and in 1967 held an estimated 2900 raptor pairs, a density of more than 19 pairs per km² (Galushin 1971). These were mainly scavenging species, such as Black Kites *Milvus migrans* (15 pairs per km²) and White-backed Vultures *Gyps bengalensis* (4 pairs per km²), but also included some other species. This high density was associated primarily with a huge amount of food within the city (mainly garbage and animal carcasses), but also with an abundance of nesting sites, and an unusual tolerance by the human population.

In all these various raptors, breeding density was broadly related to food supply, and unusual concentrations were associated with unusual food abundances. Only in very few species, however, was both breeding density and food-supply actually measured in several different areas. The implications from such natural variation are that raptors respond to the food situation, and that solitary species space themselves more widely, in larger territories, where food is sparse. An early argument concerning whether raptor breeding densities are limited by food-supply or by territorial behaviour, becomes redundant if the birds adjust their spacing to the resources available (Marquiss and Newton 1982; Village 1983). So one line of evidence that breeding density is limited in relation to food supply comes from the long-term stability of populations, but at different densities in different regions.

Annual variations in breeding density

For other species, the idea that breeding density is adjusted in relation to food supply comes, not from long-term stability in density, but from annual fluctuations in density which parallel fluctuations in food. Most species in the regions concerned have restricted diets based on cyclic

Table 1.5 Annual variation in breeding populations of raptors that exploit greatly fluctuating food sources

	Population size	Period of study	Locality	Reference
A. Species that eat rodents (approximately four-year cycles)				
Rough-legged Hawk *Buteo lagopus*	*0–9 pairs	9 years	North Norway	Hagen (1969)
	26–90 pairs	34 years	Colville River, Alaska	Mindell et al. (1987)
	10–82 pairs	5 years	Seward Peninsula, Alaska	Swartz et al. (1974)
Hen Harrier *Circus cyaneus*	10–24 females in 33 km²	22 years	Orkney, Scotland	Balfour, in Hamerstrom (1969)**
	*13–25 females in 160 km²	5 years	Wisconsin	Hamerstrom (1969)
	*0–9 pairs	6 years	North Norway	Hagen (1969)
Kestrel *Falco tinnunculus*	*35–109 clutches	4 years	Netherlands	Cavé (1968)
	*approx 20-fold fluctuation in index of no. of broods ringed, peaks every 4–5 years	42 years	Britain	Snow 1968
	*1–14 pairs	5 years	North Norway	Hagen (1969)
	1–16 nests	12 years	Germany	Rockenbauch (1968)
	*28–38 pairs	4 years	South Scotland	Village (10988)
	*9–28 pairs	6 years	Southern England	Village (1988)

	*12–22 pairs	6 years	Southern England	Korpimaki (1985)
	*3–37 pairs	8 years	South Finland	Malherbe (1963)
Black-shouldered kite *Elanus caeruleus*	*Increase of 1–8 nests during rodent plague	1 year	South Africa	Mendelsohn (1983)
	*19–35 individuals	3 years	Transvaal, South Africa	
B. Species that eat gallinaceous birds and rabbits (four- or ten-year cycles)				
Ferruginous Hawks *Buteo regalis*	*5–16 pairs	8 years	Utah	Woffinden and Murphy (1977)
	*1–8 pairs	3 years	Utah	
Goshawk *Accipiter gentilis*	0–4 nests in 100 km²	13 years	Sweden	Hoglund (1964)
	2–9 nests in 200 km²	7 years	Sweden	
	1–9 nests in 372 km²	4 years	Alaska	McGowan (1975)
Gyr Falcon *Falco rusticolus*	*13–49 pairs	5 years	Seward Peninsula, Alaska	Swartz et al. (1974)
	*19–31 occupied cliffs and 12–29 successful nests	4 years	Alaska	Platt (1977)
	8–19 pairs	27 years	Colville River, Alaska	Mindell et al. (1987)
	19–31 pairs	4 years	Brooks Range, Alaska	Platt (1977)

* Prey population also assessed and related to raptor numbers
** Excluding one year when population dropped from DDT poisoning

prey. Two main cycles are involved: (a) an approximate 4-year cycle of rodents on northern tundras and temperate grasslands; and (b) an approximate 10-year cycle of Snow-shoe Hares *Lepus americanus* in the boreal forests of North America. Some gamebirds are also cyclic, but whereas in northern Europe they follow the 4-year rodent cycle, in North America they follow the 10-year hare cycle (Lack 1954; Keith 1963). The main raptor species involved in these fluctuations are listed in Table 1.5, and all have been found to nest at greater density in years when their food is most plentiful. Moreover, some species, such as the Goshawk *Accipiter gentilis*, fluctuate in density where they feed on cyclic prey, but remain fairly stable in density where their food-supply is more stable (often through being more varied). Such regional variations within species provides further circumstantial evidence for a link between breeding density and food supply.

Prey abundance is not always a good measure of food-supply, when other factors influence its availability. In some small raptors, notably the European Kestrel, year-to-year changes in resident breeding populations have been linked with winter weather, which influences food availability (Village 1990; Kostrzewa and Kostrzewa 1990). Marked declines in Kestrel numbers, associated with heavy mortality, occur in years of prolonged snow cover, when rodents remain hidden for long periods. Larger raptors, which can withstand longer periods on reduced rations, seem much less affected by hard winters. Some raptor species, which are normally stable in numbers, have shown a marked change in breeding density following a marked change in food-supply. An example is provided by the Buzzard in Britain, in which breeding densities fell after the viral disease myxomatosis reduced the number of rabbits that formed the main prey (Moore 1957). In one area, numbers dropped from 21 pairs to 14 between one year and the next, as rabbit numbers were reduced (Dare 1961).

Density limitation in relation to nest sites

In some districts, raptor breeding densities are held by shortage of nest sites below the level that would normally be supported by the available food supply. The evidence is of two kinds: (a) breeding raptors are scarce or absent in areas in which nesting places are scarce or absent, but which otherwise appear suitable (nonbreeders may live in such areas); (b) the provision of artificial nest sites is sometimes followed by an increase in breeding density. Kestrel numbers increased from a few pairs to more than 100 pairs when nesting boxes were provided in a Dutch polder area with few natural sites (Cavé 1968). Similar results were obtained with other populations of Kestrels, and also with American Kestrels *Falco sparverius*,

Ospreys *Pandion haliaetus* and Prairie Falcons *F. mexicanus* (Hammerstrom *et al.* 1973; Reese 1970; Rhodes 1972; R. Fife, unpublished). The results from a controlled experiment on European Kestrels are given in Table 1.6.

Sometimes the provision of nesting sites has facilitated an extension of breeding range, as exemplified by the Mississippi Kite *Ictinia mississippiensis* and other species which nest in tree plantations on North American grasslands (Parker 1974). Likewise, nesting on buildings has allowed Peregrines and Lanners *F. biarmicus* to breed in areas otherwise closed to them through lack of cliffs. On the other hand, the destruction of nesting sites has sometimes led to reductions in breeding density, as in certain eagles when large free-standing trees were felled (Bijleveld 1974), and in Peregrines when cliffs providing nest sites were destroyed by mining (Porter and White 1973). Lastly, when nest sites are scarce, species may compete for them, and the presence of one dominant species may restrict the number and distribution of another (Newton 1979).

Table 1.6 Results of experimental provision of nest sites for European Kestrels in areas previously devoid of breeding Kestrels. From Village (1983)

	Number of nest sites provided	Number of pairs nesting in same year*
Experimental areas	17	8
Control areas	0	0

* All birds in the boxes provided, and include both first-year and adult individuals.

In conclusion, to judge from available evidence, the carrying capacity of any habitat for raptors is set by two main resources, food and nest sites, and which ever is most restricted is likely to limit breeding density. On this basis, much of the natural variation in breeding densities can probably be explained. However, much of the relevant research has been done on resident populations, and it is possible that some migrant populations may be limited on winter quarters, and so be unable to occupy their breeding habitat to the full. The Honey Buzzard *Pernis apivorus* in Europe provides a possible example.

Most of the evidence presented above on the role of food and nest sites is based on correlations, but experiments have confirmed the existence of surplus birds in some populations, and the provision of nest sites has confirmed their importance as limiting factors for some species in certain areas. So far, to my knowledge, no proper experiments have been made that involve manipulation of the food supply. This is an obvious gap in the study of raptor populations, the results of which could have applications in management; for however good a habitat in other respects, raptors cannot occupy it without the prey to support them.

There have, however, been some attempts to provide extra food for raptors. At one period during the management of the California Condor *Gymnogyps californianus* population deer carcasses were provided as food (Wilbur *et al*. 1974). This exercise helped to confirm that food was not limiting, for Condor numbers continued to fall despite the handouts (Snyder 1986). Subsequent work confirmed that other factors were involved. Similarly, the carcasses of domestic animals have been provided on a large scale for White-tailed Eagles *Haliaetus albicilla* in Sweden (Helander 1978). The main aim was to provide food relatively free of the contaminants known to be causing decline, but feeding also resulted in more birds overwintering near the feeding stations than previously.

Winter densities

Some raptors stay on their breeding sites all year, but others spread over a wider area after breeding, or migrate to a completely different area. Outside the breeding season, nest sites are unimportant, so in theory the birds have greater freedom to move around, and can exploit temporary food sources in a way not possible while breeding. Some Palearctic raptors, which winter in Africa, may spend the whole period between breeding seasons on the move, following rain belts, and exploiting temporary abundances of termites, locusts, and other prey (Newton 1979). At this season, therefore, food can take over as the all-important limiting factor.

The few studies that have been made of species that remain in an area all winter again imply that food supply has a major influence on density. For example, Craighead and Craighead (1956) compared the raptor population in an area of Michigan over 2 years, in one of which there was a high vole population and in the other a low one. In the good vole year, 96 raptors of seven species were present in the area, but in the poor year only 27 individuals of five species were counted. Other studies involve individual species over longer periods (Village 1988). Evidently, within the habitats that are occupied, food supply can be an important limiting factor at all seasons.

Human impact on habitat

Probably the main way in which human activities over the years have reduced raptor populations, in Europe and elsewhere, is through destruction of habitat and associated food supplies. Sometimes a habitat may be removed completely, as for example when a forest is felled or a marsh is drained and replaced by crop land. At other times a habitat may be degraded, so as to reduce its carrying capacity for raptors, through loss of prey or nest sites. Examples include the removal by human hunters of the small game animals otherwise available to raptors; the conversion of natural mixed forest to even-aged monocultures; or the intensive grazing by domestic animals of grassland, so as to reduce the numbers of rabbits and rodents available to the raptors as prey. In such cases the habitat may look superficially similar but lack the food.

For raptors limited by habitat, an increase in population could be achieved either by: (1) increasing the area of habitat available; or (2) increasing the carrying capacity of existing habitat patches, through increasing the food-supply or nest sites. Management by increasing the food-supply may be difficult where it entails changing the land use, because this may often conflict with other human interest. On the other hand, providing nest sites in areas short of suitable sites is an easy and less controversial way of increasing the breeding density of species that will accept artificially provided sites.

One of the problems of conserving raptors is the large areas of land that are required, compared to many other animals. This presumably results from their predatory behaviour, and their position near the tops of food chains. The amount of land needed to conserve raptors varies with the body size of the species concerned, and large raptors require very large areas. Small species, such as Kestrels, live at densities of 1–3 km² per breeding pair, while some large eagles extend over more than 100 km² per pair (Newton 1979). Perhaps the extreme is shown by the Bearded Vulture *Gypaetus monachus*. In a recent South African study, pairs were found during the course of a year to range over areas of around 4000 km², though these ranges overlapped greatly between neighbouring pairs (Brown 1988).

This relationship between body size and range among raptors provides further indirect evidence for a link between density and food-supply. In general, large raptors eat larger prey-species than do small raptors, and large prey animals generally occur at lower densities than do small ones. Some National Parks are large enough to sustain isolated populations of large raptors, but in the long-term, the survival of such species over much of the world will depend on land areas outside parks. In effect this means sea coast and grazing or forest land where wild or domestic animals are available as prey, and where no poisoning of predators is done. The presence of large raptors often conflicts with stock-rearing interests, so not surprisingly they have been eliminated from large parts of their former range.

Another factor that acts against the large raptors is their long-deferred maturity and low breeding rates. This means that, after a decline in population, it may take many years for numbers to recover.

Other limiting factors

Habitat and food-supply are not the only factors limiting raptors in the modern world. Some populations are held below the level that resources would support, either by human persecution or by toxic chemicals. Particularly in Europe, many raptors owe their present scarcity to persecution in the past (Bijleveld 1974; Newton 1979). Populations of many species were reduced either by shooting or by poisoning. A more recent problem is disturbance of nesting birds, which may prevent them from hunting effectively or from breeding successfully. In most European countries the problem of persecution has been much reduced because of a change in human attitudes and the enactment of protective legislation.

In much of Europe, as elsewhere, organochlorine pesticides have caused widespread declines in several raptor species, notably Peregrine and Sparrowhawk (Newton 1979; Ratcliffe 1980; Cade *et al*. 1988). Besides being toxic, these chemicals have three main properties that contribute to their effects on raptors and other wildlife. First, they are extremely stable, so that they can persist more or less unchanged in the environment for many years. Second, they dissolve readily in fat, which means that they accumulate in animal bodies, and on passing from prey to predator become concentrated at successive steps in the food chain. Predatory birds, near the tops of food chains, are especially liable to accumulate large amounts. Thirdly, at sub-lethal levels of only a few p.p.m. in tissues, some organochlorines can disrupt the breeding of certain species. By dispersal in the bodies of migratory birds and insects, or in wind and water currents, organochlorine pesticides may thus reach remote regions. None of the large numbers of bird species analysed at Monks Wood Experimental Station in the 1960s was found to be free of organochlorine residues, but the biggest concentrations were found in predatory birds, especially in fish-eating and bird-eating species (Prestt and Ratcliffe 1972). These findings were mirrored elsewhere in Europe and in North America, wherever studies were made.

Different organochlorines affect populations in different ways (Fig. 1.2). DDT, and its metabolite DDE, are of relatively low toxicity to birds and their main impact is on reproduction. DDE causes shell thinning (leading to egg breakage) and embryo deaths, thus lowering the breeding rate. If this reduction in breeding is sufficiently marked, too few young are produced to offset the usual adult mortality, and the population declines. The cyclodiene compounds, such as aldrin and dieldrin, are highly toxic to birds, in fact up to several hundred times more toxic than DDE (Hudson *et al*. 1984). Population declines were primarily caused by increasing the mortality rate above the natural level. The relative contributions made by reduced production (from DDE) and increased mortality [from HEOD

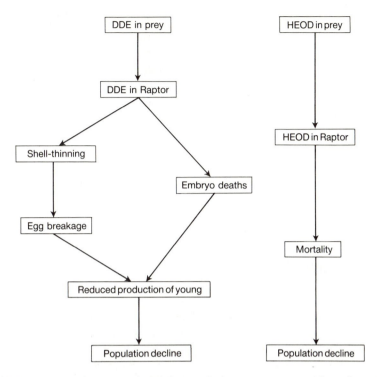

Fig. 1.2. Main mechanisms of population decline in raptors resulting from DDE (derived from the insecticide DDT) and HEOD (derived from the insecticides aldrin and dieldrin). DDE causes decline mainly by reducing the breeding rate, and HEOD mainly by increasing mortality. The relative importance of these two systems varied in different parts of the world, depending on usage and exposure patterns (from Newton 1986).

(aldrin/dieldrin)] to population declines in birds of prey have evidently varied from one region to another, depending on usage patterns. In much of Europe, declines in the Sparrowhawk and Peregrine occurred so rapidly in the late 1950s that they must have been due mainly to additional mortality (Newton 1979; Ratcliffe 1980). In parts of North America, however, regional extinctions of the Peregrine were probably due largely (or entirely) to a reduction in breeding rate caused by DDE (Cade *et al*. 1988). This is apparently the only case of a chemical known to cause population decline by means other than mortality.

As the use of DDT, aldrin, and dieldrin has been progressively curtailed, residues of DDE and HEOD in raptor eggs and tissues have declined, and populations have largely recovered (Newton 1986; Cade *et al*. 1988). This is the case over much of Europe and North America, but not in some other parts of the world where large-scale use of these chemicals continues.

Large carrion-feeding raptors have proved vulnerable to such chemicals as strychnine, 1080, or mevinphos, which are commonly used on meat baits in mammalian control programmes. The disappearance of vultures from large parts of their range has been linked with widespread attempts to get rid of wolves, foxes, or jackals (Bijleveld 1974). Carrion-eaters have also proved vulnerable to lead poisoning. This metal is ingested as bullets or shot, in the scavenged carcasses of deer or waterfowl, which are shot but not retrieved by hunters. Lead poisoning may have contributed to the decline of the California Condor (Snyder 1986). The only long-term solution to all these problems is to reduce the use of the offending chemicals, and ensure that they are not replaced by others equally detrimental to raptors.

Summary

1 Bird of prey populations are normally regulated, rather than fluctuating at random. In the absence of human intervention, the regulation in many species comes through competition for breeding space, and is helped by the presence of surplus adults, which breed only when an existing nesting territory becomes vacant.

2 In habitat where nest sites are freely available, breeding density is limited by food supply. This may be inferred from: (a) species that exploit fairly stable (often varied) food sources show fairly stable densities, which differ between regions where food abundance differs; and (b) species that exploit annually fluctuating (often restricted) food supplies show fluctuating densities. In other areas, however, breeding density may be restricted by shortage of nest sites to a lower level than would normally occur with the available food-supply. This may be inferred from: (a) the absence of breeding pairs in areas that lack nest sites but are otherwise apparently suitable; and (b) the provision of artificial nest sites is sometimes followed by a big increase in breeding density. Hence, in the habitat available in any one region, breeding density is naturally limited by food supply or nest sites, whichever is most restricting.

3 Many modern populations are below the level that would be permitted by the habitat because of pesticide use or other human action. Chemicals of importance include the various biocides used in vertebrate control programmes, organochlorine pesticides, and lead which is obtained from the scavenged carcasses of shot game not recovered by hunters.

References

Beebe, F. C. (1960). The marine Peregrines of the northwest Pacific coast. *Condor*, **62**, 145–89.

Bijleveld, M. (1974). *Birds of prey in Europe*. MacMillan Press, London.

Bowman, R. and Bird, D. M. (1986). Ecological correlates of male replacement in the American Kestrel. *Condor*, **88**, 440–5.

Brown, C. J. (1988). Study of the Bearded Vulture *Gypaetus barbatus* in southern Africa. Ph.D. Thesis, University of Natal, Pietermaritzburg.

Brown, L. H. and Watson, A. (1984). The Golden Eagle in relation to its food supply. *Ibis*, **106**, 78–100.

Cade, T. J. (1960). Ecology of the Peregrine and Gyrfalcon populations in Alaska. *University of California Publications in Zoology*, **63**, 151–290.

Cade, T. J., Enderson, J. H., Thelander, C. G., and White, C. M. (1988). Peregrine Falcon populations: their management and recovery. The Peregrine Fund Inc., Boise.

Cavé, A. J. (1968). The breeding of the Kestrel, *Falco tinnunculus* L., in the reclaimed area Oostelijk Flevoland. *Netherlands Journal of Zoology*, **18**, 313–407.

Craighead, J. J. and Craighead, F. C. (1956). *Hawks, owls and wildlife*. Stackpole Co., Pennsylvania.

Dare, P. (1961). Ecological observations on a breeding population of the Common Buzzard *Buteo buteo*. Ph.D. Thesis, Exeter University.

Galushin, V. M. (1971). A huge urban population of birds of prey in Delhi, India (preliminary note). *Ibis*, **113**, 522.

Gargett, V. (1975). The spacing of Black Eagles in the Matopos, Rhodesia. *Ostrich*, **46**, 1–44.

Gargett, V. (1977). A 13-year population study of the Black Eagles in the Matopos, Rhodesia, 1964–1976. *Ostrich*, **48**, 17–27.

Hagar, J. A. (1969). History of the Massachusetts Peregrine Falcon population, 1935–57. In *Peregrine Falcon populations: their biology and decline* (ed. J. J. Hickey), pp. 123–32. University of Wisconsin Press, Madison, Milwaukee and London.

Hagen, Y. (1969). Norwegian studies on the reproduction of birds of prey and owls in relation to micro-rodent population fluctuations. *Fauna*, **22**, 73–126.

Hamerstrom, F. (1969). A harrier population study. In *Peregrine Falcon populations: their biology and decline* (ed. J. J. Hickey), pp. 367–85. University of Wisconsin Press, Madison, Milwaukee and London.

Hamerstrom, F., Hamerstrom, F. H., and Hart, J. (1973). Nest boxes: an effective management tool for Kestrels. *Journal of Wildlife Management*, **37**, 400–3.

Helander, B. (1978). Feeding White-tailed Sea Eagles in Sweden. In *Birds of prey management techniques* (ed. T. A. Geer), pp. 47–56. British Falconers Club, Oxford.

Henny, C. J. (1988). Large Osprey colony discovered in Oregon in 1899. *Murrelet*, **69**, 33–36.

Holdsworth, M. W. (1971). Breeding biology of Buzzards at Sedbergh during 1937–67. *British Birds*, **64**, 412–20.

Hoglund, N. (1964). Der Habicht *Accipiter gentilis* Linné in Fennoscandia. *Viltrevy*, **2**, 195–270.

Hudson, R. H., Tucker, R. K., and Haegele, M. A. (1984). Handbook of toxicity of pesticides to wildlife, 2nd edn. Fish and Wildlife Service, Resource Publication **153**, Washington.

Keith, L. B. (1963). *Wildlife's ten-year cycle*. University of Wisconsin Press, Madison.

Korpimaki, E. (1985). Rapid tracking of microtine populations by their avian predators: possible evidence for stabilising predation. *Oikos*, 45, 281–4.

Kostrzewa, R. and Kostrzewa, A. (1990). Effects of winter weather on spring and summer density and subsequent breeding success of Kestrel *Falco tinnunculus*, Common Buzzard *Buteo buteo* and Goshawk *Accipiter gentilis*. *Ornis Scandinavica* (in press).

Lack, D. (1954). *The natural regulation of animal numbers*. Oxford University Press, Oxford.

Lack, D. (1966). *Population studies of birds*. Clarendon Press, Oxford.

Malherbe, A. P. (1963). Notes on birds of prey and some others at Boshock north of Rustenburg during a rodent plague. *Ostrich*, 34, 95–6.

Marquiss, M. and Newton, I. (1982). A radio-tracking study of the ranging behaviour and dispersion of European Sparrowhawks *Accipiter nisus*. *Journal of Animal Ecology*, 51, 111–33.

McGowan, J. D. (1975). Distribution, density and productivity of Goshawks in interior Alaska. Report, Alaska Department of Fish and Game.

Mendelsohn, J. J. (1983). Social behaviour and dispersion of the Black-shouldered Kite. *Ostrich*, 54, 1–18.

Mindell, D. P., Albuquerque, J. L. B., and White, C. M. (1987). Breeding population fluctuations in some raptors. *Oecologia*, 72, 382–8.

Monneret, R-J. (1988). Changes in the Peregrine Falcon populations of France. In *Peregrine Falcon populations: their management and recovery* (eds. T. J. Cade, J. H. Enderson, C. G. Thelander, and C. M. White), pp. 201–13. The Peregrine Fund Inc., Boise.

Moore, N. W. (1957). The past and present status of the Buzzard in the British Isles. *British Birds*, 50, 173–97.

Newton, I. (1979). *Population ecology of raptors*. Poyser, Berkhamsted.

Newton, I. (1986). *The Sparrowhawk*. Poyser, Calton.

Newton, I. (1988). A key factor analysis of a Sparrowhawk population. *Oecologia*, 76, 588–96.

Newton, I., (1988a). Individual performance in Sparrowhawks: the ecology of two sexes. *Proceedings 19th International Ornithological Congress, Ottawa*, pp. 125–54.

Newton, I., Davis, P. E., and Davis, J. E. (1982). Ravens and buzzards in relation to sheep-farming and forestry in Wales. *Journal of Applied Ecology*, 19, 681–706.

Newton, I., Wyllie, I., and Mearns, R. M. (1986). Spacing of Sparrowhawks in relation to food supply. *Journal of Animal Ecology*, 55, 361–70.

Parker, J. W. (1974). Populations of the Mississippi Kite in the Great Plains. Raptor Research Foundation. *Raptor Research Report*, 3, 159–72.

Platt, J. (1977). The breeding behaviour of wild and captive Gyrfalcons in relation to their environment and human disturbance. Ph.D. Thesis, Cornell University, Ithaca, New York.

Porter, R. D. and White, C. M. (1973). The Peregrine Falcon in Utah, emphasising ecology and competition with the Prairie Falcon. *Brigham Young University, Science Bulletin*, 18, 1–74.

Prestt, I. and Ratcliffe, D. A. (1972). Effects of organochlorine insecticides on European birdlife. Proceedings 15th International Ornithology Congress, pp. 407–27.

Ratcliffe, D. A. (1962). Breeding density in the Peregrine *Falco peregrinus* and Raven *Corvus corax*. *Ibis*, 104, 13–39.

Ratcliffe, D. A. (1972). The Peregrine population of Great Britain in 1971. *Bird Study*, **19**, 117–56.

Ratcliffe, D. A. (1980). *The Peregrine Falcon*. Poyser, Calton.

Reese, J. G. (1970). Reproduction in a Chesapeake Bay Osprey population. *Auk*, **87**, 747–59.

Rhodes, L. I. (1972). Success of Osprey nest structures at Martin National Wildlife Refuge. *Journal of Wildlife Management*, **36**, 1296–9.

Rockenbauch, D. (1968). Zur Brutbiologie des Turmfalken (*Falco tinnunculus* L.) *Anzeiger der Ornithologischen Gesellschaft in Bayern*, **8**, 267–76.

Rowan, W. (1921–2). Observations on the breeding habits of the Merlin. *British Birds*, **15**, 122–9, 194–202, 222–31, 246–53.

Snow, D. W. (1968). Movements and mortality of British Kestrels *Falco tinnunculus*. *Bird Study*, **15**, 65–83.

Snyder, N. F. R. (1986). California Condor Recovery Program. *Raptor Research Reports*, **5**, 56–71.

Swartz, L. G., Walker, W., Roseneau, D. G., and Springer A. M. (1974). Populations of Gyrfalcons on the Seward peninsula, Alaska, 1968–1972. *Raptor Research Report*, **3**, 71–5.

Titus, K. and Mosher, J. A. (1981). Nest-site habitat selected by woodland hawks in the Central Appalachians. *Auk*, **98**, 270–81.

Tubbs, C. R. (1974). *The Buzzard*. David and Charles, London.

Village, A. (1982). The home range and density of Kestrels in relation to vole numbers. *Bird Study*, **28**, 215–24.

Village, A. (1983). The role of nest-site availability and territorial behaviour in limiting the breeding density of Kestrels. *Journal of Animal Ecology*, **52**, 635–45.

Village, A. (1988). Factors limiting European Kestrel numbers in different habitats. In *Raptors in the Modern World* (ed. R. Chancellor) pp. 193–302. World Working Group on Birds of Prey and Owls, London.

Village, A. (1990). *The Kestrel*. Poyser, London.

Watson, A. (1970). Work on Golden Eagles. In *Research on vertebrate predators in Scotland*, pp. 14–18. Nature Conservancy Progress Report, Edinburgh.

Wendland, V. (1952–3). Populationstudien on Raubvogeln 1 and 2. *Journal of Ornithology*, **93**, 144–53; **94**, 103–13.

White, C. M. and Cade, T. J. (1971). Cliff-nesting raptors and ravens along the Colville River in arctic Alaska. *Living Bird*, **10**, 107–50.

Wilbur, S. R., Carrier, W. D. and Borneman, J. C. (1974). Supplemental feeding program for California Condors. *Journal of Wildlife Management*, **38**, 343–46.

Woffinden, N. D. and Murphy, J. R. (1977). Population dynamics of the Ferruginous Hawk during a prey decline. *Great Basin Naturalist*, **37**, 411–25.

2 Demography of Passerines in the temperate southern hemisphere

IAN ROWLEY and ELEANOR RUSSELL

Introduction

Most ecological theory has evolved in the northern hemisphere, largely because that is where most ecologists live, learn, and look. Although some of these theories have tended to achieve the status of universally applicable dogmas, they appear to be inadequate in southern hemisphere situations which differ in many important aspects. This paper attempts to review the ecological patterns found in temperate parts of the southern hemisphere, and to contrast these with patterns found elsewhere.

Demography is concerned with births and deaths—two events largely regulated by the timing of a season of abundance and the hazards of existence. For most species, to reproduce successfully the parents need to gather sufficient food, quickly, in order to fuel themselves, to grow their progeny to independence, and to launch them into an environment with plenty of easily attainable food. At the other end, the hazards of life range from predation to starvation and disease and physical exhaustion. All these parameters vary with geography and climate, and the pattern of variation of these features between the northern and southern hemispheres is likely to be significant.

This paper is concerned with the demography of passerines and since no passerines occur on the continent of Antarctica, that landmass can be ignored here. Apart from Antarctica, the most southerly point reached by continental land masses are Tierra del Fuego at 55 °S, South Africa 35 °S, Tasmania 42 °S, and New Zealand 47 °S. These compare with Scandinavia, Siberia, Greenland, Canada, and Alaska, all of which extend north of the Arctic Circle (66.5 °N) and in consequence 'enjoy' a long, very cold winter, the evasion of which has stimulated the evolution of extensive migrations. The rewards for birds are long daylight hours, throughout which the parents can forage and raise a lot of nestlings in a brief but bountiful breeding season. There are disadvantages of such behavioural adaptation and northern passerines tend to be relatively short-lived (Dobson 1987; Lack 1954).

The extreme contrast may be found in the tropics where seasonality is minimal, clutches tend to be small, and many species continue to breed all the year. Although many of these attempts fail due to heavy predation, replacement or repeat breeding make up for both the failures and the small brood size so that populations tend to be stable and long-lived (Bell 1982; Fogden 1972; Snow and Lill 1974). 'Because ornithology was born in the north temperate zone where broods tend to be large, we ask why the broods of tropical birds are so small. If more ornithologists had grown up in the tropics, we would be asking why birds at high latitudes lay so many eggs—a question easier to answer.' (Skutch 1985).

The temperate regions of the southern hemisphere appear to be inter-mediate between these two extremes. Demographic data for small passerines are few from South America, South Africa, or New Zealand, and so this paper concentrates on temperate Australia. It is now widely accepted that the southern continents derived from a super-continent, Gondwanaland, which was fragmented by continental drift during the last 200 million years (MY). Until quite recently it was assumed that Australia was colonized by birds from the Asian mainland to the north in successive waves via island land bridges. Recent studies of DNA (Sibley and Ahlquist 1985) have suggested a Gondwanan origin for the passerines. When Australia became separated from Antarctica 40–50 MY ago, it contained the ancestors of a major passerine radiation which occurred during the 20 MY or so of the Tertiary when Australia was drifting north from its connection with Antarctica and before it came within reach of Asia about 20 MY ago. Since then members of Australian lineages have dispersed outwards and some passerines of Afro-Eurasian origin have dispersed in. The result of this complex history may be summarized roughly as follows: of the 322 endemic, naturally occurring species of passerine found in Australia, 258 almost certainly belong to lineages that originated and radiated in Australia and New Guinea. The remaining 64 species represent later immigrants and their descendants, to which must be added the 16 exotic species 'introduced' during the past 200 years (Schodde 1975). Overall, the avifauna shows more than 30 per cent endemism at the family level and some 75 per cent at the species level. Since Australia is not the recipient of significant large-scale passerine migration from the north, the life histories of the majority of Australian passerines have evolved in Australia, in response to past and present environmental conditions.

Palaeoclimatic data for Australia (Frakes, McGowran, and Bowler 1987; Kemp 1981) indicates that the seasonal climatic extremes have never been such as to initiate major migrations out of Australia, although considerable seasonal movement over shorter distances does occur (Keast 1968; Rowley 1975). The remoteness of Australia from Asia for so long suggests that this is one reason why it does not receive a major influx of passerine immigrants from northern Asia—it wasn't there at the appro-priate time.

The first person to draw attention to the different demography of many temperate Australian passerines was Fry (1980) who presented a paper at the 4th Pan African Congress in 1976 reviewing evidence for high rates of annual survival in tropical land birds. However, the Australian data that Fry included came from the south-eastern temperate parts of the continent where most banding has been done. That review was largely concerned with longevity and survival; little data on productivity was available to him in 1976. Further studies have been published over the past decade, which not only confirmed Fry's thesis but provided sufficient information on productivity to prompt several important reviews (Ford *et al*. 1988; Woinarski 1985*a*, 1989; Yom-Tov 1987). The present review is in three parts: productivity, longevity, and survival.

Productivity

The annual productivity of a breeding unit is the sum of young reared from successive breeding episodes, the number of which depends on the length of the breeding season, the duration of an episode, and the ability of a species to nest several times in succession.

Breeding season

The number of breeding episodes, if all were successful, is limited by the length of the breeding season, which in turn is restricted by the number of months in which parents can find sufficient food to build eggs and to raise young. Yom-Tov (1987) calculated the average breeding season for Australian passerines as 5.5 months, and Woinarski (1985*a*, 1989) showed that Australian leaf-gleaners had much longer breeding seasons than their north temperate equivalents (Fig. 2.1).

Most of the new data that has become available since 1976 comes from the evergreen eucalypt forests of temperate southern Australia. Woinarski and Cullen (1984) in Victoria, Bell (1985), Pyke (1983), and Recher *et al*. (1983) in New South Wales have shown that foliage insects remain available throughout the year without the severe peaks and troughs found in the northern hemisphere. Similarly, Calver and Wooller (1981) and Majer and Koch (1982) found that litter arthropods in eucalypt forests in south-western Australia, although subject to fluctuations with a spring peak and a decline in late summer, were always available.

Of course not all Australian passerines are insectivores, many are nectarivores and again the milder climate, compared with that of the northern hemisphere, permits nectar-providing species to flower throughout the year (Collins and Rebelo 1987). Thus some birds can maintain resident

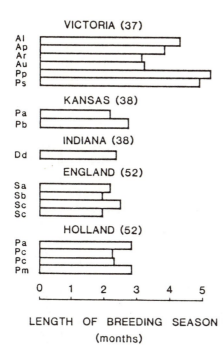

Fig. 2.1. Comparison of lengths of
breeding seasons between species of
Australian and north temperate leaf-
gleaning birds (after Woinarski
1985*a*).
Species and symbols:
Victoria: Al, *Acanthiza lineata*; Ap,
A. pusilla; Ar, *A. reguloides*; Au,
A. uropygialis; Pp, *Pardalotus
punctatus*; Ps, *P. striatus*
Kansas: Pa, *Parus atricapillus*; Pb,
P. bicolor
Indiana: Dd, *Dendroica discolor*
England: Sa, *Sylvia atricapilla*; Sb,
S. borin; Sc, *S. communis*; Sc,
S. curruca
Holland: Pa, *Parus ater*; Pc,
P. caeruleus; Pc, *P. cristatus*; Pm,
P. major
Number in parentheses following area
is the number of degrees latitude from
the equator.

status (helped along by a little part-time insectivory), while many others
have developed extensive nomadism, as indeed do our human beekeepers
(Tennant 1956) who travel long distances to harvest massive blossomings
(Keast 1968; Pyke 1983).

Breeding episode

The length of time occupied by a breeding episode is the sum of the number
of eggs laid, the interval between eggs, the duration of incubation, the
length of the nesting period, and the period of dependence on their parents
once the young have left the nest.

Clutch size

The number of young raised in each episode is limited by the number of
eggs laid, clutch size being usually a species characteristic that may vary
with latitude (Lack 1954). Yom-Tov (1987) compared Australian passerines with those from North Africa, an area of similar size and latitude but
drier and with fewer passerine species (299:119). Figure 2.2 shows the
smaller clutch size of Australian species (mean 2.7:4.3), particularly from
the old endemic families ('Maluridae' = Maluridae and Acanthizidae;

PASSERINES

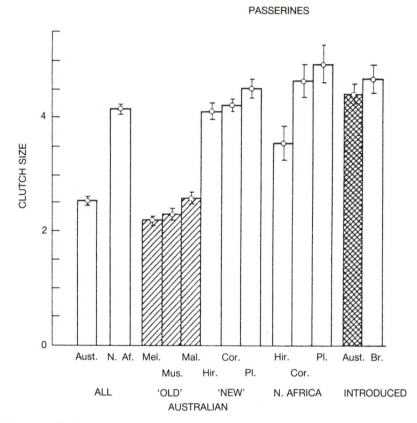

Fig. 2.2. Clutch size in various groups of passerines. Australia (299 spp.) compared with North Africa (119 spp.). Australian 'old endemics' [Meliphagidae, 67 spp.; Muscicapidae (= Pachycephalidae), 55 spp.; Maluridae (= Maluridae + Acanthizidae), 58 spp.] compared with 'new invaders' (Hirundinidae, 4 spp.; Corvidae, 5 spp.; Ploceidae, 21 spp.) and with the same three families (6, 8, and 6 spp. respectively) in North Africa. Eight European species in Britain and introduced into Australia (after Yom-Tov 1987).

Meliphagidae and 'Muscicapidae' = Pachycephalidae). This is further illustrated by the data of Table 2.1.

The newer 'invaders' tend to have clutches of comparable size to those in North Africa and the recent introductions to Australia have not yet changed their clutch size since 'transportation'. However, in New Zealand, Niethammer (1970) demonstrated that six out of ten introduced passerines had significantly lower clutch size than the same species in England, which he attributed to the increased population density in New Zealand and not to differences in latitude. The broad analyses of Thomas

(1974) and Yom-Tov (1987) suggest that in contrast to the northern hemisphere, there is little change in clutch size with increasing latitude in Australia. Analysis of clutch sizes from Nest Record Scheme data for two species with wide distribution, *Rhipidura leucophrys* and *Hirundo neoxena*, found no variation of clutch size with latitude (Marchant 1974; Marchant and Fullagar 1983).

Clutch sizes in other southern hemisphere temperate regions are similarly low. Mean clutch size for passerines in southern Africa was 2.8 ± 0.7 calculated for 353 species from data in Maclean (1985). New Zealand has few native passerines, and mean clutch size calculated from data in Falla *et al.* (1978) is 3.2 ± 0.7.

Laying interval

While a small clutch size certainly reduces the demands on the female, these demands are further lessened in several endemic Australian passerines which lay their eggs at 2-day intervals (Marchant 1986). Thomas (1974) suggested that this was due to food being inadequate for the female to form eggs more quickly, but there is no evidence to support this.

Incubation and nestling period

Woinarski (Fig. 1, 1985*a*, 1989) plotted data for leaf-gleaners with roughly similar ecology, from the Australian Nest Record Scheme and for northern temperate regions, and showed that the Australian species (Acanthizidae and Pardalotidae) had significantly longer incubation and nestling periods. However the Maluridae and Pachycephalidae do not appear to follow this trend, which therefore cannot be regarded as a general southern hemisphere adaptation. The Acanthizidae and Pardalotidae include many hole-nesting species and species that use domed nests and that might be expected, for this reason, to have longer incubation and nestling periods, but the Malurids, which do not, also have domed nests.

Dependence

As elsewhere in the world, data on the length of time fledglings remain dependent on their parents are scarce in Australia, but such data that do exist suggest that in many species parental care is more prolonged; 'weaning' is not precipitated by the need to avoid competition or for the adults to complete a moult and to build up massive reserves before migrating, as happens in much of the northern hemisphere.

Multiple broods

Most single-brooded species are capable of laying a replacement clutch when their first is lost (Perrins 1985). Multibrooded species are those that

Table 2.1 Productivity of southern temperate passerines. Clutch size ± standard deviation. R = resident; N = nomad; M = migrant

Species	Body mass (g)	Status	Clutch size	Repeat breed	Success (fledglings/egg laid)	Fl /year	Source
Australia							
Eopsaltria georgiana	17	R	1.9 ± 0.3	+	.81	3.1	Brown and Brown (1980; unpublished)
E. australis	22	R	2.3 ± 0.5	+	.28	2.1	Marchant (1984, 1985)
Pomatostomus temporalis	40	R	2.4	+	.60	2.3	Moffatt (1982) Brown and Brown (1981)
Malurus splendens	10	R	2.9 ± 0.3	+	.47	2.8	Rowley and Russell (1989) Rowley (unpublished)
M. cyaneus	10	R	3.4 ± 0.5	+	.53	3.9	Rowley (1965)
M. elegans	9	R	2.4 ± 0.6	+	.52	2.5	Rowley et al. (1988)
Acanthiza reguloides	8	R	3.4	−	.20	1.1	Bell and Ford (1986) Woinarski (1985b)
A. pusilla	7	R	2.8	−	.19	c. 1.0	Woinarski (1985b); Bell and Ford (1986)

A. chrysorrhoa	7	R	3.2 ± 0.5	+	.53	3.7	Ford (1963)
Manorina melanophrys	32	R	1.9 ± 0.1	+	.33	3.7	Clarke (1988; personal communication)
Phylidonyris novaehollandiae	22	R/N	2.0 ± 0.1	+	.76	2.8	Rooke (1977) Paton (1985)
Zosterops lateralis	11	R	3.0 ± 0.2	+	.44	3.1	Kikkawa and Wilson (1983)
Corvus mellori	541	N	4.2 ± 0.8	–	.36	1.5	Rowley (1973)
C. coronoides	645	R	4.4 ± 0.9	–	.35	1.5	Rowley (1973)
New Zealand							
Petroica australis	22	R	2.7 ± 0.5	+	.27	2.5	Powlesland (1983)
Gerygone igata	6	R	3.9 ± 0.3	+	.38	4.0	Gill (1982)
Mohoua albicilla	18	R	2.8 ± 0.5	+	.35	1.1	McLean and Gill (1988)
South Africa							
Hirundo spilodera	21	M	2.4	+	.57	1.8	Earlé (1986)
Pogonocichla stellata	21	R	3.0 ± 0.2	–	.51	c.1.5	Oatley (1982)
Motacilla clara	20	R	2.0	+	–	1.7	S. Piper (unpublished)
Myrmecocichla formicivora	46	R	3.7 ± 0.7	+	–	3.2	Earlé and Herdolt (1988)

repeat nest after successfully raising an earlier brood to fledging or independence. Good data on this aspect of breeding biology are scarce; incidence and frequency are seldom separated and the basic demographic figure of production per female per year is usually obscure. Valid data can only result from intensive banding studies spanning the entire breeding season. Many northern temperate species are capable of producing a second brood, but the number of such attempts varies considerably with latitude and from year to year. In many northern species, only a small percentage of females initiate a second brood and these tend to be less successful than the first. Data are accumulating that many Australian passerines regularly renest after a successful breeding episode and often raise a second or third brood. Table 2.1 includes only species where data are sufficient to calculate a value for production per female per year; the data illustrate the wide spread of genera that are multi-brooded. Repeat clutches were no different in size from first clutches in *Malurus splendens*, *M. elegans*, and *Eopsaltria georgiana* and nor were they any less successful.

In predominantly single-brooded species with large clutches, clutch size appears to vary considerably with time of year, from year to year, with latitude, with age, and from individual to individual (Perrins 1985). In many multibrooded Australian species clutch size shows little variation (Table 2.1). For example, in *M. splendens*, of 277 clutches, 30 were of 2 eggs, 244 of 3 eggs and 3 of 4 eggs; in *M. elegans*, of 117 clutches, 4 were of 1 egg, 58 of 2 and 55 of 3 eggs; in *E. georgiana*, of 145 clutches, 133 were of 2 eggs and 12 of 1 egg. Variation in seasonal reproductive effort arises not through clutch size, but through variation in the number of nests, which in part depends on the length of the breeding season.

Annual productivity

The way in which productivity data are presented leaves much to be desired when it comes to trying to draw comparisons. Most of the information comes from single nesting attempts which form the bulk of the Nest Record Scheme data banks. While this provides valuable data on clutch size, incubation, and nestling periods, and even breeding seasons, it is inadequate for giving annual productivity. A commonly used figure calculated from this data is 'breeding success'—the number of fledglings produced per egg laid. The breeding success for 19 of the southern hemisphere temperate species listed in Table 2.1 averaged 44.8 per cent; for temperate northern hemisphere open-nesting altricial species, Lack 1954) calculated a breeding success of 45 per cent and Nice (1957) gives a similar value of 46 per cent.

While the figure for breeding success may be adequate for comparing the productivity of single-brooded species or those in which a repeat nesting is infrequent and unsuccessful, it does not present a true picture for regularly multibrooded species. For the latter, the figure of demographic significance is the number of fledglings produced per female each year (FFY) and this figure is remarkably difficult to extract from the literature. In Table 2.1, only 1 of the 21 Australian species has an FFY value greater than 4. Our attempt to find such data for the northern hemisphere unearthed 35 species, 19 of which had FFY greater than 4. These values do not approach a level of great precision, but they do point towards a relatively low productivity in Australian passerines compared with that of their northern counterparts, although breeding success was similar. Despite this, Australian populations appear to maintain their numbers satisfactorily.

Longevity

Longevity represents the opposite end of the demographic yardstick from productivity. The maximum recorded longevity of a species is the longest elapsed time from ringing to recovery. As a species characteristic, longevity is a rather inexact and unscientific measurement which, because it is easily available and has striking publicity value, is much quoted. While maximum longevity is partly dependent upon the number of individuals ringed and the length of time over which the species has been exposed as ringed birds, there obviously is an ecological or species component in the figure which, it has been claimed, shows differences between continents, habitats, and life-styles. For example, Fry (1980) compared the number of species with recorded maximum longevity of more than 7 years in North America (53 spp.), Europe (54 spp.), and Australia (53 spp.), pointing out that the North American and European banding schemes had each ringed more than 100 million birds over some 50 years whereas the Australian scheme had only ringed one million birds over less than 20 years. Fry concluded that this represented greater average longevity by Australian passerines.

Recent analysis of American data (Clapp *et al.* 1983; Klimkiewicz *et al.* 1983; Klimkiewicz and Futcher 1987) has shown that 130 species of North American passerines (of some 300 possibles) have now passed the '7-year-mark', from 20 million passerines ringed since about 1926. South African data shows that 61 species (of some 400 possibles) have longevities greater than 7 years, from only 465 099 passerines banded since 1950 (T. Oatley, personal communication).

The recent analysis of 20 years' data from a long-term ringing study in the Brindabella Ranges of the Australian Capital Territory (Tidemann, Wilson and Marples 1988) has shown that of 23 species for which more than 40 birds were banded, 15 species had maximum longevity greater than 7 years, and 8 species greater than 10 years. These data are significant indicators that a long life is possible for many Australian resident passerines (see also Table 2.3). However it is survival data that are really convincing and which are not dependent upon a single, possibly aberrant, long-lived individual.

Survival

Survival figures are difficult to compare because they are calculated in different ways, using very diverse data, which come from three different types of study:

(a) Studies of colour-ringed populations in which individuals were followed from ringing to disappearance (death or dispersal);
(b) intensive ringing studies at one location where most birds were ringed and survival data resulted from regular retrapping; and
(c) general ringing scheme data.

Three types of survival estimates are made:

1. Survival of known individuals based on repeated sightings or recaptures. Disappearance may be due to death or dispersal; once the age of dispersal is past, disappearance usually means death and is assumed to be such. Only long-term intensive studies of single species can give data on the survival of birds known to be breeding adults and of birds of known age (nestlings or juveniles with distinctive plumage, eye colours, or mouth markings) which can be followed throughout their lives to give age-specific survival. Data of this type are now available for a few Australian species (Table 2.2).

2. Data based on the recovery of marked individuals (death or recapture). There are many methods for estimating survival probabilities from this type of data, as discussed by Chapman and Robson (1960), Eberhardt (1972), and Seber (1973) and a variety of methods have been used to obtain the figures quoted in Table 2.3.

3. Fry (1980) used the 'S_{10}' value in order to compare survival between species in many different places from general banding data. The S_{10} value is calculated by considering all recoveries or recaptures of a given species; the proportion surviving for 1, 2, . . . x years since ringing are plotted and the age to which 10 per cent of the retraps survived is measured. This is the S_{10} value and we have calculated it for the Australian species (Table 2.3).

Table 2.2 presents data on survival for 17 Australian resident species, 14 of them from 'old endemic' families and 3 (*Zosterops* and 2 *Corvus*) that are relatively 'new'; the limited data for New Zealand and South Africa are also included. In only 2 of these 22 studies was annual survival less than 60 per cent and 8 had survival greater than 80 per cent. The European Starling

Table 2.2 Survival of adult southern temperate passerines. Data for breeding adults unless marked *, where the data is for all adults

Species	Body mass (g)	Percentage survival	Source
Australia			
Eopsaltria georgiana	17	78	R. and M. Brown (unpublished)
E. australis	22	75*	Marchant (1985)
Malurus cyaneus	10	66	Rowley (1965)
M. splendens	10	72	Rowley (unpublished)
M. coronatus	11	69	Rowley (1988)
M. elegans	9	81	Rowley *et al.* (1988)
Acanthiza pusilla	7	87	Bell and Ford (1986)
A. reguloides	7	58	Bell and Ford (1986)
A. inornata	7	62	M. G. Brooker (unpublished)
Climacteris erythrops	23	83	Noske (1983)
C. picumnus	33	77	Noske (1983)
C. leucophaea	22	86	Noske (1983)
Manorina melanophrys	32	63	Clarke (unpublished)
Phylidonyris novaehollandiae	22	55*	Paton (1985)
Zosterops lateralis	11	60	Kikkawa and Wilson (1983)
Corvus coronoides	645	77	Rowley (1971)
C. mellori	541	85	Rowley (1971)
New Zealand			
Petroica australis	22	80	Powlesland (1983)
Gerygone igata	6	82	Gill (1982)
Sturnus vulgaris	80	72	Flux and Flux (1981)
South Africa			
Pogonocichla stellata	20	80	Oatley (1982)
Motacilla clara	20	75	S. Piper (unpublished)

Sturnus vulgaris introduced to New Zealand more than 100 years ago shows minimum survival of breeding adults of 72 per cent (Flux and Flux 1981). However, all these studies were of colour-ringed birds and calculations based on resighting of identifiable individuals can give a higher survival rate (and a better estimate) than recovery or recapture data (Brown *et al*. 1990; Elder and Zimmerman 1983; Elder 1985; Clobert and Lebreton, this volume).

Table 2.3 is based on recapture data and does not include resightings. Even non-resident honeyeaters show survival rates greater than 55 per cent (*Lichenostomus*, *Lichmera*, *Melithreptus*, and *Acanthorhynchus*). The calculated S_{10} values indicate that for almost all species in Table 2.3, a substantial minority of birds survive for several years. In 13 of 22 species, 10 per cent of adults survive for at least 4 years after banding, and six species for more than 5 years. Fourteen of the species have maximum longevity of more than 10 years. These data and those of Table 2.2 are

Table 2.3 Survival and longevity of Australian passerines from ringing studies

Species (status)*	Body mass (g)	Maximum longevity (years, months)	Survival (method)	Number recaptured (ringed)	S_{10} (years)
(a) Manjimup, W.A.					
Eopsaltria georgiana (R)	17	>12 y	72 (1)	216 (365)	5.2
Pachycephala pectoralis (R)	25	>12 y	78 (1)	130 (287)	6.8
Pomatostomus superciliosus (R)	40	8 y 9 m	61 (1)	40 (76)	4.6
Malurus elegans (R)	9	>12 y	68 (1)	276 (406)	5.9
Sericornis frontalis (R)	14	11 y	77 (1)	189 (288)	5.7
Colluricincla harmonica (R)	70	7 y 11 m	68 (1)	40 (74)	5.8
(b) New Chum's Road, A.C.T.					
Zoothera dauma (R/A)	70	7 y 1 m	53 (2)	77 (283)	3.8
Petroica rosea (R/A)	8	5 y 11 m	49 (2)	80 (310)	4.0
Eopsaltria australis (R)	22	12 y 10 m	56 (2)	237 (534)	3.8
Pachycephala pectoralis (R/N)	25	10 y 11 m	62 (2)	203 (576)	4.8
Sericornis frontalis (R)	14	14 y 5 m	62 (2)	850 (1472)	4.0
Acanthiza pusilla (R)	7	13 y 5 m	60 (2)	526 (1507)	3.9
A. lineata (R)	7	9 y 6 m	62 (2)	407 (928)	5.4
Lichenostomus chrysops (R/N)	17	12 y 5 m	59 (2)	402 4988)	4.0
Melithreptus lunatus (R/N)	15	10 y	58 (2)	309 (1705)	4.5
(c) Miscellaneous					
Pomatostomus ruficeps (R) (Boehm 1974)	45	>7 y	78 (1)	80 (329)	5.6
Climactris picumnus (R) (Boehm 1982)	34	11 y 4 m	70 (1)	80 (171)	4.2
Lichenostomus melanops (R/N) (Morris 1975)	22	>7 y	57 (4)	244 (1139)	3.4
L. ornatus (R/N) (Boehm 1978)	18	>9 y	57 (2)	156 (716)	2.8
Lichmera indistincta (R/N) (Robertson and Woodall 1987)	11	10 y 11 m	56 (2)	248 (1759)	3.9
Acanthorhynchus tenuirostris (R/N) (Macfarland and Ford 1987)	11	10 y 6 m	60 (3)	229 (2597)	3.4
Strepera graculina (R/N) (Nicholls and Woinarski 1988)	350	14 y	75 (1)	332 (1910)	3.8

(a) Manjimup, Western Australia. Wet eucalypt forest; unpublished data from R. and M. Brown.
(b) New Chum's Road, Australian Capital Territory. Wet eucalypt forest; data from Tidemann, Wilson and Marples (1988) and Marples (personal communication).
(c) Miscellaneous (from sources quoted).
All data are from birds ringed as adults. Methods of calculating survival rates: (1) Weighted mean annual survival, after Caughley (1977, p. 104). (2) After Chapman and Robson (1960), using captures 1961–71 and recaptures to 1979. (3) Fisher-Ford method, after Southwood (1966). (4) After Lack (1954). Survival was calculated by the present authors unless indicated otherwise. S_{10}, time to which 10 per cent of birds survived after banding, after Fry (1980).
* Status: R = resident; N = nomad; A = altitudinal migrant.

combined in Table 2.4 to compare with figures from Europe and North America. Although the values lack precision because they include migrants and residents from widely different latitudes, the data do suggest that Australian passerines tend to survive better than American ones, which in turn may survive better than European ones. Few species from Europe or America appear to reach survival greater than 70 per cent, in contrast to 15/38 from Australia.

Table 2.4 Annual survival of adults
Summary of available data for northern and southern temperate species, taken from the literature. Survival rates as calculated in the original studies have been used. This summary includes more than one value for a species where survival has been estimated in different places. Where a species spans temperate and tropical regions, only studies in temperate areas have been included

Region	Annual survival (%)					
(No. of studies)	<40	40–49	50–59	60–69	70–79	80–89
Australia (37)	—	2	10	11	9	6
New Zealand (3)	—	—	—	1	—	2
South Africa (8)		3	2	1	2	—
Europe (69)	19	24	21	5	—	—
North America (36)	3	3	15	13	1	1

Tables 2.2–2.4 are concerned with adult survival. In Table 2.5 we include data on juvenile survival—the gap between fledging and reaching adulthood—from three well-known species; juvenile survival ranges from 28–41 per cent, which appears similar to that of northern passerines, however space prohibits a detailed comparison here. Few Australian studies have continued for long enough to show whether or not mortality is density-dependent. For *Malurus splendens* over 16 years, there is no evidence of density-dependent mortality in adults or juveniles. Figure 2.3 shows that the mortality of breeding adults in *M. splendens* was not correlated with offspring production in the previous year; similarly there was no correlation between production of independent young and their survival through the first winter.

Discussion

The Australian situation appears to have led to a proliferation of resident insectivores and mobile nectarivores, characterized by low reproductive rates spread over a long breeding season, and by high adult survival. Table 2.5 summarizes life history data for three well-known resident insecti-vores, with breeding seasons of longer than 3 months. With breeding episodes lasting 60 days, this allows at least two successful nesting

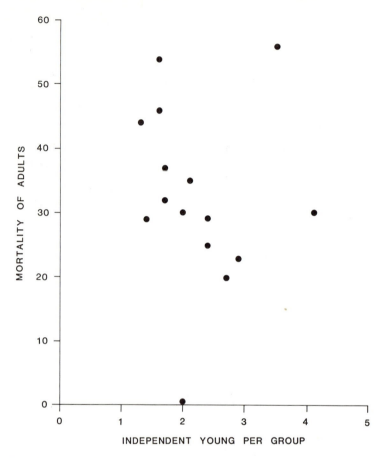

Fig. 2.3. Plot of mortality of *Malurus splendens* breeding adults, during the year from the start of one breeding season (1 September) to the next, against production of independent young during the first breeding season. If adult mortality were density dependent, we would expect a positive correlation ($r = -0.08$).

attempts in a season. Each of these three species is a cooperative breeder, and in each case the presence of a number of 'helpers' shortens the time between laying successive clutches; helpers take over raising the young thus allowing the female to relay sooner than would otherwise be possible. Clutches are small and relatively invariable. Even with repeat breeding, fledgling production per year is low, yet their survival is such that approximately one fledgling per pair reaches at least 1 year old. Bearing in mind that the survival of adult breeders is high (66–83 per cent) and that more than 10 per cent of adults survive for more than 5 years after ringing, vacancies for breeders are rare. Age at first breeding is therefore often more

Table 2.5 Demographic parameters for three Australian passerines

	Malurus *splendens*	*Malurus* *elegans*	*Eopsaltria* *georgiana*
Breeding season	Sept.–Dec.	Oct.–Dec.	July–Dec.
Clutch size	2.9	2.4	1.9
Breeding success (%)	47	52	81
Production/nest (fledglings)	1.4	1.3	1.5
Breeding episode (days)	55–60	55–60	55–60
Nests/female per year	2.1	1.5	2.0
Fledglings/female per year	2.8	2.5	3.1
Independent Y/female per year	1.8	1.9	?
Survival (%)			
Fledgling to 1 year	34	41	28
Independence to 1 year	53	54	?
Breeding male	71	82	83
Breeding female	66	80	73
S_{10}	5.5	5.9	5.2
Age at first breeding (years)			
Male	1–2	2+	2+
Female	1–2	2	2
Maximum longevity (years)	12	>12	>11
Body mass (female)	10 g	9 g	17 g

All species are capable of breeding at 1 year old but it rarely occurs.
> = Still alive.
Data for *Malurus* spp. from Rowley *et al*. (1988) and Rowley (unpublished); Data for *Eopsaltria georgiana* from R. and M. Brown (unpublished).

than 2 years and this in species such as *Malurus* where adult females weigh only 10 g.

In Australia, most breeders are likely to reproduce through several seasons. For historical reasons species do not have to share their winter resources with an influx of migrating refugees from the northern winter and many species retain their progeny in the family social group long after they have achieved 'independence' in foraging. Nor, as Skutch (1985) pointed out, do they suffer the occasional high mortality of severe northern winters or long migrations, reducing populations to a low level that may be compensated for by a high reproductive rate. Droughts do occur but have probably been overestimated as major causes of mortality in eucalypt forests and woodlands, and although breeding may be deferred or curtailed, the cost in terms of lifetime reproductive success is not great. Because food is rarely scarce in most Australian temperate habitats (Ford *et al*. 1988), species tend to remain as residents and this situation has led to resident insectivores frequently adopting the cooperative breeding strategy (22 per cent of 258 old endemic species; Russell 1988).

Most temperate northern hemisphere studies have been made at higher latitudes, and few population studies have been completed at latitudes strictly comparable with those of the southern studies we have quoted. The continuing studies of Blondel (Blondel 1985; Blondel *et al*. 1987; Blondel, this volume) on Corsica and the Mediterranean mainland provide a significant exception. These data illustrate significant variation in demographic parameters in relation to habitat variability, but do not reach the levels of productivity and survival found in southern hemisphere species at similar latitudes.

Summarizing the characteristics of Australian passerine demography, it would appear that small relatively invariable clutch size is more similar to the tropical situation reviewed by Skutch (1985) than that of the northern temperate region. Repeat breeding is commonplace and as successful as first attempts, which means that annual productivity per season is the significant demographic parameter. Adjustment of annual reproductive effort is achieved by variation in the length of the breeding season and number of nesting attempts, rather than by variation in clutch size. Density-dependent variation in reproductive rate or survival has not so far been demonstrated in any species studied in depth. Survival of adults is higher than for most northern temperate species, and survival of independent juveniles to 1 year may also be higher, especially in comparison with species that migrate. This pattern of covarying life history traits, longer life, lower reproductive rate, and later maturity is usually accepted as typical of tropical passerines. It is clearly also possible in southern temperate latitudes, in situations where adequate food is available throughout the year, winters are not severe, and long migrations across water or deserts do not occur.

Studies of patterns of covariation in life history traits in birds (Saether 1989) show that much of the variation is explained by differences present above the level of family, suggesting that variation in reproductive rates is strongly influenced by differences evolved early in the species' evolutionary history. If that is so, then the life history characteristics exhibited by the old endemic Australian passerine families were probably acquired a very long time ago.

The implications for management of the reproductive strategy shown by Australian passerines are clear. Most species are not capable of a rapid reproductive effort to re-establish a devastated local population after a major disaster or to recolonize an area from which the population has been eliminated. Furthermore, most populations will contain a large proportion of older experienced breeders that are not only more efficient per nesting attempt, but are also more likely to repeat nest; they are the high producers amongst a low-producing species. Mismanagement that alters this age structure will lower reproductive output and slow population recovery still further. For effective management, which in most cases means application of some sort of fire regime, it must be recognized that most Australian passerines reproduce at a low rate spread over a long breeding season. Fire needs to be patchy so that adequate cover for nesting remains, and needs to be timed so that minimum interference with nesting is achieved, i.e. autumn burns would be best *if* they could be controlled sufficiently to prevent wholesale conflagrations after the summer. Available evidence suggests that while adults and even independent juveniles can survive an intense fire, the effects of fire on productivity may only become obvious several years afterwards (Rowley and Brooker 1987; unpublished). Fire frequency needs to be adjusted to allow productivity and age-structure to be maintained.

The complex balance between lower reproductive rates and high survival in some tropical and temperate environments is not yet sufficiently understood, especially with regard to levels of predation, food availability, population density, environmental variability, and the relative merits of one large versus several small reproductive attempts. Lower reproductive rates do not necessarily mean lower costs of reproduction. The length of time that a female is exposed to the hazards of nesting is increased in multi-brooded species; she builds several nests per year and feeds more than one brood in an environment where the peak of food available is probably below that of a northern temperate habitat. Seen from this perspective, the preoccupation of northern temperate ornithologists with clutch size and the factors that determine it may mean that the other important aspects of life history strategies tend to be ignored.

Summary

Temperate areas of the southern hemisphere do not extend into the high latitudes of their northern equivalents. Southern temperate species for which demographic data are available are mainly Australian. A major passerine radiation occurred in Australia during the 20 million years that the continent was drifting north after separating from Antarctica and before it came within reach of Asia. As a result 75 per cent of Australian passerines are of endemic origin and evolved in response to past and present environmental conditions. In temperate non-arid regions, seasonal climatic extremes have never been severe.

Breeding seasons of 4–5 months are not shortened by the need to migrate. Although the main food resources of insects and nectar fluctuate, with a spring peak and decline in late summer/winter, food is always available and resident species are common. Clutch size (mean 2.7) of the old endemic Australian families is lower than that of passerines from equivalent latitudes in North Africa (4.3). Many Australian passerines are regularly multibrooded and variation in reproductive effort occurs through the number of breeding attempts rather than in clutch size; even so, annual productivity is low, generally fewer than 4 fledglings per female per year.

Adult survival is higher than in most northern hemisphere species; only 2 of 22 species had annual survival below 60 per cent, and 8 of these species had survival of 80 per cent or more. For many species, a significant proportion of adults survive for a long time becoming the experienced, productive breeders; 14 of 22 species showed maximum longevity in excess of 10 years. The longer life, lower reproductive rate, and delayed sexual maturity of tropical passerines appear to be also characteristic of temperate southern hemisphere passerines. Management strategies need to bear in mind the inability of such populations to recover rapidly following a disaster.

Acknowledgements

Much of the data used in this review is not yet published and we thank Michael Brooker, Dick and Molly Brown, Stephen Marchant, Tim Marples, Terry Oatley, and Steven Piper for their cooperation and patience in the face of our delving, and Hugh Ford for helpful comments on the manuscript.

References

Bell, H. L. (1982). Survival among birds of the understorey in lowland rainforest in Papua New Guinea. *Corella*, **6**, 77–82.

Bell, H. L. (1985). Seasonal variation and the effects of drought on the abundance of arthropods in savanna woodland on the northern tablelands of New South Wales. *Australian Journal of Ecology*, **10**, 207–21.

Bell, H. L. and Ford, H. A. (1986). A comparison of the social organization of three syntopic species of Australian Thornbill, *Acanthiza*. *Behavioural Ecology and Sociobiology*, **19**, 381–92.

Blondel, J. (1985). Comparative breeding ecology of the Blue Tit and the Coal Tit in mainland and island Mediterranean habitats. *Journal of Animal Ecology*, **54**, 531–56.

Blondel, J., Clamens, A., Cramm, P., Gaubert, H., and Isenmann, P. (1987). Population studies on tits in the Mediterranean region. *Ardea*, **75**, 21–34.

Boehm, E. (1974). Results from banding Chestnut-crowned Babblers. *Australian Bird Bander*, **12**, 76–8.

Boehm, E. (1978). Banding the Yellow-plumed Honeyeater on the Mount Mary Plains, South Australia. *Corella*, **2**, 65–8.

Boehm, E. F. (1982). Results from banding Brown Treecreepers. *Corella*, **6**, 16–17.

Brown, J. L. and Brown, E. R. (1981). Kin selection and individual selection in babblers. In *Natural selection and social behaviour* (eds. R. D. Alexander and D. W. Tinkle), pp. 244–56. Chiron Press, New York.

Brown, R. J. and Brown, M. N. (1980). Cooperative breeding in robins of the genus *Eopsaltria*. *Emu*, **80**, 89.

Brown, R. J., Brown, M. N., and Russell, E. M. (1990). Survival of four species of passerine in karri forests in south-western Australia. *Corella*, **14**, 69–78.

Calver, M. C. and Wooller, R. D. (1981). Seasonal differences in the diets of small birds in the karri forest understorey. *Australian Wildlife Research*, **8**, 653–57.

Caughley, G. C. (1977). *Analysis of vertebrate populations*. John Wiley, London.

Chapman, D. G. and Robson, D. S. (1960). The analysis of a catch curve. *Biometrics*, **16**, 354–68.

Clapp, R. B., Klimkiewicz, M. K., and Futcher, A. G. (1983). Longevity records of North American birds: Columbidae through Paridae. *Journal of Field Ornithology*, **54**, 123–37.

Clarke, M. F. (1988). The reproductive behaviour of the Bell Miner *Manorina melanophrys*. *Emu*, **88**, 88–100.

Collins, B. G. and Rebelo, T. (1987). Pollination biology of the Proteaceae in Australia and southern Africa. *Australian Journal of Ecology*, **12**, 387–421.

Dobson, A. P. (1987). A comparison of seasonal and annual mortality for both sexes of fifteen species of common British birds. *Ornis Scandinavica*, **18**, 122–8.

Earle, R. A. (1986). The breeding biology of the South African Cliff Swallow. *Ostrich*, **57**, 138–56.

Earle, R. A. and Herholdt, J. J. (1988). Breeding and moult of the Anteating Chat *Myrmecocichla formicivora*. *Ostrich*, **59**, 155–61.

Eberhardt, L. L. (1972). Some problems in estimating survival from banding data. In *Population ecology of migratory birds*, pp. 153–71. Bureau of Sport Fisheries and Wildlife, Wildlife Research Report, No. 2.

Elder, W. H. (1985). Survivorship in the Tufted Titmouse. *Wilson Bulletin*, **97**, 517–24.

Elder, W. H. and Zimmerman, D. (1983). A comparison of recapture *versus* resighting data in a 15-year study of survivorship of the Black-capped Chickadee. *Journal of Field Ornithology*, **54**, 138–45.

Falla, R. A., Sibson, R. B., and Turbot, E. G. (1978). *The new guide to the birds of New Zealand*. Collins, Auckland.

Flux, J. E. C. and Flux, M. M. (1981). Population dynamics and age structure of Starlings (*Sturnus vulgaris*) in New Zealand. *New Zealand Journal of Ecology*, **4**, 65–72.

Fogden, M. P. L. (1972). The seasonality and population dynamics of equatorial forest birds in Sarawak. *Ibis*, **114**, 307–43.

Ford, H. A., Bell, H., Nias, R., and Noske, R. (1988). The relationship between ecology and the incidence of cooperative breeding in Australian birds. *Behavioural Ecology and Sociobiology*, **22**, 239–49.

Ford, J. R. (1963). Breeding behaviour of the Yellow-tailed Thornbill in south-western Australia. *Emu*, **63**, 185–200.

Frakes, L. A., McGowran, B., and Bowler, J. M. (1987). Evolution of Australian environment. In *Fauna of Australia*, Vol. 1A (eds. G. R. Dyne and D. W. Walton), pp. 1–16. Australian Government Publishing Service, Canberra.

Fry, C. H. (1980). Survival and longevity among tropical land birds. *Proceedings of the Fourth Pan-African Ornithological Congress* (1976), pp. 334–43.

Gill, B. J. (1982). Breeding of the Grey Warbler *Gerygone igata* at Kaioura, New Zealand. *Ibis*, **124**, 123–47.

Keast, J. A. (1968). Seasonal movements in the Australian honeyeaters (Meliphagidae) and their ecological significance. *Emu*, **67**, 159–209.

Kemp, E. M. (1981). Tertiary palaeogeography and the evolution of Australian climate. In *Ecological biogeography of Australia*, Vol. 1 (ed. A. Keast), pp. 33–49. W. Junk, The Hague.

Kikkawa, J., and Wilson, J. M. (1983). Breeding and dominance among the Heron Island Silvereyes. *Zosterops lateralis chlorocephala*. *Emu*, 83, 181–98.

Klimkiewicz, M. K. and Futcher, A. B. (1987). Longevity records of north American birds: Coerebinae through Estrildidae. *Journal of Field Ornithology*, 58, 318–33.

Klimkiewicz, M. K., Clapp, R. B., and Futcher, A. B. (1983). Longevity records of North American birds: Remizidae through Parulinae. *Journal of Field Ornithology*, 54, 287–94.

Lack, D. (1954). *The natural regulation of animal numbers*. Oxford University Press, Oxford.

McFarland, D. C. and Ford, H. A. (1987). Aspects of population biology of the Eastern Spinebill *Acanthorhynchus tenuirostris* (Meliphagidae) in New England National Park, NSW. *Corella*, 11, 52–8.

Maclean, G. L. (1985). *Roberts' birds of Southern Africa*. John Voelcker Book Fund, Cape Town.

McLean, I. G. and Gill, B. J. (1988). Breeding of an island-endemic bird: the New Zealand Whitehead *Mohoua albicilla*; Pachycephalinae. *Emu*, 88, 177–82.

Majer, J. D. and Koch, L. E. (1982). Seasonal activity of hexapods in woodland and forest leaf litter in the south-west of Western Australia. *Journal of the Royal Society of Western Australia*, 65, 37–45.

Marchant, S. (1974). Analyses of nest-records of the Willie Wagtail. *Emu*, 74, 149–60.

Marchant, S. (1984). Nest records of the Eastern Yellow Robin *Eopsaltria australis*. *Emu*, 84, 167–74.

Marchant, S. (1985). Breeding of the Eastern Yellow Robin, *Eopsaltria australia*. In *Birds of eucalypt forest and woodlands: ecology, conservation, management* (ed. A. Keast, H. F. Recher, H. Ford, and D. Saunders), pp. 231–40. Surrey Beatty and Sons, Sydney.

Marchant, S. (1986). Long laying intervals. *Auk*, 103, 247.

Marchant, S. and Fulagar, P. J. (1983). Nest records of the Welcome Swallow. *Emu*, 83, 66–74.

Moffatt, J. D. (1982). Territoriality and use of space in Grey-crowned Babblers, *Pomatostomus temporalis*. Unpublished Ph.D. Thesis, University of Queensland, Brisbane, Australia.

Morris, A. K. (1975). Results from banding Yellow-tufted Honeyeaters. *Australian Bird Bander*, 13, 3–8.

Nice, M. M. (1957). Nesting success in altricial birds. *Auk*, 74, 305–21.

Nicholls, D. G. and Woinarski, J. Z. (1988). Longevity of Pied Currawongs at Timbertop, Victoria. *Corella*, 12, 43–7.

Niethammer, G. (1970). Clutch sizes of introduced European Passeriformes in New Zealand. *Notornis*, 17, 214–42.

Noske, R. A. (1983). Comparative behaviour and ecology of some bark-foraging birds. Unpublished Ph.D. Thesis, University of New England, Armidale, Australia.

Oatley, T. B. (1982). The Starred Robin in Natal, Part 3: Breeding, populations and plumages. *Ostrich*, 53, 206–21.

Paton, D. C. (1985). Food supply, population structure and behaviour of New Holland Honeyeaters *Phylidonyris novaehollandiae* of woodland near Horsham, Victoria. In *Birds of eucalypt forests and woodlands: ecology, conservation, manage-*

ment (ed. A. Keast, H. F. Recher, H. Ford, and D. Saunders), pp. 219–30. Surrey Beatty and Sons, Sydney.

Perrins, C. M. (1985). Clutch size. In *A dictionary of birds* (ed. B. Campbell and E. Lack). T. and A. D. Poyser, Calton, England.

Powlesland, R. G. (1983). Breeding and mortality of the South Island Robin in Kowhai Bush, Kaikoura. *Notornis*, 30, 265–82.

Pyke, G. H. (1983). Seasonal pattern of abundance of honeyeaters and their resources in heathland areas near Sydney. *Australian Journal of Ecology*, 8, 217–33.

Recher, H. F., Gowing, G., Kavanagh, R., Shields, J., and Rohan-Jones, W. (1983). Birds, resources and time in a tableland forest. *Proceedings of the ecological Society of Australia*, 12, 101–23.

Robertson, J. S. and Woodall, P. F. (1987). Survival of Brown Honeyeaters in south-east Queensland. *Emu*, 87, 137–42.

Rooke, I. (1977). The social behaviour of the honeyeater *Phylidonyris novaehollandiae*. Unpublished Ph.D. Thesis, University of Western Australia.

Rowley, I. (1965). The life history of the Superb Blue Wren. *Malurus cyaneus*. *Emu*, 64, 251–97.

Rowley, I. (1971). Movement and longevity of ravens in south-eastern Australia. *CSIRO Wildlife Research*, 16, 49–72.

Rowley, I. (1973). The comparative ecology of Australian corvids. IV. Nesting and the rearing of young to independence. *CSIRO Wildlife Research*, 18, 91–129.

Rowley, I. (1975). *Bird Life*. Collins, Sydney.

Rowley, I. (1988). The Purple-crowned Fairy-wren—an RAOU Conservation Statement. *RAOU Report* No. 34. RAOU, Melbourne.

Rowley, I. and Brooker, M. G. (1987). The response of a small insectivorous bird to fire in heathlands. In *Nature conservation: the role of remnants of native vegetation* (ed. D. A. Saunders, G. W. Arnold, A. A. Burbidge, and A. J. M. Hopkins), pp. 211–18. Surrey Beatty and Sons, Sydney.

Rowley, I. and Russell, E. M. (1989). The Splendid Fairy-wren *Malurus splendens*. In *Cooperative breeding in birds: long-term studies of ecology and behaviour* (eds. P. B. Stacey and W. D. Koenig). Cambridge University Press.

Rowley, I., Russell, E., Brown, R., and Brown, M. (1988). The ecology and breeding biology of the Red-winged Fairy-wren *Malurus elegans*. *Emu*, 88, 161–76.

Russell, E. M. (1988). Cooperative breeding—a Gondwanan perspective. *Emu*, 89, 61–2.

Saether. B.-E. (1989). Survival rates in relation to body weight in European birds. *Ornis Scandinavica*, 20, 13–21.

Schodde, R. (1975). *Interim list of Australian songbirds. Passerines*. Royal Australian Ornithologists Union, Melbourne.

Seber, G. A. F. (1973). *The estimation of animal abundance*. Griffin, London.

Sibley, C. G. and Ahlquist, J. E. (1985). The phylogeny and classification of the Australo-Papuan passerine birds. *Emu*, 85, 1–14.

Skutch, A. F. (1985). Clutch size, nesting success, and predation on nests of neotropical birds, reviewed. In *Neotropical ornithology* (ed. P. A. Buckley, M. S. Foster, E. A. Morton, R. S. Ridgely, and F. G. Buckley). *Ornithological Monographs*, 36, 575–94.

Snow, D. W. and Lill, A. (1974). Longevity records for some neotropical land birds. *Condor*, 76, 262–67.

Southwood, T. R. E. (1966). *Ecological methods: with particular reference to the study of insect populations*. Methuen, London.

Tenannt, K. (1956). *The honey flow*. Angus and Robertson, Sydney.

Thomas, D. G. (1974). Some problems associated with the avifauna. In *Biogeography and ecology of Tasmania* (ed. W. D. Williams), pp. 339–65. W. Junk, The Hague.

Tidemann, S. C., Wilson, S. J., and Marples, T. G. (1988). Some results from a long-term bird-banding project in the Brindabella Range, A.C.T. *Corella*, **12**, 1–6.

Woinarski, J. C. Z. (1985*a*). Breeding biology and life history of small insectivorous birds in Australian forests: response to a stable environment. *Proceedings of the ecological Society of Australia*, **14**, 159–68.

Woinarski, J. C. Z. (1985*b*). Foliage-gleaners of the treetops, the pardalotes. In *Birds of eucalypt forests and woodlands: ecology, conservation, management* (ed. A. Keast, H. F. Recher, H. Ford, and D. Saunders), pp. 165–75. Surrey Beatty and Sons, Sydney.

Woinarski, J. C. Z. (1989). Some life history comparisons of small leaf gleaning bird species of south-eastern Australia. *Corella*, **13**, 73–80.

Woinarski, J. C. Z. and Cullen, J. M. (1984). Distribution of invertebrates on foliage in forests of south-eastern Australia. *Australian Journal of Ecology*, **9**, 207–32.

Yom-Tov, Y. (1987). The reproductive rates of Australian passerines. *Australian Wildlife Research*, **14**, 319–30.

3 Birds in biological isolates

JACQUES BLONDEL

Introduction

The way in which the number and identity of species inhabiting an island, a set of islands, or any other discrete patch of habitat varies with such parameters as size, degree of isolation, and habitat heterogeneity has been an important question in ecology, evolutionary biology, and more recently in conservation biology. In the late 1960s and 1970s, a tremendous amount of work was generated by the model of island biogeography of Preston (1962) and MacArthur and Wilson (1963, 1967). This model and its predictions have been investigated on real oceanic islands (e.g. Hamilton *et al*. 1963; Terborgh 1971, 1973; Diamond *et al*. 1976; Diamond and Mayr 1976; Mayr and Diamond 1976), and also on mountaintops (Cook 1974; Johnson 1975), caves (Culver *et al*. 1973; Vuilleumier 1973), ponds (Hubbard 1973), and such patches of habitat as isolated fragments of forest (e.g. Moore and Hooper 1975; Galli *et al*. 1976; Burgess and Sharpe 1981; Pickett and White 1985; Soulé 1986).

Although the model of island biogeography has been extremely useful in revitalizing island studies and, more generally, ecological biogeography, it must be recognized that many of its predictions are too general or too simplistic to produce sound working hypotheses for specific situations. MacArthur and Wilson themselves were quite aware of the major limitations of their model. They wrote in the preface of their book: 'We do not seriously believe that the particular formulations advanced in the chapters to follow will fit for very long the exacting results of future empirical investigations'. I shall not discuss the advantages, drawbacks, and shortcomings of this theory; many recent papers deal with them, especially the hypothesis of a turnover of species that is a keystone of the theory (see for instance Connor and McCoy 1979; Blondel 1979, 1986, 1987; Abbott 1980; F. S. Gilbert 1980; Simberloff 1980, 1983; Williamson 1981; Connor and Simberloff 1983).

The aims of this chapter are:

(1) to give an insight into island patterns at the three levels of regional faunas, communities, and populations, using the Mediterranean region and its islands as an example;

(2) to review some major points on the biology of isolates, especially of habitat patches resulting from forest fragmentation; and
(3) to draw conclusions for conservation.

Island biology in the Mediterranean region

One of the most stimulating trends in biology during the last decade has been the recognition that the organization of living systems is basically hierarchical and that processes in nature are sensitive to the scale at which they are considered (Allen and Starr 1982): and hence scale has a powerful influence on the conclusions that are reached (Karr and Freemark 1983; Wiens 1985). Because the entire field of island biogeography is, in a sense, a matter of scale, I shall investigate several levels of biological organization by zooming along the scales of space and time.

My approach will be to consider island patterns by switching from one level of organization to the next lower level using comparative studies in mainland and island situations. Although the spatial scales for investigtion form a continuum, I shall identify four levels along this continuum (Fig. 3.1):

(1) the whole Mediterranean;
(2) a region

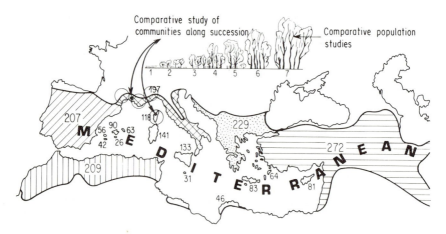

Fig. 3.1. Biogeographical processes and island patterns studied at four levels of investigation using a comparative approach whenever possible: (1) faunas in the Mediterranean region as a whole (figures indicate the numbers of bird species breeding in each region delimited on the map); (2) the level of a region, for instance Corsica by comparison to three regions of similar size on the mainland; (3) communities in habitat gradients on Corsica *versus* the mainland; and (4) populations in similar island and mainland habitats (after Blondel 1986).

(3) a mosaic of habitat patches, and
(4) local patches of habitats.

These points roughly correspond respectively to a fauna, a subset of a fauna, a set of communities, and finally a community and a population.

How history can help to explain distributional patterns

The unusual geotopographical and geobotanical diversity of habitats around or on the islands within this 'sea-amongst-land ('Medi-terranean') has precipitated an extreme diversification of the Mediterranean biotas (see Blondel 1986, 1988). Three hundred and forty-three bird species breed currently in the Mediterranean region as a whole (Blondel 1988), a number to be compared with the 419 species that breed in the whole of Europe (Voous 1960). Most of these species belong either to a fauna of temperate Eurasia or to a southern fauna which evolved in the semi-arid margins of the Mediterranean. An unexpected feature is the scarcity of birds of Mediterranean origin, although the Mediterranean region has long been recognized as a well-defined biogeographical subunit of the Palaearctic region: no more than 47 species, i.e. 14 per cent of the fauna speciated within the limits of the Mediterranean region. Among them, 15 (i.e. 32 per cent) are species of open and semi-open habitats (e.g. *Alectoris* spp., *Oenanthe* spp.), 21 (45 per cent) are shrubland species (e.g. *Hippolais* spp., *Sylvia* spp.) and only 8 (17 per cent) are forest species (e.g. *Sitta* spp., *Columba* spp., *Parus*). Although there is a great variety of forests in the Mediterranean with no less than 100 endemic tree species (Quézel 1985), no more than 2 per cent of the woodland bird fauna speciated there. As a result, old forests dominated by Mediterranean sclerophyllous trees do not include a single bird species of Mediterranean origin (Blondel and Farré 1988).

Any explanation of the dominance of Eurasian temperate forest birds and of the scarcity of Mediterranean species needs to turn to the historical background of the development of the bird faunas during the Pleistocene. Palaeoclimatic, palaeobotanical, and palaeontological findings indicate that all the forest types of Europe and their associated faunas had to find refugia in the Mediterranean during the most severe phases of the glacial periods. These forest types and their associated faunas were more or less continuously distributed at mid-altitudes along the Mediterranean mountains (Beug 1975). The absence of geographical isolation among European forests that is a prerequisite for allopatric speciation has prevented differentiation of forest birds in the Mediterranean belt of Eurasia, at least down to species level. This explains why the bird faunas of European forests are now so similar everywhere in Europe, including the Mediterranean region.

On the other hand, most cases of speciation among birds of open or shrubby habitats, for instance the *Sylvia* warblers, probably occurred during the Pleistocene in the few patches of shrubby vegetation that persisted in the Mediterranean region even during the most severe climatic phases (Pons 1981; Reille 1984). The present relative extension of Mediterranean-type shrublands is a modern feature due to human influence—large-scale deforestation of landscapes that were formerly dominated by forests, except in some localized patches of matorral (see below).

Given this historical background, an interesting question regarding island biology is: do species such as tits, that evolved in large tracts of forest, and species that evolved in patches of matorral, such as Mediterranean warblers, have the same probabilities of colonizing islands and living in isolation? If not, what will be the consequences of island colonization and then, going down through the different levels of biological organization, to community structure and population biology?

In order to answer this question I shall focus on several scales of investigation, taking Corsica as an example.

Island communities: a case study in Corsica

Corsica is a rather large island of 8700 km² which rises to 2700 m; it is 80 km from the coast of Italy and has not been connected to the mainland since the Miocene-Pliocene. It presents a large diversity of habitats from sea-level up to the subalpine zone.

In the past, too many studies of island biogeography have focused either on species lists at the level of whole islands or on comparing single communities between island and mainland habitats which are supposed to resemble each other. Such studies often yielded ambiguous results because processes that operate at one particular level do not necessarily operate in the same way at other levels (Wright 1980; Wiens 1985; Blondel *et al*. 1988). Because quantitative investigations have been underexploited in island biology (Haila and Järvinen 1981), detailed faunistic surveys and censuses have been conducted in Corsica at the level of the island as a whole and for a large series of habitats. More specifically, an experiment was designed which defined two similar habitat gradients each with 6 habitats that match each other reasonably well, one on the mainland (Provence) and the other on Corsica (see Fig. 3.2). Habitats were arranged from a very low matorral, 0.5 m high, to a mature forest of Holm oak, *Quercus ilex*, 25 m high. The bird communities of each habitat have been carefully censused (see Blondel *et al*. 1970, 1988 for the methods and experimental design).

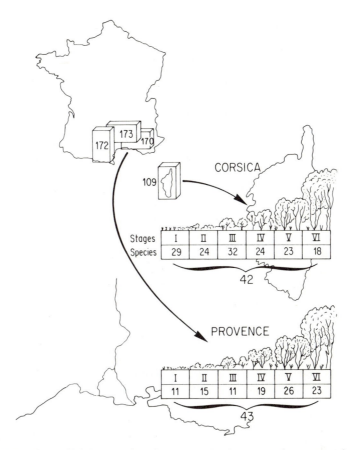

Fig. 3.2. Numbers of bird species breeding in (a) Corsica *versus* three areas of similar size on the mainland; (2) two sets of six habitats along habitat gradients of similar structure on the island and the mainland. Notice the similar number of species along the scale of the gradient but a different allocation of species per habitat in Corsica and in Provence.

Species impoverishment and the scales of investigation

Results of these surveys and censuses have shown that species impoverishment in Corsica varies according to the size of the area considered. Species richness may even be higher in some island habitats than in their mainland counterparts (see below and Fig. 3.2).

1. At the level of the whole island, species impoverishment is about 37 per cent since 109 species regularly breed in Corsica against 170–3 in three areas of similar size on the mainland. The missing species are not a random sample of the source mainland fauna (Blondel 1985a, 1986): species likely

to be missing from Corsica are (a) species for which there is a lack of suitable habitats; (b) the larger species, especially raptors; (c) species that are rare and localized on the mainland; and especially (d) forest species.

2. At the level of the combined series of 6 habitats in the gradients, there is hardly any impoverishment since there are 43 species in the mainland gradient and 42 in the island one.

3. Finally, at the habitat level, results are more complex. Some habitats, especially old forests, are heavily impoverished on the island whereas other habitats, such as shrublands, have more species than their mainland counterparts. For instance, there are 22 per cent fewer species in the old Holm Oak stand in Corsica than in its mainland counterpart and up to 66 per cent more species in a Corsican shrubby habitat than in its mainland counterpart (Fig. 3.2). Using a correspondence analysis on the data sets of each region separately and then on the combined data sets of the two regions, Blondel *et al*. (1988) have demonstrated the process of habitat-niche expansion in Corsica. They found that the structure of the habitat gradient is much looser in Corsica, i.e. dispersion ellipses which display each community on the multivariate space of the analysis overlap much more on the island than on the mainland. This is because, on the average, species on the island occupy a larger spectrum of habitats, i.e. enlarge their habitat-niche (see Blondel *et al*. 1988 for more details).

Such results clearly show that the process of species impoverishment is more complicated than the classical assumptions of island biogeography would suggest. At the level of the island as a whole, there is a clear relationship between species impoverishment and a broad range of quite different habitat types. At a smaller scale, however, the relationship between birds and habitats becomes different and some patterns vanish, while other patterns emerge. Some of them are said to be a legacy of the historical development of bird faunas, and will now be investigated at the species-specific level.

Species-specific responses to isolation

Given these different patterns of species allocation to habitats on the mainland and in Corsica, one may wonder whether there are differences between species in habitat utilization. A species-specific examination shows that species impoverishment is much more severe among forest birds than among birds of shrublands. Since forest birds have presumably evolved in large tracts of habitat over Eurasia, one may hypothesize that they have evolved life history traits, especially dispersal patterns, which are adapted to large areas. Survival of such species in biological isolates should require the evolution of ecological traits allowing them to cope with conditions of isolation. On the other hand, since shrubland species evolved in more or less isolated patches of matorral, their survival on islands should not require any further particular adaptation. If this is true, in contrast to forest birds that are poor island colonizers, there should be no species impoverishment of the guild of warblers on Mediterranean islands. This is indeed the case: there are on average as many *Sylvia* species in Mediterranean islands as there are on the mainland nearby. Therefore the two categories of birds should react differently to isolation.

Niche enlarging and density inflation, which are parts of the insular syndrome (MacArthur *et al.* 1972; Williamson 1981; Crowell 1983; Stamps and Buechner 1985; Blondel 1986) are actually much more pronounced for forest species, for example the *Parus* tits, than for the shrubland species, for example the *Sylvia* warblers (Blondel 1987). Patterns of habitat utilization support this hypothesis. Figure 3.3 shows that forest species such as tits occur in additional habitats in Corsica, whereas matorral species such as warblers do not exhibit any conspicuous change in their distributional profiles on the island (see also Table 3.1). Aside from habitat-niche expansion, the total density of forest species is much higher in the insular gradient than in its mainland counterpart (see table 3.1), a process named density compensation by MacArthur *et al.* (1972) and density inflation by Crowell (1983).

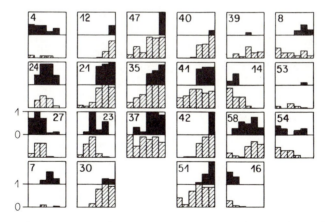

Fig. 3.3. Patterns of habitat utilization by the species over the six habitats (horizontal axis = low matorral to the left; old mature forest to the right) of the mainland gradient (black) and the Corsican gradient (hatched). Vertical axis = per cent of occurrence of the species. Species are indicated by their code numbers as given in Table 3.1. Forest species such as the three species of tits (*Parus ater* 40, *P. major* 41, and *P. caeruleus* 42) expand in additional habitats in Corsica whereas matorral species such as the Mediterranean warblers (*Sylvia melanocephala* 23, *S. cantillans* 24, and *S. undata* 27) do not (from Blondel *et al.* 1988).

Table 3.1 Mean composition of the bird communities in habitats 2, 4 and 6 of Provence and Corsica. Figures are densities in breeding pairs per 10 Ha. Arrows indicate the trend of habitat shifts of Corsican populations. Notice that these shifts involve only forest species. (From Blondel *et al*. 1988)

Species	Provence			Corsica		
	P2	P4	P6	C2	C4	C6
Accipiter nisus 2		0.1	0.1			
Buteo buteo 3						0.1
Alectoris rufa 1	0.2	0,1			0.1	
Columba palumbus 6			0.1			0.1
Streptopelia turtur 7		1.2				
Cuculus canorus 8		0.1	0.1	0.1	0.1	0.1
Picus viridis 11			0.3			
Picoides major 12			1.8		0.1 ←	1.1
Picoides minor 13			0.2			
Lullula arborea 14	1.0			0.9	0.1	
Alauda arvensis 15				0.1		
Anthus campestris 16	0.7			0.1		
Lanius excubitor 17	0.3					
Lanius collurio 19				0.2	0.1	
Sylvia atricapilla 21		6.1	5.5	0.3 ←	7.1 ←	6.7
Sylvia hortensis 22		0.3				
Sylvia melanocephala 23	0.4	0.3		1.8	0.2	
Sylvia cantillans 24	1.8	8.3		1.0	2.0	
Sylvia conspicillata 25	0.1					
Sylvia sarda 26				3.9		
Sylvia undata 27	4.4	0.1		1.4	0.1	
Phylloscopus collybita 28		0.7				
Phylloscopus bonelli 29		0.2	1.0			
Regulus ignicapillus 30			1.5		7.3 ←	9.0
Muscicapa striata 31						1.8
Saxicola torquata 32				0.3		
Oenanthe hispanica 34	2.8					
Erithacus rubecula 35		2.3	6.1	1.7 ←	4.4 ←	5.7
Luscinia megarhynchos 36	0.2	4.5	1.4			
Turdus merula 37	1.1	8.0	3.8	1.5	1.4	1.1
Turdus viscivorus 38			0.2			
Aegithalos caudatus 39		0.3		1.1 ←	1.9 ←	1.1
Parus ater 40			0.2		1.2 ←	4.1
Parus major 41		2.2	3.2	1.7 ←	3.6 ←	4.7
Parus caeruleus 42			11.5	0.5 ←	3.3 ←	14.1
Parus cristatus 43			1.8			
Sitta europaea 44			2.1			
Certhia brachydactyla 45			6.8			
Certhia familiaris 46						1.6
Troglodytes troglodytes 47			5.0	2.8 ←	6.3 ←	5.0
Emberiza cirlus 49				0.4	0.1	
Emberiza hortulana 50	2.7					
Fringilla coelebs 51		0.2	7.7	0.6 ←	6.9 ←	6.4
Carduelis carduelis 52				1.6	0.4	

Species	Provence			Corsica		
	P2	P4	P6	C2	C4	C6
Carduelis chloris 53					0.1	
Carduelis cannabina 54	0.7	0.2		1.6		
Serinus citrinella 56				5.3	1.7	0.4
Oriolus oriolus 57			0.1			
Garrulus glandarius 58	0.2	1.0	0.6	0.1	1.9	1.0
Pica pica 59	0.5					
Corvus cornix 60				0.1	0.1	
Number of species	15	19	23	24	24	18
Total density (pairs/10 ha)	17.1	36.2	61.1	29.1	49.5	64.1
Average population density	1.1	1.9	2.6	1.2	2.1	3.6
Total biomass (g)	1192	3030	2942	1156	2079	2602
Average weight of the species (g)	34.8	41.8	24.1	19.9	21.0	20.3

Specific responses to insularity and the structure of communities

Such changes in the partitioning of species within a mosaic of habitats result in changes in the structure of each particular community. Table 3.1 gives the composition and some basic parameters of three stages of the succession. Although the Corsican forest contains five fewer species than its mainland counterpart, the total population density is slightly higher. This means that average population densities of the species present are higher on the island. The total biomass is slightly lower in Corsica than in Provence because birds are on average smaller on the island. The species that have greatly increased their densities in the species-poor Corsican habitats are those forest species that have invaded matorrals (marked with an arrow on Table 3.1) (see Blondel *et al*. 1988).

An explanation of the causal mechanisms that result in such changes in community structure cannot be achieved simply at the community level. An insight into such mechanisms requires an analysis of life histories of selected populations, especially of warblers and of tits because of their different reactions to isolation. For reasons of time and research effort only the tits have been studied in detail.

Population biology of tits in Corsica and on the mainland

Population biology of tits has been studied in the habitats where community changes are the largest, i.e. in old forest stands. Study methods and a description of the sites are given by Blondel (1985*b*).

Comparative demography of the Blue Tit

Life history traits of the Blue Tit *Parus caeruleus* are strikingly different in Corsica and on the mainland (Blondel 1985*b*, Fig. 3.4). On average,

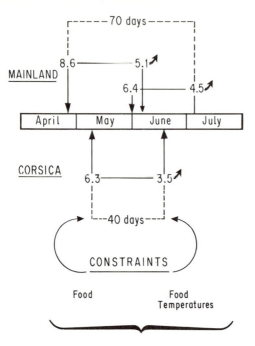

Fig. 3.4. Differences in breeding patterns between a mainland and an island population of Blue Tit. Figures indicate clutch size and numbers of fledglings. Mainland tits breed earlier, have a larger clutch size, and fledge more young than the island birds. Changes in breeding pattern on the island are due to trophic and climatic constraints. Hence the breeding cycle is much shorter on the island that on the mainland.

mainland tits start to breed about the 25th of April, lay 8.6 eggs, fledge 5.1 young, and 15.2 per cent of them attempt second broods with a clutch size of 6.4 eggs that produce 4.5 fledglings. On the other hand, Corsican tits start to lay, on average, about the 9th of May, lay 6.3 eggs, and fledge no more than 3.5 young. They never attempt second broods. Thus the breeding cycle is much more contracted and the breeding commitment much lower in Corsica than on the mainland (Fig. 3.4).

These results were first interpreted in the context of a trade-off between life history traits, especially between clutch size and survival (Cody 1966, 1971; Ricklefs 1980). High population densities in buffered and predictable island environments were supposed to result in high intraspecific competition, high adult survival, and low clutch size. Broadly speaking, mainland birds were thought to be selected for productivity in uncrowded populations because of environmental instability (r-selection), whereas island birds were hypothesized to be selected for efficiency in crowded populations because of environmental stability (K-selection) (MacArthur et al. 1972; Crowell and Rothstein 1981; Blondel 1985b).

However, the prediction of higher adult survival in Corsica than on the mainland, which is a keystone of the hypothesis stated above, has not been supported by further investigation. Estimates of survival over a 10-year period using the capture-recapture model of Jolly-Seber (Cormack 1964; Clobert *et al.* 1985) demonstrated that adult survival is of the same order of magnitude and even slightly lower in Corsica than on the mainland (Table 3.2).

This result contradicts classical theories of population biology in equitable island habitats because it does not support the contention of a direct inverse relationship between clutch size and survival. Indeed high adult survival is a component of Ashmole's and Cody's hypotheses on breeding strategies in stable tropical and island bioclimates (Ricklefs 1980; Crowell and Rothstein 1981).

Table 3.2 Survival rates of the Blue Tit in Mont-Ventoux (mainland) and in Corsica (Blondel *et al.* in preparation)

	Mont Ventoux		Corsica	
	Males	Females	Males	Females
1979/80	.56	.58	.67	.40
1980/81	.88	.68	.81	.92
1981/82	1.00	.64	.82	.60
1982/83	.74	.36	.69	.35
1983/84	.25	.29	.51	.50
1984/85	.55	.54	.35	.43
1985/86	.73	.83	.61	.64
1986/87	.65	.73		.64
Mean:	.67	.58	.64	.56

Population biology and environmental constraints

Because classical models of island biology were unable to explain changes in population biology of tits in Corsica, further investigations had to be made into the cost of reproduction and survival in relation to environmental constraints, especially climatic and trophic (see Drent and Daan 1980; Calow 1984; Sibly and Calow 1986; Martin 1987; Nur 1988).

Contrary to mid-European ecosystems, the most critical season for survival of land-birds in the Mediterranean region is probably not the winter, which is mild and moist in the Mediterranean region, but the summer which is dry and hot. High temperatures associated with drought are thought to limit the breeding season.

As to the trophic constraints, investigations of the leafing phenology of the dominant trees and of their associated insect fauna have highlighted two points of interest.

(1) The spring flush of insects in spring occurs later and more slowly in sclerophyllous oaks than in deciduous ones (Isenmann *et al*. 1987);

(2) According to preliminary studies of the caterpillar biomass in the foliage of the trees in Corsica (Zandt *et al*. in preparation), the productivity of caterpillars in sclerophyllous oaks is very low.

This late and low availability of food in the insular sclerophyllous forest forces the birds to start to breed very late, which brings the breeding event dangerously close to the unfavourable summer season. In the second half of June, high temperatures are probably a limiting factor because hyperthermia and water balance create problems for the young in the nest. Moreover the low availability of caterpillars may force the birds to switch to other prey items of lower quality. A study of the diet of the young by automatic photography inside nest boxes confirms this hypothesis (Blondel *et al*. in preparation). There are many fewer caterpillars in the diet of the Corsican nestlings and a greater diversity of prey than on the mainland (Fig. 3.5). On the whole the diversity of the diet, roughly expressed by the Shannon function H', is higher in Corsica than in Provence, even within the main prey taxa. The feeding niche of the Blue Tit in Corsica is definitely larger than on the mainland, but for other reasons than those usually given. These trophic constraints in Corsica have two implications.

(1) According to optimal foraging theory (Stephens and Krebs 1986) it should be more time-consuming, hence more costly, to search for a great many types of prey than to concentrate on just a few;

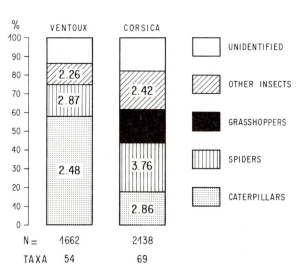

Fig. 3.5. Composition of the diet of nestling Blue Tit in Mont-Ventoux (Provence) and in Corsica. N = sample size (number of photographic registrations inside the nestboxes), taxa = numbers of different taxa in the diet, H' = Shannon's diversity index; figures in the columns are diversity indices (H') within the main food categories.

(2) The quality of the prey is not the same for all categories, especially its water content, which is highest in caterpillars and lowest in spiders with intermediate values for grasshoppers.

A high diversity of prey items that are poor in water content constitutes a strong constraint especially when temperatures are very high in the second half of the breeding cycle. In these conditions the only way to optimize the reproductive commitment, that is to optimize both the quality and survival of offspring and of the female, is to reduce clutch size.

Adult survival must be interpreted in relation to reproductive effort. Adult survival is of the same order of magnitude on the mainland and on the island (Table 3.2) despite a lower reproductive commitment on the latter. This suggests that the cost of reproduction is higher in Corsica. The summer is just beginning at a time when the birds have used much energy to reproduce and have yet to complete their moult, which is also energy consuming. These events occur just before or during the most unfavourable period of the year. It is hypothesized that a higher reproductive effort in Corsica would result in a higher adult mortality. However, no data are available to test this hypothesis. If this is true, a higher summer mortality in Mediterranean habitats should result in selective pressures against large clutch size. This conclusion supports the model of Ricklefs (1977) which tests the relationship between optimization of fecundity and components of mortality: 'Fecundity is determined primarily by density-dependent response to adult mortality mediated through resources available for reproduction' (Ricklefs 1977).

Although this very schematically summarized story of Corsican tits is far from fully understood, the best explanation of their biology is that a particularly unfavourable combination of trophic and climatic constraints squeezes the breeding event between two impassable limits (see Fig. 3.4). These limits are much closer to each other in Corsica than in any mainland habitat. An important point in this context of island studies is that local constraints that have nothing to do with community changes on species-poor island habitats explain the shape of life history strategies of Corsican tits better than any prediction from island models. This is an illustration of the observation that ecological processes at one level do not necessarily operate in a way that is predicted from investigations at other levels.

In conclusion, these case studies in Mediterranean islands provide three messages:

(1) Colonization and distributional patterns on islands must be studied in the light of their history;
(2) communities should not be investigated regardless of their neighbourhood because biological processes work at the level of a mosaic of habitat patches between which there may be exchanges; and

(3) population studies must take into account the complicated network of interactions that operate within each species-specific 'population-environment' system. Such case studies are an illustration of the limits to generalization resulting from theories that are too broad such as that of island biogeography and of demographic strategies.

The biology of isolates and conservation

Following the impetus of island studies in the late 1960s and 1970s, many attempts have been made to extract from island models principles for conservation and the design of nature reserves (e.g. Terborgh 1974, 1976; Diamond 1975a, 1976; Wilson and Willis 1975; Soulé and Wilcox 1980; Soulé 1986 etc.). This approach is fully justified and perfectly valid in theory because the process of isolate formation through habitat fragmentation is a process of insularization (Wilcox 1980). However, a pertinent critique of this approach has been made by several authors (e.g. Simberloff and Abele 1976; Abele and Connor 1979; Margules et al. 1982; Boecklen and Gotelli 1984), especially because the basic mechanisms underlying extinction and immigration are far from understood.

This section will mainly emphasize the problem of habitat islands in mainland areas. From a conservation viewpoint this is all the more important because among many ecological disasters, ecosystem fragmentation and especially massive deforestation in the tropics will probably entail a mass extinction crisis in the decades to come. Therefore since we are concerned with problems of scale, diversity, and scarcity, the challenge for conservationists is to predict the nature of the changes in community composition and structure and the likely processes of extinction as a consequence of decreases in habitat size and fragmentation.

The basic assumption of the theory of island biogeography is that any variation in the rate of change of the number of species S (ds/dt) on an island depends on the rate of immigration I into a patch of size A and the rate of extinction E within this patch. In the equilibrium hypothesis of island biogeography, I and E lead to a balance law of the form

$$ds/dt = I - E \qquad (1)$$

with $I = E$.

If I and E are different from zero, then there is a turnover of species, which is a keystone of the theory. If one were able to express accurately I and E as functions of the characters of the habitat (size and habitat heterogeneity) and the number S of the species present, then the equilibrium equation $I = E$ and, more generally, the integrative equation (1) would give a relation between S and A (Coleman et al. 1982). We are far from this goal because the theory entails several important shortcomings:

(1) it says nothing about the characterization of I and E as functions of the characters of the island and of the attributes of the species already present;

(2) it considers all the species as equivalent in their probabilities of immigration and extinction;

(3) it assumes that the location and magnitude of sources of immigration are known;

(4) our knowledge of population dynamics in natural populations is not sufficiently advanced to test the hypothesis that extinction is a function of the number of species already present on the island; and

(5) it does not take into account the history of the island and that of its species.

More generally, the theory does not take evolutionary processes into account. Nevertheless a substantial body of theory as well as empirical evidence support the existence of many changes in community composition and structure in species-poor isolates. The main changes may be brought about together in the so-called 'insular syndrome'.

The insular syndrome

It is beyond the scope of this paper to discuss in detail the components of the insular syndrome because for that an entire book would be necessary (see for instance discussions and references in MacArthur *et al.* 1972; Diamond 1975*b*; Abbott 1980; Williamson 1981; Bengtson and Enckell 1983; Haila 1983; Stamps and Buechner 1985; Blondel 1986). This syndrome involves ecological adjustments as well as evolutionary changes that are dissociated for didactic reasons only because ecology and evolution operate together in the process of biological integration on islands. Ecological adjustments include five main features:

(1) selection among the candidates to island colonization;
(2) density inflation;
(3) niche expansion;
(4) the development of sedentarity; and
(5) changes in territorial behaviour.

Evolutionary changes include modifications in body size (see Van Valen 1973) and shifts in life history traits, as shown for instance by the Corsican tits discussed in the previous section. Evolutionary changes have been extensively studied on oceanic islands. However, because strong differential selection can operate within a few generations on ecologically important life-history traits (Van Noordwijk *et al.* 1980; Van Noordwijk 1987), microevolutionary processes may occur very rapidly with important consequences to the structure and functioning of populations and communities. Therefore the increasing habitat fragmentation of forest ecosystems may have consequences for the genetic structure of populations in biological isolates.

Habitat fragmentation, species diversities, and extinction

Patterns and processes of community changes are particularly interesting to study in situations where a habitat becomes isolated, for instance as a result of the fragmentation of forest blocks. In such situations one can observe the process of relaxation from a state of supersaturation to a new equilibrium determined by the species-area relationship. Examples of losses of species as a result of fragmentation have been provided from British woodlots (Moore and Hooper 1975), the deciduous forest biome in the north-east United States (Forman *et al*. 1976; Galli *et al*. 1976; Robbins 1979, 1980; Whitcomb *et al*. 1981; Wilcove *et al*. 1986), and the newly created island of Barro Colorado, Panama (Willis 1974; Karr 1982).

It follows from the components of the insular syndrome that deforestation and ecosystem fragmentation are not only a matter of conservation of species, but also a concern of structures and systems. We must be aware of the fact that in temperate regions, especially in western Europe, most studies of populations and communities have been conducted in ecosystems so heavily modified by man for so many centuries that important ecological and evolutionary changes may have occurred in biological systems. Long ago, David Lack (1965) pointed out that it could be misleading to study evolutionary ecology in man-made habitats because they have been so greatly modified that adaptations may be hard to recognize. Bird communities are widely believed to be more resistant to fragmentation in the temperate zone than in the tropics. This is because species seem to occur in higher densities, to be more widely distributed, and to have better dispersal abilities (Wilcove *et al*. 1986). However, this is probably a fallacy. Habitat fragmentation seems to be less severe in the temperate zone because damage to the area occurred at a time long before the present stage of human ecological awareness.

The structure of communities in primaeval habitats

An insight into the changes in community structure as a result of isolation in western Europe may be obtained by a study of the composition and structure of the bird communities in the last remnants of a primaeval lowland forest of Europe, for example that of Bialowieza, Poland (Tomialojc *et al*. 1984). Of course, one must be cautious with such comparisons because bird faunas are on average richer in central than in western Europe. However, many community attributes that are more or less independent of the species richness should be the same in both regions under primaeval conditions. One of the most conspicuous features of community structures in the primaeval forest is its extreme richness of birds, but with the same or even lower overall densities as those of western bird communities (Table 3.3). This results in large differences in the distribution of individuals among species. Densities of individual species are, on average, very low in primaeval conditions: 38 species out of a total of 56 have densities lower than one breeding pair per 10 ha. For instance the density of the Blue Tit is only 2 breeding pairs per 10 ha in the Bialowieza forest as compared with 10–30 pairs in similar western isolated forest habitats (Dhondt and Eyckerman 1980). Consequently, at the scale of plots of 25 ha, the frequency of local extinction-recolonization events is especially high in the primaeval forest. For instance during a 5-year study of plots of 25 ha censused each year, Tomialojc *et al* (1984) found that no more than 18 species out of the 56 that were recorded bred at least once every year. Thus the local turnover of species is extremely high (Table 3.3) although the forest is supposed to be a stable and well-buffered environment. Such patterns as well as the structure of the forest are very similar to tropical patterns (Tomialojc *et al*. 1984).

Table 3.3 Attributes of a bird community in a plot of 25 ha in the primaeval forest of Bialowieza, Poland (after Tomialojc *et al*. 1984)

Dominant tree species	Oak, Hornbeam, Lime
Number of years of study	5
Total number of species having bred at least once	56
Mean annual number of breeding species	41
Average total density (pairs/10 ha)	75.9
Average species-specific density	1.37
Average number of species with density >10 pairs	1.0
Average number of species with density <1 pair/10 ha	38
Turnover rate	.25

Many ecological, evolutionary, and conservation conclusions may be drawn from such patterns:

(1) In large tracts of natural habitat such as the deciduous temperate forest or tropical forests, it is a mistake to believe that the size of any population is at an equilibrium determined by local habitat and resource conditions (Wiens 1983).

(2) Species distributions appear to be broken into a mosaic of 'sink' and 'source' subpopulations which exchange propagules according to local demographic situations. What is a disequilibrium at a very local

scale is actually a dynamic equilibrium at the larger scale of a landscape (Blondel 1986).

(3) The prime factor affecting transfers between subpopulations of the same neighbourhood is the specific regime of disturbance, both biotic and abiotic. In the forest of Bialowieza, predation pressures exerted by no less than 30 species of mammals and birds, appear to be the major force in the regulation of distributions and abundances at the scale of a large forest block (Tomialojc *et al*. 1984).

Species losses as a result of habitat fragmentation

Some generalizations and recommendations may be drawn from studies in habitat fragments. The survival of the pool of species at a regional level, which is a legacy of history, depends on three factors: habitat size, habitat heterogeneity, and changes in the structure of communities and populations which result from the first two factors.

Habitat size

The question of minimum viable population size has been omitted from many island models as a variable (Schoener 1976; Williamson 1981). Populations have demographic and genetic thresholds below which stochastic and rare events may lead to extinction. When the size of a habitat patch becomes too small to contain a sustainable population, stochastic extinction may result from normal random changes or environmental perturbation. This explains why large species that need, on average, more space per individual than small ones are more vulnerable to extinction than are smaller species; hence the disharmony of size classes in island communities. Expected lifetimes of small populations have been simulated by several authors (see for instance Richter-Dyn and Goel 1972; Leigh 1975; Shaffer 1981; Gilpin and Soulé 1986). Besides the demographic aspect, another point in closed populations is the relation between population size and genetic diversity. There are critical factors of population size and structure below which inbreeding and loss of selectable variation lead a population to extinction (Frankel and Soulé 1981; Schoenwald-Cox *et al*. 1983; Soulé 1987). This fact could have been responsible for the extinction of a small isolated population of Middle Spotted Woodpecker *Picoides medius* in Sweden (Pettersson 1985). Much more theoretical and empirical work is needed on this particular point, and it is discussed in other chapters of this book.

Habitat heterogeneity

Probability of extinction is not a simple function of fragment size; it depends also on habitat heterogeneity and on environmental stochasticity

(Gilpin and Soulé 1986). Even apparently uniform tracts of habitat such as the temperate primaeval forest are a mosaic of habitat patches, created mainly by the disturbance regime. The disturbance regime that is a specific attribute for each region (Loucks 1970; Sousa 1984; Pickett and White 1985) includes both abiotic events such as treefalls, storms, fires, droughts, cold spells, or floods, and biotic events such as disease, parasitism, and predation pressures of which the intensities vary in space and time. Habitat heterogeneity generated by disturbances guarantees the full array of habitat patches and is necessary for the survival of all species at a regional level (Blondel 1986). Assuming that different sets of species within the same forest block are adapted to the different types of habitat generated by disturbance within a mosaic, such as seral stages of successions, their survival requires the simultaneous existence at a regional level of all these habitats. Because of the random occurrence of disturbances in space and time, the full array of all habitat patches, the so-called 'metaclimax' (Blondel 1986, 1987), requires areas large enough for this regime to be fully applied. Levin (1976), Chesson (1978), Pickett and Thompson (1978), White (1979), Sousa (1984), Pickett and White (1985) among others have repeatedly pointed out the importance of disturbances for community ecology and the regulation of diversities at a regional level. Disturbances are unpredictable in space and time over ecological time, but their recurrent occurrence makes them predictable over evolutionary time, so that they are integrated in the evolution of species-specific life histories at a regional level.

Effects of rare climatic events such as unusually dry seasons that are critical for many species (Wiens 1977) may be overcome thanks to habitat heterogeneity, which buffers environmental stochasticity. Mobile animals such as birds shift their distribution in space as a response to short-term climatic variability provided that microhabitat use is possible on a larger geographic scale than just a single habitat patch. Part of the species losses in Barro Colorado Island have been attributed to this factor (Karr 1982). When the island was broadly connected to surrounding forests, the avifauna of what became Barro Colorado Island must have shifted somewhat from season-to-season and year-to-year during particularly severe dry seasons. Prevention of such movements as a result of isolation resulted in many extinctions. Since sedentarity is a component of the insular syndrome (Blondel 1986), migrant and mobile species should be especially vulnerable to loss of heterogeneity. Species that are the most vulnerable to forest fragmentation in eastern North America have been proved to be long-distance neotropical migrants. This is because of the unsuitability to forest fragments of several of their life history attributes: low annual reproductive effort, habitat specialization, low densities, dispersal strategies, nest type, and location that make them vulnerable to predation and give poor competitive ability against an influx of birds usually associated with more open habitats (edge effect, see below) (Robbins 1979, 1980; Whitcomb *et al*. 1981; Wilcove 1985*a*).

Changes in the structure of communities

Biological isolates are supposed to be composed of a small number of mostly sedentary species with high population densities. This results in changes in interspecific interactions, especially competition and predation. Competition is classically claimed to be a factor of prime importance for extinction on true islands (MacArthur 1972; Diamond 1975*b*). However, very few studies have demonstrated this to be the case. The fact that species expand their habitat niche in the absence of competitors on islands (MacArthur 1972) is not proof of competition, as shown for instance by our studies in Corsica. The same is true on habitat islands: for the 50 to 60 species that became extinct within 50 years on Barro Colorado, inter-specific competition has been ruled out by Karr (1982) as a causal mechanism.

On the other hand, because of very high population densities, new patterns of intraspecific competition probably emerge in biological isolates. These could result in changes in life histories as well as in behavioural and spacing patterns. For instance Dhondt and Schillemans (1983) have shown that in oak woodlots in Belgium, Blue Tits accept subordinates as a result of high intrusion rates in years of high population densities. Intruding pairs attempt to nest inconspicuously within the territories of established pairs. Moreover, in very high populations of Blue Tit in Belgium, there is an important proportion of polygynous birds (Dhondt 1987*a*, *b*). Such features that have never been found in less dense populations are presumably secondary adaptations such as those described in true islands by Stamps and Buechner (1985). Indeed, changes in territorial behaviour on islands include reduced territory size, increased territory overlap, acceptance of intruders within territories, and reduced situation-specific aggressiveness. Thus biological isolates probably evolve traits that are typical of insular situations.

Patterns of predation are more complex. Tomialojc *et al*. (1984) gave arguments supporting the view that predation is one of the main factors responsible for the regulation and the spacing patterns of birds in the primaeval forest of Bialowieza. Release from predation pressures is a classical feature in biological isolates with many consequences on populations. Predation within natural communites is, so to speak, endogenous and contributes to their normal functioning, but in isolated habitat fragments, another type of predation that could be called exogene-ous may cause extinctions. In forested fragments that are not isolated by an impassable barrier like true oceanic islands, the man-made matrix between fragments can be a barrier for inter-fragments colonists, but not for some predators that may become harmful to species within fragments (Wilcove *et al*. 1986). This is the so-called edge effect. For instance the number of species of songbirds sharply declines in small woodlots in the eastern United States as a result of excessive nest predation by exogenous predators (Robbins 1979; Wilcove 1985*b*) and by brood parasitism exerted by the Brown-headed Cowbird *Molothrus ater* (Mayfield 1977). A similar explanation has been given to account for many avian extinctions on Barro Colorado Island (Terborgh 1974; Karr 1982). This is an edge effect that gradually corrodes the communities on small habitat islands (see Wilcove *et al*. 1986).

Finally, besides classical interspecific interactions, more subtle mechanisms may be disrupted. These include some kinds of predator-prey, parasite-host, plant-pollinators, plant-dispersers, and mutualistic interactions (L. E. Gilbert 1980; Terborgh and Winter 1980). Extinctions of the so-called keystone species in complex webs of interactions may lead to additional and unexpected extinctions (Foster 1980; L. E. Gilbert 1980; Terborgh 1986).

Concluding remarks

Changes in community structure in forest fragments have had important consequences for ecological theory. Most studies in community and population ecology have been conducted in impoverished man-made forest fragments in western Europe and North America. In such habitats, the disturbance regime has been largely obliterated through forest management and the guild of predators dramatically reduced. Consequently, high and stable population densities in species-poor communities characterized by a low turnover rate have led to the concept of community equilibrium with populations near the carrying capacity. According to the Hutchinson–MacArthur theory of the niche, this equilibrium was believed to be mediated mainly by interspecific competition. In fact, this monolithic and deterministic view of community organization was at least partly due to the fact that communities and populations were studied in secondary man-modified habitats. Modern geographical ecology is more cautious, less dogmatic, and biologically more realistic because of the awareness that each island is a singular case with its own history and also because one takes into account the species-specific attributes of its elements. The process of extinction in biological isolates with which conservation is concerned is far more complex than can be predicted from simple habitat attributes such as size and degree of isolation. Species extinction cannot be pegged to one factor, because it is a matter of systems involving the interaction of many processses and states. This emphasizes the need to evaluate population and community dynamics under nonequilibrium conditions (Connell 1978) in both continental and island situations and carries an important message for studies on conservation and the design of nature reserves for the benefit of biological isolates.

Summary

Using the Mediterranean region and its bird faunas as a model, an insight is given to island patterns and processes at the three levels of regional faunas, communities, and populations. Because the organization of living systems is basically hierarchical, island patterns are sensitive to the scale of time and space at which they are considered. As a consequence, colonization and distributional patterns on islands must be studied in the light of their past history. Species impoverishment in Corsica is shown to vary according to the size of the area considered. Missing species are not a random sample of the species pool of the source on the mainland. For historical reasons, woodland faunas are much more impoverished in Corsica than shrubland faunas. The insular syndrome is discussed at both the community level and that of species-specific responses to insularity. Hypotheses raised at the community level were tested at the population level using long-term population studies of the Blue Tit in mainland and island habitats. Large differences in life history traits between the two populations do not fit classical models of population biology on islands. The shaping of life history strategies is better explained by local constraints than by any prediction from broad theories on island biogeography and/or demographic strategies.

Finally an insight is given to problems of conservation in habitat islands on mainland areas. It follows from the components of the 'insular syndrome' that habitat fragmentation is not only a matter of conservation of species, but also a concern of structures and systems. Examples are given comparing patterns in the primaeval forest of Bialowieza, Poland, to those reported in small woodland areas in western Europe. Some generalizations are drawn from studies in habitat fragments.

Acknowledgements

I am greatly indebted to C. M. Perrins who made many constructive comments and suggestions on a first draft of the manuscript and who improved the English.

References

Abbott, I. (1980). Theories dealing with the ecology of landbirds on islands. *Advances in Ecological Research*, **11**, 329–71.

Abele, L. G.and Connor, E. F. (1979). Application of island biogeography to refuge design: making the right decision for the wrong reasons. In *Proceedings first Conference on Scientific Research in the National Parks* (ed. R. M. Linn), pp. 89–94. U.S. Dept. of Interior, National Park Service Transactions and Proceedings Series 5.

Allen, T. H. and Starr, T. B. (1982). *Hierarchy: perspectives for ecological complexity*. Chicago University Press, Chicago.

Bengston, S-A. and Enckell, P. H. (1983) (ed.). Island ecology. *Oikos*, **41**, 296–547.

Beug, H. J. (1975). Changes of climate and vegetation belts in the mountains of Mediterranean Europe during the Holocene. *Bulletin Geology*, **19**, 101–10.

Blondel, J. (1979). *Biogéographie et écologie*. Masson, Paris.

Blondel, J. (1985*a*). Habitat selection in island versus mainland birds. In *Habitat selection in birds* (ed. M. L. Cody), pp. 477–516. Academic Press, New York.

Blondel, J. (1985*b*). Comparative breeding ecology of the Blue Tit and the Coal Tit in mainland and island mediterranean habitats. *Journal of Animal Ecology*, **54**, 531–56.

Blondel, J. (1986). *Biogéographie évolutive*. Masson, Paris.

Blondel, J. (1987). From biogeography to life history theory: a multithematic approach. *Journal of Biogeography*, **14**, 405–22.

Blondel, J. (1988). In *Acta XIX Congressus Internationalis Ornithologici* (ed. H. Ouellet), pp. 155–88.

Blondel, J. and Farré, H. (1988). The convergent trajectories of bird communities in European forests. *Oecologia (Berlin)*, **75**, 83–93.

Blondel, J., Ferry, C., and Frochot, B. (1970). La méthode des indices ponctuels d'abondance (I.P.A.) ou des relevés d'avifaune par 'stations d'écoute'. *Alauda*, **38**, 55–71.

Blondel, J., Chessel, D., and Frochot, B. (1988). Bird species impoverishment, niche expansion, and density inflation in Mediterranean island habitats. *Ecology*, **69**, 1899–1917.

Boecklen, W. J. and Gotelli, N. J. (1984). Island biogeography theory and conservation practice: species-area or specious area relationships? *Biological Conservation*, **29**, 63–80.

Burgess, R. L. and Sharpe, D. M. (ed.) (1981). *Forest island dynamics in man-dominated landscapes*. Springer-Verlag, New York.

Calow, P. (1984). Economics of ontogeny-adaptational aspects. In *Evolutionary ecology* (ed. B. Shorrocks), pp. 81–104. Blackwell, Oxford.

Chesson, P. (1978). Predator-prey theory and variability. *Annual Review of Ecology and Systematics*, **9**, 323–47.

Clobert, J., Lebreton, J. D., Clobert-Gillet, M., and Coquillart, H. (1985). The estimation of survival in bird populations by recaptures or sightings of marked animals. In *Statistics in ornithology* (ed. B. J. T. Morgan and P. M. North), pp. 197–213. Springer Verlag, Berlin.

Cody, M. L. (1966). A general theory of clutch-size. *Evolution*, **20**, 174–84.

Cody, M. L. (1971). Ecological aspects of reproduction. In *Avian biology*, vol. 1 (eds. D. S. Farner and J. R. King), pp. 462–503. Academic Press, New York.

Coleman, B. D., Mares, M. A., Willig, M. R., and Hsieh, Y-H. (1982). Randomness, area, and species richness. *Ecology*, **63**, 1121–33.

Connell, J. H. (1978). Diversity in tropical rain forests and coral reefs. *Science*, **199**, 1302–10.

Connor, E. F. and McCoy, E. D. (1979). The statistics and biology of the species-area relationship. *American Naturalist*, **113**, 791–833.

Connor, E. F. and Simberloff, D. S. (1983). Interspecific competition and species co-occurrence patterns on islands: null models and the evaluation of evidence. *Oikos*, **41**, 455–65.

Cook, R. E. (1974). Origin of the highland avifauna of southern Venezuela. *Systematic Zoology*, **23**, 257–65.

Cormack, R. M. (1964). Estimates of survival from the sighting of marked animals. *Biometrika*, **51**, 429–38.

Crowell, K. L. (1983). Islands—insight or artifact? Population dynamics and habitat utilization in insular rodents. *Oikos*, **41**, 442–54.

Crowell, K. L. and Rothstein, S. I. (1981). Clutch-sizes and breeding strategies among Bermudan and North American passerines. *Ibis*, **123**, 42–50.

Culver, D. C., Holsinger, J. R., and Baroody, R. (1973). Toward a predictive cave biogeography: the Greenbriar Valley as a case study. *Evolution*, **27**, 689–95.

Dhondt, A. A. (1987*a*). Reproduction and survival of polygynous and monogamous Blue Tits. *Ibis*, **129**, 327–34.

Dhondt, A. A. (1987*b*). Blue tits are polygynous but Great Tits monogamous: does the polygyny threshold model hold? *American Naturalist*, **129**, 213–20.

Dhondt, A. and Eyckerman, R. (1980). Competition and the regulation of numbers in Great and Blue Tit. *Ardea*, **68**, 121–32.

Dhondt, A. A. and Schillemans, J. (1983). Reproductive success of the Great Tit in relation to its territorial status. *Animal Behaviour*, **31**, 902–12.

Diamond, J. M. (1975*a*). The island dilemma: lessons of modern biogeographical studies for the design of natural reserves. *Biological Conservation*, **7**, 129–46.

Diamond, J. M. (1975*b*). Assembly of species communities. In *Ecology and evolution of communities* (eds. M. L. Cody and J. M. Diamond), pp. 342–444. Harvard University Press, Cambridge, Massachusetts.

Diamond, J. M. (1976). Island biogeography and conservation: Strategy and limitations. *Science*, **193**, 1027–9.

Diamond, J. M. and Mayr, E. (1976). Species-area relation for birds of the Solomon Archipelago. *Proceedings National Academy of Science U.S.A.*, **73**, 262–6.

Diamond, J. M., Gilpin, M. E., and Mayr, E. (1976). Species-distance relation for birds of the Solomon archipelago, and the paradox of the great speciators. *Proceedings National Academy of Sciences U.S.A.*, **73**, 2160–4.

Drent, R. H. and Daan, S. (1980). The prudent parent: energetic adjustments in avian breeding. *Ardea*, **68**, 225–52.

Forman, R. T., Galli, A. E., and Leck, Ch. F. (1976). Forest size and avian diversity in New Jersey woodlots with some landuse implications. *Oecologia (Berlin)*, **26**, 1–8.

Foster, R. B. (1980). Heterogeneity and disturbance in tropical vegetation. In *Conservation biology: an evolutionary-ecological perspective* (eds. M. E. Soulé and B. A. Wilcox), pp. 75–92. Sinauer, Sunderland, Massachusetts.

Frankel, O. H. and Soulé, M. E. (1981). *Conservation and evolution*. Cambridge University Press, New York.

Galli, A. E., Leck, C. F., and Forman, R. T. (1976). Avian distribution patterns within sized forest islands in central New Jersey. *Auk*, **93**, 356–65.

Gilbert, F. S. (1980). The equilibrium theory of island biogeography: fact or fiction? *Journal of Biogeography*, **7**, 209–35.

Gilbert, L. E. (1980). Food web organization and the conservation of neotropical diversity. In *Conservation biology: an evolutionary-ecological perspective* (eds. M. E. Soulé and B. A. Wilcox), pp. 11–34. Sinauer, Sunderland, Massachusetts.

Gilpin, M. E. and Soulé, M. (1986). Minimum viable populations: processes of species extinction. In *Conservation biology* (ed. M. Soulé), pp. 19–34. Sinauer, Sunderland, Massachusetts.

Haila, Y. (1983). Ecology of island colonization by northern land birds: a quantitative approach. Department of Zoology, University of Helsinki.

Haila, Y. and Järvinen, O. (1981). The underexploited potential of bird censuses in insular ecology. In *Estimating numbers of terrestrial birds* (eds. C. J. Ralph and J. M. Scott), pp. 559–65. Studies in Avian Biology No. 6.

Hamilton, T. H., Rubinoff, I., Barth, R. H. Jr., and Bush, G. L. (1963). Species abundance: natural regulation of insular variation. *Science*, **142**, 1575–77.

Hubbard, M. D. (1973). Experimental insular biogeography: ponds as islands. *Florida Scientist*, **36**, 132–41.

Isenmann, P., Cramm, P., and Clamens, A. (1987). Etude comparée de l'adaptation des mésanges du genre *Parus* aux différentes essences forestières du bassin méditerranéen occidental. *Revue d' écologie (Terre et Vie)*, Suppl. 4, 17–25.

Johnson, N. K. (1975). Controls of number of bird species on montane islands in the Great Basin. *Evolution*, **29**, 545–74.

Karr, J. R. (1982). Avian extinction on Barro Colorado island, Panama: a reassessment. *American Naturalist*, **119**, 220–39.

Karr, J. R. and Freemark, K. E. (1983). Habitat selection and environmental gradients: dynamics in the 'stable' tropics. *Ecology*, **64**, 1481–94.

Lack, D. (1965). Evolutionary ecology. *Journal of Ecology*, **53**, 237–45.

Leigh, E. G. (1975). Population fluctuations, community stability and environmental variability. In *Ecology and evolution of communities* (eds. M. L. Cody and J. M. Diamond), pp. 51–73. Harvard University Press, Cambridge, Massachusetts.

Levin, S. A. (1976). Population dynamics models in heterogeneous environments. *Annual Review of Ecology and Systematics*, **7**, 287–310.

Loucks, O. L. (1970). Evolution of diversity, efficiency and community stability. *American Zoologist*, **10**, 17–25.

MacArthur, R. H. (1972). *Geographical ecology*. Harper and Row, New York.

MacArthur, R. H. and Wilson, E. O. (1963). An equilibrium theory of insular zoogeography. *Evolution*, **17**, 373–87.

MacArthur, R. H. and Wilson, E. O. (1967). *The theory of island biogeography*. Princeton University Press, Princeton, New Jersey.

MacArthur, R. H., Karr, J. R., and Diamond, J. M. (1972). Density compensation in island faunas. *Ecology*, **53**, 330–42.

Margules, C., Higgs, A. J., and Rafe, R. W. (1982). Modern biogeographic theory: are there any lessons for nature reserve design? *Biological Conservation*, **24**, 115–28.

Martin, T. E. (1987). Food as a limit on breeding birds: a life history perspective. *Annual Review of Ecology and Systematics*, **18**, 453–87.

Mayfield, H. (1977). Brown-headed cowbird: agent of extermination? *American birds*, **31**, 107–13.

Mayr, E. and Diamond, J. M. (1976). Birds on islands in the sky: Origin of the montane avifauna of Northern Melanesia. *Proceedings of the National Academy of Science U.S.A.*, **73**, 1765–9.

Moore, N. W. and Hooper, M. D. (1975). On the number of bird species in British woods. *Biological Conservation*, **8**, 239–50.

Noordwijk, A. van (1987). Quantitative ecological genetics of Great Tit *Parus major*.

In *Avian genetics* (eds. F. Cooke and P. A. Buckley), pp. 363–80. Academic Press, London.

Noordwijk, A. van, Balen, J. H. van, and Scharloo, W. (1980). Heritability of ecologically important traits in the Great Tit. *Ardea*, **68**, 193–203.

Nur, N. (1988). The cost of reproduction in birds: An examination of the evidence. *Ardea*, **76**, 155–68.

Pettersson, B. (1985). Extinction of an isolated population of the Middle Spotted Woodpecker *Dendrocopos medius* (L.) in Sweden and its relation to general theories of extinction. *Biological Conservation*, **32**, 335–53.

Pickett, S. T. A. and Thompson, J. N. (1978). Patch dynamics and the design of Nature Reserves. *Biological Conservation*, **13**, 27–37.

Pickett, S. T. A. and White, P. S. (ed.) (1985). *The ecology of natural disturbance and patch dynamics*. Academic Press, New York.

Pons, A. (1981). The history of the Mediterranean shrublands. In *Maquis and chaparrals, coll. Ecosystems of the world* (eds. F. di Castri, D. W. Goodall and R. L. Specht), pp. 131–8, UNESCO, Elsevier, Amsterdam.

Preston, F. W. (1962). The canonical distribution of commonness and rarity. *Ecology*, **43**, 185–215; 410–32.

Quézel, P. (1985). Definition of the mediterranean origin of its flora. In *Plant Conservation in the Mediterranean area* (ed. C. Gomez-Campo), pp. 9–24. Dr W. Junk, Dordrecht.

Reille, M. (1984). Origine de la végétation actuelle de la Corse sud-orientale; analyse pollinique de cinq marais côtiers. *Pollen et Spores*, **26**, 43–60.

Richter-Dyn, N. and Goel, N. S. (1972). On the extinction of a colonizing species. *Theoretical Population Biology*, **3**, 406–33.

Ricklefs, R. E. (1977). A note on the evolution of clutch size in altricial birds. In *Evolutionary ecology* (eds. B. Stonehouse and C. M. Perrins), pp. 193–214. MacMillan, London.

Ricklefs, R. E. (1980). Geographical variation in clutch-size among passerine birds: Ashmole's hypotheses. *Auk*, **97**, 38–49.

Robbins, C. S. (1979). Effect of forest fragmentation on bird populations. *USDA Forest Service, General Technical report* NC-51, pp. 198–212.

Robbins, C. S. (1980). Effect of forest fragmentation on breeding bird populations in the piedmont of the Mid-Atlantic region. *Atlantic Naturalist*, **33**, 31–6.

Schoener, T. W. (1976). The species-area relation within archipelagoes: models and evidence from island land birds. *Proceedings 16th International Ornithological Congress*, pp. 629–42.

Schoenwald-Cox, C. M., Chambers, S. M., MacBryde, F., and Thomas, L. (eds.) (1983). *Genetics and conservation: a reference for managing wild animal and plant populations*. Benjamin Cummings, Menlo Park, California.

Shaffer, M. L. (1981). Minimum population sizes for species conservation. *Bio-Science*, **31**, 131–4.

Sibly, R. M. and Calow, P. (1986). *Physiological ecology of animals*. Blackwell, Oxford.

Simberloff, D. S. (1980). A succession of paradigms in ecology: Essentialism to materialism to probabilism. *Synthèse*, **43**, 3–39.

Simberloff, D. S. (1983). Biogeography: the unification and maturation of a science. In *Perspectives in ornithology* (eds. A. H. Brush and G. A. Clark, Jr.), pp. 411–55. Cambridge University Press, Cambridge, Massachusetts.

Simberloff, D. S. and Abele, L. G. (1976). Island biogeography theory and conservation practice. *Science*, **191**, 285–7.

Soulé, M. E. (1986) (ed.). *Conservation biology*. Sinauer, Sunderland, Massachusetts.

Soulé, M. E. (1987). *Viable populations for conservation*. Cambridge University Press, Cambridge, Massachusetts.

Soulé, M. E. and Wilcox, B. A. (1980). *Conservation biology: An evolutionary-ecological perspective*. Sinauer, Sunderland, Massachusetts.

Sousa, W. P. (1984). The role of disturbance in natural communities. *Annual Review of Ecology and Systematics*, **15**, 353–91.

Stamps, J. A. and Buechner, M. (1985). The territorial defense hypothesis and the ecology of insular Vertebrates. *Quarterly Review of Biology*, **60**, 155–81.

Stephens, D. W. and Krebs, J. R. (1986). *Foraging theory*. Princeton University Press, Princeton, New Jersey.

Terborgh, J. W. (1971). Distribution on environmental gradients: theory and a preliminary interpretation of distributional patterns in the avifauna of the Cordillera Vilcabamba, Peru. *Ecology*, **52**, 23–40.

Terborgh, J. W. (1973). Chance, habitat and dispersal in the distribution of birds in the West Indies. *Evolution*, **27**, 338–49.

Terborgh, J. W. (1974). Preservation of natural diversity: the problem of extinction prone species. *BioScience*, **24**, 715–22.

Terborgh, J. W. (1976). Island biogeography and conservation: strategy and limitations. *Science*, **193**, 1029–30.

Terborgh, J. W. (1986). Keystone plant resources in the tropical forest. In *Conservation biology* (ed. M. E. Soulé), pp. 330–44. Sinauer, Sunderland, Massachusetts.

Terborgh, J. W. and Winter, B. (1980). Some causes of extinction. In *Conservation biology. An evolutionary-ecological perspective* (eds. M. E. Soulé and B. A. Wilcox), pp. 119–33. Sinauer, Sunderland, Massachusetts.

Tomialojc, L., Wesolowski, T., and Walankiewicz, W. (1984). Breeding bird community of a primaeval temperate forest (Bialowieza National Park, Poland). *Acta Ornithologica*, **20**, 241–310.

Valen, L. van. (1973). Pattern and the balance of nature. *Evolutionary Theory*, **1**, 31–49.

Voous, K. H. (1960). *Atlas of european birds*. Nelson, Edinburgh.

Vuilleumier, F. (1973). Insular biogeography in continental regions. II. Cave faunas from Tesin, Southern Switzerland. *Systematic Zoology*, **22**, 64–76.

Whitcomb, R. F., Robbins, C. S., Lynch, J. F., Whitcomb, B. L., Klimkiewicz, M. K., and Bystrak, D. (1981). Effects of forest fragmentation on avifauna of the eastern deciduous forest. In *Forest island dynamics in man-dominated landscapes* (eds. R. L. Burgess and D. M. Sharpe), pp. 125–205. Springer-Verlag, New York.

White, P. S. (1979). Pattern, process and natural disturbance in vegetation. *Botanical Review*, **45**, 229–99.

Wiens, J. A. (1977). On competition and variable environments. *American Scientist*, **65**, 590–7.

Wiens, J. A. (1983). Avian community ecology: an iconoclastic view. In *Perspectives in ornithology* (eds. A. H. Brush and G. A. Clark) pp. 355–403. Cambridge University Press, Cambridge.

Wiens, J. A. (1985). Vertebrate responses to environmental patchiness in arid and semi-arid ecosystems. In *Natural disturbance: the patch dynamics perspective* (eds. S. T. A. Pickett and P. S. White), pp. 169–93. Academic Press, New York.

Wilcove, D. S. (1985*a*). Forest fragmentation and the decline of migratory songbirds. D.Phil. Thesis. University of Princeton.

Wilcove, D. S. (1985*b*). Nest predation in forest tracts and the decline of migratory songbirds. *Ecology*, **66**, 1211–4.

Wilcove, D. S., McLellan, C. H., and Dobson, A. P. (1986). Habitat fragmentation in the temperate zone. In *Conservation biology* (ed. M. E. Soulé), pp. 237–56. Sinauer, Sunderland, Massachusetts.

Wilcox, B. A. (1980). Insular ecology and conservation. In *Conservation biology: An evolutionary-ecological perspective* (eds. M. E. Soulé and B. A. Wilcox), pp. 95–117. Sinauer, Sunderland, Massachusetts.

Williamson, M. (1981). *Island populations*. Oxford University Press.

Willis, E. O. (1974). Populations and local extinctions of birds on Barro Colorado. *Ecological Monographs*, **44**, 153–69.

Wilson, E. O. and Willis, E. O. (1975). Applied biogeography. In *Ecology and evolution of communities* (eds. M. L. Cody and J. M. Diamond), pp. 522–34. Harvard University Press, Cambridge, Massachusetts.

Wright, S. J. (1980). Density compensation in island avifaunas. *Oecologia*, **45**, 385–9.

Estimating the Parameters

4 Estimation of demographic parameters in bird populations

J. CLOBERT and J.-D. LEBRETON

Introduction

Describing the population dynamics of a group of individuals requires some knowledge of its density, the number of births and deaths, and the number of individuals moving in or out (Lack 1954; Von Haartman 1971; Ricklefs 1973; Eberhardt 1985). Population biologists who wish to measure these processes must focus part of their attention on how to estimate demographic parameters (Lincoln 1930; Jackson 1933, 1939; Ricker 1956; De Lury 1947; Deevey 1947; Bailey 1951). However, it is extremely difficult to obtain comprehensive data on all aspects of the demography of a population. Usually only partial information about the life and death of a variable proportion of the population can be gathered. Thus in practice, demographic parameters are often estimated by sampling the population (involving some assumptions) and using appropriate statistical techniques.

A review of all the demographic parameters and of their associated methods of estimation is outside the scope of this paper. Several recent books and meetings have reviewed the methodology (Cormack *et al.* 1979; Amlaner and Macdonald 1980; Ralph and Scott 1981; Manly 1985*a*; Brownie *et al.* 1985; Morgan and North 1985; Seber 1986; Burnham *et al.* 1987; North 1987; Blondel and Frochot 1987). We focus on the following basic demographic parameters: fecundity; age-specific proportion of breeders; immigration and emigration rates; survival rates and population density. These parameters were chosen on the following basis. Firstly, knowledge of them makes it possible to describe fully the population dynamics of a species in terms of flows. Secondly, most of the other parameters can be obtained by a combination of these. Not all the parameters will be treated in the same way; some, such as fecundity, are easier to estimate than others, but can raise problems of definition. On the other hand, survival rate, age-specific proportions of breeders, and emigration are technically difficult to estimate. Some parameters are poorly studied (e.g. age-specific proportions of breeders, emigration, immigration). Some methods or type of data which relate to very specific situations have been deliberately omitted from this review.

Most of the biological significance of a demographic parameter is determined by the population studied. In particular, the definition of a population will depend on its geographical context (whether isolated or not . . .), the investigator's interest (genetical, demographic . . .), and the methodology used, or will vary from one species to another. In this paper, 'population' most of the time refers to a group of reproducing individuals at some point in time and space which has on average closer interactions between them than with members of other groups. This definition is obviously wrong in several cases (the chapters on the survival estimation using recoveries and in some parts of the chapter on population size estimation), but it allows a clear understanding of most of the other parameters.

Lastly, it would have been more logical to treat the estimation of density first. Most applied studies are concerned first with this parameter. However some of the techniques required to estimate our stated parameters are common to the estimation of some other parameters, and can be more easily introduced when the elementary parameters and their associated techniques of estimation have been considered.

Fecundity

Definitions of fecundity are numerous and sometimes contradictory (Krebs 1972; Southwood 1978; Ricklefs 1980; Begon *et al*. 1986; see Table 4.1). In life table analysis, age-specific fecundities are currently defined as: the expected number of 'daughters' produced by a female of age x at time t (Leslie 1945, 1966; Ricklefs 1980; Charlesworth 1980; Rose 1987). The term 'daughter' is taken as the number of female zygotes by Charlesworth (1980) but as the number of fledgings by Leslie (1966). For theoretical purposes, the first definition seems preferable but the second one is more relevant on practical grounds. Also from a functional point of view, the number of fledged young per breeding pair is better related to body size than any other measure of breeding success (Allainé *et al*. 1987). Egg and nestling survival contributes to little variation in

Table 4.1 Some definitions of fecundity

Andrewartha and Birch (1954): number of offspring produced in unit time by a female aged x.

Krebs (1972): potential capability of an organism to produce reproductive units such as eggs and sperms.

Charlesworth (1980): expected number of female zygotes produced by a female aged x in year t (age-specific fecundities).

Ricklefs (1980): rate at which an individual produces offspring, usually expressed only for females.

Begon *et al*. (1986): total number of eggs deposited during each stage (fecundity schedule).

---

lifetime reproductive success (Van Balen *et al*. 1987; McCleery and Perrins 1988). Estimating the number of fledings requires regular inspection of nest activity. From nest initiation this can be achieved in small-scale studies using nest boxes or in species where nests can be easily found and followed (e.g. in raptors where incubation and nestling periods are long). When information about parents is collected at the same time, age-specific fecundity and the percentage of second or replacement clutches can be obtained. Most of the time, the analysis of such data uses standard statistics.

The use of nest boxes can lead to an overestimation of the fledging success. The differential use of next boxes by different classes of individuals can also lead to a misleading estimate of the mean age of the breeders (Van Balen *et al*. 1982; Van Balen 1984; Nilsson 1984; Gustafsson and Nilsson 1985; Slagsvold 1987). Overestimation can also arise from the fact that small-scale studies are often done in good habitats, with associated problems of pseudoreplication (Hurlbert 1984). When the observer interferes with the reproductive activity, underestimation is expected (Willis 1973; Strang 1980; Nichols *et al*. 1984; Vacca and Handel 1988).

Studies of fecundity on a medium or large scale, or on difficult species or rare species usually means that each nest is followed less intensively and that information about parents cannot be gathered. Most nests are not found at the same stage and when nesting success is calculated this must be taken into account. Mayfield (1961, 1975) devised a method for estimating daily nest survival rates of the various stages of reproduction. Early developments of this approach were aimed at providing a better statistical framework (e.g. maximum likelihood, variance estimates: see Johnson 1979; Hensler and Nichols 1981) and to provide tests of basic model assumptions (for example, time of loss of the unknown nest, Johnson 1979; variable daily survival rate, Johnson 1979; Bart and Robson 1982; Hensler 1985; not constant sampling, Bart and Robson 1982). A new promising extension to the method estimates nest survival probabilities taking into account age at discovery of successful nests (Pollock and Cornelius 1988). Furthermore, data of this type can be analysed using the computer program SURVIV (White 1983) which allows greater flexibility in model testing (for a review, see Cornelius and Pollock, submitted). Further developments should allow stage length to be variable according to clutch size or number of offspring. Unfortunately, these methods cannot provide estimates of age-specific fecundity. Nevertheless when sampling is random with respect to the age of parents, the mean value obtained can be used for all age categories in Leslie matrix models (Lebreton 1981).

It is sometimes very difficult to follow chicks of a particular brood up to fledging (e.g. nidifugous species). In such cases, it is practical to define fecundity as the number of hatchlings per breeding pair rather than using ratio of young to adults, which relies on the assumption that young and adults have equal detection probabilities. In some situations (nidification areas not accessible for political or logistical reasons), the latter method can be useful (e.g. geese and waders), but only when applied on a wide scale, to avoid the problem of differences in the geographical and temporal distribution of adults and young within the wintering zone. When fecundity is calculated as hatchlings per breeding pair, most of the mortality arising during the 'nestling' period is incorporated into the juvenile survival rate. Clearly it is important to be consistent in the definition of fecundity.

In the formal definition of fecundity, the number of female offspring is divided by the total number of females of age x, irrespective of whether they reproduce or not. Techniques for estimating age-specific proportions of breeders are quite different from those for estimating fecundity and, in birds, it is easier to use the number of breeding females when measuring fecundity.

Generally speaking, fecundity is a demographic parameter which does not raise major problems in its estimation. Providing sampling designs take into account habitat and/or seasonal variability, long-term trends can be assessed. Nevertheless, the efficiency with which observers find nests and the possibility of induced additional mortality may present problems in interpreting observed changes. In all cases, accuracy of fecundity estimation is especially important when population growth rate is particularly sensitive to fecundity parameters as in many Galliformes and Anseriformes (see Lebreton and Clobert, this volume).

Age-specific proportion of breeders

There is much evidence that breeding densities are partly limited either by the availability of nests sites or by spacing behaviour (Lack 1954; Tinbergen 1957; Wynne-Edwards 1962; Brown 1969; Klomp 1972; Newton 1976; and this volume). The presence of a surplus of potential breeding adults (mainly young) has been demonstrated experimentally by removal of breeding birds (Hensley and Cope 1951; Stewart and Aldrich 1951; Watson 1967; Krebs 1977; Potts *et al.* 1980; Village 1983; Gauthier and Smith 1987; Pedersen 1988), although in some of these experiments we do not know whether the replacement birds were non-breeders or whether they moved from other territories. Other assessment is by comparing densities of breeders in areas with and without nest-boxes (Von Haartman 1971; Van Balen *et al.* 1982; Brawn and Balda 1988).

Although age at first breeding is a potentially important factor in population regulation (Birkhead and Furness 1985; Porter and Coulson

1987; Brawn and Balda 1988) as well as in life history theory (Stearns and Crandall 1981; Prout 1984; Stearns and Koella 1986), very few attempts have been made to estimate age-specific proportions of breeders (Barrat *et al*. 1976; Chabrzyk and Coulson 1976). Methods are usually crude, using the distribution of ages when birds are seen to breed for the first time, without correcting for variation in survival rates and catching effort (Wooler and Coulson 1977; Duncan 1978; Finney and Cooke 1978; Coulson *et al*. 1982; Weimerskirch and Jouventin 1987; for measurements of life-time reproductive success, see also Brown 1988).

This lack of formal analyses arises because it is difficult to obtain separate estimates of the proportion of breeders and the survival/return rate, requiring complex statistical modelling. Very recently, attempts have been made partly to fill this gap. As it is rarely possible to estimate the non-breeding part of the adult population (i.e. birds having bred once), all the models built up to now assume that there is an age at which all the birds are breeders (see Lebreton and Clobert, this volume). A first approach (Lebreton *et al*. in press) develops the estimation of age-specific proportions of breeders using data in which birds born in the same year are all marked with the same type of tags, in a way similar to that of Chapman and Robson (1960) for estimating survival rates (Table 4.2). Assuming that survival rates of breeders and non-breeders are the same, that survival rates are not year specific and that resighting probabilities are not age specific, various models (different full breeding ages, equality or not of recruitment rates per cohort, equality or not of resighting probabilities) can be fitted using the statistical package GLIM (Baker and Nelder 1978).

When individual capture-recapture histories are available some of the assumptions (e.g. constancy of survival rate) can be relaxed (Clobert *et al*. in preparation). As these models (easily built using SURGE version 4.0, Pradel *et al*. 1990) have not yet been applied to many data, it is difficult at this stage to know the importance of the various assumptions. One assumption, full breeding, has a potentially large effect. The model assumes that all birds, or at least a constant proportion of them, become breeders from a particular age whereas it is known that in some species breeding activity decreases in the oldest birds (Dhondt 1989). These estimates are all based on resightings or recaptures of young which return to their natal area to breed. The fact that favourable habitats are very often chosen for such studies, may introduce bias.

Much remains to be done in this field. There is an urgent need to quantify the nonbreeding part of populations and to develop models allowing a greater flexibility in the estimation of these parameters. Further analysis of existing data and appropriate sampling designs (different interconnected populations followed up in the same time) are some immediate requirements. Obviously, the estimation of these parameters

Table 4.2 Probability of resighting or of recapture rate

C	N	Years of resightings (recapture as breeders)		
		1	i	m
1	N_1	$N_1(s_{11} \times \pi_{11})\alpha_{11}p_{11}$	$N_1(s_{11} \times \pi_{11})\cdots(s_{12} \times \pi_{12})(S_{1i} \times \pi_{1i})\alpha_{1i}p_{1i}$	$N_1(S_{11} \times \pi_{11})(s_{12} \times \pi_{12})\cdots(s_{1m} \times \pi_{1m})\alpha_{1m}\ p_{1m}$
i	N_i		$N_i(S_{ii} \times \pi_{ii})\alpha_{ii}p_{ii}$	$N_i(s_{ii} \times \pi_{ii})(s_{i,i+1} \times \pi_{i,i+1})\cdots(s_{im} \times \pi_{im})\alpha_{im}\ p_{im}$
				$\alpha_{im} - i + 1\quad p_{im}$
m	Nm			$N_m(s_{mm} \times \pi_{mm})(s_{12} \times \pi_{12})\alpha_{m1}\quad p_{mm}$

In addition to classical hypotheses in C-M-R as independence between individuals:

Set of hypotheses for cohort data analysis	For all i, j:	$s_{ij}\,\pi_{ij} = \phi_{ij}$
	For all $j > i$:	$\phi_{ij} = \phi$
	For all $j - i + 1 > k$	$\alpha_{ij} = 1$
Set of hypotheses for individual marking data	For all $j - i + 1 > k$	$\alpha_{ij} = 1$

C, cohort marking; N, number ringed; s, survival rate; π, return rate; $s\pi = \emptyset$ apparent survival rate; α, age-specific survival rate; ρ, capture, recapture, or resighting rate; k, full breeding age.

requires long-term studies. This will probably lead us to reconsider the role of age-specific proportions of breeders in short-lived species, which has been largely neglected up to now.

Survival rate

The literature on survival estimation based on recoveries and on live recaptures has increased markedly in the past 15 years (Seber 1986; Brownie *et al*. 1985; Burnham *et al*. 1987). The resulting increase in the relevance of the methods used is well illustrated in Fig. 4.1, where much of the trend observed is confounded with the different types of methodology (recoveries or capture-recapture) employed. In birds, two main types of data are used to estimate survival rates:

(a) ringing recoveries of individuals usually recorded as dead, shot, or found dead; and
(b) live recaptures of an individual either one or several times.

Until recently, these two types of data have generated relatively independent methodological developments (Burnham *et al*. 1987).

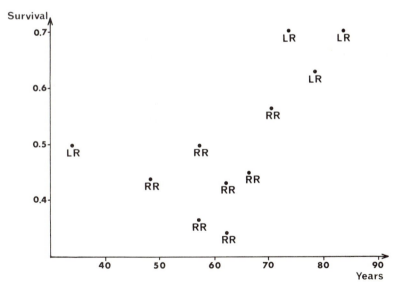

Fig. 4.1. Increase of survival rate through time in Starling, as a probable result of increase in the refinement of the methodology. Much of the trend is confounded by the different methodologies used.
LR, estimation from live recapture; RR, estimation from ringing recoveries.

Ringing recoveries

A schematic illustration for modelling data from recoveries of dead birds is given in Fig. 4.2A. In the fifties, simple models assumed that all birds were dead at the end of the period of analysis (Lack 1951) and that survival was constant over time and age (Lack 1951; Haldane 1955). Similar to Lack's method was the so-called composite dynamic life table (Hickey 1952, age-dependent survival rate). Although the Haldane model would probably be fairly robust if all that is required is an average adult survival rate of a species or population, the strong assumptions that underlie these methods are rarely met (Burnham and Anderson 1979; Anderson *et al.* 1981, 1985; Seber 1986). Subsequently, models involving age-dependent and time-dependent survival rates as well as time-dependent reporting rates were developed (age-dependent survival rate, Cormack 1970; Seber 1971; Lebreton 1977; time-dependent survival rate, Seber 1970; Aebischer 1987; mixed, Brownie *et al.* 1985). However, problems of identifiability which are not always recognized (Piper *et al.* 1981), particularly in age-dependent models (Burnham and Anderson 1979; Lakhani and Newton 1983) are important when data on birds ringed as young are used. Goodness-of-fit and likelihood ratio tests do not help much in this context (North 1987). To overcome such problems, Lakhani and Newton (1983) and North and Cormack (1981) have

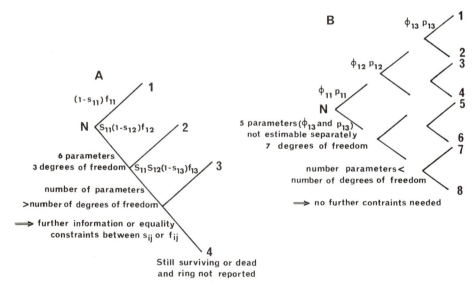

Fig. 4.2. Data presentation for ringing recoveries (A) and recaptures of live birds (B). f, reporting rate; s, survival rate; p, recapture or resighting rate; ϕ, apparent survival rate.

advised the use of independent information (e.g. live recaptures). Two other approaches are:

(a) The use of data from birds ringed as young, as 1-year-olds, and adults (Johnson 1974; Brownie and Robson 1976; Brownie *et al*. 1985).

This seems the most advisable although problems of identifiability can still be present (Lakhani 1987; but see North 1987, pp. 151–69), and also augmentation of the data is impossible for some species. Freeman and Morgan (personal communication) suggest that the identifiability problem in the age-specific model can be solved by making first year survival rates time-dependent.

(b) To analyse variation in survival rates in relation to external variables (North and Morgan 1979).

Building intermediate models with constraints on subsets of parameters and then comparing them is made easy by the development of flexible software, especially when numerical procedures are needed to estimate parameters (Lebreton 1977; White 1983; Conroy and Williams 1984).

Sources of variation in survival rates (age, time, sex) and particularly in reporting rates (age, time, localities, delay in reporting rates) are numerous (Brownie *et al*. 1978; Anderson and Burnham 1980; Clobert 1982; Nichols and Hines 1983). This heterogeneity can cause important biases (Johnson *et al*. 1986). Some methods to detect or account for them have been developed (Pollock and Raveling 1982; Nichols *et al*. 1982; Schwarz *et al*. 1988). However, dealing with ringing recoveries is still difficult and requires a very careful and sophisticated analysis. Furthermore, when individuals are reported dead from shooting, problems of compensatory mortality may further complicate the analysis (Nichols *et al*. 1984; see this volume). Nevertheless, the availability of software (SURVIV: White 1983; Brownie *et al*. 1985) makes it possible to take into account most of these problems, by comparing models with various constraints.

The development of powerful goodness-of-fit tests (see Brownie *et al*. 1985; and computer program ESTIMATE) increases also the possibility of dealing with such problems. In this context, having the information on the numbers ringed is essential. Finally, we have to remember that these data are often the only way to estimate survival rates of young.

Live recaptures

Survival rate estimation using multiple recaptures of live birds (Fig. 4.2B) first arose as a correction factor for population size estimates (Clobert and Lebreton 1987) in the so-called 'open population' case, i.e. population submitted to death-emigration, and birth-immigration (Jolly 1965; Seber

1965). One year earlier, Cormack (1964) developed a method using capture-resighting data which was subsequently shown to be the same as the survival part of the Jolly–Seber model (Seber 1973; Begon 1983; Brownie and Robson 1983; Brownie 1987; Clobert and Lebreton 1987).

These original models consider time-dependent survival rate and capture rate, plus time-dependent population sizes and recruitment rates, in the case of Jolly-Seber model. They rely on the following set of assumptions (from Seber 1973):

(a) every animal, whether marked or not, has the same probability of being caught on one particular occasion (no trap response, no age variation);
(b) every marked animal has the same probability of surviving between two sampling occasions (no age variation, no heterogeneity);
(c) marked animals do not lose their marks; and
(d) all samples are instantaneous.

A large number of parameters are also necessitated, sometimes to the detriment of precision (for a critical review of its use see Cormack 1979; Begon 1983).

In putting emphasis on the modelling of survival, population size estimation has become a by-product. Part of the recent effort aimed at developing models that reduce the set of assumptions concerning age-dependent survival and/or capture rate (Pollock 1981*b*; Stokes 1984; Clobert *et al*. 1987); models including trap response (Sandland and Kirkwood 1981; Cormack 1981); or models with a reduced number of parameters (survival rate and/or recapture rate constant: Jolly 1979, 1982; Sandland and Kirkwood 1981; Crosbie and Manly 1985; Clobert *et al*. 1985; Brownie *et al*. 1986). As with recoveries, the pursuit of generality (less constraining assumptions) and parsimony (models that describe the biological situation well using the fewest number of parameters) has led to the development of computer programs for fitting a wide range of models (Jolly and Dickson 1980; Arnason and Baniuk 1980; Arnason and Schwarz 1986; Crosbie and Manly 1985; Burnham *et al*. 1987; Lebreton and Clobert 1986). In some cases, models can be fitted using existing statistical software (GLIM: Cormack 1981, 1985; SAS: Burnham 1989). Crosbie and Manly (1985), Clobert and Lebreton (1985) and Clobert *et al*. (1987) analyse survival rates in relation to some external variables. Refined ways to compare models and compute goodness-of-fit tests for heterogeneity in survival and capture rate (including trap response) have also been developed by Brownie and Robson (1976), Pollock (1981*b*); Pollock *et al*. (1985); Brownie *et al*. (1986); and see particularly Burnham *et al*. 1987. In these developments, most of the models have no explicit analytic solution and require the use of numerical procedures. Estimates outside of the range [0.1] can arise and cause problems for the estimation of sampling variances. Techniques to improve variance estimates are provided by Buckland (1980, 1984, and in preparation), using bootstrap and Monte Carlo confidence intervals. Survival estimates from telemetry and radiotracking data (Heisey and Fuller 1985) can be easily obtained using existing models and corresponding computer packages with capture rates set equal to one, or by using continuous models and proportional hazards models available from standard statistical packages (Bunck 1987; Pollock *et al*. submitted).

Unfortunately, these methods are scattered over a number of different areas and use different notations and computer programs, making them unavailable to many biologists and even some specialists in these fields. Evidently, some unifying developments are needed (Seber 1986); recently Burnham *et al*. (1987) have provided both a good update of the notation and theory and a comprehensive computer program (RELEASE) for goodness-of-fit testing. Most of the models described can be fitted using program SURVIV (White 1983), JOLLY (Brownie *et al*. 1986) or SURGE (Lebreton and Clobert 1986; Pradel *et al*. 1990).

Although methods to estimate survival rate, using live recaptures are well-developed, some weaknesses remain. In particular, the necessary intensive data collection is often carried out on a relatively small scale. As a result:

(a) the survival rate estimates do not necessarily apply to the entire population, particularly if biologists choose good areas to carry out their studies.

(b) survival rates are nearly always underestimated (see Fig. 4.1) by an unknown factor because of birds moving permanently out of the study area. The problem is acute for birds living in non-permanent habitats, like the Bee-Eaters (*Merops* sp.), the Kingfisher (*Alcedo atthis*), and the Bank Swallow (*Riparia riparia*).

(c) juvenile survival rates from birth to first breeding are always underestimated because of confounding with return rate to the study area (Greenwood 1980).

Live recaptures and ringing recoveries: a unified approach

In Table 4.3, the main characteristics of the two types of data are summarized. Both approaches have different relative strengths and weaknesses and so combining the information in some way is warranted. On the statistical side, the structure underlying the two types of data is quite similar (Brownie *et al*. 1985; Brownie and Pollock 1985; Burnham *et al*. 1987) and some important work has already been done on this aspect (Buckland

Table 4.3 Main characteristics of ringing recoveries data and live recapture data compared

Level of comparison	Ringing recoveries	Live recaptures
Data available	Numerous	Less numerous
Scale of application	Wide	Narrow
Precision	Weak	Medium to high
Biases	Many	Few
Juvenile survival rate estimation	Possible but often biased	Rarely possible
Models available	Many	Many
Possibility of model development	Weak to medium	Important
Use in experiments	Weak	Strong
Link with other parameters	Weak	Important

1980; Burnham, in preparation). Experimental designs using live recaptures and ringing recoveries to estimate survival rate on a wide scale have also been proposed (Buckland and Baillie 1987) and may provide a basis for a co-ordinated approach by different ringing centres. Such studies of subdivided populations (set of colonies, populations in patchy environments . . .) are badly needed in order to estimate the juvenile survival rates under good conditions (see Manly 1985*b* for some methodological developments on this point) (see also Lebreton and Clobert, this volume). Extra information on winter recapture or resightings is worth including in the analysis to increase sample size and provide better coverage over critical periods. Finally, the specific problem of density-dependent mortality needs to be studied carefully. Correlations between survival and densities in the presence of abiotic variables require careful interpretation and further experimental study.

Immigration and emigration rates

The scope of this paper is to focus our attention on parameters directly linked to changes in population size from one life cycle to the next. Movements associated with territory defence, the search for food in the home range, or migration between breeding sites and wintering grounds are not considered here, although there can be a major effect on immigration-emigration processes (Krebs 1972).

To estimate these rates, one needs to know something about the pattern of movements between breeding sites or between natal sites and breeding sites. An important part of this section is devoted to ways of estimating this kind of dispersal.

For technical reasons, emigration and immigration rates are not estimated in the same way. While (age-specific) emigration rates are expressed as the percentage of surviving birds which definitely leave the study site, immigration is measured as the number of birds coming into the study site (Jolly 1965; Seber 1965).

Immigration

There are very few definitions of immigration rates in textbooks on population ecology (see Krebs 1972; Begon *et al*. 1986) which may be partly explained by the difficulties of measurement that are encountered. By definition, immigrants are coming from other populations and a proper understanding of immigration patterns requires sampling of several neighbouring populations at the same time. Very few attempts have been made in modelling immigration patterns (see Crosbie and Manly 1985).

When birth cannot be separated from immigration, the joint effect is often called recruitment rate. Jackson (1939) provides a way of separating the two effects which we discuss later in the section on emigration.

Emigration rates

Natal dispersal can be defined as the movement of individuals from their birthplace to their first breeding area (Berndt and Stenberg 1968; Greenwood 1980; Shields 1983; Von Bauer 1987). Breeding dispersal between two breeding sites can occur subsequently.

Dispersal is a central theme (Gadgil 1971; Begon *et al*. 1986) and has generated a lot of theoretical work (Horn 1983) in life history theory (for example Hamilton and May 1977; Comins *et al*. 1980; Stenseth 1983) and in population regulation theory and conservation management (Lidicker 1962; Gadgil 1971; Levin 1976; Soulé and Simberloff 1986; Gilpin 1987). However, despite numerous hypotheses and partial analysis of existing data (Greenwood 1980; Greenwood and Harvey 1982; Horn 1983; Shields 1983; Moore and Ali 1984; Dobson and Jones 1985; Pusey 1987), little is known about patterns and processes of dispersal.

Also, there have been few methodological developments in data analysis (Levin 1985). Two types of ringing data can be analysed: (i) location of ringing recoveries and (ii) location of live recaptures. Data provided by radio-tracking is another possibility but is currently limited to short-term studies.

Ringing recoveries

One can use locations of ringing recoveries to study dispersal distances of birds ringed as breeders or nestlings and found dead during one of the following breeding seasons. Very few statistical developments have been made using these data (for a review see Von Bauer 1987). One approach is to model dispersal distances using some *a priori* distribution such as an exponential (for a review see North 1978, 1985), or some composite distribution if dispersal distances are generated by two types of dispersal behaviour. North (1978, 1985) proposed a more refined approach using random walk and stepping-stone models with some position dependence in the probability of moving. Using these models the proportion of birds remaining on their marking, natal, or breeding area could be calculated according to age. Unfortunately, available data rarely offer such a possibility. Firstly, the precision of the location is often poor. Birds marked or recovered within about 5 km are often considered as not having moved. As the distribution of dispersal distances is often highly skewed towards the origin (Greenwood 1980; Shields 1983), most of the movements are not adequately measured. Although dispersal over larger distances are relevant from a conservation standpoint (recoveries are important data in this respect), the shape of the dispersal curve within 5 km of the place of ringing can be of some importance in landscape ecology. Secondly, recovery rates must be constant with respect to age, time, and more critically to area. At the same time, the probability of death has also to be constant with area which is not often the case (Cavé 1968). Lastly, as the breeding period is usually the period in which the fewest number of birds die (at least from natural causes), data are scarce. At the present stage, there is therefore little point in developing sophisticated models for describing the shape of the dispersal curve from recoveries. However they can be useful for comparing the dispersal of different groups, provided they are analysed with caution. In order to estimate dispersal curves, more precise location of ringing and recovery places is obviously required. Models incorporating time between ringing and recovery (Manly 1977*a*), together with the use of live recapture data with careful planning are also needed.

Live recapture

We have seen that survival rate estimates using live recaptures are always minimal estimates because study areas are often severely limited in size. Some breeders can permanently leave the study site, but this is much more important when juvenile survival rates are considered. There are several ways of estimating the intensity of dispersal, which have not yet been fully developed. Comparing observed distances with possible distances (Greenwood *et al*. 1979; but see Van Noordwijk 1984) can give useful information about the magnitude of the phenomenon. One problem is that birds around the border are more likely to settle outside the experimental area. In order to take this into account, one has to assume that the distribution of potential breeding sites is similar in and out of the study site. If the assumption holds, one can evaluate the emigration rates by using the observed distances to estimate the proportion of missing birds. This approach does not apply to population in nest boxes or in areas chosen for their high breeding potential.

Few methods propose simultaneous estimation of survival and emigration rates using the distribution of dispersal distances. Jackson (1939), Darroch (1961), Manly (1985*b*) and a few others (see Seber 1973) propose a method which divides the study area into several parts and estimates movements between sub-areas. However, the assumptions about equality of breeding potential in and out of the study site is critical (and partially excludes the study of nest box populations). Furthermore, any polymorphism in dispersal behaviour implies a mixture of distributions of dispersal distances. In such a case, estimating the proportion of emigrating birds using the dispersal distances observed *in* the study area is likely to lead to wrong conclusions (Fig. 4.3).

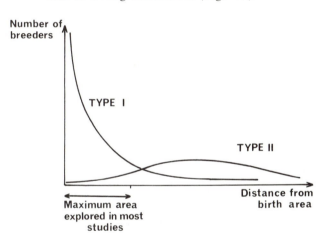

Fig. 4.3. Polymorphism in dispersal behaviour, here two types characterized by different dispersal distance distributions.

It is obvious that much remains to be done to estimate these demographic parameters both in collecting more relevant data and in developing appropriate statistical tools. Some progress could be made using medium-scale studies on carefully selected species (e.g. colonial birds). Increased precision in the location of ringing recovery data (at least from ringers) together with some changes in the experimental planning (encourage trapping during breeding seasons) can increase the value of the ringing centre data. In the meantime, one has to rely mainly on theoretical work.

Population size estimation

Literature on bird censusing is abundant (Taylor *et al*. 1985; Blondel and Frochot 1987; Ralph and Scott 1981; Verner 1985). There is no universal method for estimating bird densities and appropriate methods vary according to species, time, and place. Recent reviews are given by Seber (1986) and Verner (1985). We merely give an outline focusing on problems of relevance-efficiency and management requirement.

Management policy requires the estimation of population size not only to evaluate the number of individuals in a particular species, but to follow their fluctuations through time and space to analyse the causes and consequences of these trends. Methods of censusing birds can be ranked by their ability to answer such questions. We can split them into three main categories:

(a) those that give relative abundance;
(b) those that give absolute abundance, without individual marking; and
(c) those that give absolute abundance with individual marking.

Methods that allow the detection of population changes are not considered in this paper (see Lebreton and Clobert, this volume).

Relative abundance

Relative abundance methods are 'quick' and 'easy' techniques which are often chosen for large-scale survey. They can involve direct indices such as counts of singing birds, and numbers of birds seen or trapped from points or lines (Blondel *et al.* 1970; Verner 1985) or indirect indices such as tracks, feeding activities, aerial photographs, nest counts (for a review see Bull 1981). The use of indices relies on the assumption that the index measures a constant but unknown proportion of the population. Calibration is therefore a critical step, particularly to ensure that relative measures are comparable. Very few, if any, indices have a direct proportional relationship with density. For example, counting singing males ignores the variation in the non-breeding populations and in the sex ratio, i.e. non-singing part of the population. The magnitude of the biases is obviously dependent on the species. For example, male density will reflect the state of the population more closely in monogamous than in polygamous species. Observers are limited in the number of birds they can hear in a particular unit of time (Mayfield 1981; Bart 1985). However, the idea of paired between-year comparisons, as used by the BTO in the Common Bird Census, can remove a significant part of the observer effect. Other problems also arise when counting by viewing, e.g. double counting, birds in flocks (Verner 1985). Ideally a calibrated index should be used by a wide range of observers, on a wide range of species, habitats, seasons, and years to measure the magnitude of the different biases (Burnham *et al.* 1980, 1981). Very few attempts have been made to do this (see De Santé 1981; Svensson 1981; Hamel 1984; Bollinger *et al.* 1988) and we agree with

Pollock's (1981*a*) comment which states that some biologists have been too optimistic about the use of indices of abundance.

Absolute abundance without individual or cohort marking

To complement calibrated indices and provide data for modelling, we require methods measuring absolute densities. Count of birds seen or heard can be in variable-circular plots (Reynolds *et al*. 1980) and/or in line transects (Burnham *et al*. 1980). The area sampled can be fixed in advance or determined from the observations. A variety of methods have been derived from these different designs to record density, and a detailed review is not given here (see Ralph and Scott 1981; Taylor *et al*. 1985; Verner 1985; Seber 1986). On statistical grounds, mapping techniques differ from distance techniques in that the former uses estimation of bird territory size, implying several visits to the same area, while the latter uses the distance at which the observation is recorded to estimate the sampled area. Statistical methods to determine territory size (North 1977, 1978, 1979; Scheffer 1987) are well developed but critically dependent on the number of points (Worton 1987) and on the way birds are detected. Detection by singing can miss birds (Mayfield 1981; see above) whereas methods involving sighting may be misleading when there is overlap of home ranges (Frochot *et al*. 1977), with the exception of some colonial birds (Rothery *et al*. 1988). Mixing methods of recording birds is obviously worse. Statistical models based on line transect data or variable circular plot data are now well developed (Burnham *et al*. 1980; Buckland 1985), but these methods critically depend upon the shape of the detection function (how our ability to detect a bird decreases with the distance). Other models allowing flexibility in the shape of this function have been developed (Buckland 1985). The following important assumptions need to be made:

(a) all the birds on the transect centre are detected;
(b) no birds are counted twice; and
(c) sightings are independent events. Often for birds (Verner 1985) these assumptions are violated and such methods also need to be calibrated.

Absolute abundance with individual marking

Some of the above hypotheses can be avoided when birds are individually marked. Using mist-nets or some other way to trap birds (Karr 1981; Baillie and Marchant 1986), one can estimate population size using information about the changing proportion of newly trapped birds to already marked birds (C-M-R. methods; Nichols *et al*. 1981). Statistical models are now very well developed (Seber 1986) for closed (White *et al*. 1982) and open populations (Burnham *et al*. 1987; and see section on estimation of survival rates). They account for a wide range of possible sampling designs (Burnham *et al*. 1987). However, although survival rate and capture rate modelling are well developed, it remains to develop models for estimating population size in the same way. In particular, large variances of population size estimates can be reduced by the use of parsimonious models (Jolly 1982) although variances are not easily derived in all cases (Brownie *et al*. 1986). One of the main problems is the determination of transient birds and Manly (1977*b*) proposes a way to tackle the problem.

Because they are time-consuming and their use is restricted to some species, such techniques have rarely been used in monitoring bird populations (Verner 1985). In particular, very few comparisons between C-M-R and other techniques have been carried out (Frochot *et al*. 1977; De Santé 1981; Conner *et al*. 1983; Bollinger *et al*. 1988).

What to choose

The choice of a suitable method for estimating population size should be determined by the purpose of the study and the species considered. We restrict ourselves to some general considerations. Managers and conservationists need methods of estimating population size of a particular species that work well on a medium to large scale. Techniques have to be applicable to this scale. Furthermore the methods should have a sound statistical basis so that sampling errors can be assessed. Ideally, they should be robust against violations of underlying assumptions. Line transect and variable circular-plot methods appear to be good in this respect.

Nevertheless, because of observer variation, problems of double counting, effects of time of day, habitat type, . . . calibration may still be required against methods that provide more precise estimation of population size on a small scale using individually marked birds. There is often a trade-off between quantity and quality of the data and the associated methods. Many methods can be easily applied on a wide scale, but do not provide data that are adequate for analysing the mechanisms of population changes. Therefore, detailed population studies are crucial not only for providing a way to calibrate other methods for estimating density, but also for analysing mechanisms of population changes. Generally, both types of approaches are required for successful study of bird population dynamics.

Conclusions

1. Some parameters, although central to our understanding of population dynamics are poorly known. For immigration, emigration, and age-

specific proportion of breeders, not only are relevant data difficult to collect but appropriate existing data are rare and suitable statistical models virtually non-existent. Carefully planned sampling designs on well chosen species, using techniques that give a maximum of information about each individual, are probably the only way to gather relevant data. For example, increasing the scale of existing long-term studies to include other populations at various distances from the main area is a way to achieve this goal. This will require increased effort in manpower which is unavoidable if precise questions concerning these parameters are to be answered.

2. Many models or methods are presented in the context of a particular situation, without merging them in the broader context of other connected methods (relationships with other methods, use of similar notations . . .). Good advances in these areas have been made on survival rate (Brownie *et al*. 1985; Burnham *et al*. 1987; Clobert *et al*. 1987). Hierarchy of models, tests between models, numerical estimation and associated goodness-of-fit tests underpin the philosophy underlying future developments in parameter estimation, model-building, and testing.

3. Appropriate statistical techniques are not used as widely in research on bird populations as they could be. A typical example is the way survival rates are estimated. Many studies of survival rate still use ratios of survivors at time $t + 1$ among those present at time t, or worse age-ratios, although more refined models are available. Furthermore, assumptions often go untested (Begon 1983) and we are tempted to generalize Burnham's (1981) remark, on censusing, that at least a 10-year gap exists between the state-of-the-knowledge and the state-of-the-practice. On the one hand, biologists and managers do not often appreciate that population dynamics, because of problems of sampling, requires a high level of sophisticated statistics and modelling. On the other hand, biometricians and statisticians sometimes write in such a way and in such journals that biologists and managers are effectively denied access to their methods. For example, up until a few years ago, the use of statistical packages for analysing capture-recapture data which allow the non-specialist access to these methods, have been largely obscured by a mass of algebra. In addition to this, some model development is not in the direction that biologists really want. Communication in both directions has to be encouraged. Also the use of methods from other fields such as epidemiology for survival analysis, or agronomy for sampling design may be fruitful.

4. No techniques or methods are universal. If managers and conservationists wish to know the level of the population, the annual changes, the causes of these changes, and what to do to modify them, quick and easy methods will not generally be sufficient. A broad approach, combining information collected on a wide scale, with small-scale, more detailed data

is likely to be more successful. Some steps are being made in this direction (e.g. Buckland and Baillie 1987) but there is still a long way to go in:

(a) integrating the existing long-term studies;
(b) building appropriate statistical models to analyse data;
(c) developing large-scale survey using properly calibrated indices together with a few intensively studied sites using more detailed methods; and
(d) using designed experiments to identify causal mechanisms (e.g. compensatory mortality). Within this framework, sampling design (Hurlbert 1984) is a central problem.

5. Estimating demographic parameters of shy or rare species, even if their population size can be estimated using appropriate techniques or evaluated habitat requirements is very difficult because of the small sample size. If crude analyses can help somehow, the comparison with species that are ecologically, behaviourally, or taxonomically similar (see this volume) can also be valuable. In the case of species with small populations, high quality data acquisition (which is possible for industrial countries) should be promoted to help locating the species main characteristics and requirements. The use of theoretical or predictive models (Soulé 1987; Lebreton and Clobert, this volume) is also advisable.

Acknowledgements

We are deeply indebted to Dr P. M. North who critically read a first draft of the manuscript. We also greatly acknowledge two anonymous referees for their very constructive comments and their improvements to the English.

Summary

This paper reviews methods of estimation of demographic parameters in animal populations. If fecundity is clearly defined with respect to the other demographic parameters, its estimation does not raise major problems. Data and methods for analysing survival rate are numerous although problems still exist in obtaining survival rates of young individuals. Data on age-specific proportions of breeders and immigration-emigration processes are scarce. Suitable methods for estimating them are not well developed. Methods of population size estimation are divided into two main groups, those that give a relative estimation and those that give an absolute one. The reliability of the former depends on a variety of assumptions which clearly need to be tested more fully than hitherto. Calibration with methods giving estimates of absolute density (CMR technique, direct counts, mapping techniques . . .) is needed.

The screening of the literature leads to the following conclusions:

1. More data have to be collected for the estimation of some parameters (immigration-emigration, age-specific proportion of breeders).
2. The philosophy of the methodological development seems to be driven by the following principles: hierarchy of models, test between models, numerical estimations, and powerful goodness-of-fit tests.
3. Communication between biologists and statisticians has to be increased as well as transfer of techniques from other fields.
4. The sampling design and the methodology chosen need to be in agreement with managers' and conservationists' requirements.
5. Rare or shy species offer little possibility to get precise estimation of demographic parameters. The acquisition of high quality data (behaviour, habitat requirements . . .), comparison with similar species and the exploration of theoretical or predictive models may be good ways of overcoming this difficulty.

References

Aebischer, N. J. (1987). Estimating time-specific survival rates when the reporting rate is constant. *Acta Ornithologica*, **23**, 35–40.

Allainé, D., Pontier, D., Gaillard, J. M., Lebreton, J. D., Trouvilliez, J., and Clobert, J. (1987). The relationship between fecundity and adult body weight in Homeotherms. *Oecologia*, **73**, 478–80.

Amlaner, A. J. and McDonald, D. W. (eds) (1980). *A handbook on biotelemetry and radiotracking*. Pergamon Press, Oxford.

Anderson, D. R. and Burnham, K. P.. (1980). Effect of delayed reporting of band recoveries on survival estimates. *Journal of Field Ornithology*, **51**, 244–7.

Anderson, D. R., Wywialowski, A. P., and Burnham, K. P. (1981) Tests of the assumptions underlying life table methods for estimating parameters from cohort data. *Ecology*, **62**, 1121–4.

Anderson, D. R., Burnham, K. P., and White, G. C. (1985). Problems in estimating age-specific survival rates from recovery data of birds ringed as young. *Journal of Animal Ecology*, **54**, 89–98.

Andrewartha, H. G. and Birch, L. C. (1954) *The distribution and abundance of animals*. University of Chicago Press, Chicago.

Arnason, A. N. and Baniuk, L. (1980). A computer system for mark-recapture analysis of open population. *Journal of Wildlife Management*, **44**, 325–32.

Arnason, A. N. and Schwartz, C. J. (1986). *Popan-3 Extended analysis and testing features for Popan-2*. CBRC publication, Manitoba.

Bailey, N. T. J. (1951). On estimating the size of mobile populations from capture-recapture data. *Biometrika*, **38**, 293–306.

Baillie, S. and Marchant, J. (1986). Constant effort ringing. A new approach to bird population studies. *BTO News*, **143**.

Baker, R. J. and Nelder, J. A. (1978). *The GLIM system, release 3, generalized linear interactive modelling*. Numerical algorithm group, Oxford.

Barrat, R., Barré, H., and Mougin, J. L. (1976). Données écologique sur les grands. Albatros *Diomedea exulans* de l'île de la Possession (archipel Crozet). *L'oiseau et la Revue Français d'Ornithologie*, **46**, 143–55.

Bart, J. (1985). Causes of recording errors in singing bird surveys. *Wilson Bulletin*, **97**, 161–72.

Bart, J. and Robson, D. S. (1982). Estimating survivorship when the subjects are visited periodically. *Ecology*, **63**, 1078–90.

Begon, M. (1983). Abuses of mathematical techniques in ecology: applications of Jolly's capture-recapture method. *Oikos*, **40**, 155–8.

Begon, M., Harper, J. L., and Townsend, C. R. (1986). *Ecology: individuals, populations and communities*. Blackwell, Oxford.

Berndt, R. and Sternberg, H. (1968). Terms, studies and experiments on the problems of bird dispersion. *Ibis*, **110**, 256–69.

Birkhead, T. R. and Furness, R. W. (1985). Regulation of seabirds populations. In *Behavioural ecology, ecological consequences of adaptative behaviour* (ed. R. M. Sibly and R. H. Smith), pp. 145–67. Blackwell, Oxford.

Blondel, J. and Frochot, B. (eds.) (1987). *Bird census and atlas studies*. Proceedings of the IXth International Conference on Bird Census and Atlas Work. Université de Dijon, France, 2–6 Septembre 1985.

Blondel, J., Ferry, C., and Frochot, B. (1970). La méthode des indices ponctuels d'abondance (IPA) ou des relevés d'avifaune par station d'écoute. *Alauda*, **38**, 55–71.

Bollinger, E. K., Gavin, T. A., and McIntyre, D. C. (1988). Comparison of transects and circular-plots for estimating bobolink densities. *Journal of Wildlife Management*, **52**, 777–86.

Brawn, J. D. and Balda, R. P. (1988). Population biology of cavity nesters in northern Arizona: do nest sites limit breeding densities? *Condor*, **90**, 61–71.

Brown, D. (1988). Components of lifetime reproductive success. In *Reproductive success. Studies of individual variation in contrasting breeding systems* (ed. T. H. Clutton-Brock), pp. 439–53. Chicago University Press, Chicago.

Brown, J. L. (1969). Territorial behaviour and population regulation in birds. *Wilson Bulletin*, **81**, 293–329.

Brownie, C. (1987). Recent models for mark recapture and mark resighting data. *Biometrics*, **43**, 1017–19.

Brownie, C. and Pollock, K. H. (1985). Analysis of multiple capture-recapture data using band recovery methods. *Biometrics*, **41**, 411–20.

Brownie, C. and Robson, D. S. (1976). Models allowing for age-dependent survival rates for band-return data. *Biometrics*, **32**, 305–23.

Brownie, C. and Robson, D. S. (1983). Estimation of time-specific survival rates from tag-resighting samples: a generalization of the Jolly-Seber model. *Biometrics*, **39**, 437–53.

Brownie, C., Hines, J. E., and Nichols, J. D. (1986). Constant parameter capture-recapture models. *Biometrics*, **42**, 561–74.

Brownie, C., Anderson, D. R., Burnham, K. P., and Robson, D. S. (1978). *Statistical inference from band recovery data—a handbook* (1st edn.). Fish and Wildlife Service, Resource Publication, no. 131, Washington.

Brownie, C., Anderson, D. R., Burnham, K. P., and Robson, D. S. (1985). *Statistical inference from band recovery data—a handbook* (2nd edn.). Fish and Wildlife Service, Resource Publication, no. 156, Washington.

Buckland, S. T. (1980). A modified analysis of the Jolly-Seber capture-recapture model. *Biometrics*, **36**, 419–35.

Buckland, S. T. (1984). Monte Carlo confidence intervals. *Biometrics*, **40**, 811–17.

Buckland, S. T. (1985). Perpendicular distance models for line transect sampling. *Biometrics*, **41**, 177–95.

Buckland, S. T. and Baillie, S. R. (1987). Estimating bird survival rates from organized mist-netting programmes. In *Ringing recovery analytical methods* (ed. P. M. North). Proceeding of Euring Technical Conference and meeting of the Mathematical Ecology Group of the Biometric Society (British region) and British Ecological Society. *Acta Ornithologica*, **23**, 89–100.

Bull, E. L. (1981). Indirect estimates of abundance in birds. In *Estimating number of terrestrial birds* (eds. C. J. Ralph, and J. M. Scott), pp. 76–80, Studies in Avian Biology, **6**. Cooper Ornithological Society, Lawrence.

Bunck, C. M. (1987). Analysis of survival data from telemetry projects. In *Experimental design and data analysis for telemetry projects: summary of a workshop* (ed. V. J. Meretsky). *Journal of Raptor Research*, **21**, 132–4.

Burnham, K. P. (1981). Summarizing remarks: environmental influences. In *Estimating number of terrestrial birds* (eds. C. J. Ralph and J. M. Scott), pp. 324–5. Studies in Avian Biology, **6**. Cooper Ornithological Society, Lawrence.

Burnham, K. P. (1989). Numerical survival rate estimation for capture recapture models using SAS PROC NLIN. In *Estimation and analysis of insect populations* (ed. L. L. MacDonald, B. F. J. Manly, J. Lockwood, and J. Logan), pp. 416–35. Lecture Notes in Statistics, no. 55, Springer Verlag, New York.

Burnham, K. P. and Anderson, D. R. (1979). The composite dynamic method as evidence for age-specific waterfowl mortality. *Journal of Wildlife Management*, **43**, 356–66.

Burnham, K. P., Anderson, D. R., and Laake, J. L. (1980). *Estimation of density from line transect sampling of biological populations*. Wildlife Monographs, no. 72.

Burnham, K. P., Anderson, D. R., and Laake, J. L. (1981). Line transect estimation of bird population density using a Fourier series. In *Estimating numbers of terrestrial birds* (eds. C. J. Ralph and J. M. Scott), pp. 466–82. *Studies in Avian Biology*, **6**.

Burnham, K. P., Anderson, D. R., White, G. C., Brownie, C., and Pollock, K. H. (1987). *Design and analysis methods for fish survival experiments based on release-recapture*. American Fisheries Society, Monograph 5, Bethesda, Maryland.

Cavé, A. J. (1968). The breeding of the Kestrel, *Falco tinnunculus* L., in the reclaimed area Oostelijk Flevoland. *Netherlands Journal of Zoology*, **18**, 313–407.

Chabrzyk, G. and Coulson, J. C. (1976). Survival and recruitment in the Herring Gull *Larus argentatus*. *Journal of Animal Ecology*, **45**, 187–203.

Chapman, D. G. and Robson, D. S. (1960). The analysis of a catch curve. *Biometrics*, **16**, 354–68.

Charlesworth, B. (1980). *Evaluation in age-structured populations*. Cambridge University Press, Cambridge.

Clobert, J. (1982). Etude des taux de survie chez l'Etourneau, calculés à partir de données de capture-marquage-recapture. *Le Gerfaut*, **72**, 255–85.

Clobert, J. and Lebreton, J. D. (1985). Dépendance de facteurs de milieu dans les estimations de taux de survie par capture-recapture. *Biometrics*, **41**, 1031–7.

Clobert, J. and Lebreton, J. D. (1987). Recent models for mark-recapture and mark-resighting data: A response. *Biometrics*, **43**, 1019–22.

Clobert, J., Lebreton, J. D., and Allainé, D. (1987). A general approach to survival rate estimation by recaptures or resightings of marked birds. *Ardea*, **75**, 133–42.

Clobert, J., Lebreton, J. D., Clobert-Gillet, M., and Coquillart, H. (1985). The estimation of survival in bird populations by recaptures or sightings of marked individuals. In *Statistics in ornithology* (ed. B. J. T. Morgan, and P. M. North), pp. 197–213. Springer Verlag, New York.

Comins, H. N., Hamilton, W. D., and May, R. M. (1980). Evolutionarily stable dispersal strategies. *Journal of Theoretical Biology*, **82**, 205–30.

Conner, R. N., Dickson, J. G., and Williamson, J. H. (1983). A comparison of breeding census techniques with mist netting results. *Wilson Bulletin*, **95**, 276–80.

Conroy, M. J. and Williams, B. K. (1984). A general methodology for maximum likelihood inference from band-recovery data. *Biometrics*, **40**, 739–48.

Cormack, R. M. (1964). Estimates of survival from the sighting of marked animals. *Biometrika*, **51**, 429–38.

Cormack, R. M. (1970). The construction of life tables from the recovery of dead animals. Statistical appendix. In: *Mortality and population change of dominican gulls in Wellington, New Zealand* (ed. R. A. Fordham). *Journal of Animal Ecology*, **39**, 13–27.

Cormack, R. M. (1979). Models for capture-recapture. In *Sampling biological populations* (ed. R. M. Cormack, G. P. Patil, and D. S. Robson), pp. 217–55. International Cooperative Publishing House, Fairland, Maryland.

Cormack, R. M. (1981). Loglinear models for capture-recapture experiments on open populations. In *The mathematical theory of the dynamics of biological populations* (ed. R. W. Hiorns and D. Cooke). Academic Press, London.

Cormack, R. M. (1985). Examples of the use of GLIM to analyse capture-recapture studies. In *Statistics in Ornithology* (ed. B. J. T. Morgan and P. M. North), pp. 243–73. Springer Verlag, New York.

Cormack, R. M., G. P. Patil, and Robson, D. S. (ed.) (1979). *Sampling biological populations*. Statistical ecology series no. 5, International Cooperative Publishing House, Fairland, Maryland.

Coulson, J. C., Duncan, N., and Thomas, C. (1982). Changes in the breeding biology of the Herring Gull (*Larus argentatus*) induced by reduction in the size of the colony. *Journal of Animal Ecology*, **51**, 739–56.

Crosbie, S. F. and Manly, B. F. J. (1985). A new approach for parcimonious modelling of capture-mark-recapture experiments. *Biometrics*, **41**, 385–98.

Darroch, J. N. (1961). The two-sample capture-recapture census when tagging and sampling are stratified. *Biometrika*, **48**, 241–60.

Deevey, E. S. Jr. (1947). Life tables for natural populations of animals. *Quarterly Review in Biology*, **22**, 283–314.

De Lury, D. B. (1947). On the estimation of biological populations. *Biometrics*, **3**, 145–67.

De Santé, D. F. (1981). A field test of the variable circular-plot censusing technique in a California coastal scrub bird community. In *Estimating numbers of terrestrial birds* (eds. C. J. Ralph and J. M. Scott), pp. 177–85. Studies in Avian Biology, **6**. Cooper Ornithological Society, Lawrence.

Dhondt, A. A. (1989). The effect of old age on the reproduction of Great and Blue Tit. *Ibis* **131**, 268–80.

Dobson, F. S. and Jones, W. T. (1985). Multiple causes of dispersal. *American Naturalist*, **126**, 855–8.

Duncan, N. (1978). The effects of culling Herring Gulls (*Larus argentatus*) on recruitment and population dynamics. *Journal of Applied Ecology*, **15**, 697–713.

Eberhardt, L. L. (1985). Assessing the dynamics of wild population. *Journal of Wildlife Management*, **49**, 997–1012.

Finney, G. and Cooke, F. (1978). Reproductive habits in the Snow Goose. The influence of female age. *The Condor*, 80, 147–58.

Frochot, B., Reudet, D., and Leruth, Y. (1977). A comparison of preliminary results of three census methods applied on the same population of forest birds. *Polish Ecological Studies*, 3, 71–5.

Gadgil, M. (1971). Dispersal: population consequences and evolution. *Ecology*, 52, 253–61.

Gauthier, G. and Smith, J. N. M. (1987). Territorial behaviour, nest-site availability, and breeding density in buffleheads. *Journal of Animal Ecology*, 56, 171–84.

Gilpin, M. E. (1987). Spatial structure and population vulnerability. In *Viable populations for conservation* (ed. M. E. Soulé), pp. 125–39. Cambridge University Press, Cambridge.

Greenwood, P. J. (1980). Mating systems, philopatry and dispersal in birds and mammals. *Animal Behaviour*, 28, 1140–62.

Greenwood, P. J. and Harvey, P. H. (1982). Natal and breeding dispersal of birds. *Annual Review of Ecology and Systematics*, 13, 1–21.

Greenwood, P. J., Harvey, P. H., and Perrins, C. M. (1979). The role of dispersal in the Great Tit (*Parus major*): the causes, consequences and heritability of natal dispersion. *Journal of Animal Ecology*, 48, 123–42.

Gustafsson, L. and Nilsson, S. G. (1985). Clutch size and breeding success of Pied and Collared Flycatchers *Ficedula* spp. in nest-boxes of different sizes. *Ibis*, 127, 380–5.

Haldane, J. B. S. (1955). *The calculation of mortality rates from ringing data*. Proceedings of the XI International Ornithological Congress, pp. 454–8. Basel.

Hamel, P. B.(1984). Comparison of variable circular plot and spot-mapping censusing methods in temperate deciduous forest. *Ornis Scandinavica*, 15, 266–74.

Hamilton, W. D. and May, R. M. (1977). Dispersal in stable habitats. *Nature*, 269, 578–81.

Heisey, D. M. and Fuller, T. K. (1985). Evaluation of survival and cause-specific mortality rates using telemetry data. *Journal of Wildlife Management*, 49, 668–74.

Hensler, G. M. (1985). Estimation and comparison of functions of daily nest survival probabilities using the Mayfield method. In *Statistics in Ornithology* (eds. B. J. T. Morgan and P. M. North), pp. 289–301. Lecture Notes in Statistics, no. 29, Springer-Verlag, Berlin.

Hensler, G. M. and Nichols, J. D. (1981). The Mayfield method of estimating nesting success: a model, estimators and simulation results. *Wilson Bulletin*, 93, 42–53.

Hensley, M. M. and Cope, J. B. (1951). Further data on removal and repopulation of the breeding birds in a spruce-fir forest community. *Auk*, 68, 483–93.

Hickey, J. J. (1952). Survival studies of banded birds. Special scientific report. *Wildlife*, 15, 1–177.

Horn, H. S. (1983). Some theories about dispersal. In: *The ecology of animal movement* (eds. I. A. Swingland and P. J. Greenwood), pp. 54–62. Clarendon Press, Oxford.

Hurlbert, S. H. (1984). Pseudoreplication and the design of ecological field experiments. *Ecological Monographs*, 54, 187–211.

Jackson, C. H. N. (1933). On the true density of Tsetse flies. *Journal of Animal Ecology*, 2, 204–9.

Jackson, C. H. N. (1939). The analysis of an animal population. *Journal of Animal Ecology*, 8, 238–46.

Johnson, D. H. (1974). Estimating survival rates from banding of adult and juvenile birds. *Journal of Wildlife Management*, 38, 290–7.

Johnson, D. H. (1979). Estimating nest success: the Mayfield method and an alternative. *Auk*, **96**, 651–61.

Johnson, D. H., Burnham, K. P., and Nichols, J. D. (1986). *The role of heterogeneity in animal population dynamics*. Proceedings of the thirteenth International Biometrics Conference. Biometric Society, University of Washington, Seattle.

Jolly, G. M. (1965). Explicit estimates from capture-recapture data with both death and immigration-stochastic model. *Biometrika*, **52**, 225–47.

Jolly, G. M. (1979). A unified approach to mark-recapture stochastic models, exemplified by a constant survival rate model. In *Sampling biological populations* (eds. R. M. Cormack, G. P. Patil, and D. S. Robson), pp. 277–82. International Cooperative Publishing House, Maryland.

Jolly, G. M. (1982). Mark-recapture models with parameters constant in time. *Biometrics*, **38**, 301–21.

Jolly, G. M. and Dickson, J. M. (1980). Mark-recapture suite of programs. In *Compstat 1980: Proceedings in computational statistics*, **4** (eds. M. M. Barriet and D. Wishart), pp. 570–6. Physica-Verlag, Berlin.

Karr, J. R. (1981). Surveying birds with mist nets. In *Estimating numbers of terrestrial birds* (eds. C. J. Ralph and J. M. Scott), pp. 62–7. Studies in Avian Biology, **6**. Cooper Ornithological Society, Lawrence.

Klomp, H. (1972). Regulation of the size of bird populations by territorial behaviour. *Netherlands Journal of Zoology*, **22**, 456–88.

Krebs, C. J. (1972). *Ecology. The experimental analysis of distribution and abundance*. Harper International Edition, New York.

Krebs, J. R. (1977). Song and territory in the Great Tit. In *Evolutionary Ecology* (eds. B. Stonehouse and C. M. Perrins), pp. 47–62. Macmillan, London.

Lack, D. (1951). Population ecology in birds. Proceedings of the Xth International Ornithological Congress, Uppsala, pp. 409–48.

Lack, D. (1954). *The natural regulation of animal numbers*. Clarendon Press, Oxford.

Lakhani, K. H. (1987). Efficient estimation of age-specific survival rates from ring recovery data, of birds ringed as young, augmented by further field information. *Journal of Animal Ecology*, **56**, 969–87.

Lakhani, K. H. and Newton, I. (1983). Estimating age-specific bird survival rates from ring recoveries—can it be done? *Journal of Animal Ecology*, **52**, 83–91.

Lebreton, J. D. (1977). Maximum likelihood estimations of survival rates from bird band returns: some complements to age-dependent methods. *Biométrie-Praximétrie*, **17**, 145–61.

Lebreton, J. D. (1981). Contribution à la dynamique des populations d'oiseaux. Modèles mathématiques en temps discret. Unpublished Thesis, Université de Lyon.

Lebreton, J. D. and Clobert, J. (1986). *User's manual for program Surge*, version 2.0. CEPE/CNRS, Montpellier.

Lebreton, J. D., Hemery, G., Clobert, J., and Coquillart, H. (1990). The estimation of age-specific breeding probabilities from recaptures or resightings in vertebrate populations. I. Transversal models. *Biometrics*, **46**, in press.

Leslie, P. H. (1945). On the use of matrices in certain population mathematics. *Biometrika*, **33**, 213–45.

Leslie, P. H. (1966). The intrinsic rate of increase and the overlap of successive generations in a population of Guillemots (*Uria aalge*). *Journal of Animal Ecology*, **35**, 291–301.

Levin, S. A. (1976). Population dynamics models in heterogeneous environments. *Annual Review in Ecology and Systematics*, **7**, 287–310.

Levin, S. A. (1985). Ecological and evolutionary aspects of dispersal. In *Mathematics tropics in population biology, morphogenesis and neurosciences* (ed. C. Teramoto, and S. Yamaguti), pp. 80–1. Springer Verlag, Berlin.

Lidicker, W. Z., Jr. (1962). Emigration as a possible mechanism permitting the regulation of population density below carrying capacity. *American Naturalist*, **96**, 29–33.

Lincoln, F. C. (1930). *Calculating waterfowl abundance on the basis of banding returns*, pp. 1–4. U.S. Department of Agriculture, Circular no. 118.

Manly, B. J. F. (1977*a*). A model for dispersion experiments. *Oecologia*, **31**, 119–30.

Manly, B. J. F. (1977*b*). The analysis of trapping records for birds in mist-nets. *Biometrics*, **33**, 404–10.

Manly, B. F. J. (1985*a*). *The statistics of natural selection*. Population and Community Biology Series, Chapman and Hall, London.

Manly, B. F. J. (1985*b*). A test of Jackson's method for separating death and emigration with mark-recapture data. *Research on Population Ecology*, **27**, 99–109.

Mayfield, H. F. (1961). Nesting success calculated from exposure. *Wilson Bulletin*, **73**, 255–61.

Mayfield, H. F. (1975). Suggestions for calculating nest success. *Wilson Bulletin*, **87**, 456–66.

Mayfield, H. F. (1981). Problems in estimating population size through counts of singing males. In *Estimating numbers of terrestrial birds* (eds. C. J. Ralph and J. M. Scott), pp. 220–24. Studies in Avian Biology, **6**. Cooper Ornithological Society, Lawrence.

McCleery, R. H. and Perrins, C. M. (1988). Life-time reproductive success of the Great Tit, *Parus major*. In *Reproductive success. Studies of individual variation in contrasting breeding systems* (ed. T. H. Clutton-Brock), pp. 136–53. Chicago University Press.

Moore, J. and Ali, R. (1984). Are dispersal and inbreeding avoidance related? *Animal Behaviour*, **32**, 94–112.

Morgan, B. J. T. and North, P. M. (eds.) (1985). *Statistics in Ornithology. Lectures notes in statistics*, **29**. Springer Verlag, New York.

Newton, I. (1976). Population limitation in diurnal raptors. *Canadian Field Naturalist*, **90**, 274–300.

Nichols, J. D. and Hines, J. E. (1983). The relationships between harvest and survival rates of mallards: a straightforward approach with partitioned data sets. *Journal of Wildlife Management*, **47**, 334–48.

Nichols, J. D., Noon, B. R., Stokes, S. L., and Hines, J. E. (1981). Remarks on the use of mark-recapture methodology in estimating avian population size. In *Estimating numbers of terrestrial birds* (eds. C. J. Ralph and J. M. Scott), pp. 121–36. Studies in Avian Biology, **6**. Cooper Ornithological Society, Lawrence.

Nichols, J. D., Stokes, S. L., Hines, J. E., and Conroy, M. J. (1982). Additional comments on the assumption of homogeneous survival rates in modern bird banding estimation models. *Journal of Wildlife Management*, **46**, 953–62.

Nichols, J. D., Percival, H. F., Coon, R. A., Conroy, M. J., Hensle, G. L., and Hines, J. E. (1984). Observer initiation frequency and success of mourning dove nests: a field experiment. *Auk*, **101**, 398–402.

Nilsson, S. G. (1984). Clutch-size and breeding success of the Pied Flycatcher *Ficedula hypoleuca* in natural tree-holes. *Ibis*, **126**, 407–10.

North, P. M. (1977). A novel clustering method for estimating numbers of bird territories. *Applied Statistics*, **26**, 149–55.

North, P. M. (1978). Statistical methods in Ornithology. Unpublished Ph.D. Thesis, University of Kent, Canterbury.

North, P. M. (1979). A novel clustering method for estimating numbers of bird territories: an addendum. *Journal of the Royal Statistical Society*, Series C, **28**, 300–1.

North, P. M. (1985). Models to describe razorbill movements. In *Statistics in ornithology* (eds. B. J. T. Morgan and P. M. North), pp. 77–84. Lecture Notes in Statistics, **29**. Springer Verlag, New York.

North, P. M. (ed.) (1987). *Ringing recovery analytical methods*. Proceedings of the Euring Technical Conference and Meeting of the Mathematical Ecology Group of the Biometric Society (British region) and British Ecological Society, Wageningen 4–7 March 1986. *Acta Ornithologica*, 23(1).

North, P. M. and Cormack, R. M. (1981). On Seber's method for estimating age-specific bird survival rates from ringing recoveries. *Biometrics*, **37**, 103–12.

North, P. M. and Morgan, B. J. T. (1979). Modelling heron survival using weather data. *Biometrics*, **35**, 667–81.

Pedersen, H. C. (1988). Territorial behaviour and breeding numbers in Norwegian Willow Ptarmigan: a removal experiment. *Ornis Scandinavica*, **19**, 81–7.

Piper, S. E., Mundy, P. J., and Ledger, J. A. (1981). Estimates of survival in the Cape Vulture *Gyps coprotheres*. *Journal of Animal Ecology*, **50**, 815–25.

Pollock, K. H. (1981*a*). Summarizing remarks: data analysis. In *Estimating numbers of terrestrial birds* (eds. C. J. Ralph and J. M. Scott), pp. 509–10. Studies in Avian Biology, **6**. Cooper Ornithological Society, Lawrence.

Pollock, K. H. (1981*b*). Capture-recapture models allowing for age-dependent survival and capture rates. *Biometrics*, **37**, 521–9.

Pollock, K. H. and Cornelius, W. L. (1988). A distribution-free nest survival model. *Biometrics*, **44**, 397–404.

Pollock, K. H. and Raveling, D. G. (1982). Assumptions of modern band-recovery models, with emphasis on heterogeneous survival rates. *Journal of Wildlife Management*, **46**, 88–98.

Pollock, K. H., Hines, J. E., and Nichols, J. D. (1985). Goodness-of-fit tests for open capture-recapture models. *Biometrics*, **41**, 399–410.

Porter, J. M., and Coulson, J. C. (1987). Long-term changes in recruitment to the breeding group, and the quality of recruits at a kittiwake *Rissa tridactyla* colony. *Journal of Animal Ecology*, **56**, 675–89.

Potts, G. R., Coulson, J. C., and Deans, I. R. (1980). Population dynamics and breeding success of the shag, *Phalacrocorax aristotelis*, on the Farne Islands, Northumberland. *Journal of Animal Ecology*, **49**, 465–84.

Pradel, R., Clobert, J., and Lebreton, J. D. (1990). Recent developments for the analysis of multiple capture-recapture data sets. An example concerning two Blue Tit populations. *Ring*, (in press).

Prout, T. (1984) The delayed effect on adult fertility of immature crowding: population dynamics. In *Population Biology and Evolution* (ed. K. Wochrmann and L. Loeschke), pp. 83–6, Springer Verlag, Berlin.

Pusey, A. E. (1987). Sex-biased dispersal and inbreeding avoidance in birds and mammals. *Trends in Ecology and Evolution*, **2**, 295–9.

Ralph, C. J. and Scott, C. M. (eds.) (1981). *Estimating numbers of terrestrial birds*, Studies in Avian Biology, **6**. Cooper Ornithological Society, Lawrence, USA.

Reynolds, R. T., Scott, J. M., and Nussbaum, R. A. (1980). A variable circular-plot method for estimating bird numbers. *Condor*, **82**, 309–13.

Ricker, N. E. (1956). Uses of marking animals in ecological studies: the marking of fish. *Ecology*, **37**, 665–70.

Ricklefs, R. E. (1973). Fecundity, mortality and Avian demography. In *Breeding biology of birds* (ed. D. J. Farner), pp. 366–447. Natural Resources Council, Washington DC.

Ricklefs, R. E. (1980). *Ecology* (2nd edn). Nilson. Simbury-on-Thames.

Rose, M. R. (1987). *Quantitative ecological theory. An introduction to basic models*. Croom Helm, London.

Rothery, P., Wanless, S., and Harris, M. P. (1988). Analysis of counts from monitoring guillemots in Britain and Ireland. *Journal of Animal Ecology*, 57, 1–19.

Sandland, R. L. and Kirkwood, G. P. (1981). Estimation of survival in marked populations with possibly dependent sighting probabilities. *Biometrics*, 68, 531–41.

Scheffer, M. (1987). An automated method for estimating the number of bird territories from an observation map. *Ardea*, 75, 231–6.

Schwarz, C. J., Burnham, K. P., and Arnason, A. N. (1988). Post-release stratification in band-recovery models. *Biometrics*, 44, 765–85.

Seber, G. A. F. (1965). A note on the multiple-recapture census. *Biometrika*, 52, 249–59.

Seber, G. A. F. (1970). Estimating time-specific survival and reporting rates for adult birds from band returns. *Biometrika*, 58, 313–18.

Seber, G. A. F. (1971). Estimating age-specific survival rates for birds from bird-band returns when the reporting rate is constant. *Biometrika*, 58, 491–7.

Seber, G. A. F. (1973). *The estimation of animal abundance and related parameters* (1st edn). Griffin, London.

Seber, G. A. F. (1986). A review of estimating animal abundance. *Biometrics*, 42, 267–92.

Shields, W. M. (1983). Optimal inbreeding and the evolution of philopatry. In *The ecology of animal movement* (eds. I. R. Swingland and P. J. Greenwood), pp. 132–59. Clarendon Press, Oxford.

Slagsvold, T. (1987). Nest site preference and clutch size in the Pied Flycatcher *Ficedula hypoleuca*. *Ornis Scandinavica*, 18, 189–97.

Soulé, M. E. (ed.) (1987). *Viable populations for conservation*, pp. 1–189. Cambridge University Press.

Soulé, M. E. and Simberloff, D. (1986). What do genetics and ecology tell us about the design of nature reserves? *Biological Conservation*, 35, 19–40.

Southwood, T. R. E. (1978). *Ecological methods* (2nd edn.). Chapman and Hall, London.

Stearns, S. C. and Crandall, R. E. (1981). Quantitative predictions of delayed maturity. *Evolution*, 35, 455–63.

Stearns, S. C. and Koella, J. C. (1986). The evolution of phenotypic plasticity in life-history traits: predictions of reaction norms for age and size at maturity. *Evolution*, 40, 893–913.

Stenseth, N. C. (1983). Causes and consequences of dispersal in small mammals. In *The ecology of animal movements* (eds. I. A. Swingland and P. J. Greenwood), pp. 63–101. Clarendon Press, Oxford.

Stewart, R. E. and Aldrich, J. M. (1951). Removal and repopulation of breeding birds in a spruce-fir forest community. *Auk*, 68, 471–82.

Stokes, S. L. (1984). The Jolly-Seber method applied to age-stratified populations. *Journal of Wildlife Management*, 48, 1053–9.

Strang, C. A. (1980). Incidence of avian predators near people searching for waterfowl nests. *Journal of Wildlife Management*, 44, 220–4.

Svensson, S. E. (1981). Do transect counts monitor abundance trends in the same way as territory mapping in study plots? In: *Estimating numbers of terrestrial birds* (eds.

C. J. Ralph and J. M. Scott), pp. 209–14. Studies in Avian Biology, **6**. Cooper Ornithological Society, Lawrence.

Taylor, K., Fuller, R. J., and Lack, P. C. (eds.) (1985). *Birds census and atlas studies.* Proceedings VIII International Conference on Bird Census and Atlas Work. B.T.O., Tring.

Tinbergen, N. (1957). The functions of territory. *Bird Study*, **4**, Vacca, M. M. and Handel, C. M. (1988). Factors influencing predation associated with visits to artificial Goose nests. *Journal of Field Ornithology*, **59**, 215–23.

Van Balen, J. H. (1984). The relationship between nest-box size, occupation and breeding parameters of the Great Tit *Parus major* and some other hole-nesting species. *Ardea*, **72**, 163–75.

Van Balen, J. H., Van Noordwijk, A. J., and Visser, J. (1987). Life time reproductive success and recruitment in two Great Tit populations. *Ardea*, **75**, 1–12.

Van Balen, J. H., Booy, C. J. H., Van Franeker, J. A., and Osieck, E. R. (1982). Studies on hole-nesting birds in natural nest sites. I. Availability and occupation of natural nest sites. *Ardea*, **70**, 1–24.

Van Noordwijk, A. J. (1984). Problems in the analysis of dispersal and a critique on its heritability in the Great Tits. *Journal of Animal Ecology*, **55**, 331–50.

Verner, J. (1985). An assessment of counting techniques. *Current Ornithology*, **2**, 247–302.

Village, A. (1983). The role of nest site availability and territorial behaviour in limiting the breeding density of Kestrels. *Journal of Animal Ecology*, **52**, 635–45.

Von Bauer, H. G. (1987). Geburstsortstreue und Streuungsverhalten junger Singvögel. *Die Vogelwarte*, **34**, 15–32.

Von Haartman, L. (1971). Population dynamics. In: *Avian Biology*, Vol. 1 (eds. D. S. Farmer and J. R. King), pp. 391–459. Academic Press, Orlando.

Watson, A. (1967). Population control by territorial behaviour in red grouse. *Nature*, **215**, 1274–5.

Weimerskirch, H. and Jouventin, P. (1987). Population dynamics of the Wandering Albatross, *Diomedea exulans*, of the Crozet Islands: causes and consequences of the population decline. *Oikos*, **49**, 315–22.

White, G. C. (1983). Numerical estimation of survival rates from band-recovery and biotelemetry data. *Journal of Wildlife Management*, **47**, 716–28.

White, G. C., Anderson, D. R., Burnham, K. P., and Otis, D. L. (1982). *Capture, recapture and removal methods for sampling closed populations.* Los Alamos National Laboratory, Los Alamos.

Willis, E. O. (1973). Survival rates for visited and unvisited nests of Bicolored Antbirds. *Auk*, **90**, 263–7.

Wooler, R. D. and Coulson, J. C. (1977). Factors affecting the age of first breeding of the kittiwake *Rissa tridactyla*. *Ibis*, **119**, 339–49.

Worton, B. J. (1987). A review of models of home range for animal movement. *Ecological Modelling*, **38**, 277–98.

Wynne Edwards, V. C. (1962). *Animal dispersion in relation to social behaviour.* Hafner, New York.

5 Bird population dynamics, management, and conservation: the role of mathematical modelling

J.-D. LEBRETON and J. CLOBERT

Introduction

In the real world animal populations rarely show simple patterns of growth such as the exponential or logistic model: most of the time complex mechanisms interact over time in an intrinsically multiplicative process. This complexity limits the value of intuitive reasoning or of simple models. In a parallel way, biological goals in studies of the dynamics of bird populations are very diverse, ranging e.g. from deciding if a population is stable in numbers, to assessing the effects of dispersal at various levels in time and space. This might be used equally to fight a pest as to predict the risk of extinction of an endangered population.

Thus, it would be a formidable task to attempt to review all the modelling approaches that might be valuable in such studies. For example, the modelling of species-environment relationships using multivariate methods is an important field which has been the subject of several recent reviews (see e.g. Verner *et al*. 1986; Ter Braak 1986, 1988; Sabatier *et al.* 1989). Therefore, we will confine ourselves to demographic models concerned with flows and rates of change rather than on population size *per se*: 'Since population is a changing entity, we are interested not only in its size and composition at any one moment, but also how it is changing. A number of important population characteristics are concerned with rates' (Odum 1971). From this point of view, as noted by Reddingius (1971*a*), 'there is no sharp borderline' between models built for either theoretical or practical purposes. We will however emphasize realism and precision as well as generality, three classical attributes for model judgment (Levins 1968; Walters, in Odum 1971, p. 278).

Wherever possible, we will cite primarily papers in biological rather than mathematical or statistical journals. For the sake of clarity we will cite for biological examples reviews rather than particular papers. Estimation of demographic parameters is reviewed in another paper (Clobert and Lebreton, this volume).

Initially we consider models with constant parameters. Then we introduce environmental variability over time and density dependence. Finally we consider complex approaches incorporating density dependence, spatial effects, and various kinds of stochasticity.

Models of population dynamics with constant parameters

The Leslie matrix model

Sizeable differences in demographic parameters according to age seem to be so general in birds that their incorporation in population models should be given priority (survival: see e.g. Coulson and Wooler 1976, for *Rissa tridactyla*; fecundity: see Ricklefs 1973). Demographic models arising from age-specific demographic rates considered as constant over time represent a starting point in population modelling. In the case of seasonal breeding it seems natural to consider a discrete time scale to model the long-term behaviour of the population: the formulation most commonly used is that of matrix models (Leslie 1945, 1948). The Leslie matrix model (for a detailed study, see Cull and Vogt 1973) is presented in Table 5.1 with a parameterization particularly suitable for birds, incorporating age-specific probabilities of breeding.

It is well known that, under mild conditions, this model leads to asymptotically exponential growth: exponential growth is reached only simultaneously to a stable age structure. This is the price to pay for taking into account differences in demographic parameters according to age. The asymptotic population *multiplication* rate is obtained as the largest eigenvalue of the Leslie matrix, traditionally denoted as λ.

Table 5.1 The Leslie matrix model with a parameterization suitable for bird population

$$
\begin{bmatrix} N_1 \\ N_2 \\ N_3 \\ \cdot \\ \cdot \\ \cdot \\ N_m \end{bmatrix}_{t+1}
=
\begin{bmatrix}
0 & \cdots & & s_1 a_n f_n & \cdots & s_1 a_m f_m \\
s_2 & 0 & 0 & \cdots & & 0 \\
0 & s_3 & 0 & \cdots & & 0 \\
\cdot & & & & & 0 \\
\cdot & & & & & 0 \\
\cdot & & & & & 0 \\
0 & \cdots & & 0 & s_m & s_{m+1}
\end{bmatrix}
\begin{bmatrix} N_1 \\ N_2 \\ N_3 \\ \cdot \\ \cdot \\ \cdot \\ N_m \end{bmatrix}_{t}
$$

N, age at first reproduction; m, age of stabilization of parameters or maximum age; a_i, age-specific probability of reproduction; f_i, reproductive output (in *females per breeding female*); s_i, age dependent survival, S_{m+1}, 0 if m is maximum age; $s_{m+1} >$ if survival remains constant for individuals older than m; s_1 and f_i should have coherent definitions so that there is no gap in the life cycle. For example if f_i is expressed as a number of newly hatched young, s_1 must include fledging success.

$r = \text{Log}_e \lambda$ is the asymptotic *growth* rate of the modelled population. The ratio of the modulus of the second largest eigenvalue of the Leslie matrix to the first one provides a straightforward index of convergence speed to this asymptotic age distribution. Lewis (1976) illustrates this underused approach for the Herring Gull *Larus argentatus*. In this instance, the value of this ratio (0.536/1.0886) indicated a fairly rapid convergence, as is usually found.

Reasoning directly from flows of individuals leads to a renewal equation for λ, known as the Lotka equation (for some particular cases see Capildeo and Haldane 1954):

$$\Sigma a_i f_i s_1 \ldots s_i / \lambda^i = 1$$

The renewal equation and the matrix approaches are strictly equivalent formulations of the (discrete time) stable population theory (Keyfitz 1968), which would more appropriately be called stable <u>structure</u> population theory.

Various age structures arise from the model:

(a) the longitudinal one, i.e. that of a cohort over time, proportional to $(s_1, s_1 s_2, s_1 s_2 s_3, \ldots)$;
(b) the transversal one, i.e. the stable age structure of the model, proportional to $(s_1 / \lambda^1, s_1 s_2 / \lambda^2, s_1 s_2 s_3 / \lambda^3, \ldots)$;
(c) the transversal structure of breeders, proportional to $(a_1 s_1 / \lambda^1, a_2 s_1 s_2 / \lambda^2, a_3 s_1 s_2 s_3 / \lambda^3, \ldots)$;
(d) the transversal structure of breeders, corrected for fecundity, proportional to $(a_1 f_1 s_1 / \lambda^1, a_2 f_2 s_1 s_2 / \lambda^1, a_3 f_3 s_1 s_2 s_3 / \lambda^3, \ldots)$.

The last structure is made up of the terms of the Lotka equation, which sum up to 1: considered as a frequency distribution, it is the asymptotic distribution of the ages of mothers of newly born young, taking into account multiple births. Its mean $\Sigma i a_i f_i s_1 \ldots s_i / \lambda^i = T$ is a particularly meaningful definition of generation time (Leslie 1966). The stable age structure of the whole population and that of breeders of a Black-headed Gull *Larus ridibundus* population (Lebreton 1987) are shown in Fig. 5.1.

Ratios of numbers observed in the age classes can provide estimates of survival rates for the longitudinal age structure: this is the basis for the longitudinal life table techniques. This is also possible for the transversal (stable) age structure if and only if λ is equal to 1, i.e. if the population is stable (see Chapman and Robson 1960). Moreover, for many birds, only counts of breeders will be available, and nearly always relative to original numbers ringed since ringing is usually the only way to determine the age of a bird. Specific models are needed (Clobert and Lebreton, this volume).

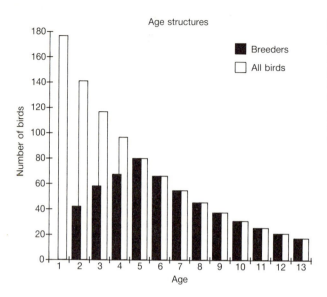

Fig. 5.1. Asymptotic age structures of the total population and of breeders, predicted by a Leslie matrix model of a Black-headed Gull *Larus ridibundus* population (parameter values from Lebreton 1987).

Sensitivity of population multiplication rate to parameters

Various authors noticed the strong sensitivity of the multiplication rate to adult survival and the limited role of fecundity in simulations of populations of long-lived species (Lebreton and Isenmann 1976: *Larus ridibundus*; Eberhardt and Siniff 1977: marine mammals; Kosinski and Podolski 1979: *Rissa tridactyla*). The same conclusion, contrasted with heavy sensitivity to fecundity of short-lived species had been obtained by Young (1968) considering hypothetical populations of Robin *Turdus migratorius* and Bald Eagle *Haliaetus leucocephalus*. Explicit results on sensitivity for individual parameters have been given by a number of authors (e.g. Goodman 1971). The generation time T (see above and Leslie 1966) plays a central role which is particularly striking when sensitivity is expressed as the relative sensitivity to changes in all fecundity rates, or all survival rates after age one (Fig. 5.2, from Lebreton 1981; see also Houllier and Lebreton 1986). This role has not been emphasized by human demographers, working on populations with fairly homogeneous generation time. However, it is of primary importance in our case since it means that in any sharp change of population growth rate for a long-lived species, one should first suspect a change in adult survival. Among outputs of the model, generation time is therefore as important as the multiplication rate.

Stable age structures are much less sensitive than multiplication rate to permanent changes in parameters. The most extreme case is that an overall change of all survival rates, s_i, in a same proportion: λ will also change in the same proportion, and the stable age structures will not be affected at

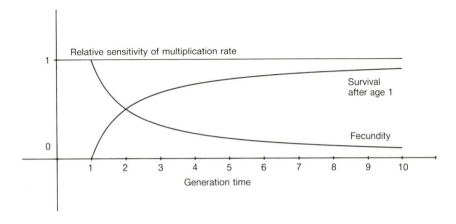

Fig. 5.2. Relative sensitivity of the asymptotic population multiplication rate to survival after 1st year and to fecundity, as functions of generation time (see text for details).

all. This is the first reason which limits possible inferences on population dynamics from estimates of age structure (Caughley 1974). Slow changes in demographic rates have been considered by Artzrouni (1986) who gives approximations to the transient age structures occurring during such changes.

Uncertainty in parameters

In the ideal case where estimates of all parameters are available, the population multiplication rate λ obtained from them can be viewed as an estimate of a true unknown λ. A linear approximation of $\hat{\lambda}$ as a function of parameters θ_i leads to an approximation of its sampling variance (Daley 1979): $\text{var}(\hat{\lambda}) = \Sigma(\partial\lambda/\partial\theta_i)(\partial\lambda/\partial\theta_j) \, S_{ij}$, where $S_{ii} = \text{var}(\theta_i)$, $S_{ij} = \text{cov}(\theta_i, \theta_j)$. This approximation includes the simultaneous influence of the uncertainty of parameter estimates on λ through the sensitivity coefficients. The role of the various parameters can be examined separately as components of the variance. Although this expression slightly underestimates the true variance, a cumbersome second order approximation does not improve it appreciably (Houllier *et al*. 1989).

The multiplication rate can also be estimated from counts, under the hypothesis of exponential growth and of measurement error; various estimates are possible, according to the structure of the error variance (Lebreton 1982a). Most can be considered as weighted means of the indices of population changes in successive years N_{t+1}/N_t. If the imprecision in counts is expressed as var(N_{t+1}) and var(N_t), standard approximation formulas lead to an approximation of the variance of N_{t+1}/N_t (Sen 1983) and then to an approximation of the variance for the other estimates (Lebreton, unpublished). Another good estimate of $\log \lambda$, especially in the case of irregularly distributed censuses is the slope of a regression of $\log N_t$ over t. Plotting $\log N_t$ against t is always useful. When stochasticity is more complex (see below) and causes autocorrelation between counts, fitting a similar equation taking account of the autocorrelation can provide a more satisfactory estimate (see an example in Binkley and Miller 1988).

There are thus two kinds of estimates of the multiplication rate: those arising from estimates of parameters through a demographic model, and those arising from counts under the hypothesis of exponential growth. As noted by Nur (1987) these two kinds of estimates have not always been clearly separated in the literature. When both are available and accompanied by estimates of their variance, a formal comparison is possible under the hypothesis of normality of these estimates, by comparing $(\lambda_1 - \lambda_2)/\sqrt{(S_1^2 + S_2^2)}$ to a standardized normal random deviate. It tests consistency between the external approach (counts) and the internal one (demographic mechanisms). The demographic estimate can also be compared with the theoretical value 1, to test stability. For example Lande (1988) shows for *Strix occidentalis* that the observed $\hat{\lambda} = .961$, with a standard error of .029, does not depart significantly from stability (1 is within the interval $.961 \pm 2_*.029$). Because of possible skewness and bias it might be useful to work out such comparisons after transformation (e.g. to log) (Sen 1983, p. 711). In the first test, rejection of the hypothesis of equality indicates a bias in demographic estimates (e.g. an underestimated survival; overestimated proportions of breeders . . .), a failure in assumptions of the model (e.g. variable parameters, immigration, and emigration . . .), or a bias in counts.

The application of LESLIE matrix models

Possible strategies for using a Leslie matrix model range from such a formal comparison to simply running scenarios when estimates of some of the required parameters are missing (e.g. Mertz 1971). In the latter case, it is still worthwhile to calculate the generation time and to perform sensitivity analyses. Between these two extremes, many intermediate uses are possible.

Whatever the degree of sophistication, the model should be used as a tool, and not as a perfect representation of the population under study. From this point of view, even running it with estimates from other populations, neighbouring species, or parameter predictions obtained from comparative studies may be worthwhile. Estimates of unknown parameters can also be deduced from a matrix model by equating the multiplication rate to a fixed value (see e.g. Henny *et al.* 1970; O'Neill *et al.* 1981). This leads to moderately biased estimates because of the nonlinearity of λ as a function of the unknown parameters. In many cases such estimates will nevertheless provide an interesting order of magnitude, e.g. for juvenile survival.

As noted by Henny *et al.* (1970), Lebreton and Isenmann (1976), and Eberhardt (1985), matrix models for seasonally breeding populations appear to be valid on their own without having to be considered as approximations of homogeneous continuous time models. The only correct continuous time approach would involve cumbersome periodic rates: 'the matrix approach is both compact and efficient and permits a great deal of exploration and special analysis not so readily done in other ways' (Eberhardt 1985). On the other hand, biases and shortcomings of matrix models are obvious. For example, the population is supposed not to be affected by emigration and immigration; the scale in space should thus account for recruitment dispersal. The parameters are supposed to be constant; this might be an acceptable hypothesis for a short time-scale for some populations. Although conclusions in terms of sensitivity make these models a necessary—and underused—step, they are clearly inadequate for any precise assessment of population resilience (in the meaning of Holling 1973) because they do not consider stochasticity and density dependence.

More generally, any graphical model of flows over seasons between classes of individuals leads to a generalized matrix model if parameters are assumed to be constant (Lebreton and Isenmann 1976; Lewis 1976); this is another underused possibility. Rogers and Castro (1973) gives examples of such models for human populations interconnected by migration. In species with demographic rates dependent on the duration of breeding experience or the time since the pair-bond was established (e.g. *Rissa tridactyla*, Coulson 1966) models with stages rather than age classes might be preferable (Houllier and Lebreton 1986; Crouse *et al.* 1987).

Effect of variations of parameters over time

Models for random environment

Year-to-year random environmental variation can be modelled in a straightforward way as a succession of random Leslie matrices (Sykes

1969). Since these matrices are positive, population numbers are positive; this kind of model does not consider population extinction. Numerical simulation has been and still is heavily used. However, under mild conditions log population size becomes asymptotically normal and grows or decreases linearly with an expected rate, a (Tuljapurkar and Orzack 1980); population growth is then best characterized by $\Lambda = \exp(a)$. When environmental variations are independent over the years this multiplication rate can be approximated as (Tuljapurkar 1982):

$\Lambda \approx \lambda \exp(-\frac{1}{2}\mathrm{var}(\lambda)/\lambda^2) \approx \lambda (1 - \frac{1}{2}\mathrm{var}(\lambda)/\lambda^2)$, where the previous linear approximation for $\mathrm{var}(\lambda)$ can be used; Lande (1988) show for *Strix occidentalis* that reasonable environmental variation leads to $\Lambda = .98\lambda$.

If variability in a single parameter π is considered, $\Lambda/\lambda \approx 1 - \frac{1}{2}(\partial\lambda/\partial\pi)^2 \mathrm{var}(\pi)/\pi^2$. The effect of the variability in π can be viewed in terms of the increase $\delta\pi$ in the average π needed to compensate for variability, given by $\delta\pi/\pi = \frac{1}{2}(\partial\log\lambda/\partial\log\pi) \mathrm{var}(\pi)/\pi^2$. The relative increase required to compensate for variability is thus higher for sensitive parameters. Reducing the variability is thus a way to reach a higher average multiplication rate.

Environmental variability and population counts

Simultaneously to detecting variability over time, a common goal is to attribute its origin to some external variables (weather . . .). Regression of any index of population change, such as N_{t+1}/N_t, on an environmental variable x_t (for example see Hafner *et al*. 1987) should be used with extreme caution. While the usual estimator of slope is not biased, its variance is; usual tests of significance for the regression will thus be biased, in a direction depending on the sign of the autocorrelation of the environmental variable. This effect jeopardizes all classical key-factor analyses, and will be even more dramatic when density dependence is considered (see next paragraph). If key-factor analysis is defined as looking for parameters of which variability induces the largest changes in numbers (Manly 1977), it reduces to sensitivity analysis.

Density dependence

After the early discussions on density dependence (for review see Royama 1977), the problem is still challenging because, as in general in cybernetics (Wilbert 1970), the existence and intensity of negative feedbacks governs the stability of the system (Holling 1973) and its ability to withstand perturbations.

Modelling density dependence

Modelling techniques are straightforward: one or several parameters in the Leslie matrix are considered as functions of one or several components of population numbers. After an early proposal by Leslie (1948), Beddington (1974) gave a general formulation and studied dependence on the *total* number in the population. A stable equilibrium is reached under mild conditions. For this equilibrium level, the largest eigenvalue of the Leslie matrix is equal to 1. Density-dependent Leslie matrix models can thus be considered as multivariate discrete time analogues to logistic growth. Beddington's results can be easily extended to dependence on any linear compound of components of population size (Lebreton 1981): this covers dependence on total breeding population (cf. Fig. 5.1) or on total biomass as particular cases. In a particularly clear paper Smouse and Weiss (1975) investigate density dependence in fecundity, prereproductive survival, or adult survival. The differences obtained in the stable age-structure under these different regimes of regulation are limited if the intrinsic growth rate ($\log \lambda$ at density 0) is low, which is the case for bird populations. The stable age structure is reached earlier than the plateau in numbers; both results cast further doubt on inferences about population mechanisms drawn from age structures. Although discrete time density-dependent models exhibit complex behaviour (periodic, chaotic; see May 1975), especially in multivariate models, such effects are mostly irrelevant here, again because of low maximum growth rates of bird populations.

Detecting density dependence

The unreliability of naive statistical approaches based on regression in detecting density dependence has been emphasized by many authors (Salt 1966; Eberhardt 1970; St-Amant 1970; Ito 1972; Slade 1977). Considering for example the regression of $\log N_{t+1}/N_t$ on N_t, one can see that errors in $Y_t = \log N_{t+1}/N_t$ and $X_t = N_t$ are not independent, as well as those in $Y_t = \log N_{t+1}/N_t$ and $Y_{t-1} = \log N_t/N_{t-1}$. Estimates of slope and of its variance are biased. All other kinds of stochasticity will affect the slope in the same way (St-Amant 1970). All these methods tend to detect density dependence too often. Despite contrary claims by their authors, the same problem exists in recently proposed *ad hoc* methods and reviews (Vickery and Nudds 1984; Pollard *et al*. 1987; Cuperus and de Bruyn 1987). Time-series techniques are the logical tool (Bulmer 1975; Lebreton 1989). However, measurement error should be accounted for by the Kalman filter technique as noted by Poole (1978) and Brillinger (1981), and proposed in the second model of Bulmer (1975, p. 903). Both Poole and Brillinger review time-series approaches potentially relevant for population studies, with an emphasis on forecasting in Poole's paper. Although appealing, especially with reference to management, such techniques will rarely be powerful with the number of points usually available in population dynamics, especially because the effect of environmental variables has to be accounted for simultaneously (Lebreton 1982*b*). Their best use seems to be as approximations of mechanistic models rather than as descriptive forecasting models on their own.

Fortunately enough, the problem is much more to measure the intensity of density dependence, locate the concerned parameters and assess the consequences, rather than simply detect density dependence. Our advice is to look for effects of population density on parameters such as survival or fecundity (see Clobert and Lebreton, this volume), and to include environmental variables and density as explanatory variables to limit confounding between sources of variation. To avoid the pitfalls of key-factor analysis, the estimates of parameters used in such an approach should be independent of the estimates of density.

Model use

In view of these problems, using density-dependent models is clearly difficult. Evidence for density dependence is usually limited, although it is supposed to play a major role in determining the stability of bird populations. Numerical simulation, easy with a desk computer, is of great help, particularly to achieve a sensitivity analysis of equilibrium level. Patterns of sensitivity are roughly similar to those of multiplication rate in density-independent models with a main role of survival in long-lived species (Lebreton 1981).

Density dependence and seasonality

Up to now seasonality and age structure seem not to have affected the general pattern of growth. They seem to have been included for the sake of realism only. However results depart from the standard logistic growth when we look at the consequences of simultaneous seasonality (or random environment) and density dependence. Figure 5.3 gives numbers over time in a model alternating a good and a poor season (or year). The population never reaches at time t the carrying capacity for the conditions prevailing at this time, but an equilibrium level depending on the overall dynamics of the system. Although this role of seasonality in regulated populations has been well studied (Fretwell 1976; May 1976), is not usually realized how weak it makes the concept of carrying capacity (see e.g. Caughley 1976). This paradoxical effect is more pronounced for populations with low maximum growth rate, i.e. in practice for long-lived species. The numbers

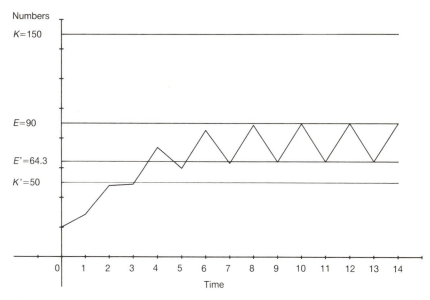

Fig. 5.3. Population size over time in a model alternating good and poor seasons (or years), according to equations: $N_{t+1} = 2N_t/(1 + N_t/150)$ (good) and $N_{t+1} = 2N_t/(1 + N_t/50)$ (poor). Asymptotic population size does not oscillate between the poor year model ($K' = 50$) and the good year model ($K = 150$) carrying capacities, but between two equilibrium levels ($E' = 64.3$ and $E = 90$), both well below the average carrying capacity ($(K + K')/2 = 100$).

in populations of such species do not track environmental variability as narrowly as those of shorter-lived species. This has two consequences. Firstly, it is better to examine the influence of environment on population dynamics at the level of population mechanisms, i.e. at the parameter level, because species–environment relationships may appear to be weak when considered on the basis of population numbers. Secondly, a regulating factor in spring may determine the *existence* of an equilibrium level, while the *value* of this equilibrium level will be determined by internuptial survival. For migratory species in particular, ecological factors on the wintering grounds may determine what is usually called the carrying capacity of the breeding habitat (see Lebreton 1981; *Larus ridibundus*).

Harvesting

Another approach to model stability which is relevant to conservation and management is concerned with optimizing harvesting. The basic literature on the subject (see full account by Getz and Haight 1989) comes from exploited fish populations (e.g. Deriso 1980; Reed 1980) where the goal is to maximize the sustainable yield. One could also look for maximum efficiency against a pest (Murton and Westwood 1976), with guidelines from sensitivity analysis. Doubleday (1975) studied optimal harvesting in the density-independent case, which is however of restricted practical interest, as noted by Mendelssohn (1976).

Optimal harvesting is in general studied in models with density dependence in fecundity (or survival in the first year or any other 'immature' parameter) (density-dependent recruitment). The shape of the 'stock-recruitment curves', i.e. of the density-dependent relationship (see e.g. Eberhardt 1977), and the sustainable yield (Winters 1978; Eberhardt and Siniff 1977) are concepts of general validity. However, they appear difficult to use with realism and precision for bird populations in the present state of the art, since they suppose widescale information on mechanisms of density dependence. In exploited fish populations it is commonly found or assumed that the proportion of immature animals that recruit to the breeding part of the population (called the breeding stock) is density dependent. The extent to which this is true in bird populations is unclear (Duncan 1978; Coulson *et al*. 1982; Brown 1969; Klomp 1972; Patterson 1980). This question requires a discussion of more complex phenomena such as spatial heterogeneity and dispersal.

Incorporating more spatial and temporal heterogeneity and stochasticity

Variability between habitats and dispersal

Building population models that include dispersal does not raise particular technical difficulties. Spatial cells can be considered to have the same status as age classes. Such models have been developed in both density-independent (Rogers and Castro 1976) and density-dependent cases. Gurney and Nisbet (1975), Kot and Schaffer (1986), and Den Boer (1971) show that oriented dispersal from high densities to low densities tends to stabilize numbers.

This is probably the best example of a problem for which a good array of modelling techniques exist, but for which data are extremely difficult to obtain because of the complexity of the time and space scales in bird populations. Developing studies integrating modelling of meta-populations and field evidence is obviously one of the present challenges in population dynamics.

Demographic stochasticity

Demographic stochasticity, or within individual stochasticity (Chesson 1978) is the residual variability that affects the demographic results of a given individual, under fixed conditions: for example, death with probability p is viewed as a coin-tossing process. This type of variability, both in survival and in fecundity, can easily be incorporated in a process together with age structure and seasonality. Such 'multitype discrete time branching processes' have been heavily studied by mathematicians (see e.g. Jagers 1975). Average numbers are described by a Leslie matrix model and the probabilistic structure relies on the hypothesis of independence between individuals. Although the relative role of demographic stochasticity is negligible in a large homogeneous population, it may be important in a small or subdivided one (Chesson 1978; Mode and Pickens 1986). Models including subunits and dispersal should thus consider demographic as well as environmental stochasticity.

Extinction

With branching processes, possible extinction is now under concern: the ultimate probability of extinction is lower than 1 if and only if the population increases ($\lambda > 1$), i.e. goes to infinity rapidly enough to avoid the trap of extinction. The next step considers simultaneously demographic and environmental stochasticity with parameters of the branching process varying from year to year in a random way. Such models (Branching Processes in Random Environment = BPRE) have been used for theoretical purposes (Heyde 1978; Mountford 1973). Moreover, BPRE can incorporate realistic features of seasonal age-structured populations (Mode and Jacobson 1987a), since they have a Leslie matrix model as mean equation. They constitute a natural tool to study extinction phenomena.

Decreasing populations coming close to extinction present, as modelled by branching processes, a particular behaviour of convergence to what is called a *quasi-stationary distribution*. Once reduced to a few individuals, the population can become extinct next year, or survive. In the latter case, it will include on average more than one individual; conditional on non-extinction, there is thus a small residual population. This conditional residual population tends to stabilize. This means that when this quasi-stationary distribution is reached, the population behaves in terms of survival like a single individual, dying with probability $1 - p$, or surviving next year identical to itself (on the average in the case of the population) with probability p. Extinction time then follows a geometric distribution. A striking result is that $p = \Lambda$, which depends not only on the average parameters but also strongly on the environmental process, in particular on its autocorrelation (Mode and Jacobson 1987b). In the case of the decreasing population of White Stork of the Alsace area (Lebreton 1982b), a BPRE leads to a quasi-stationary distribution with about 7 individuals (about 1.5 breeding pair), with a yearly probability of extinction of .15.

Results become scarcer when density dependence is mixed with seasonality, age structure, discrete time, demographic stochasticity, and environmental stochasticity. In some simple models it is possible to prove the certainty of extinction (Lebreton 1981). When an equilibrium distribution is reached, extinction in a single step of time is always possible with an extremely small probability (Lebreton 1982*b*), and such an equilibrium distribution can only be a quasi-stationary distribution, of which formal existence is difficult to prove (see however De Angelis 1976; Lebreton 1981; Lande and Orzack 1988); simulation is often the only possible tool. De Angelis (1976; Canada Goose *Branta canadensis* population) and Lebreton (1981; Great Tit *Parus major*) provide examples of quasi-stationary distributions for discrete time density-dependent models.

The convergence between studies of mathematical processes (Branching Processes in Random Environment have mainly been developed since 1970) and the concept in conservation biology of Minimum Viable Populations (MVP) (Soule 1986, 1987; Frankel and Soule 1981) is thus very striking, and is explicitly pointed out by Mode and Jacobson (1987*a*, *b*). One can speak with Gilpin and Soule (1986) and Burgman *et al*. (1988) of extinction models. Extinction models in continuous time (Pettersson 1985) are highly questionable because they neglect the seasonal structure of the population (De Angelis 1976). The variations are buffered by the continuous role of density dependence, and the predictions of extinction seem likely to be optimistic. Burgman *et al*. (1988), who emphasized the theoretical character of extinction models, focused on continuous time models (e.g. pp. 17–18) and neglected the wide literature on discrete time branching processes published mainly in probability journals. Discrete time branching models built as generalizations of demographic models bring some more realism to the general concept of MVP. However, the sensitivity of predictions to variations in the environmental process and to the amount of density dependence makes them very difficult to validate.

It seems striking to us that the effects of parameter uncertainty, on which rapid progress has been made for the Leslie model in the last ten years, have still to be investigated in extinction models. This difficulty must always be borne in mind for small populations where a reasonable precision on demographic parameters is unattainable. More realistic models would also include genetic (Schonewald-Cox *et al*. 1983) and behavioural aspects (e.g. mate selection, social stimulation). As a rule, it seems wise in the present state of the art to consider MVP obtained from models as being underestimated.

In conclusion, such complex phenomena are probably easier to model than to assess. Models can be used with caution to test some hypotheses in an informal way, but the robustness of the conclusions will usually be limited by the absence of replication (Hurlbert 1984). It seems dangerous for conservation and management purposes to use complex models developed in an *ad hoc* way by incorporating all potential mechanisms and effects (see e.g. McKelvey *et al*. 1980).

Discussion

Modelling of population dynamics for seasonal populations with constant parameters has reached a stage of maturity. Demographic stochasticity, environmental stochasticity (Chesson 1978), and density dependence have progressively been brought together in parallel in a common probabilistic framework. Such a framework, already advocated by Reddingius (1971*b*) provides a sound basis for studies of regulation, harvesting, and extinction, and more generally of resilience of populations. The application of these approaches to practical problems will take time, as is to be expected for a multidisciplinary endeavour. There are no major differences in modelling approaches for fundamental research or conservation and management goals. It seems likely that in the future the diversity of interest in population dynamics modelling will maintain the same balance between theoretical and practical approaches.

General models of population regulation have been developed less extensively and less rapidly for bird populations than for exploited populations such as fish, marine mammals (Eberhardt and Siniff 1977), or large terrestrial mammals (Caughley 1976). However the information on demography and on mechanisms of regulation is usually much greater for birds. This has probably inhibited the broad generalizations which have been made, perhaps unjustifiably, for fish populations.

While some generality and realism have already been reached, precision will frequently remain out of reach, for reasons of cost, or for intrinsic reasons in the case of small populations. Models of adaptive management (Holling 1978) might in this case be helpful, as well as arrays of methods in monitoring programs (Aebischer 1986).

The 'new frontier' of population modelling—which hopefully will develop simultaneously with refined designs in field studies—is in modelling meta-populations *sensu lato*, i.e. in focusing on dispersal phenomena. We also emphasize the need for flexibility: models should be seen as constantly developing tools, which should be periodically re-evaluated.

Summary

This paper reviews, in four steps, mathematical models of the dynamics of bird populations which are relevant to conservation and management.

(1) demographic models with constant parameters, with an emphasis on sensitivity analysis and on the comparison of model results with population censuses;
(2) environmental variability and its effect on population growth rate;
(3) density-dependent models, with comments on the difficulties to assess density dependence using censuses;
(4) models incorporating density dependence, spatial aspects, and various kinds of stochasticity.

Acknowledgements

We thank R. Green and R. Lande for useful comments on a previous version of the manuscript.

References

Aebischer, N. (1986). Retrospective investigation of an ecological disaster in the Shag, *Phalacrocorax aristotelis*. *Journal of Animal Ecology*, **55**, 613–29.

Artzrouni, M. (1986). On the dynamics of a population subject to slowly changing vital rates. *Mathematical Biosciences*, **80**, 265–90.

Beddington, J. R. (1974). Age distribution and the stability of simple discrete time population models. *Journal of Theoretical Biology*, **47**, 65–74.

Binkley, C. S. and Miller, R. S. (1988). Recovery of the Whooping Crane *Grus americana*. *Biological Conservation*, **45**, 11–20.

Brillinger, D. R. (1981). Some aspects of modern population mathematics. *The Canadian Journal of Statistics*, **9**, 173–94.

Brown, J. L. (1969). Territorial behavior and population regulation in birds. A review and re-evaluation. *Wilson Bulletin*, **81**, 283–329.

Bulmer, M. G. (1975). The statistical analysis of density-dependence. *Biometrics*, **31**, 901–11.

Burgman, M. A., Akcakaya, H. R., and Loew, S. S. (1988). The use of extinction models for species conservation. *Biological Conservation*, **43**, 9–25.

Capildeo, R. and Haldane, J. B. S. (1954). The mathematics of bird population growth and decline. *Journal of Animal Ecology*, **23**, 215–23.

Caughley, G. (1974). Interpretation of age ratios. *Journal of Wildlife Management*, **38**, 557–62.

Caughley, G. (1976). Wildlife management and the dynamics of ungulate populations. In *Applied biology* (ed. T. H. Coaker), pp. 183–246, Academic Press, London.

Chapman, D. G. and Robson, D. S. (1960). The analysis of a catch curve. *Biometrics*, **16**, 354–68.

Chesson, P. (1978). Predator-prey theory and variability. *Annual Review of Ecology and Systematics*, **9**, 223–347.

Coulson, J. C. (1966). The influence of pair-bond and age on the breeding biology of the Kittiwake *Rissa tridactyla*. *Journal of Animal Ecology*, **35**, 269–79.

Coulson, J. C. and Wooler, R. D. (1976). Differential survival rate among breeding Kittiwake Gulls. *Journal of Animal Ecology*, **45**, 205–13.

Coulson, J. C., Duncan, N., and Thomas, C. (1982). Changes in the breeding biology of the Herring Gull (*Larus argentatus*) induced by reduction in the size and density of the colony. *Journal of Animal Ecology*, 51, 739–56.

Crouse, D. T., Crowder, L. B., and Caswell, H. (1987). A stage-based population model for Loggerhead Sea-Turtles and implications for conservation. *Ecology*, 68, 1412–23.

Cull, P. and Vogt, A. (1973). Mathematical analysis of the asymptotic behaviour of the Leslie population model. *Bulletin of Mathematical Biology*, 35, 645–61.

Cuperus, R. and De Bruyn, G. J. (1987). Statistical impacts on the detection of natural regulation in bird population on account of census data. *Ardea*, 75, 237–43.

Daley, D. J. (1979). Bias in estimating the malthusian parameter for Leslie matrices. *Theoretical Population Biology*, 15, 257–63.

De Angelis, D. L. (1976). Application of stochastic models to a wildlife population. *Mathematical Biosciences*, 31, 227–36.

Den Boer, P. J. (1971). Stabilization of animal numbers and the heterogeneity of the environment, the problem of the persistence of sparse populations. In *Proceedings Advanced Institute on the 'Dynamics of Numbers in Populations'* (Oosterbeek, 1970), (ed. P. J. den Boer and G. R. Gradwell), pp. 77–97.

Deriso, R. B. (1980). Harvesting strategies and parameter estimation for an age-structured model. *Canadian Journal of Fisheries and Aquatic Sciences*, 37, 268–82.

Doubleday, W. G. (1975). Harvesting in matrix population models. *Biometrics*, 31, 189–200.

Duncan, N. (1978). The effects of culling Herring Gulls (*Larus argentatus*) on recruitment and population dynamics. *Journal of Applied Ecology*, 15, 697–713.

Eberhardt, L. L. (1970). Correlation, regression, and density dependence. *Ecology*, 51, 306–10.

Eberhardt, L. L. (1977). Relationship between two stock-recruitment curves. *Journal of The Fisheries Research Board of Canada*, 34, 425–8.

Eberhardt, L. L. (1985). Assessing the dynamics of wild populations. *Journal of Wildlife Management*, 49, 997–1012.

Eberhardt, L. L. and Siniff, D. B. (1977). Population dynamics and marine mammal management policies. *Journal of Fisheries Research Board of Canada*, 34, 183–90.

Frankel, O. H. and Soule, M. E. (1981). *Conservation and evolution*. Cambridge University Press, Cambridge.

Fretwell, S. D. (1976). *Populations in a seasonal environment*. Princeton University Press, Princeton.

Getz, W. M. and Haight, R. G. (1989). *Population harvesting*. Princeton University Press, Princeton.

Gilpin, M. E. and Soule, M. E. (1986). Minimum viable populations: processes of species extinction. In *Conservation biology: the science of scarcity and diversity* (ed. M. E. Soule), pp. 19–34, Sinauer, Sunderland.

Goodman, L. A. (1971). On the sensitivity of the intrinsic growth rate to changes in the age-specific birth and death rates. *Theoretical Population Biology*, 2, 339–54.

Gurney, W. S. C. and Nisbet, R. M. (1975). The regulation of inhomogeneous populations. *Journal of Theoretical Biology*, 52, 441–57.

Hafner, H., Wallace, J. P., and Dugan, P. J. (1987). *Tour du Valat Heron Programme: Progress Report 1987*. Mimeographed paper, Station Biologique de la Tour du Valat, Arles, France.

Heyde, C. C. (1978). On an explanation for the characteristic clutch size of some bird species. *Advances in Applied Probability*, 10, 723–5.

Henny, C. J., Overton, W. S. and Wight, H. M. (1970). Determining parameters for populations by using structural models. *Journal of Wildlife Management*, **34**, 690–703.

Holling, C. S. (1973). Resilience and stability of ecological systems. *Annual Review of Ecology and Systematics*, **4**, 1–23.

Holling, C. S. (ed.) (1978). *Adaptive environmental assessment and management*. I. I. A. S. A. Laxenburg, Austria.

Houllier, F. and Lebreton, J. D. (1986). A renewal equation approach to the dynamics of stage-grouped populations. *Mathematical Biosciences*, **79**, 185–97.

Houllier, F., Lebreton, J. D., and Pontier, D. (1989). Sampling properties of the asymptotic behaviour of age- or stage-grouped population models. *Mathematical Biosciences*, **95**, 161 77.

Hurlbert, S. H. (1984) Pseudoreplication and the design of ecological field experiments. *Ecological Monographs*, **54**, 187–211.

ITO, Y. (1972). On the methods for determining density-dependence by means of regression. *Oecologia (Berlin)*, **10**, 347–72.

Jagers, P. (1975) *Branching processes with biological applications*. Wiley, London.

Keyfitz, N. (1968). *Introduction to the mathematics of population*. Addison Wesley, Reading, Mass.

Klomp, H. (1972). Regulation of the size of bird populations by means of territorial behaviour. *Netherlands Journal of Zoology*, **22**, 456–88.

Kosinski, R. J. and Podolski, R. H. (1979). An analysis of breeding and mortality in a maturing Kittiwake colony. *Auk*, **96**, 537–43.

Kot, M. and Schaffer, W. M. (1986). Discrete-time growth-dispersal models. *Mathematical Biosciences*, **80**, 109–36.

Lande, R. (1988). Demographic models of the northern spotted owl (*Strix occidentalis caurina*). *Oecologia (Berlin)*, **75**, 601–7.

Lande, R. and Orzack, S. H. (1988). Extinction dynamics of age-structured populations in a fluctuating environment. *Proceedings National Academy of Sciences, U.S.A.*, **85**, 7418–21.

Lebreton, J. D. (1981). Contribution à la dynamique des populations d'oiseaux. Modèles mathématiques en temps discret. Unpublished thesis, Université Lyon I, Lyon.

Lebreton, J. D. (1982*a*). Modèles dynamiques déterministes définis par des équations de récurrence. In *Modèles dynamiques déterministes en biologie* (ed. J. D. Lebreton and C. Millier), pp. 58–98. Masson, Paris.

Lebreton, J. D. (1982*b*). Application of discrete time branching processes to bird population dynamics modelling. In *Anais 10a Conferência Internacional de Biometria*, pp. 115–33. Embrapa, Brasilia.

Lebreton, J. D. (1987). Regulation par le recrutement chez la Mouette rieuse *Larus ridibundus*. *Revue d'Ecologie (Terre Vie)*, Suppl. **4**, 173–82.

Lebreton, J. D. (1989). Statistical methodology for the study of animal populations. *Bulletin International Statistical Institute*, **53**, 267–82.

Lebreton, J. D. and Isenmann, P. (1976). Dynamique de la population camarguaise de Mouette rieuse, un modèle mathématique. *La Terre et La Vie (Revue d'Ecologie)*, **30**, 529–49.

Leslie, P. H. (1945). On the use of matrices in population mathematics. *Biometrika*, **33**, 183–212.

Leslie, P. H. (1948). Some further notes on the use of matrices in population mathematics. *Biometrika*, **35**, 213–45.

Leslie, P. H. (1966). The intrinsic rate of increase and the overlap of successive generations in a population of Guillemots (*Uria aalge* Pont.). *Journal of Animal Ecology*, **35**, 291–301.

Levins, R. (1968). *Evolution in changing environments*. Princeton University Press, Princeton.

Lewis, E. R. (1976). Applications of discrete and continuous network theory to linear population models. *Ecology*, **57**, 33–47.

Manly, B. J. F. (1977). The determination of key factors from life table data. *Oecologia (Berlin)*, **31**, 111–7.

May, R. M. (1975). Biological populations obeying difference equations: stable points, stable cycles, and chaos. *Journal of Theoretical Biology*, **51**, 11–524.

May, R. M. (1976). Models for single populations. In *Theoretical ecology, principles and applications* (ed. R. M. May), pp. 4–25. Blackwell, Oxford.

McKelvey, R., Hankin, D., Yanosko, K., and Snygg, C. (1980). Stable cycles in multi-stages recruitment models: an application to the Northern California Dungeness crab (*Cancer magister*) fishery. *Canadian Journal of Fisheries and Aquatic Sciences*, **37**, 2323–45.

Mendelssohn, R. (1976). Optimization problems associated with a Leslie matrix model. *American Naturalist*, **110**, 339–49.

Mertz, D. (1971). The mathematical demography of the California Condor population. *American Naturalist*, **105**, 437–53.

Miller, R. S., Botkin, D. B., and Mendelssohn, R. (1974). The Whooping Crane (*Grus americana*) population of North America. *Biological Conservation*, **6**, 106–11.

Mode, C. J. and Jacobson, M. E. (1987*a*). A study of the impact of environmental stochasticity on extinction probabilities by Monte Carlo integration. *Mathematical Biosciences*, **83**, 105–25.

Mode, C. J. and Jacobson, M. E. (1987*b*). On estimating critical population size for an endangered species in the presence of environmental stochasticity. *Mathematical Biosciences*, **85**, 185–209.

Mode, C. J. and Pickens, G. T. (1986). Demographic stochasticity and uncertainty in population projections—a study by computer simulation. *Mathematical Biosciences*, **79**, 55–72.

Mountford, M. D. (1973). The significance of clutch size. In *The mathematical theory of the dynamics of biological populations* (ed. M. S. Bartlett and R. W. Hiorns), pp. 315–23. Academic Press, London.

Murton, R. K. and Westwood, N. J. (1976). Birds as pests. In *Applied biology*, Vol. 1 (ed. T. H. Coaker), pp. 89–181. Academic Press, London.

Nisbet, R. M. and Gurney, W. S. C. (1986). The formulation of age-structure models. In *Mathematical ecology, an introduction* (Biomathematics Vol. 17) (ed. T. Ghallam and S. A. Levin), pp. 95–115. Springer-Verlag, Berlin.

Nur, N. (1987). Population growth rate and the measurement of fitness: a critical reflection. *Oikos*, **48**, 338–41.

Odum, E. P. (1971). *Fundamentals of ecology*, 3rd edn. Saunders, Philadelphia.

O'Neill, R. V., Gardner, R. H., Christensen, S. W., Vanwinkle, W., Carney, J. H., and Mankin, J. B. (1981). Some effects of parameter uncertainty in density-independent and density-dependent Leslie models for fish populations. *Canadian Journal of Fisheries and Aquatic Sciences*, **38**, 91–100.

Patterson, I. J. (1980). Territorial behaviour and the limitation of population density. *Ardea*, **68**, 53–62.

Pettersson, B. (1985). Extinction of an isolated population of the middle spotted woodpecker *Dendrocopos medius* (L.) in Sweden and its relation to general theories on extinction. *Biological Conservation*, **32**, 335–53.

Pollard, E., Lakhani, K. H., and Rothery, P. (1987). The detection of density-dependence from a series of annual censuses. *Ecology*, **68**, 2046–55.

Poole, R. W. (1978). The statistical prediction of population fluctuations. *Annual Review of Ecology and Systematics*, **9**, 427–48.

Reddingius, J. (1971*a*). Models as research tools. In *Proceedings Advanced Institute on the Dynamics of Numbers in Populations*, pp. 64–76 (Oosterbeek, 1970).

Reddingius, J. (1971*b*). Gambling for existence. *Acta Biotheoretica*, **20** (Suppl.), 1–208.

Reed, W. J. (1980). Optimum age-specific harvesting in a non-linear population model. *Biometrics*, **36**, 579–93.

Ricklefs, R. E. (1973). Fecundity, mortality and avian demography. In *Breeding biology of birds* (ed. D. S. Farmer), pp. 366–447. National Academy of Sciences, Washington, DC.

Rogers, A. and Castro, L. J. (1976). Model multiregional life tables and stable populations. IIASA, Laxenburg, Austria.

Royama, T. (1977). Population persistence and density-dependence. *Ecological Monographs*, **47**, 1–35.

Sabatier, R., Lebreton, J. D., and Chessel, D. (1989). Principal components with instrumental variables as a tool for modelling composition data. In *Multiway data analysis* (ed. R. Coppi and S. Bolasco), pp. 341–52. North Holland, Amsterdam.

Salt, G. W. (1966). An examination of logarithmic regression as a measure of population density response. *Ecology*, **47**, 1035–9.

Schonewald-Cox, C. M., Chambers, S. M., MacBryde, B., and Thomas, W. L. (ed.) (1983). *Genetics and conservation*. Benjamin-Cummings, Menlo park, Calif.

Sen, A. R. (1983). Review of some important techniques in wildlife sampling and sampling errors. *Biometrical Journal*, **7**, 699–715.

Slade, N. A. (1977). Statistical detection of density dependence from a series of sequential censuses. *Ecology*, **58**, 1094–102.

Smouse, P. E. and Weiss, K. M. (1975). Discrete demographic models with density-dependent vital rates. *Oecologia (Berlin)*, **21**, 205–18.

Soule, M. E. (ed.) (1986). *Conservation biology: the science of scarcity and diversity*. Sinauer, Sunderland, Mass.

Soule, M. E. (ed.) (1987). *Viable populations for conservation*. Cambridge University Press, Cambridge, U.K.

St-Amant, J. L. S. (1970). The detection of regulation in animal populations. *Ecology*, **51**, 823–8.

Sykes, Z. M. (1969). Some stochastic versions of the matrix model for population dynamics. *Journal of the American Statistical Association*, **64**, 111–30.

Ter Braak, C. J. F. (1986). Canonical correspondence analysis: a new eigenvector technique for multivariate direct gradient analysis. *Ecology*, **67**, 1167–79.

Ter Braak, C. J. F. (1988). Partial canonical correspondence analysis. In *Classification and related methods of data analysis* (ed. H. H. Bock), pp. 551–8. North Holland, Amsterdam.

Tuljapurkar, S. D. (1982). Population dynamics in variable environments. III. Evolutionary dynamics of r-selection. *Theoretical Population Biology*, **21**, 141–65.

Tuljapurkar, S. D. and Orzack, S. H. (1980). Population dynamics in variable environments. I. Long-run growth rates and extinction. *Theoretical Population Biology*, **18**, 314–42.

Verner, J., Morrison, M. L., and Ralph, C. J. (1986). *Wildlife 2000: modeling habitat relationships of terrestrial vertebrates*. University of Wisconsin Press, Madison.

Vickery, W. L. and Nudds, T. D. (1984). Detection of density-dependent effects in annual duck censuses. *Ecology*, 65, 96–104.

Wilbert, H. (1970). Cybernetic concepts in population dynamics. *Acta Biotheoretica*, 19, 54–81.

Winters, G. H. (1978). Production, mortality, and sustainable yield of northwest atlantic Harp Seals (*Pagophilus groenlandicus*). *Journal of the Fisheries Research Board of Canada*, 35, 1249–61.

Young, H. (1968). A consideration of insecticide effects on hypothetical avian populations. *Ecology*, 49, 991–3.

The Species Approach

6 Effects of predation on the numbers of Great Tits Parus major

R. H. McCLEERY and C. M. PERRINS

The Great Tit, *Parus major* is one of the most intensively studied small birds in the world and, more importantly perhaps, has been the subject of a number of long-term studies from which some conclusions can be drawn about the dynamics of populations. The longest running study was that started in the Netherlands by H. Wolda (Kluijver 1951) and continued in the Hoge Veluwe (van Balen 1980) and on the island of Vlieland. The Oxford study was started in Wytham Woods by Drs D. Lack and J. A. Gibb in 1947 (e.g. Lack 1966); initially this study was confined to a relatively small area of the woods, Marley Wood, but was extended to cover the whole of Wytham Woods during the years 1958–65. Few breeding adults were trapped before 1960 and from 1960–4 only females were caught, so for studies such as this one, involving survival rates of individuals, 1965 marks the beginning of the usable data set.

Since the beginning of the study both Great Tits and Blue Tits have been subject to predation during and outside of the breeding season. Weasels (*Mustela nivalis*) enter nest boxes where they take whole broods and sometimes the incubating female as well. Dunn (1977) showed that such predation was density dependent and that it tended to be more prevalent when the populations of rodents were lower. Losses increased dramatically after about 1955, when myxomatosis reduced the availability of rabbits, apparently because there was no alternative mammalian prey. Faced with such extensive destruction of the study population, the nest boxes were changed from a flush-mounted wooden pattern to a hanging concrete one which is more or less weasel-proof. Since 1976, when this change was completed, weasel predation has been insignificant.

Sparrowhawks *Accipiter nisus* may take a large proportion of newly fledged young, especially in the first 5 days after leaving the nest, and take adults both in the breeding season and at other times (Geer 1978; Perrins and Geer 1980; Gray 1987). From 1959 until the early 1970s the Sparrowhawk was largely absent from southern England as a result of almost total breeding failure due to the effects of DDT and its derivatives (e.g. Newton 1979, chapter 13). During this period losses of Great Tits to Sparrowhawks were negligible. However during the early to middle 1970s,

Sparrowhawks made a dramatic come-back and started to breed again in Wytham Woods. From corpses intercepted at nests and remains found at plucking posts, Geer (1978) and Gray (1987) concluded that breeding adult Great Tits and newly-fledged young are an important component of the diet of Sparrowhawks. However, they both thought that the Sparrowhawk predation largely removed some of the 'doomed surplus' of young and that the effect of overall numbers of breeding tits in Wytham was not marked.

The aim of this study was to collate evidence about the effects of predation on this population and to present this in the context of other factors affecting the fluctuations in numbers. In theory, the presence and absence of weasel predation and the presence-absence-presence of Sparrowhawk predation, give us two 'natural' experiments which should enable us to assess the importance of predation to Great Tit numbers. However, this approach is complicated by other factors which demonstrably affect the population and which are not necessarily consistent across periods when predators were either present or absent. What we have tried to do here is to analyse both types of factor and model their joint effects on the population.

General methods

The data referred to in this paper have been collected routinely by researchers working in Wytham Wood since 1947. Nest boxes are visited weekly from mid-April until the young have hatched; since normally one egg is laid per day, the date of laying of the first egg by each pair can be calculated from a weekly visit. Thereafter each nest is visited as necessary to establish success, to ring the chicks, and to weigh them on the 15th day after hatching. The adults are also trapped using a spring-loaded trap-door on the box, ideally between the 7th and 12th day after hatching. We currently catch 76 per cent of the males and 86 per cent of the females (means for 1987–9), but the rate was lower in the early years of the study partly because the nest losses due to weasel predation denied us the opportunity to trap the adults.

We assume that the number of pairs is equal to the number of occupied nest boxes (allowance has been made for pairs that lose a nest at an early stage and lay a replacement clutch); the number of pairs that nest outside nest boxes is small and detailed studies of small areas have shown there to be a very close correlation between the numbers of territorial males and the number of nests in boxes (East and Perrins 1988). Unlike most continental populations, Wytham Great Tits rarely attempted to raise more than a single brood in a season.

In interpreting adult survival and nesting success in different years, two further difficulties arise. Measurement of adult survival is complicated by the fact that the proportion of adults trapped has differed between years. In order to overcome this problem we have, where necessary, multiplied observed survival rates by the retrap efficiency in order to provide an estimate of true survival as follows:

$$\text{estimated survivors} = \text{observed survivors} \times \frac{\text{Number of occupied boxes}}{\text{Number of birds trapped.}}$$

This was done separately for each sex. Clobert *et al*. (1988) investigated the recapture rate using an extension of Cormack's (1964) method for estimating survival from mark-release-recapture data with repeated recaptures. Allowing for differential trapability between sexes and between birds born in Wytham and those born outside and immigrating into the breeding population, they found that the simple estimates used here usually agreed closely with the survival estimates made in a more rigorous way.

A more difficult problem is emigration from the wood. We assume that this is negligible for established breeders, but we do not know how many Wytham-born birds leave the wood. It is clear from the life tables, calculated on the basis of young that are found entering the breeding population (recruits) that individual parents do not produce enough recruits to the Wytham Wood population to maintain a stable population. Using uncorrected retraps (see above) the mean number of recruits produced in their whole breeding life is about 0.80 per female and about 0.89 per male (McCleery and Perrins 1989; the reason males produce slightly more on average is that those that survive to breed live slightly longer than the females). Since we find many unringed birds (immigrants) breeding in the nest boxes in Wytham, we have no doubt that a proportion of the Wytham-born young emigrate, but we cannot estimate the proportion that does so. This may not matter for many purposes since it seems that the number of recruits each year is a reflection of survival rate rather than a variable emigration rate. Survival rates are not correlated with the mobility of the British population as a whole (using data from the British Trust for Ornithology), neither do we find that low local recruitment occurs in years when the average distance moved in the wood is high as might be expected if we were really measuring the movement of individuals out of the study area.

It is instructive to consider these interpretational problems at the outset, since most population studies are likely to encounter similar difficulties, unless they happen to be on an island. In spite of the problems, we feel that we have a reasonably accurate estimate of the survivorship of individuals which can be related to the changes in overall numbers.

Factors other than predation affecting population fluctuations

Before looking at the effect of predation on the population it is necessary to review the other factors that are known to affect it.

Extrinsic factors

It has been known for quite a long time that fluctuations in the Wytham population were related to the food supply in winter (Perrins 1966). The quality of winter food is expressed by whether or not beech trees *Fagus sylvatica* produce a seed crop. Such 'Mast' crops never occur in two successive winters, but may be absent in more than one. Both adult and juvenile survival from one summer to the next are lower when there is no beech crop in the intervening winter, with the result that the total population drops. The situation is presumably more complex than this since most of the mortality probably occurs before the mast becomes available and the fact that the beech index is related to tit population fluctuations in places where there are no beech trees (Perrins 1966). Within the population, the biggest fluctuations in survival are found in the females and juveniles, apparently because the more dominant adult males displace them from insect foods when beech is not available (Gosler 1987).

Intrinsic factors

The dynamics of the Great Tit population in Wytham Wood have been the basis of several studies in the past. Lack (1966) claimed on the basis of total population and mean clutch sizes for the period 1947 to 1962 that there was a density-dependent control on fecundity in Marley woods. Krebs' (1970) study using key-factor analysis was consistent with this; he found that shortfall in egg production was negatively correlated with population size and also that the factor contributing most to changes in total population was survival from fledging to recruitment as breeding adults; the latter is hardly surprising since this period covers three-quarters of the year.

More recently calculations using the data up to 1983 confirm the general result, although altering the details a little (McCleery and Perrins 1985). As before, the key factor explaining most of the changes in total numbers was the survival rate of newly fledged offspring. When this survival rate was split into factors for early winter and a factor for later winter, using a mark-release-recapture estimate for the mid-winter population, it appears that the key factor is early winter survival. The recruitment rate of those juveniles that are still alive in mid-winter is negatively correlated with their density, and the k-factor describing their survival is negatively correlated with the breeding population in the previous year. There are several problems with calculating the adult mid-winter population from recapture data, so we were unable to relate recruitment rate to total density. However, adult overwinter survival is quite well correlated with that of juveniles, so it seems likely that density of juveniles gives an index of total density, and that we can conclude there is density-dependent regulation of the number of potential breeders that actually obtain territories each spring.

Predation

Figure 6.1 shows the proportion of nests considered to have been predated, or which failed for other reasons at the egg (Fig. 6.1b, *c*) or chick (Fig. 6.1*d*, *e*) stage. For the purposes of this study 'predation', which was scored from evidence for death and destruction in the nest, was usually attributable to weasels. Failures due to desertion are probably due to the death of one or both parent since Great Tits normally have a high level of nesting success and do not commonly desert nests when they are still alive. As can be seen from Fig. 6.1, nest losses attributed to predation were highest between 1963 and 1976, and have fallen to a low level subsequently.

Losses not attributed to predation were also highest in the period 1963–76, but have subsequently remained at a noticeable level. At least some of this mortality is due to Sparrowhawk predation, though it may have other causes. To assess the validity of this argument it is necessary to consider the effects sparrowhawks are known to have on tit populations.

Effects of Sparrowhawk predation

Reduction in the number of breeding pairs recorded

Perrins and Geer (1980) showed that there was a lower than normal nesting density of tits in the immediate vicinity of Sparrowhawk nests. This might be due to tits avoiding settling in such places, but since the Sparrowhawks build their nests after the tits have taken up territories, it seems rather more likely that the individuals whose territories end up close to a hawk nest have a high probability of being predated before they themselves nest, so leading to the observed 'spaces'. However this would be hard to observe in the total population measured since the 6–8 pairs of hawks breeding in the wood would only remove about 25 nests.

Losses of nests due to Sparrowhawks

Almost certainly, many of the tit nests which fail for no apparent reason (i.e. excluding 'predation', which we have taken to be weasel predation) do

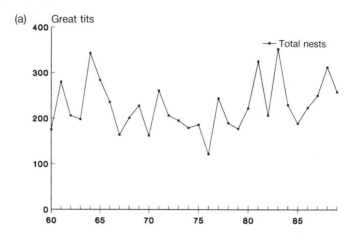

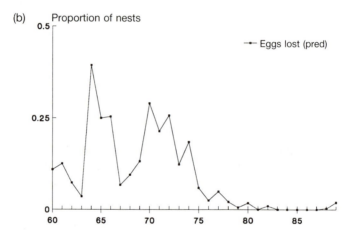

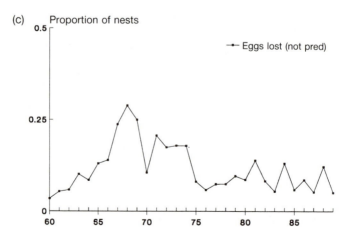

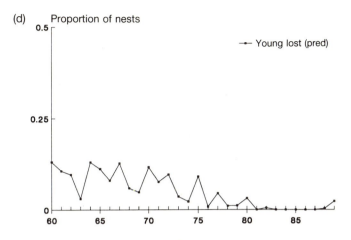

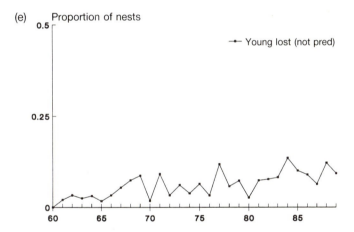

Fig. 6.1. Time series plots of (a) total numbers of Great Tit pairs in Wytham; (b) proportion of nests failing before hatching scored as due to predation; (c) proportion of nests failing before hatching not attributed to predation; (d) nests where chicks hatched but none fledged due to predation; and (e) nests where no chicks fledged, not attributed to predation.

so as a result of the Sparrowhawks taking one of the parent tits. Geer (1978) showed that the incidence of 'total nest failures' was higher when Sparrowhawks were present in the wood than when they were absent. There is a slight increase in the incidence of chick death not attributed to predation by weasels from 1970 onwards (Fig. 6.1e).

Losses of fledglings due to Sparrowhawks

Table 6.1 shows the number of fledgling Great Tits estimated to have been taken in the 3 years of Geer's study and the 2 years of Gray's work; Blue Tit numbers are also shown for comparison.

The figures given by Gray are calculated by averaging Geer's percentage mortality figures, since Gray did not have sufficiently precise data on the numbers of fledglings brought to individual Sparrowhawk nests. Note that in 1977 approximately the same number of Great Tits were killed as in 1976, even though there were more fledglings available in 1977. In 1978 however, nearly twice as many Great Tits were killed even though the population of fledglings was only slightly more than in 1977. It is difficult to interpret this result precisely given the available information. The Sparrowhawks may have had an alternative food which they used in 1976 and 1978 but not in 1977, but if so it was not Blue Tits (Table 6.1). Alternatively, fewer of the estimated 8 pairs of Sparrowhawks may have successfully raised chicks in 1976 and 1978 than in 1977. Geer and Perrins (1980) thought that tit chicks might be most vulnerable in the first 5 days after fledging, so possibly the predators were 'swamped' in 1977.

Table 6.1 Estimated mortality of juvenile tits during the period May 1st to Sept. 1st (condensed from Geer 1978 and Gray 1987)

Year	Great Tits			Blue Tits		
	Juveniles	Estimated no. killed	Per cent taken	Juveniles	Estimated no. killed	Per cent taken
1976	746	254	34	2030	548	27
1977	1125	203	18	2544	458	18
1978	1280	422	33	2248	607	27
1983	2376	672	28.3[a]	2468	591[b]	24
1984	1204	341	28.3[a]	2493	591[b]	24

[a] This figure is the average of the figures for 1976–8 (see text).
[b] These figures are derived in the same way as those for Great Tits.

Losses to Sparrowhawks at other times of year

Gray (1987) used radio-tracking and other observation of individual Sparrowhawks to estimate their daily food requirements, and the proportion of their diet that was comprised of Great Tits. Table 6.2 (modified from Gray 1987) shows the total Great Tits taken in 1983/4 and 1984/5 were 1382 and 1090, respectively, representing 44 per cent and 66 per cent of the total birds fledged and their parents in those years. It is clear

Table 6.2 Estimated annual deaths in Great Tits due to Sparrowhawk predation in 1983/4 and 1984/5 (from Gray 1987)

Year	Total available adults + juveniles	Losses during			Total losses
		May–Aug.	Sept.–Oct.	Nov.–Apr.	
1983/4	3084	774	219	389	1382
1984/5	1646	405	219	466	1090

that such losses are substantial, but with only two seasons' data and with the Sparrowhawk numbers remaining fairly constant, we cannot relate variations in Sparrowhawk predation rate outside the breeding season to actual Great Tit numbers.

The effect of predation on the population

In this section, we try to assess the effects of these levels of predation on the Wytham Great Tit population. The loss of territorial pairs in spring is very small as is the loss due to failure by hawk predation of the breeding parents; these are not considered further. As we have indicated, insufficient information exists to consider the effects of losses outside the breeding season in any detail.

For the purposes of analysis we have treated losses due to 'predation', that is, cases where nest destruction was thought to be due to weasels, with 'total failures', where a clutch or brood was abandoned for no apparent reason, as being due to Sparrowhawks. In fact most of the variation between years in total failures is due to losses of eggs rather than young (Fig. 6.1).

The losses of broods due to weasel predation was high in some years, reaching a peak of 52 per cent in 1964. It seems as if, at least in those years with such high predation, the losses might lead to a lowering of the breeding population the following year. Since the purpose of this study is to examine the effect of predation on population processes, it is of interest to relate predation to the key-factor analysis for this population (McCleery and Perrins 1985). Fledging failure, k_3, is significantly related to number of nests scored as having been predated (Fig. 6.2). Number scored as predated is a minimum estimate of predation loss since some nests which failed due to predation may not have been recognized as such; a similar relationship holds between k_3 and number of 'total losses', i.e. nests where no young were fledged, which would represent a maximum estimate of predation. However, k_3 does not seem to be significantly correlated with overall population change (K), so there is no evidence on a year by year basis for a determining effect of chick loss on the overall population.

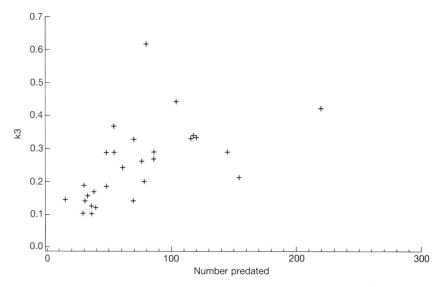

Fig. 6.2. Relationship betwen k3 and number of nests scored as predated. The significant positive relationship indicates that predation, mainly attributable to weasels, accounts for a significant proportion of fledging failure.

However, this method of analysis does not take into account factors external to the population that may influence its numbers.

Relative importance of predation

In order to assess how strongly predation influences the changes in numbers of Great Tits in relation to other factors it is desirable to use a statistical model to assess the joint significance of factors known individually to affect the population. One problem with this approach is that eventually one ends up doing simple arithmetic. Since we claim to be able to count most of the offspring individually, we do not have to worry about the fact that sample errors in successive periods of the life cycle may not be independent, but at the limit of accuracy the change in population is exactly the sum of adult survivors and new juveniles subtracted from the previous population, with no residual errors. The problem is precisely analogous to that involved in the estimation of variance in life-time reproductive success (Brown 1989). There are no generally agreed methods for multivariate statistical modelling of factors external to the population in such a situation. Moller (1989) used bivariate analysis of k-factors of interest with candidate environmental variables. We have already shown a relationship between k3 and our measure of total predation losses, but concluded that since k3 is not correlated with K such losses do not account for fluctuations in the population. However, it would obviously be neater to use a multivariate approach in which some factors could be added and removed to see whether they account for population changes when other factors are taken into account.

Table 6.3 shows the result of such an approach, using the numerical change in population as the dependent variable and fitting a linear regression-like model. Inspection of Figure 6.3a suggests that large increases in population often follow years in which large clutches are laid. This is confirmed by the fit of the parameter egg/nest in the model, and is consistent with the fact that k1 in the k-factor analysis is significantly correlated with K. Beech mast winters tend to be followed by increases in population due to improved overwinter survival of both adults and

Table 6.3 Regression model for prediction of population change in Wytham Great Tits. The dependent variable is the numerical difference between current population and next year's population

Parameter	Prediction	Standard error	F	d.f.	P
Mean cl.	17.28	7.93	4.75	1,14	<0.05
Mast	13.98	21.21	0.43	1,14	n.s.
Adult survivors	171.4	58.66	8.54	1,14	<0.05
Juvenile survivors	599.1	211.5	8.03	1,14	<0.05
No. fail	−0.36	0.13	7.58	1,14	<0.05
Population	−0.02	0.16	0.01	1,14	n.s.

F calculated by dropping variables/factors in turn from the model.

juveniles; this is reflected in the fact that a factor for beech is significant on its own, but not significant when variables for fledgling survival and adult survival from the end of the breeding season to the next is added; in other words these two variables contain the same information as beech mast with regard to population fluctuations, doubtless because both adult and juvenile survival are strongly dependent on beech crop. Having taken these factors into account, it then appears that the occurrence of total failures in nests is a significant negative correlate of the changes in population, so the population tends to drop after years when total failures were high. None of these variables have very highly significant coefficients, so the effects they represent are not very strong but, as can be seen in Fig. 6.4, the overall fit of the model containing as parameters mean clutch size, juvenile and adult survival, and nest failures (the upper estimate of predation) is quite good. After fitting all these factors, population size itself is not a significant predictor of population change. This suggests that the apparent relationship between population size and population change is explicable in terms of fecundity, nest predation, and overwinter survival rate. It is possible that

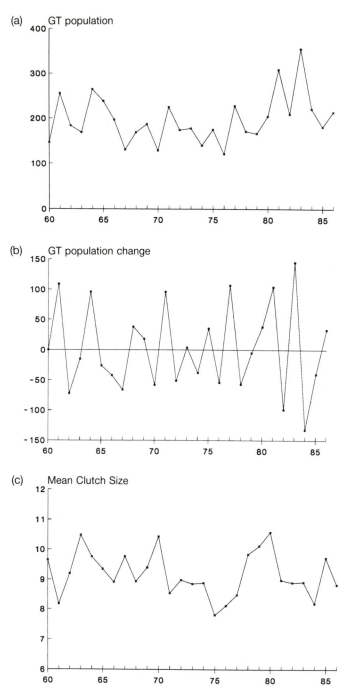

Fig. 6.3. Time series of (a) Great Tit population; (b) difference between current population and that of the previous year; and (c) mean clutch year.

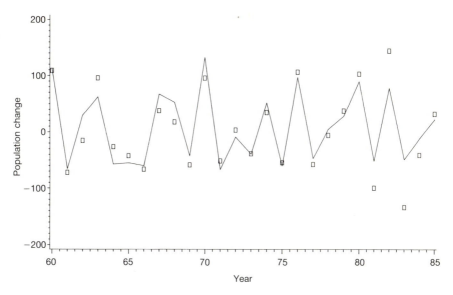

Fig. 6.4. Fit of the regression model described in Table 6.3. The squares show the difference between this year's and next year's population, the line joins the fitted values for the model containing the following significant parameters: mean clutch size, proportion of adult breeders breeding again, proportion of fledged young recruited to breeding population, and number of nests raising no young.

the dependence on clutch size represents the mechanism of density dependence, with low clutches occurring in years of high population, as proposed by Lack (1966). On the other hand, the factor for mean clutch size fits much better than that for population, which is indeed not significant when the parameters of adult and juvenile survival are in the model. This is consistent with the fact that fecundity appears to be negatively dependent on population size at some times and not at others (Perrins and McCleery 1989), while the density dependence of juvenile survival (i.e. k4 in the key-factor analysis) is consistent throughout the study.

Long and short-term fluctuations

One of the assumptions made in both the above analyses is that the processes determining the population act independently in successive years. However it seems quite likely (and inspection of Fig. 6.1a suggests) that at least some of the processes involved act over more than one year, so that the population level in any year depends on more than just factors related to the previous year. Because tit populations tend to show quite marked variations between years, such long-term changes are more difficult to unravel. In this section we look at the longer term changes that seem to have occurred in the population.

Inspection of Fig. 6.1a suggests that there are short-term fluctuations in Great Tit numbers of the order of a year or two combined with long-term trends. The population seems to have decreased a bit during the period 1964 to 1976 and then increased up to about 1982. One way of extracting such trends from data without, as autocorrelational methods do, assuming a strictly periodic structure, is to use running means or medians to smooth the data. The basic idea is that underlying long-term trends are character- ized by the extent to which neighbouring points are similar to one another. A simple smoother would be a running median of three points, where each point is replaced by the median of itself, the previous and the next one. Such simple methods are prone to instability however, and a considerable body of experience has been built using 'compound' or 'resistant' smoothers (Velleman and Hoaglin 1981).

Figure 6.5a shows a resistant smoother applied to the Great Tit numbers. In a sense this procedure simply confirms the results of eyeballing, though it may reveal patterns that were not obvious. The smoothed curve seems to confirm that Great Tit numbers dropped some- what between 1965–8, but also reveals a slight further dip in 1974–6. We can now ask whether there is any sign of a relationship between the longer term (smoothed) population level and the presence or absence of predators. The drop in 1965–8 more or less coincides with the peak of weasel predation recorded in 1964–6 (Fig. 6.1b). The second peak is more of a problem. It follows the second peak of weasel predation shown in Fig. 6.1b (1971–2), but two factors lead one to doubt a connection. Firstly the drop in weasel predation in 1967–9 is mirrored by a rise in nest failures *not* attributed to predation. As Sparrowhawks were absent during this period it seems possible that nests which were, in fact, predated by weasels were not recorded as such during these years. On the other hand, the further drop in population in 1974–6 appears to correspond well with the period during which both Sparrowhawks *and* weasels were causing substantial losses (bearing in mind that some of the decrease is due to adult breeders being removed after nesting but before their nests had been recorded). After about 1976 numbers seem to increase again, at least up to 1982, corresponding with the drop in weasel predation to a low level. If we assume that all age classes of breeder are equally likely to be caught, then we can see (Fig. 6.5b) that the proportion of known birds that are at Wytham born 1st years increases, whereas the proportion that are immigrant 1st years decreases slightly (Fig. 6.5c). The proportion that are new adults (not shown), and the survival rate of adults from the previous year (Fig. 6.5d), remain fairly constant throughout this period though there are short-term fluctuations and at times adult survival and recruitment of juveniles seem to compensate. Finally, it is noticeable that the Blue Tit population has dropped slightly in concert with the rise in Great Tits (Fig. 6.5a, e).

Although some caution is needed in interpreting the results of such smoothing exercises, especially as there are no agreed methods for testing the statistical significance of the effects revealed, they do show features of the situation that are not apparent from other methods. From the evidence presented above it looks as though the Great Tit population of Wytham did decline somewhat in the face of predation by weasels and when this was prevented by means of weasel-proof boxes the population increased again. The most obvious change in the structure of the population is the increase in the proportion that are Wytham-born breeders. It is possible that the combined impact of Sparrowhawks and weasels was also noticeable in the total population, though the impact of Sparrowhawks alone does not seem to be very large, in spite of evidence that the actual numbers of birds they take can be large. One problem in interpreting this is the fact that when they were present the population of Sparrowhawks was rather stable, at about 8 pairs each year. It also appears from this study that between-year fluctuations are strongly affected by the number of eggs laid. This may depend on population density, and also on the timing of the season which we have not examined here, but the relationship is complex and requires further investigation. Furthermore, although survival of juveniles through the winter is perhaps the most important factor in numerical fluctuations, it seems more likely to depend on food supply than on predation.

General conclusion

It appears that the effects of predation on small bird populations are often smaller than one might expect. Even the combined effect of weasels and Sparrowhawks at most reduced the running median of the population to about 71 per cent of its value in the peak year of 1964; weasels alone may have reduced it by no more than 25 per cent. This may seem surprising in view of the fact that at times both prey species have removed as much as one-third of the young. A simple way of looking at this is to imagine that each pair of tits lays ten eggs and hatches nine young. If weasel predation removed one-third of these, an average of six young per pair would fledge; if Sparrowhawks then removed a further third, there would be four fledglings per pair alive by September. On average, one adult per pair dies each year, requiring one juvenile per pair to survive if the population is to remain stable. Hence, even after such heavy predation, a further three-quarters of the young that reach the autumn will die before the next breeding season.

Great Tit Population

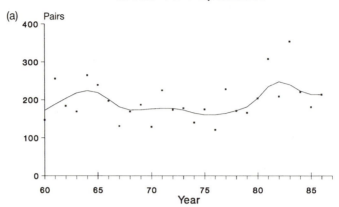

Wytham born

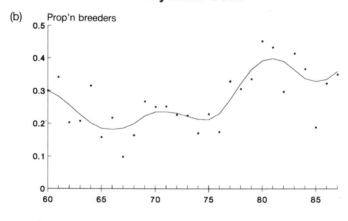

Immigrants

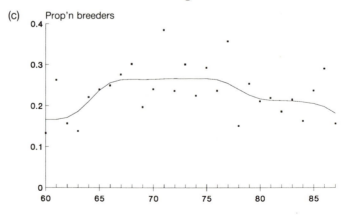

Adult survivors

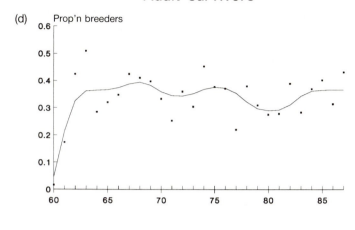

Blue Tit Population

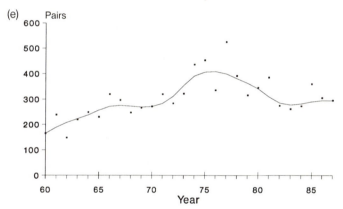

Fig. 6.5. Time series plots of (a) Great Tit population; (b–d) proportion of breeders that were (b), 1st years raised in Wytham, (c) 1st year breeders that were not raised in Wytham, (d) adults surviving from last year; (e) population of Blue Tits.

It then becomes important to know whether the numbers of birds taken in winter are sufficient to lower the population. We have insufficient quantitative data on this point, though *prima facie* the relative stability of numbers during the study suggests that winter mortality due to predation does not consistently lower breeding numbers.

We suggest that the data presented here gives some idea of how very resilient a population of small birds is to the pressure of predation, but it would be foolish to conclude from this that predation is never important. Predation on birds with lower reproductive rates would obviously be more serious, and an introduced predator might push the system beyond its maximum sustainable yield, at which point a sudden collapse would ensue. This non-linear behaviour means that the gravity of the effect of predation on populations can only be determined with the help of detailed knowledge.

Acknowledgement

We thank Jacquie Clark of the B.T.O. for data on the movements of the general Great Tit population of the U.K.

References

Brown, D. (1989). Components of lifetime reproductive success. In: *Reproductive success* (ed. T. H. Clutton-Brock), pp. 439–53. Chicago University Press.

Clobert, J., Perrins, C. M., McCleery, R. H., and Gosler, A. G. (1988). Survival rate in the great tit *Parus major* in relation to age, sex and immigration status. *Journal of Animal Ecology*, 57, 287–306.

Cormack, R. M. (1964). Estimates of survival rate from sightings of marked animals. *Biometrika*, 51, 429–38.

Dunn, E. K. (1977). Predation by weasels (*Mustela nivalis*) on breeding tits (*Parus* spp.) in relation to density of tits and rodents. *Journal of Animal Ecology*, 46, 633–52.

East, M. L. and Perrins, C. M. (1988). The effect of nestboxes on breeding population of birds in broad leaved temperate woodlands. *Ibis*, 130, 393–401.

Geer, T. A. (1978). Effects of nesting Sparrowhawks on nesting tits. *Condor*, 80, 419–22.

Gosler, A. G. (1987) Some aspects of bill morphology in relation to ecology in the Great Tit *Parus major*. Unpublished D.Phil. thesis, University of Oxford.

Gray, I. (1987). The feeding ecology of the Sparrowhawk *Accipiter nisus* outside the breeding season. Unpublished D.Phil. Thesis, University of Oxford.

Kluijver, H. N. (1951). The population ecology of the Great tit, *Parus major*. *Ardea*, 39, 1–135.

Krebs, J. R. (1970). Regulation of numbers in the Great tit. *Journal of the Zoological Society of London*, 162, 317–33.

Lack, D. (1966). *Population studies of birds*. Clarendon Press, Oxford.

McCleery, R. H. and Perrins, C. M. (1985). Territory size, reproductive success and population dynamics in the Great tit, *Parus major*. In *Behavioural ecology: Ecological consequences of adaptive behaviour* (eds. R. M. Sibly and R. H. Smith), pp. 353–74. Blackwell, Oxford.

McCleery, R. H. and Perrins, C. M. (1989). Lifetime reproductive success of the Great tit, *Parus major*. In *Reproductive success* (ed. T. H. Clutton-Brock), pp. 136–53. Chicago University Press.

Moller, A. P. (1989). Population dynamics of a declining Swallow *Hirundo rustica* population. *Journal of Animal Ecology*, 58, 1051–64.

Newton, I. (1979). *Population ecology of raptors*. Poyser, Berkhamsted.

Perrins, C. M. (1966). The effect of beech crops on Great tit populations and movements. *British Birds*, 59, 419–32.

Perrins, C. M. and Geer T. A. (1980). The effects of Sparrowhawks on nesting tit populations. *Ardea* **68**, 133–42.

Perrins, C. M. and McCleery R. H. (1989). Laying dates and clutch size in the Great tit. *Wilson Bulletin*, **101**, 236–53.

van Balen, J. H.(1980). Population fluctuations of the Great tit and feeding conditions in winter. *Ardea*, **68**, 143–64.

Velleman, P. F. and Hoaglin, D. C. (1981). *Applications, basics and computing of exploratory data analysis*. Duxbury, Boston, Mass.

7 Social behaviour and population regulation in insular bird populations: implications for conservation

JAMES N. M. SMITH, PETER ARCESE, and WESLEY M. HOCHACHKA

Introduction

To conserve any population in a natural setting, it is useful to understand the factors that determine average abundance and changes in numbers. Compared with other vertebrates, birds are easy to mark and count; thus, fluctuations in the numbers of birds are well described (Lack 1966; Pimm and Redfearn 1988) and there is in the literature considerable information about the processes that regulate bird populations (Lack 1966; Newton 1980; Patterson 1980; Moss and Watson 1985; McCleery and Perrins 1985 and this volume; Hannon 1988).

In birds, two forms of density-dependent regulation have been discussed extensively: (1) regulation via intraspecific competition for food (Lack 1966; Martin 1987); and (2) regulation below food limits by social intolerance (Patterson 1980; Moss and Watson 1985; Hannon 1988). The second topic is the principal subject of this paper. We first briefly review the evidence for territorial regulation of bird populations, and then present a case history of regulation in an island population of Song Sparrows. We also show how the mechanisms by which populations are regulated can affect the options for active management of a population.

Social behaviour and population regulation

Territorial behaviour is almost ubiquitous in birds, and confers several benefits to individuals (Davies 1978; Moss and Watson 1985; Gauthier 1987a). Howard (1920) first proposed that bird numbers are regulated by territorial behaviour; early empirical studies were done by Stewart and Aldrich (1951), Hensley and Cope (1951), Kluyver and Tinbergen (1953), and Tompa (1962, 1963, 1964, 1971). The idea of population regulation by social factors has been reviewed by Wynne-Edwards (1962), Lack (1966), Chitty (1967), Brown (1969), Watson and Moss (1970), Klomp (1972), Davies (1978), Krebs and Perrins (1978), Murray (1979), and

Patterson (1980). Two related forms of social behaviour, territoriality and dominance, can affect survival and reproduction. We first discuss territoriality.

Territoriality affects the number of individuals when the most suitable habitats become saturated. Excluded individuals must then live and breed in habitats where food is poor or predation rates are high (Kluyver and Tinbergen 1953; Fretwell and Lucas 1970; Krebs 1971; Watson and Jenkins 1968; Krebs *et al*. this volume), or live as non-territorial floaters on the territories of breeders (Smith 1978). Territoriality thus often indirectly influences reproduction and survival, although territorial fights can cause the death of participants (Arcese 1987). Heritable aggressive behaviour (Chitty 1967; Krebs 1985) may alter the composition of populations as density changes and cyclic population dynamics may be promoted (Moss *et al*. 1984; Hannon 1988).

Watson and Moss (1970) presented a useful list of conditions that should be satisfied if a population is being limited by territoriality. A population is *limited* by a particular factor when removal or amelioration of that factor causes the population to increase. We also use the term *regulation* as defined by Begon and Mortimer (1986). Effects of density-dependent regulation combine with other influences on populations to determine bird numbers (Begon and Mortimer 1986, p. 176). Watson and Moss' list of conditions was extended by Smith and Arcese (1986) and is modified further in Table 7.1. Conditions 1 and 2 state that populations should contain non-breeders that can breed if territory owners die or are removed. Non-breeding individuals are commonly revealed by removal experiments (Patterson 1980; Hannon 1983; Wesolowski *et al*. 1987; Martin 1989; Newton, this volume), and can occasionally be observed directly (Smith 1978; Birkhead *et al*. 1986; Arcese 1987).

Establishment of other individuals after removal of territorial birds is

Table 7.1 Necessary conditions for: (a) the territorial limitation of the numbers of breeders, and (b) social regulation of total population size

Type	Condition	Source[a]
1(a)	Individuals cannot breed because of the territorial intolerance of others	A
2(a)	Excluded individuals breed if they can gain a territory, and are of the limiting sex	A, C
3(a)	The population is not limited by a resource such as food or nest sites	A
4(a)	The size of the pool of non-territorial individuals during the breeding season depends on the density of territory owners	B
5(a, b)	The proportion of young individuals that take up breeding territories each year declines with increasing density of territory owners	B

[a] A = Watson and Moss (1970); B = Smith and Arcese (1986); C = Hannon (1983).

not, however, sufficient proof of the limitation of breeding numbers by territorial behaviour. Male replacements are usually more common than female replacements (Hannon 1988), probably because adult sex ratios in birds are often biased towards males (Breitwisch 1989). Since females are often limiting, male territoriality may not affect the size of the breeding population, even though some males are excluded from territory owner-ship (see below).

If other regulating factors operate instead of social behaviour (condition 3, Table 7.1), removal or reduction of these factors should increase density, survival, or reproductive success (Watson and Moss 1970). A shortage of nest sites regulates some populations of birds (Lack 1966; Patterson 1980; Haramis and Thompson 1985), while others may be regulated by predation (Marcström *et al*. 1988; Martin 1988), or parasites and/or diseases (van Riper *et al*. 1986; Hudson and Dobson, this volume). The best studied alternative regulatory factor is food supply (Martin 1987; Gauthier 1987*b*). We shall examine the interaction of food and social regulation in the next section. If territoriality acts in a density-dependent manner, both the size and proportion of the non-territorial component of the population should be a function of the density of territory holders (conditions 4, 5 in Table 7.1).

Interactions of social behaviour and food shortage

Early studies of population regulation (see Lack 1966) emphasized the regulatory potential of single factors such as food, predators, and parasites. It is now clear (Watson and Moss 1970; Hilborn and Stearns 1982; Sinclair 1990) that several factors often interact to affect animal numbers. Food and territorial behaviour are almost inseparable because the very existence of territoriality depends on the distribution of the food supply (Zahavi 1971; Rubenstein 1981). Territory size is often negatively correlated with food abundance (Enoksson and Nilsson 1983; Gauthier 1987*b*; but see Moss and Watson 1985; Gauthier 1987*a*). Experimentally increasing the food abundance often reduces territory size (Enoksson and Nilsson 1983; Enoksson 1988), perhaps due to increased intrusion pressure on food-rich territories (Norton *et al*. 1982). Supplemental feeding can also affect the two sexes differentially. In Dunnocks *Prunella modularis*, female territory sizes are reduced more than those of males when food is supple-mented (Davies and Lundberg 1984). In Great Tits *parus major*, pairs with small territories reproduced poorly (McCleery and Perrins 1985), presu-mably because small territories contain insufficient food. Thus, territoriality and food can interact intimately to affect population size.

Interdependence may also apply to other combinations of regulating factors. Food supplies of herbivorous birds are affected by successional changes in vegetation (Zwickel and Bendell 1985; Boag and Schroeder 1987), and food shortage may alter the behaviour of prey so that they are more likely to be eaten by predators (McNamara and Houston 1987).

Non-territorial components of territorial populations

One consequence of regulation by territoriality is the temporary or permanent presence of non-territorial individuals in populations. These include young that have yet to contest for breeding territories, and older 'floaters' excluded from territory ownership (e.g. Smith 1978). To date, little quantitative information of the numbers of floaters has been available. Such knowledge might be useful for two reasons: the presence and numbers of non-territorial birds could give clues to the demographic health of a population and such individuals could be translocated in active conservation programs.

Social dominance and population regulation

Social dominance is universal in the many species and populations that live in stable or permanent groups. In most birds, social groups form during the non-breeding season, and social dominance is most likely to affect survival at that time. In temporary social groups, older birds generally dominate yearling birds (Kikkawa 1980; Smith *et al*. 1980; Catterall *et al*. 1989; Piper and Wiley 1989), which commonly survive less well than adults during the non-breeding season (Kikkawa 1980; Kikkawa and Wilson 1983; Arcese and Smith 1985; but see Greig-Smith 1985). There is an unanswered question relating dominance to population regulation: would more birds of the limiting sex survive the non-breeding season in the absence of social dominance? If the answer is yes, dominance could regulate survival in the non-breeding season and hence affect breeding density indirectly. Alternatively, dominance might result in the early death of the most subordinate group members, and thus ameliorate exploitative intra-specific competition for food. It seems more likely, however, that dominance simply determines which individuals survive, and not how many. This argument might also apply to territoriality regulating the number of breeders (Lack 1966).

In group-breeding species, dominance within the group lasts throughout the year (Stacey and Koenig 1990) and thus directly affects the number of breeders. Some aggressive interactions among individuals can be interpreted as struggles to attain dominant breeding status (Hannon *et al*. 1985). Since most species that breed in groups also defend group territories, dominance and territoriality can interact to regulate reproduction at a level that is unaffected by food limits. A problem in testing any hypothesis about the regulating effects of dominance is that it is very difficult to remove experimentally the effects of dominance.

Thus, although territoriality and dominance might regulate populations separately or in concert, these social factors usually interact closely with other density-dependent effects. Boag and Schroeder (1987) and Hannon (1988) reached similar conclusions for grouse populations. We now illustrate these ideas from our own work.

The Song Sparrow population of Mandarte Island

Active conservation of a population has two basic goals:

(a) maintenance of a substantial minimum population size; and
(b) in the case of endangered populations, the production and 'harvest' of excess individuals for captive breeding or introduction to other areas.

There are several ways of achieving these goals. We consider here the consequences of the following actions using the Song Sparrow population as a model: harvest of members of the population, control of predation and brood parasitism, supplying nest sites, and providing supplemental food.

The resident population of four to 72 breeding females and 9 to 100 adult males inhabits a small (6 ha) rocky island in the Haro Strait, about 25 km north of Victoria, British Columbia. The study site and neighbouring islands are described by Tompa (1964) and Drent *et al.* (1964). Tompa studied the population from 1960 to 1963 (Tompa 1962, 1963, 1964, 1971). We began our study late in 1974. Most adult Song Sparrows on Mandarte are territorial throughout the year, although most territorial birds spent time off the territory in a communal feeding area in winter (Table 3 in Arcese 1989*a*). Males sing in every month and changes in territory ownership occur throughout the year, with a strong peak in early spring (Arcese 1989*a*). Most changes of territory ownership among males involve aggressive contests (Arcese 1989*a*) and we suspect this also to be true for females.

The Song Sparrow is a typical passerine, in that most birds breed in monogamous pairs and surviving mates generally remain together from year to year. Most breeding pairs raise two broods of 1–4 young per year. There is some polygyny (Smith *et al.* 1982; Arcese 1989*b*) and adult sex ratios are usually male-biased (Fig. 7.1). Because only one-third of the island is used for breeding, densities reach 35 breeding pairs per ha, much higher than is reported for the species elsewhere (Tompa 1964). Because of low immigration rates, changes in population size are almost entirely driven from within; about one immigrant, usually a female, joins the population per year. Therefore the high density and isolation of the Song Sparrow population on Mandarte Island provides a model situation for endangered species that are locally abundant, but live in insular or fragmented habitat.

We colour-ring all adult individuals, map territories each spring, and monitor survival throughout the year. In analyses concerning the non-territorial part of the population, we define territorial status (owner, floater) on April 30 each year. We have studied the breeding success of all individuals in detail. Tompa's methods were similar, but he studied reproduction less intensively. Details of methods are given by Tompa (1964), Smith (1981), Arcese and Smith (1985, 1988), Arcese (1987, 1989*a*, *b*, *c*), and Smith and Arcese (1989).

The numbers of breeding males and females in late April, after general breeding has commenced, are shown in Fig. 7.1. Tompa (1962) observed stable numbers of breeders in the first 3 years of study and suggested that territoriality regulated numbers in spring. He later modified this view (1964), because of the increase in numbers that followed the deaths of some adult birds during a snowstorm in March 1962. Tompa (1971) reiterates the earlier view of the regulating effect of territoriality; he suggested that territoriality normally constrains settlement of yearlings, but that this constraint is relaxed when a number of territorial vacancies occur at the same time (Van den Assem 1967). This hypothesis was later supported in an experimental test by Knapton and Krebs (1974) on a population of Song Sparrows on the nearby mainland. Thus, the regulating effect of territorial behaviour depends on the amount and distribution of unoccupied space at the time of settlement.

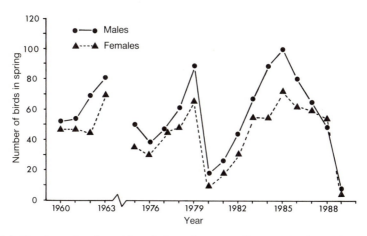

Fig. 7.1. Numbers of male and female Song Sparrows alive on Mandarte Island in late April. The values for males include territorial birds and non-territorial floaters. All females were breeders. Data from 1960–3 from Tompa (1964).

Changes in breeding population size

Densities were high to extremely high for much of the time during the course of our studies (Fig. 7.1). Peaks in numbers were reached in 1963, 1979, and 1985. The study was interrupted at the peak in 1963 and numbers were moderate when work recommenced in 1974. The peak in numbers in 1979 was followed by a severe population crash of unknown cause in the winter of 1979–80. The peak in 1985 was followed by a gradual decline until an even more severe crash took place in February 1989. This was caused by extreme winter cold, combined with high winds and was a local phenomenon, since marked populations of Song Sparrows on nearby islands did not decline precipitously in 1989 (unpublished data).

The role of social behaviour in population changes

Territoriality

For the removal of individuals for captive breeding or translocation to be most effective, there should be an excess of potentially reproducing individuals in the population (conditions 1, 2, and 4; Table 7.1). Condition 1 was clearly supported for male Song Sparrows, but not for females; there were non-breeding floater males in most years (Fig. 7.3a). There were non-territorial young females in early spring each year, but most of these either gained territories and bred with unpaired, late-settling, or already-paired males, or emigrated or died in late spring.

Condition 2 was supported only for males. Floater males took over territories and bred when territorial males were removed during the breeding season (Smith *et al*. 1982). In some years, aggressive male floaters frequently usurped territories (Arcese 1987, 1989*a*). Knapton and Krebs (1974) removed territorial Song Sparrows in autumn and early spring and found rapid replacements by young birds of both sexes, which were capable of breeding.

Although territorial behaviour prevented many males from breeding locally, it had less effect on the local number of females. Territorial fights between females were seen in early spring and female singing was associated with such contests (Arcese *et al*. 1988). Once females began to incubate eggs in April, however, time constraints probably prevented them from excluding other females from the territory (Arcese 1989*b*). In support of this idea, females occupying territories supplied with supplemental food in early spring 1985 were better able to resist settlement by non-territorial females than were control females (Table 4 in Arcese 1989*b*). Female territoriality may, however, have depressed the reproductive success of late-settling females. Thus, although excess males existed, and female territoriality affected settlement of females on territories, there was no non-breeding surplus of females with which excess males could pair. Thus, condition 2, like condition 1, was supported for males only.

Social dominance

We measured social dominance of juvenile and adult birds by observing agonistic interactions at feeders in winter in 1978–9 and in late summer in both 1983 and 1984. Birds that won over half of five or more decisive encounters were classed as dominant; birds that lost over half of their encounters were defined as subordinates (Arcese and Smith 1985).

Adult males were the most dominant class of birds at feeders in winter, and juvenile females the most subordinate (Table 1 in Smith *et al*. 1980). Juveniles survived less well than adults in winter (Smith *et al*. 1980; Arcese and Smith 1985). Juvenile survival during the non-breeding season averaged 0.37 (SE = 0.03, $n = 13$ years), compared to an adult survival of 0.65 (SE = 0.06, $n = 11$ years). Dominant juveniles of both sexes were approximately twice as likely as subordinates to settle on breeding territories as yearlings (Table 7.2). Juveniles survived much less well in the non-breeding season as population density increased (Fig. 7.2), but it is not known whether dominance influenced this result. Dominance thus played a critical role in determining whether juveniles gained breeding territories. We have no way to judge the effect of dominance on population regulation.

Nonbreeding floaters

Few other studies of pair-breeding passerine birds have been able to census the non-breeding segment of the adult population (but see Patterson 1980; Birkhead *et al*. 1986). We found, as expected from the hypothesis that

Table 7.2 Survival of juvenile Song Sparrows hatched in 1982 and 1983 during their first winter, in relation to sex and dominance status as scored at feeders in late summer. Data replotted from Arcese and Smith (1985)

	Status on April 30 in the year after hatching			
	Settled on territory	Floater	Absent	% Settling
Females				
Dominant	27	0	9	75
Subordinate	16	0	22	42
Males				
Dominant	34	11	5	68
Subordinate	17	15	19	33

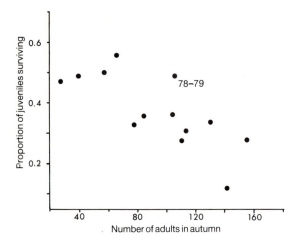

Fig. 7.2. Proportions of independent young Song Sparrows that survived on Mandarte Island until 1 year of age, in relation to population density of adults in the autumn. The data point for the 1978–9 winter, during which supplemental food was provided, is identified by a subscript.

territoriality regulates numbers of males, that both the number and proportion of non-breeding floater males present increased with population density (Smith and Arcese 1989). Floaters were less common at low population densities (Fig. 7.3a) and comprised a lower proportion of the male population at low population densities (Fig. 7.3b), thus meeting conditions 4 and 5 of Table 7.1 for males.

The proportion of the previous year's crop of independent young that gained a territory each year was also related negatively to the density of territory owners (condition 5, Fig. 7.2). This is strong evidence for regulation of juvenile survival in the non-breeding season. We believe that contests for territories in both males and females, and perhaps also food limitation, are the underlying mechanisms.

A further mechanism that might affect reproduction is that social interference by floaters could depress reproductive success of territory owners at high densities (Tompa 1964). Intrusions by floater males were common at high densities (Arcese 1987) and in the year with a very male-biased sex ratio (Tompa 1964). Intrusions led to many turnovers in territory ownership among males (Arcese 1987, 1989a). Intrusions by non-breeding females, however, were rare or went undetected by us, and disturbance by unmated males did not markedly depress female reproductive success. Thus, although there were excess males in the population at high densities, these did not interfere greatly with the production of young.

If individuals are to be translocated as part of an active conservation programme, the timing of territory establishment and the workings of social dominance make juveniles the logical choice for capture. Removals should be done as soon as possible after the end of a breeding season to minimize losses due to death or emigration among the remaining juveniles.

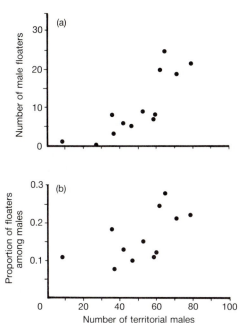

Fig. 7.3. (a) Numbers of non-territorial male Song Sparrows alive on Mandarte Island on April 30 each year, in relation to population density. (b) The proportion of the male population not defending a territory on April 30 each year. Correlation coefficients are $+0.82$ ($P < 0.001$) for (a) and $+0.65$ ($P < 0.02$) for (b). $N = 13$ for both correlations.

Other regulatory mechanisms

Predation

We now consider whether predators, brood parasites, or the availability of nest sites and food, might limit or regulate the reproductive success or population size of Song Sparrows (condition 3, Table 7.1). Predation on adults was uncommon, although visiting hawks, owls, and shrikes killed some birds in winter, gulls occasionally killed territorial adults, and deer mice *Peromyscus maniculatus* killed a few incubating females.

The main predators of eggs and nestlings were believed to be deer mice, and the proportion of nests that failed (mostly due to predation) was strongly density-dependent (Fig. 2d in Arcese and Smith 1988). North-western Crows *Corvus caurinus* preyed upon fledglings and some nestlings. Supplemental feeding in 1985 reduced predation on nest contents slightly (Arcese and Smith 1988), suggesting that predation rates are higher because food is in short supply.

Since predation may have regulated breeding success, control of predators is thus a possible tactic in an active conservation programme. However, active control measures are controversial, especially when several native predators are involved and the quantitative impact of each predatory species is unknown, as was the case on Mandarte Island. In addition, even if predator control does increase breeding success, reduced survival of juveniles in the following winter may compensate for increased production of young (see below). We therefore recommend caution in the use of predator control programmes to increase breeding success of endangered populations of insular birds, except when introduced predators are responsible for most predation, or when populations are critically endangered.

Brood parasitism

Brood parasitism can regulate reproduction in passerine birds (Walkinshaw 1983; May and Robinson 1985; Wiley 1985), but removal of Brown-headed Cowbirds *Molothrus ater* from Mandarte in 1977 had little effect on the Mandarte Song Sparrow population (Smith 1981). When food was supplemented in 1985, pairs of birds with extra food suffered a reduced frequency of brood parasitism, presumably because they were better able to exclude cowbirds from their territories (Arcese and Smith 1988, p. 127).

Cowbird removal did increase breeding production successfully in the endangered Kirtland's Warbler *Dendroica kirtlandii*, but did not increase adult populations (Walkinshaw 1983), presumably because numbers of Kirtland's Warblers are regulated in the non-breeding season.

Nest sites

Female Song Sparrows build an open nest in dense cover near the ground. There was little indication that such sites were in short supply on Mandarte, although shortages of well-concealed nest sites might affect predation rates on nests. Attempts to supply extra nest sites would therefore be of little use as tools to raise numbers of Song Sparrows on Mandarte.

Food supply

Supplying excess food to a population might improve survival or reproductive success, and provide surplus individuals for translocation to other sites. Survival of juveniles, but not adults, was density dependent during the non-breeding period. When we supplemented food in the winter of 1978–9, we found no evidence that provision of food improved adult survival (Smith *et al.* 1980); however, recruitment of juveniles, was disproportionately high in relation to adult density (Fig. 7.2). It is possible, although the suggestion is unsubstantiated, that food shortage may be involved in juvenile recruitment; food supplementation in winter may thus aid conservation in critical years.

Annual reproductive output of the population was density dependent (Fig. 2f in Arcese and Smith 1988). The probable mechanism involved was tested by two supplemental feeding experiments (Arcese and Smith 1988; Hochachka, unpublished results). Supplemental feeding from a single feeder per territory improved reproductive success four-fold at high density (Fig. 5 in Arcese and Smith 1988), although territory sizes were not affected. When the experiment was repeated at a lower density in 1988 with three feeders per territory, feeders were not monopolized by territory owners. Territories of birds adjacent to food-supplemented territories expanded slightly to encompass a feeder and reproductive success was the same as on neighbouring control territories (Hochachka, unpublished results).

These two experiments suggest that territoriality and food shortage interacted to regulate reproductive success in the population. Since predation on nest contents also regulated reproduction in Song Sparrows on Mandarte (see above), reproduction in Song Sparrows is likely to be regulated by at least three closely related factors.

Regulation, catastrophic mortality, and the determination of Song Sparrow numbers on Mandarte Island

We found strong evidence for regulation of the population in both the breeding and non-breeding seasons. Social behaviour played a role in regulation in the non-breeding season, through contests for future breeding territories, and in the breeding season, through the defendability of concentrated food sources. However, other factors, including food shortage and predation, were more important causes of regulation during the breeding season. The list of conditions that need to be satisfied for territorial behaviour to limit or regulate the numbers of breeders (Table 7.1) was satisfied only for males, while females were the limiting sex in the population in most years (Fig. 7.1).

Despite the presence of strong regulation of reproduction in summer, and recruitment of juveniles in winter, the population on Mandarte did not fluctuate closely around an equilibrium point. The reason for this was that two episodes of catastrophic mortality reduced populations to low levels (Fig. 7.1). Paradoxically, without these two density-independent events, we would not have found evidence for regulation because numbers would not have varied enough to demonstrate density dependence.

The first population crash occurred during the winter of 1979–80 when the population was not being studied intensively. The second crash occurred in February 1989. Good survival of adults and young was maintained until January 26, but about 90 per cent of the population disappeared during the next 14 days. The weather was unusually cold for the first 4 days in February 1989, with freezing temperatures (mean daily temperature of $c.-10°C$) being accompanied by strong ($c.45$ km/h) winds. We suspect that, under these weather conditions, most Song Sparrows were unable to maintain a positive energy balance.

It can be seen from Fig. 7.1 that the crashes of 1979–80 and 1989 had a dominant effect on the performance of the population. Thus, the numbers of Song Sparrows on Mandarte were determined jointly by regulatory mechanisms involving contests for territories, food shortage, and predation, and by density-independent catastrophic events, at least one of which was caused by bad weather in winter.

The dynamics of populations on small islands

Bird populations on islands often exist at higher densities than on adjacent mainland sites (MacArthur *et al*. 1972). Nevertheless, bird populations on small islands on continental shelves exhibit high local extinction rates (Williamson 1983; Pimm *et al*. 1988) and many extinctions of species have taken place on oceanic islands (King 1985; van Riper *et al*. 1986). Extinction rates are currently also high in disturbed habitats on continents (Whitcomb *et al*. 1981; Ambuel and Temple 1983; Temple 1985; Green and Hirons, this volume; Rands, this volume), and many endangered populations now occupy isolated patches of forest (Lande, this volume). Populations in such patches are likely to show demographic similarities to those occupying true islands, although models of the demography of isolated populations (e.g. Lande 1987; papers in Soulé 1987) are currently outstripping empirical data. We therefore suggest our data are relevant to the conservation of insular populations.

Although our study population maintained a high average density, and was strongly regulated during both breeding and non-breeding seasons, it approached extinction twice in 19 years of study. Since the population crash in 1989 was more severe on Mandarte than on neighbouring islands, one might have expected increased immigration in 1989. No influx of immigrants, however, had occurred by August 1989. The recovery from the population crash in 1989 may thus depend entirely on local recruitment and the population may become extinct if a further catastrophic event occurs while numbers are low.

Very similar results have been found in an insular population of Silvereyes *Zosterops lateralis* on Heron Island, Great Barrier Reef, Australia. These small passerines are also influenced strongly by climatic castrophe (Kikkawa 1977; Kikkawa and Catterall, unpublished data), by food distribution and abundance (Catterall *et al*. 1982; Catterall 1985), and by social behaviour (Kikkawa 1980, 1987; Kikkawa and Wilson 1983; Catterall *et al*. 1982, 1989). Our results and those of Kikkawa and Catterall contrast with studies of Great Tits in insular patches of forest among farmland and suburban habitats (Dhondt 1979; McCleery and

Perrins 1985). Dhondt found frequent movement between habitat patches and McCleery and Perrins found that a large and variable fraction of breeders immigrated from outside their study area each year. The difference between these two sets of studies may lie either in the ease with which birds can move between sites, or in the frequency of catastrophic mortality or both.

Other long-term population studies have also seen striking drops in insular populations, usually in association with diseases (Van Riper *et al*. 1986; Woolfenden and Fitzpatrick, this volume), or bad weather (Kikkawa 1977). Studies of bird populations in isolated habitat patches in tropical forests show initial increases in abundance as immigrants move from neighbouring disturbed areas, with pronounced edge effects and subsequent population collapses (Bierregaard and Lovejoy 1988).

Implications for conservation

Our results suggest some guidelines for the management of endangered insular populations of vertebrates. First, the unstable demography of the Mandarte Island population of Song Sparrows confirms that even dense isolated populations in optimal habitats are still at considerable risk of extinction. However, if bad weather reduces numbers and removes density-dependent controls, rapid recovery in numbers follows. The bottleneck in 1979–80 did not seem to cause permanent genetic harm to the population (Pimm *et al*. 1989). Conversely, when densities are high in such populations, reproductive rates and juvenile survival may be depressed. Thus prereproductive juveniles could be safely transplanted from dense populations to sites elsewhere, or used in captive breeding programs. There is, however, a slight risk that capturing individuals could coincide with stochastic losses and thus increase the chance of local or global extinction.

The existence of regulation involving social behaviour has both positive and negative implications for conservation. Social regulation is an indicator of high density and good demographic health of a population. If territoriality or social dominance creates non-breeders, these individuals are candidates for translocation to new sites. In the Song Sparrow population there were no female non-breeders. Thus juvenile females captured before the onset of winter would need to be used for translocation. Removal experiments have shown that non-territorial birds are quite capable of breeding, despite their low social status.

If, however, non-breeders are displaced immigrants from disturbed areas nearby, the demographic health of a host population may be poor. Although we did not find this, displaced non-territorial individuals may depress the reproduction or survival of local breeders. Bierregaard and

Lovejoy (1988) found that presumed resident birds survived poorly in isolated patches of tropical rain forest, when these were flooded by immigrants from cleared areas nearby.

Our results from feeding experiments suggest that the provision of supplemental food might be used to increase the reproductive success of endangered populations, and hence augment numbers so that part of a population could be transplanted to a suitable new site. Predation on nest contents was a density-dependent mechanism that interacted with food shortage negatively to regulate reproductive success. Programmes of predator control are thus a second way to increase the production of young in populations at high density, although we recommend caution in the use of this method because the identity of nest predators is usually unknown.

Summary

We have reviewed regulation of numbers by means of social behaviour. Bird populations meet several of the conditions necessary for regulation by territorial intolerance. However, effects of social behaviour are intimately related to regulation by food shortage and probably also to other regulating factors such as predation. The principal regulatory mechanisms in the resident Song Sparrow population on Mandarte Island are:

(a) disappearance of juveniles in winter, probably because territories cannot be obtained at the prevailing high population densities; and
(b) food shortage and predation during the breeding season.

Social regulation creates a non-breeding population of yearling males, which displace some male territory owners at high densities. Numbers of yearling females, however, are regulated before the onset of breeding, and there is no non-breeding surplus of yearling females. Despite the existence of strong density-dependence in both the breeding and non-breeding seasons, numbers in the population were also strongly influenced by catastrophic winter mortality in severe weather.

Acknowledgements

Our work was supported by grants from the Natural Sciences and Engineering Research Council of Canada. The Tsawout and Tseycum Indian Bands kindly allowed us to work on their island. We thank C. P. Catterall, S. J. Hannon, J. Kikkawa, J. H. Myers, and C. M. Rogers for criticism of the manuscript and C. P. Catterall, D. H. Chitty, A. R. E. Sinclair, J. Kikkawa, and S. J. Hannon for helpful discussion. We also thank our many colleagues and assistants for help with field work.

References

Ambuel, B. and Temple, S. A. (1983). Area-dependent changes in the bird communities and vegetation of southern Wisconsin forests. *Ecology*, **64**, 1057–68.

Arcese, P. (1987). Age, intrusion pressure, and defence against floaters by territorial male Song Sparrows. *Animal Behaviour*, **35**, 773–84.

Arcese, P. (1989*a*). Territory acquisition and loss in male Song Sparrows. *Animal Behaviour*, **37**, 45–55.

Arcese, P. (1989*b*). Intrasexual competition and the mating systems of primarily monogamous birds: the case of the Song Sparrow. *Animal Behaviour*, **38**, 96–111.

Arcese, P. (1989*c*). Intrasexual competition, mating system and natal dispersal in Song Sparrows. *Animal Behaviour*, **38**, 958–79.

Arcese, P. and Smith, J. N. M. (1985). Phenotypic correlates and ecological consequences of dominance in Song Sparrows. *Journal of Animal Ecology*, **54**, 817–30.

Arcese, P. and Smith, J. N. M. (1988). Effects of population density and supplemental food on reproduction in the Song Sparrow. *Journal of Animal Ecology*, **57**, 119–36.

Arcese, P., Stoddard, P. K., and Hiebert, S. M. (1988). The form and function of song in female Song Sparrows. *Condor*, **90**, 44–50.

Begon, M. and Mortimer, M. (1986). *Population ecology. A unified study of animals and plants*, 2nd edn. Blackwell, Oxford.

Bierregaard, R. O. Jr. and Lovejoy, T. E. (1988). Birds in Amazonian forest fragments: effects of insularization. In *Proceedings of 19th International Ornithological Congress, II* (ed. H. Ouellet), pp. 1564–79. University of Ottawa Press, Ottawa.

Birkhead, T. R., Eden, S. F., Clarkson, K., Goodburn, S. F., and J. Pellatt (1986). Social organization of a population of Magpies *Pica pica*. *Ardea*, **74**, 59–68.

Boag, D. A. and Schroeder, M. A. (1987). Population fluctuations in Spruce Grouse: what determines their numbers in spring? *Canadian Journal of Zoology*, **65**, 2430–5.

Breitwisch, R. (1989). Mortality patterns, sex ratios, and parental investment in monogamous birds. *Current Ornithology*, **6**, 1–50.

Brown, J. L. (1969). Territorial behaviour and population regulation in birds: a review and re-evaluation. *Wilson Bulletin*, **81**, 293–329.

Catterall, C. P. (1985). Winter energy deficits and the importance of fruit versus insects in a tropical island population. *Australian Journal of Ecological*, **10**, 265–79.

Catterall, C. P., Wyatt, W. S., and Henderson, L. J. (1982). Food resources, territory density, and reproductive success of an island silvereye population. *Ibis*, **124**, 405–21.

Catterall, C. P., Kikkawa, J., and Gray, C. (1989). Inter-related age-dependent patterns of ecology and behaviour in a population of silvereyes (Aves: Zosteropidae). *Journal of Animal Ecology*, **58**, 557–70.

Chitty, D. H. (1967). The natural selection of self-regulatory behavior in animal populations. *Proceedings of the Ecological Society of Australia*, **2**, 51–78.

Davies, N. B. (1978). Ecological questions about territorial behaviour. In *Behavioural ecology, an evolutionary approach* (eds. J. R. Krebs and N. B. Davies), pp. 317–50. Blackwell, Oxford.

Davies, N. B. and Lundberg, A. (1984). Food distribution and a variable mating system in the Dunnock, *Prunella modularis*. *Journal of Animal Ecology*, **53**, 895–912.

Dhondt, A. A. (1979). Summer disperal and survival of juvenile Great Tits in southern Sweden. *Oecologia (Berlin)*, 42, 137–57.

Drent, R., van Tets, G. F., Tompa, F., and Vermeer, K. (1964). The breeding birds of Mandarte Island, British Columbia. *Canadian Field Naturalist*, 78, 208–63.

Enoksson, B. (1988). Prospective resource defence and its consequences in the Nuthatch *Sitta europea* L. Thesis, University of Uppsala, Sweden.

Enoksson, B. and Nilsson, S. G. (1983). Territory size and population density in relation to food supply in the Nuthatch *Sitta europea*. *Journal of Animal Ecology*, 52, 927–35.

Fretwell, S. D. and Lucas, H. R. Jr. (1970). On territorial behavior and other factors influencing habitat distribution in birds. I. Theoretical development. *Acta Biotheoretica*, 19, 16–36.

Gauthier, G. G. (1987a). The adaptive significance of territorial behaviour in breeding Buffleheads: a test of three hypotheses. *Animal Behaviour*, 35, 348–60.

Gauthier, G. G. (1987b). Brood territories in Buffleheads: determinants and correlates of territory size. *Canadian Journal of Zoology*, 65, 1402–10.

Greig-Smith, P. (1985). Winter survival, home ranges and feeding of first-year and adult Bullfinches. In *Behavioural Ecology: Ecological Consequences of adaptive behaviour* (eds. R. M.Sibly and R. H. Smith), pp. 387–392. Blackwell, Oxford.

Hannon, S. J. (1983). Spacing and breeding density of Willow Ptarmigan in response to an experimental alteration of sex ratio. *Journal of Animal Ecology*, 52, 807–20.

Hannon, S. J. (1988). Intrinsic mechanisms and population regulation in grouse—a critique. *Proceedings of the 19th International Ornithological Congress, II* (ed. H. Ouellet), pp. 2478–89. University of Ottawa Press, Ottawa.

Hannon, S. J., Mumme, R. L., Koenig, W. D., and Pitelka, F. A. (1985). Replacement of breeders and within-group conflict in the cooperatively breeding Acorn Woodpecker. *Behavioural Ecology and Sociobiology*, 17, 303–12.

Haramis, G. M. and Thompson, D. Q. (1985). Density-production characteristics of box-nesting Wood Ducks in a northern greentree impoundment. *Journal of Wildlife Management*, 49, 429–36.

Hensley, M. M. and Cope, J. B. (1951). Further data on removal and repopulation of breeding birds in a spruce-fir forest community. *Auk*, 68, 483–93.

Hilborn, R. and Stearns, S. C. (1982). On inference in ecology and evolutionary biology: the problem of multiple causes. *Acta Biotheoretica*, 31, 145–64.

Howard, H. E. (1920). *Territory in bird life*. Murray, London.

Kikkawa, J. (1977). Ecological paradoxes. *Australian Journal of Ecology*, 2, 121–36.

Kikkawa, J. (1980). Winter survival in relation to dominance class among Silvereyes, *Zosterops lateralis chlorocephala*, of Heron Island, Great Barrier Reef. *Ibis*, 122, 437–46.

Kikkawa, J. (1987). Social relations and fitness in silvereyes. In *Animal societies, theories and facts* (eds. Y. Ito, J. L. Brown, and J. Kikkawa), pp. 253–66. Japan Scientific Societies Press, Tokyo.

Kikkawa, J. and Wilson, J. M. (1983). Breeding and dominance among the Heron Island Silvereyes *Zosterops lateralis chlorocephala*. *Emu*, 83, 181–91.

King, W. B. (1985). Island birds: will the future repeat the past? In *Conservation of island birds, case studies for the management of threatened island species* (ed. P. J. Moors), pp. 3–15. ICBP Technical Publications No. 3, Cambridge.

Klomp, H. (1972). Regulation of the size of bird populations by means of territorial behaviour. *Netherlands Journal of Zoology*, 22, 456–88.

Kluyver, H. N. and Tinbergen, L. (1953). Territory and regulation of density in titmice. *Archives Néerlandaises de Zoologie*, 10, 265–86.

Knapton, R. W. and Krebs, J. R. (1974). Settlement patterns, territory size, and breeding density in the Song Sparrow (*Melospiza melodia*). *Canadian Journal of Zoology*, **52**, 1413–20.

Krebs, C. J. (1985). Do changes in spacing behaviour drive population cycles in small mammals? In *Behavioural ecology: Ecological consequences of adaptive behaviour* (eds. R. M. Silby and R. H. Smith), pp. 295–312. Blackwell, Oxford.

Krebs, J. R. (1971). Territory and breeding density in the Great Tit, *Parus major*. *Ecology*, **52**, 2–22.

Krebs, J. R. and Perrins, C. M. (1978). Behaviour and population regulation in the Great Tit (*Parus major*). In *Population control by social behaviour* (eds. F. J. Ebling and D. M. Stoddard), pp. 23–47. Institute of Biology, London.

Lack, D. (1966). *Population studies of birds*. Clarendon Press, Oxford.

Lande, R. (1987). Extinction thresholds in demographic models of territorial populations. *American Naturalist*, **130**, 624–35.

MacArthur, R. H., Diamond, J. M., and Karr, J. R. (1972). Density compensation in island faunas. *Ecology*, **53**, 330–42.

Marcström, V., Kenward, R. E., and Engren, E. (1988). The impact of predation on boreal tetraonids during vole cycles: an experimental study. *Journal of Animal Ecology*, **57**, 859–72.

Martin, K. (1989). Pairing and adoption of offspring by replacement male willow ptarmigan: behaviour, costs and consequences. *Animal Behaviour*, **37**, 569–78.

Martin, T. E. (1987). Food as a limit on breeding birds: a life-history perspective. *Annual Review of Ecology and Systematics*, **18**, 453–87.

Martin, T. E. (1988). Habitat and area effects on forest bird assemblages: is nest predation an influence? *Ecology*, **69**, 74–84.

May, R. M. and Robinson, S. K. (1985). Population dynamics of avian brood parasitsm. *American Naturalist*, **126**, 475–94.

McCleery, R. H. and Perrins, C. M. (1985). Territory size, reproductive success and population dynamics in the Great Tit, *Parus major*. In *Behavioural ecology: Ecological consequences of adaptive behaviour* (eds. R. M. Sibly and R. H. Smith), pp. 353–73. Blackwell, Oxford.

McNamara, J. M. and Houston, A. I. (1987). Starvation and predation as factors limiting population size. *Ecology*, **68**, 1515–19.

Moss, R. and Watson, A. (1985). Adaptive value of spacing behaviour in population cycles of Red Grouse and other animals. In *Behavioural ecology: Ecological consequences of adaptive behaviour* (eds. R. M. Sibly and R. H. Smith), pp. 275–94. Blackwell, Oxford.

Moss, R., Kolb, H. H., Marquiss, M., Watson, A., Treca, B., Watt, D., and Glennie, W. (1984). Aggressiveness and dominance in captive cock Red Grouse. *Aggressive Behaviour*, **8**, 59–84.

Murray, B. G. (1979). *Population dynamics, alternative models*. Academic Press, New York.

Newton, I. (1980). The role of food in limiting bird numbers. *Ardea*, **68**, 11–30.

Norton, M. E., Arcese, P., and Ewald, P. W. (1982). Effect of intrusion pressure on territory size in Black-chinned Hummingbirds (*Archilocus alexandri*). *Auk*, **99**, 761–4.

Patterson, I. J. (1980). Territorial behaviour and the limitation of population density. *Ardea*, **68**, 53–62.

Pimm, S. L. and Redfearn, A. (1988). The variability of population densities. *Nature*, **334**, 613–14.

Pimm, S. L., Jones, H. L., and Diamond, J. (1988). On the risk of extinction. *American Naturalist*, **132**, 757–85.

Pimm, S. L., Gittelman, J. L., McCracken, G. F., and Gilpin, M. (1989). Plausible alternatives to bottlenecks to explain reduced genetic diversity. *Trends in Ecology and Evolution*, **4**, 176–8.

Piper, W. H. and Wiley, R. H. (1989). Correlates of dominance in wintering White-crowned Sparrows: age, sex and location. *Animal Behaviour*, 37, 298–310.

Rubenstein, D. I. (1981). Population density, resource patterning, and territoriality in the Everglades pygmy sunfish. *Animal Behaviour*, 29, 155–72.

Sinclair, A. R. E. (1990). Population regulation of animals. In *Ecological concepts* (ed. J. M. Cherrett). *Symposium of the British Ecology Society*, pp. 197–241. Blackwell, Oxford.

Smith, J. N. M. (1981). Cowbird parasitism, host fitness, and age of the host female in an island Song Sparrow population. *Condor*, 83, 152–61.

Smith, J. N. M. and Arcese, P. (1986). How does territorial behaviour influence breeding bird numbers? In *Behavioral ecology and population biology*, Readings from the 19th International Ethological Conference (ed. L. C. Drickamer), pp. 89–94. Toulouse.

Smith, J. N. M. and Arcese, P. (1989). How fit are floaters? Consequences of alternative territorial behaviors in a non-migratory sparrow. *American Naturalist*, **133**, 830–45.

Smith, J. N. M., Montgomerie, R. D., Taitt, M. J., and Yom-Tov, Y. (1980). A winter feeding experiment on an island Song Sparrow population. *Oecologia (Berlin)*, **47**, 164–70.

Smith, J. N. M., Yom-Tov, Y., and Moses, R. (1982). Polygny, male parental care and sex ratio in Song Sparrows: an experimental study. *Auk*, **99**, 555–64.

Smith, S. M. (1978). The 'underworld' in a territorial sparrow: adaptive strategy for floaters. *American Naturalist*, **112**, 571–82.

Soulé, M. (ed.) (1987). *Viable populations for conservation*. Cambridge University Press, Cambridge.

Stacey, P. B. and Koenig, W. D. (eds.) (1990). *Cooperative breeding in birds: long-term studies of ecology and behaviour*. Cambridge University Press, Cambridge.

Stewart, R. E. and Aldrich, J. W. (1951). Removal and repopulation of breeding birds in a spruce-fir forest community. *Auk*, **68**, 471–82.

Temple, S. A. (1985). The problem of avian extinctions. *Current Ornithology*, 3, 453–85.

Tompa, F. S. (1962). Territoriality, the main controlling factor of a local Song Sparrow population. *Auk*, 79, 687–97.

Tompa, F. S. (1963). Factors determining the numbers of Song Sparrows on Mandarte Island, B.C. Thesis, University of British Columbia, Vancouver, Canada.

Tompa, F. S. (1964). Factors determining the numbers of Song Sparrows *Melospiza melodia* (Wilson) on Mandarte Island, B.C. *Acta Zoologica Fennica*, **109**, 3–73.

Tompa, F. S. (1971). Catastrophic mortality and its population consequences. *Auk*, **88**, 733–59.

Van den Assem, J. (1967). Territory in the three-spined stickleback *Gasterosteus aculeatus* L. *Behaviour (Suppl.)*, 16, 1–164.

Van Riper, C. III, Van Riper, S. G., Goff, M. L., and Laird, M. (1986). The epizootology and ecological significance of malaria in Hawaiian land birds. *Ecological Monographs*, 56, 327–44.

Walkinshaw, L. H. (1983). *Kirtland's Warbler: the natural history of an endangered species.* Cranbrook Institute of Science, Bloomfield Hills, Michigan.

Watson, A. and Jenkins, D. (1968). Experiments on population control by territorial behaviour in Red Grouse. *Journal of Animal Ecology*, 37, 595–614.

Watson, A. and Moss, R. (1970). Dominance, spacing behaviour and aggression in relation to population limitation in vertebrates. In *Animal populations in relation to their food resources* (ed. A. Watson), pp. 167–218. Blackwell, Oxford.

Wesolowski, T., Tomialojc, L., and Stawarczyk, T. (1987). Why low numbers of *Parus major* in Bialoweza Forest—removal experiments. *Acta Ornithologica*, 23, 303–16.

Whitcomb, R. F., Robins, O. S., Lynch, J. F., Bystrak, B. L., Klimkiewitz, M. K., and Bystrak, D. (1981). Effects of forest fragmentation on avifauna of the eastern deciduous forest. In *Forest island dynamics in man-dominated landscapes* (ed. R. L. Burgess and D. M. Sharpe), pp. 125–205. Springer-Verlag, New York.

Wiley, J. W. (1985). Shiny cowbird parasitism in two avian communities in Puerto Rico. *Condor*, 87, 165–76.

Williamson, M. (1983). The land-bird community of Skokholm: ordination and turnovers. *Oikos*, 41, 378–84.

Wynne-Edwards, V. C. (1962). *Animal dispersion in relation to social behaviour.* Oliver and Boyd, Edinburgh.

Zahavi, A. (1971). The social behaviour of the White Wagtail, *Motacilla alba*, wintering in Israel. *Ibis*, 113, 203–11.

Zwickel, F. C. and Bendell, J. F. (1985). Blue Grouse—effects on, and influences of, a changing forest. *Forest Chronology*, 61, 185–8.

8 Demographic changes in a Snow Goose population: biological and management implications

E. G. COOCH and F. COOKE

Introduction

There is a threefold challenge facing biologists studying economically important species: firstly, to understand the basic demographic facts about the population; secondly, to draw the appropriate inferences from these facts; and thirdly, management implications have to be pointed out. If the factual information is flawed, it is irrelevant at best and irresponsible at worst to offer advice on management. In this paper, we report some of the essential demographic findings that have been made relating to the Lesser Snow Goose (*Anser caerulescens caerulescens*) population at La Pérouse Bay (LPB), in Northern Manitoba, Canada. We provide population estimates for fecundity (production) and survival and show how the values change with time. We use these values to show that the changes in demographic parameters can explain the changes in population size, and attempt to identify the factors responsible for the demographic changes. This information will allow managers to devise policies concerning the target population levels, which may involve either a change or maintenance of the original population level. If managers wish to change the population levels, they will have the requisite information on which to attempt to alter the demography of the population.

Methods

The population

The Snow Goose colony at La Pérouse Bay has been studied intensively since 1968 during which time the population has grown from about 2000 to more than 8000 breeding pairs. This population is not isolated from other arctic breeding colonies but mixes with birds from other colonies located in the Hudson Bay, Foxe Basin, and Central Arctic regions. This mixing occurs during migration and on the wintering grounds when birds from different breeding colonies share similar feeding locations. Since pair formation occurs away from the breeding grounds, the intermixing results in a large amount of gene flow between colonies. Rockwell and Barrowclough (1987) estimated gene flow per generation of this colony to be 0.497. This is due mainly to an influx of males, since when birds from different breeding colonies pair, they return to the natal colony of the female. The extremely high rate of gene flow through male immigration means that demographic attributes of other breeding colonies could have a considerable effect on population size and population change at the La Pérouse Bay colony (Cooke and Findlay 1982). Through the auspices of the Canadian Wildlife Service and the U.S. Fish and Wildlife Service, we have obtained data on production and population size of the birds that constitute this much larger population, which uses the Central and Mississippi Flyways for migration and winters mainly in Texas and Louisiana. Henceforth we will refer to this larger population as the global population.

Techniques

Each year since 1973, up to 600 nests have been monitored annually from the laying of the first egg to the departure of goslings. An additional 2000 nests found during the incubation period increased the sample size for some of the fecundity measures. Goslings from all monitored nests are marked with individually numbered webtags which allow their survival during the fleding period to be assessed. Just prior to fledging, approximately 1500 families are rounded up, measured, aged, sexed, ringed, and colour-ringed. Their subsequent survival is estimated by a variety of methods, involving either birds that return to the breeding colony or birds that die and whose rings are reported to federal agencies.

The size of the breeding colony has been estimated in two ways. Firstly, part of the colony area is searched in detail and all nests counted and mapped. The number of nests in the remainder of the colony is estimated by making transects and extrapolating to the extent of the segment of the colony where the transect is located. Secondly, colony size is estimated by the Jolly-Seber method (Seber 1982) utilizing data from returning females (the philopatric sex) which have been ringed at the colony as breeding adults.

Results

Population size

Jolly-Seber (J-S) capture-recapture estimates of colony size based on data from ringing drives (Sulzbach 1975; Healey 1985; Fig. 8.1) and estimates

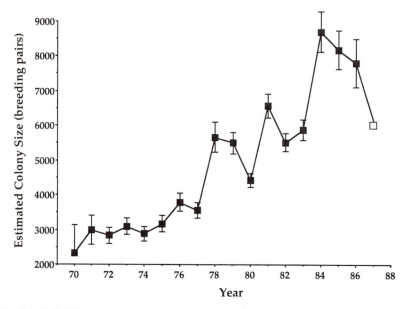

Fig. 8.1. Jolly-Seber capture-recapture estimates of colony size at La Pérouse Bay, for the years 1970–86. Data are of means ± s.d. The value of 1987 (open box) is derived from visual transect estimates (see the text for comparison of visual and J-S estimates).

from visual transects indicate that the colony has expanded in both area and numbers over the course of the study. Sulzbach (1975) and Healey (1985) showed by regression of J-S estimates of colony size against time that colony size increased at approximately 8 per cent per year. Healey (1985) commented that in recent years the visual and J-S estimates have increasingly diverged, with the visual transect estimates being consistently lower than the J-S values. The divergence may be due to violation of the assumptions inherent in one or both methods, and our estimates of colony size in recent years must be considered much more tentative.

The estimated wintering population of the combined Central and Mississippi flyways, of which birds breeding at La Pérouse Bay are a part, has also increased significantly since 1973 (Fig. 8.2).

Fecundity values

To discover whether the estimated fecundity and survival values of the La Pérouse Bay population might be expected to result in the observed population increase documented above, we have analysed these components in considerable detail. Because females usually return to their

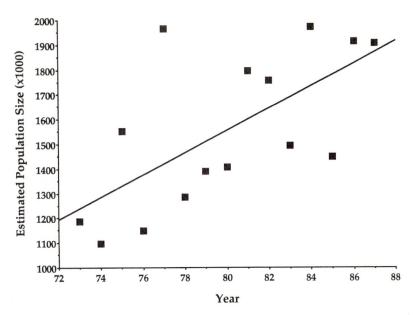

Fig. 8.2. Midwinter estimates of number of Lesser Snow Geese wintering in the combined Central and Mississippi flyway regions, 1973–87 (linear model: $F = 9.60$; d.f. = 1,14; $P = 0.009$).

natal colonies to breed, we examine demographic parameters through the females. If there is little female immigration and emigration, or if female immigration and emigration are in equilibrium, then the fecundity and survival values derived from the LPB colony should accurately predict the growth rate of the colony. To estimate the number of fledged female goslings per nesting female, we have measured the number of eggs laid, rates of partial and total clutch loss, hatching success, and total and partial brood loss. Initially these values were averaged over several seasons. Details of the method used are given by Cooke and Rockwell (1988). From these studies, the calculated average number of fledged female goslings (to the age when goslings are ringed, just prior to fledging) per nesting female is 0.95–1.10 for 2-year-olds, 1.06–1.23 for 3-year-olds and 1.17–1.38 for older females. Yearling females do not breed. The variability in these figures reflects the fact that it was difficult for us to measure accurately the total brood loss, which is however an infrequent event.

Survival values

To calculate the effect of the observed level of production on the growth rate of the colony we also calculated the age-specific probability of a female entering the breeding population, the age-specific survival rate at least to the age of four, and the average annual adult survival rate. If we have these values we can use a matrix approach (Leslie 1945) to predict whether the population is likely to be expanding, staying constant, or contracting in size.

The probability of birds entering the breeding population was calculated by comparing age-specific survival estimates with known rates of return to the breeding colony of birds of different ages (yearlings do not breed). It was found that approximately 50% of 2-year-olds and 86 per cent of 3-year-olds breed and that by the age of four years virtually all birds attempt to breed (for details see Cooke and Rockwell 1988).

Several different methods were used to estimate age-specific and adult survival rates. Each method has its own set of assumptions and there has been much debate about the reliability of the various approaches. Some of the methods calculate age-specific survival; others calculate first-year survival and adult survival, assuming that survival after the first year is independent of age.

The method of Fordham and Cormack (1970), commonly referred to as the ratio method, relies on ringing recoveries of birds marked just prior to fledging. We use data from female goslings ringed in the years 1970–81 and recovered up to 12 years later. The assumption of this method is that survival rates are the same for at least two ages (Lakhani and Newton 1983). Regardless of which pair of ages was considered, average female first-year survival rate was about 48 per cent and average adult survival rate about 75 per cent. There was no evidence of any significant changes in survival rate with age of adults, at least in the age range 2–7 (Richards 1986).

Using the method described by Brownie et al. (1985), survival during the first year and mean survival of all other age classes were calculated (yearling geese can be distinguished at ringing, but too few were caught to be included separately in the model). Survival rates were calculated for each cohort and averaged. Boyd et al. (1982) have already calculated cohort-specific survival rates for birds of both sexes combined for the LPB colony for the years 1970–7 and arrived at an average of 45.7 per cent for birds in their first year and 78 per cent for adults. Richards (1986) calculated values for the years 1970–81 and estimated an average female first-year survival of 48 per cent and an average male survival rate of 38 per cent. Adult survival for each sex was 80 per cent.

The above methods allowed us to estimate survival rates of both first-year and adult birds using ringing recovery data. We then needed to see whether the estimates obtained by these methods correspond with the number of recruits that enter the breeding population. To do this we calculated the proportion of four-year-olds in our ringing drives and assumed them to be a representative sample of the breeding birds in the colony as a whole. Averaged over 6 cohorts, approximately 20.2 per cent of female goslings ringed were alive and breeding at LPB at age four. If all surviving females return to their natal colony to breed and if we assume that first-year survival was approximately 60 per cent of that of older birds (as suggested by the above figures), we estimate first-year survival as 46 per cent and subsequent annual survival as 76 per cent. Further details of the method are described in Cooke and Rockwell (1988).

Two other methods were available for estimation of adult annual survival rates. The method of Tanner (1978) was used by Rockwell *et al.* (1985) and utilized female birds banded as goslings which had returned to breed at LPB at a minimum of 4 years of age. The method relies on the change in frequency of detection with time. Adult survival estimates averaged 79.8 per cent for the years 1976–81. The other approach is the Jolly-Seber capture-recapture method which was also used to estimate population size. This relies on females banded as adults, which return and are recaptured at the nesting colony in subsequent years. Healey (1985) calculated average annual survival rate for the years 1970–81 and obtained an estimate of 78.0 per cent. A summary of the various estimates for survival of first-year and adult Snow Geese is presented in Table 8.1.

Table 8.1 Estimates of first-year and adult survival among female Lesser Snow Geese calculated by various methods

Method	Females ringed	Re-encounter method	Years	Survival values (%)	
				First year	Adult
Tanner	Goslings	Recapture	1976–81	—	79.8
Ratio	Goslings	Recovery	1970–81	48.0	75.0
Brownie	Gos/Adults	Recovery	1970–81	48.0	79.6
Jolly–Seber	Adults	Recapture	1970–81	—	78.0
Cooke, Rockwell	Goslings	Recapture	1976–81	46.0	76.0

There is close similarity between the values shown in Table 8.1, despite the different assumptions and potential biases of the various methods of estimation. It is suggested that all the methods would be most unlikely to give the same sort of biases, indicating that some of our concerns about methods of calculating survival may be unwarranted. We conclude that average first-year survival rate is in the range of 46–48 per cent and that for adult survival is 76–80 per cent. There is no evidence that 2–4-year-old birds have lower survival rates than birds greater than 4 years of age.

Demographic change

Using the long-term average fecundity and viability estimates reported above, it is possible using the matrix model described by Geramita *et al*. (in preparation) to calculate the growth rate and stable age distribution of the colony. The model is of a type first described by Leslie (1945); the age-specific fecundity and survival values are incorporated with the variable age of first breeding and expressed as a single matrix equation. This allows the calculation of an intrinsic growth rate for the population. In a simulation run with the values obtained above and including minimum and maximum estimates for fecundity for the years 1973–81, the expected growth rate of the LPB colony was 5–12 per cent, which is in accordance with our observed rate of growth over the same time period of *c*. 8 per cent (Fig. 8.1) (Healey 1985). Both types of data, population size estimates and fecundity and viability estimates, therefore led to the same conclusion, namely that for the years 1973–81 the LPB colony was growing. The growth rate calculated by the matrix model assumed no immigration or emigration, indicating that local production of the colony itself is sufficient to account for the growth of the colony for these years. We are currently updating this analysis to include more recent years.

Long-term trends

The problem with calculating average values for fecundity and survival is that long-term trends may be obscured. If systematic changes are occurring either because of, or independently of, the increasing population size then we need to investigate annual variations in the fecundity and survival values. Such trends are of major concern to the managers and protectors of the wildlife resource. Trends are difficult to detect without an accumulation of many years of data particularly when there are likely to be effects of other environmental variables.

During the 20 years of the study there have been significant declines in clutch size, gosling fledging weight, and first-year survival. The decrease in first-year survival is not associated with increased killing by hunting (in fact the reported hunter recovery rate has declined significantly during the course of the study). All these declines should lead to a reduction in the growth rate of the population and are consistent with population regulation by density-dependent processes.

Clutch size

Annual mean clutch size, as measured by the number of eggs laid in a nest found at the one egg stage, varies for several reasons. If the season is late (i.e. if snow leaves the nesting area late) clutch size is lower (Cooch 1958;

Barry 1962; Ryder 1970; Newton 1977). Thus, annual variation in laying date will contribute significantly to variation in clutch size. Also, intra-specific nest parasitism (INP) occurs frequently in Lesser Snow Geese (Finney 1975; Lank *et al*. 1989*a*, *b*) and will inflate the recorded clutch size. Because individual females laying eggs parasitically in a given year are believed not to have a nest of their own (Lank *et al*. 1989*b*), annual variation in clutch size is affected by annual differences in INP rate. Consequently, the data of clutch sizes were adjusted to account for seasonal variations in both INP and mean laying date. We then looked at the adjusted clutch sizes over time (Fig. 8.3). During the 15 years for which clutch size data were available, mean clutch size has declined by 0.7 of an egg ($F = 39.7$; d.f. = 1,14; $P \ll 0.001$; Cooch *et al*. 1989). The rate of the long-term decline was found to be independent of female age.

Since clutch size increases with female age until the age of 5 years (and is independent for ages >5) (Finney and Cooke 1978; Rockwell *et al*. 1983), any variation in the age composition of the colony could result in variation in mean clutch size. In our samples there was a significant increase in the mean age of the nesting sample, which would be expected to lead to an increase in mean clutch size (Cooch *et al*. 1989). However, a marked decline in mean clutch size was observed.

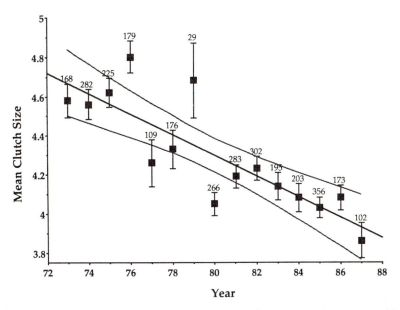

Fig. 8.3. Regression of annual mean clutch adjusted for intraspecific nest parasitism and laying date against year, for 1973–87, with 95% confidence limits of mean predicted values and sample sizes.

The considerable gene flow between Lesser Snow Goose colonies might contribute significantly to variation in clutch size (Rockwell *et al.* 1987; Rockwell and Barrowclough 1987). Because clutch size is independent of age after 4 years, variation in clutch size for individual birds aged 5+ (corrected for INP and laying date) was examined for evidence of long-term decline. A significant decline in the clutch size of individual birds was found, suggesting that the decline in population mean clutch size could not be attributed solely to gene flow.

Since natural selection favours birds laying larger clutches (Rockwell *et al.* 1987), selection is unlikely to explain the decline in clutch size. The most reasonable explanation is that there has been some systematic decline in the environment of the geese. Egg production in arctic nesting geese is closely correlated with the amount of nutrient the bird has acquired prior to egg-laying (Ankney and MacInnes 1978), so it seems likely that aspects of nutrient status at some time prior to egg production have deteriorated during the course of the study. Note that the population of geese at LPB and along the flyway as a whole has increased during the time of study. This suggests the possibility of some density-dependent reduction of egg production.

We have looked for, but failed to find, changes in other components of fecundity. There are no systematic changes in mean laying-date, INP, net failure, egg loss, or hatching success (Cooch *et al.* 1989; Lank *et al.* 1989*b*). Fledging success (for those birds that had at least one gosling with them at the time of the annual ringing drives) has declined slightly (Cooch *et al.* in preparation). The only other feature that has changed during the breeding season is gosling weight.

Gosling weight

Each year, 5000–7000 goslings are web-tagged at the time of hatching. This allows us to determine the precise age of the goslings when they are weighed and measured at ringing, which is generally 25–45 days after hatching. Since growth-rate has been shown to be approximately linear for the ranges of gosling ages encountered during the annual 8-day ringing period, it is possible to calculate a weight for each bird for some standardized age. Because there is significant annual variation in the mean age of goslings measured at ringing, gosling morphometrics were also adjusted for differences in annual mean gosling age. Figure 8.4 shows the change in standardized mean gosling weights during the years 1978–88. There has been a significant decline during the course of the study ($F = 8.60$; d.f. $= 1,10$; $P = 0.017$). From other evidence (Cooch *et al.* in preparation), genetic change in the population in respect of body size can be ruled out, so it appears that the decline in gosling weight is attributable to a deterioration in the feeding environment at LPB. Since maximal

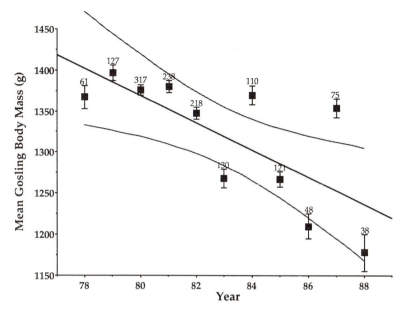

Fig. 8.4. Regression of mean gosling weight adjusted for intra- and inter-year differences in age at measurement against year (1978–88), with 95% confidence limits of mean predicted values and sample sizes.

growth rate is thought to be vital to the survival of arctic nesting geese (Cooch 1958; Barry 1962; Ryder 1970; Newton 1977), the detected decline in gosling weight suggests some adverse effects on survival. A lower gosling weight at a given age indicates that birds fledge at a lower weight or fledge later, either of which could adversely affect subsequent survival. In fact the decline in mean gosling weight of cohorts is reflected in the adult weight of those cohorts, suggesting that slower growth during the first few weeks of life results in a lower adult weight. This can only be demonstrated in females, since females if they survive, return to the natal colony to breed (Cooke *et al*. 1975) and can be remeasured as adults. Adult female body weight has declined by approximately 11.5 per cent from 1970 to 1986 ($F = 134.2$, d.f. $=1,16$; $P \ll 0.001$). Males at LPB are mostly hatched at other breeding colonies and we have detected no corresponding decline of mean weight over time. This tentatively suggests that the phenomenon of decline in gosling weight is local to LPB and not necessarily the same at other arctic breeding colonies.

Declining gosling weights suggest a deterioration in the food supply for growing goslings concomitant with the increase in population size. There are various indications in support of this suggestion. Firstly, in the early years of the study most Lesser Snow Goose families could be found on the food-rich salt-marshes around La Pérouse Bay itself. It was not difficult to round up groups of 1500 birds at one time. As time progressed, families became more scattered and were found in other feeding areas inland from the bay. The amount of salt-marsh vegetation (mainly *Puccinellia phryganodes* and *Carex subspathacea*) has declined in absolute terms due to grubbing by geese in early spring, and other plant species such as *Carex aquatilis* (Jefferies 1988) are now extensively utilized by Snow Geese. If food is limited and goslings are growing more slowly, this might affect their subsequent survival and in the next section we document such an effect.

First-year survival

First-year survival is measured from the time the birds are ringed (slightly before fledging) to the same time 1 year later. Survival was measured on an annual basis using the method of Brownie *et al.* (1978), as described earlier. Estimated annual first-year survival varies among years between 15.1 and 62.9 per cent (mean s.d. $\pm$ 40.7 $\pm$ 13.5 per cent). Sedinger and Raveling (1986) suggested that there is strong selection for early nesting in Canada Geese to synchronize with the temporal abundance of food. This is apparently supported by our finding that first-year survival is significantly lower in later years ($r = -0.683$; d.f. = 13; $P = 0.01$). However, while there is considerable annual variation in mean hatching date, the timing of ringing is generally consistent between years (late July to early August). Thus, the mean age of goslings handled during ringing shows significant variation among years ($F = 932.1$; d.f. = 10,7073; $P \ll 0.001$). First-year survival was found to be significantly higher in years when the mean age at ringing was higher ($r = 0.821$; d.f. = 13; $P < 0.001$). Because hatch-date and ringing age were highly correlated, we could not separate their relative effects on variation in first-year survival. When we controlled for differences in ringing age, which was the more highly correlated of the two factors, there was a significant decline in first-year survival over time ($F = 6.01$; d.f. = 1,12; $P = 0.032$; Fig. 8.5).

Similarly, there was considerable annual variation in adult survival, with estimates ranging from 71.9 to 100 per cent (mean $\pm$ s.d. 82.4 $\pm$ 8.3 per cent; Fig. 8.5). Although ringing age is not an appropriate variable for analysis of variation in adult survival, the time between hatching and ringing is the period during which breeding birds feed heavily to replenish reserves lost during the nesting period. There was no significant correlation between adult survival and the length of time between hatching and ringing, either alone ($r = 0.020$; d.f. = 13; n.s.) or controlling for year ($r = 0.068$; d.f. = 13, n.s.).

To determine the cause of the change in gosling mortality rates, we can partition mortality in this population into hunter kill and natural death,

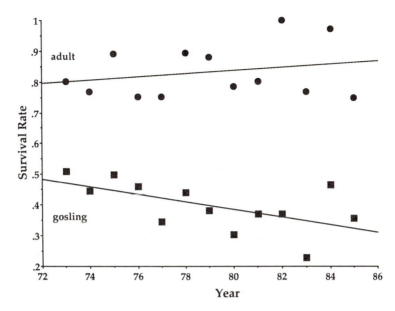

Fig. 8.5. Regression of estimated annual survival of adults (circles) and goslings (squares) against year (1973–85). Gosling survival rates adjusted for annual variation in mean age at ringing (see text).

the latter defined as all other causes of death. Natural death may occur primarily before the geese reach parts of the continent where hunters are active, or may occur at other times of year. Usually in birds, post-fledging mortality occurs soon after fledging and perhaps a reasonable hypothesis is to assume that natural death generally would precede hunter death.

Two pieces of evidence suggest that while gosling mortality has been increasing with time, hunter kill has been decreasing. Firstly, the recovery rates, defined as the probability that a bird will be shot and its ring reported to the authorities (Brownie *et al*. 1985), have declined significantly during the study (Fig. 8.6). Assuming that reporting rates, namely the probability that a hunter who shoots a ringed bird will report it to the authorities, have not changed significantly during the study, this suggests that hunters are shooting fewer birds. Secondly, the total numbers of geese in the Mississippi and Central flyways has been increasing, while the number of hunting licenses issued has declined (F. G. Cooch, personal communication). Unless hunters are shooting more geese per capita, this also suggests that hunter kill has been decreasing.

In addition, the large annual variation in first-year mortality seem more reasonably explained by large variations in natural as opposed to hunter mortality. This is because one would not expect dramatic differences in

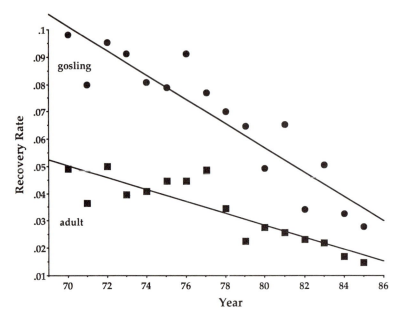

Fig. 8.6. Annual variation in recovery rates of adults (squares) and goslings (circles) recovery rates (1973–85).

continent-wide hunter activity from one year to the next. Moreover, the gradual deterioration in conditions for the geese as the population has increased provides a likely explanation for an increase in natural mortality over time. The changes in feeding pattern, the decline in gosling body weight, and the changes in vegetation (Jefferies 1988) all point to the decline in survival being a consequence of a decline in nutrient availability or nutrient quality during the course of the study.

To summarize the results of this section, we have shown that despite an overall growth of the colony, demographic changes have occurred during the course of the study which result in a decline in colony growth rate. These are (a) a decrease in clutch size, (b) a decrease in gosling weight, and (c) a decrease in first-year survival.

Discussion

Biological implications

Our findings are consistent with the regulation of population density mainly in response to food limitation; in other words as the population has increased, food availability and fecundity and survival have declined. However, until we can show the precise relationship between food availability and the effect on the geese, our suggestion of density-dependent population regulation remains somewhat speculative; food limitation is strongly implicated in all observed declines. In general, rates of protein and mineral storage of Arctic geese are highest during early migration (Alisauskas 1988). Most fat reserves are accumulated in the northern Plains states and southern regions of Manitoba and Saskatchewan. A second period of protein (but not fat) storage occurs during grazing along the James Bay-Hudson Bay coastline. Alisauskas (1988) notes that there is sufficient protein accumulated by the geese during this second phase of protein acquisition to overcome any between-year differences in body protein occurring further south along the flyway. A change in the availability of food, presumably on the Hudson Bay coastline, could result in a reduction of nutrients for egg production. This could be simply due to a larger number of birds competing for the same resources or could result from an actual reduction in the availability of resources. Another possibility is that conditions earlier in the year or in a previous year could influence the ability of the geese to gather nutrients: poor conditions on the wintering grounds might conceivably affect egg production.

A fourth and more intriguing possibility is that gosling growth may affect subsequent clutch production. Since we showed that gosling weight was closely correlated with subsequent adult weight, if adult weight influences clutch size then the decline in mean gosling weight with time may be the cause of the decline in mean clutch size. Evidence of a positive correlation between adult body weight and clutch size is controversial. Ankney and MacInnes (1978) found that larger birds (size being determined by culmen length, which correlated strongly with body weight) laid larger clutches. Davies *et al.* (1988) using principal component analysis which incorporated weight, tarsus, culmen, and head length found no such relationship. Therefore, clarification of this issue is necessary before we can implicate environmental conditions during the time of gosling growth as contributing to the decline in clutch size of those birds as adults.

Another possibility is that food itself is not limiting clutch size, but that there is an indirect effect via limitations in nest-building activity. If this were the case we might expect changes in the rate of INP or in patterns of nest initiation, but no such changes have been detected (Lank *et al.* 1989*b*; Cooch *et al.* 1989). Another possible explanation for the decline is increased predation pressures during the egg-laying period. If eggs are laid but disappear due to predators prior to the nest being visited by the scientific observers then this could result in an apparent decline in clutch size. However, we have found no systematic changes in predation rates at other times during the nesting period so this is also an unlikely explanation (Cooch *et al.* 1989). Overall the most probable explanation is food limitation and this probably occurs either in the northern prairies where birds accumulate most of their fat reserves for egg production or while the geese are 'topping up' during their stay along the Hudson and James Bay coastline. Since birds from LPB feed close to the colony for a few days prior to nesting, there is also the possibility that declines in vegetation availability or nutrient quality in the region of LPB itself could be responsible for the decline in clutch size.

The decline in gosling weight observed during the study could be the result of a continuing decline in some aspect of the environment (food, disease, or predators) or due to some change in the genetic composition of the population. A change in genetic composition could arise from directional selection for birds of small gosling weight or alternatively there could be gene flow into the population of birds producing small goslings. The available evidence does not favour a genetic explanation. There is indeed selection operating on goslings of differing weight, but large rather than small birds are recruited preferentially in some years (Cooke et al. unpublished). Thus directional selection appears to be in the opposite direction from that required to explain the decline in gosling weight. There are two potential sources of gene flow within this population. Firstly, there has been a low level of hybridization with the considerably smaller Ross' Goose Anser rossii. Since fewer than 0.1 per cent of Snow Geese in the colony are mated to Ross' Geese which are approximately 55 per cent the weight of Snow Geese, the effect of such gene flow is negligible. A second source of gene flow is potentially much more relevant. Almost all male geese breeding at LPB were hatched elsewhere (Cooke et al. 1975). If males from colonies with smaller body size have been increasing in frequency during the study, then smaller goslings could arise. There has been no change however in the structural size of males during the course of the study (Cooch et al. unpublished data), so differences due to gene flow are unlikely.

It is much more probable that decline in gosling weight reflects a decline in the quality of the environment of the gosling during the fledging period (Cooch et al. in preparation). Food supply is the obvious selecting factor and is consistent with our observations that birds are ranging more widely during the brood rearing period and that salt-marsh vegetation has declined in extent due to geese uprooting plants during the early part of the season (Jefferies 1988). Disease and increasing predation pressures also cannot be ruled out.

Decreased survival in the first year of life seems to be due to an increase in natural as opposed to hunter mortality. In years of extremely low survival, there are actually low levels of hunter recovery. This is consistent with the idea that most natural mortality occurs before the geese arrive in those parts of North America frequented by hunters. This suggestion is supported by findings for other bird species, namely that most first-year mortality occurs soon after fledging. In our study we have some evidence that the ringing activities increase subsequent gosling mortality (unpublished data); but in the course of the study we have introduced new methods which reduce the stress on the birds so it is unlikely that our method of ringing accounts for an increase in mortality. If however, the geese are stressed by natural processes, then our ringing activities may contribute over time in an increased death rate; by the same argument, the ringing activity is simply exaggerating a natural phenomenon. There is no evidence of change in ringing-induced mortality occurring during the study.

Assuming that the increased first-year mortality is a natural phenomenon, is it possible to identify the contributory factors? An attractive possibility is that the decreased gosling weight prior to fledging could itself result in higher mortality. Indeed in some years, smaller goslings are recruited at a lower rate into the breeding population. Also, years of lower mean gosling weight are generally years of higher gosling mortality ($r = 0.786$; d.f. $= 13$; $P \ll 0.001$). However it is more likely that the same environmental factors leading to lower gosling weight also lead to increased mortality. Food availability is again implicated, but the evidence is much more tenuous; indeed without additional information one cannot speculate further on the underlying cause of the mortalities.

In summary, the declines in all three life-history characteristics lead to a reduced growth rate of the population and could serve as density-dependent population regulators. Food availability is strongly implicated in reduced clutch size and gosling weight and weakly implicated in lower post-fledging survival. Food shortages could occur at several possible times in the annual cycle. A likely explanation for all three declines is that the feeding habitat near the colony (and possibly at other parts of the Hudson Bay coastline) has deteriorated due to the activity of the geese themselves. The destruction of salt-marsh habitat by geese is well-documented (Jefferies 1988). Such a reduction of food would be expected directly to influence gosling growth and subsequently gosling survival. Whether this could also effect clutch size is more uncertain, since we do not understand completely the critical dynamics essential to egg production. If clutch size is determined simply by the amount of nutrient material acquired, then it seems more likely that the food supply acquired in the prairie spring staging areas is more crucial than the Hudson Bay coastal marshes. If however, some essential ingredient(s) for egg production is acquired at the coastal marshes, then a deterioration in that habitat may effect a reduction in clutch size. It is crucial that research be continued to elucidate the times of year at which food availability is affecting the life history parameters.

Management implications

If we assume that the biological deductions that we have drawn from the facts we have discovered are correct, there is a further question of relevance to a broad management policy for Snow Geese. How representative of the overall population of Snow Geese is the colony at La Pérouse Bay? If each breeding population is subject to its own local demographic characteristics then a management policy suitable for the LPB colony may be totally invalid elsewhere. No other colony has been studied in as much detail for as long a period of time as the LPB colony, but some facts are clear. Firstly there has been a widespread increase in the continent-wide Snow Goose population. Increases have been documented in the mid-continent wintering population of which the LPB birds are a tiny proportion. Nesting increases have been documented in the Central Arctic, in the Queen Maud Gulf area, and large wintering increases have been noted in New Mexico (F. G. Cooch, personal communication). It seems likely therefore that other breeding colonies may be subject to the same demographic changes as those demonstrated at LPB, reflecting an increase in the global population. Indeed at the much larger and earlier established breeding colony at McConnell River, N.W.T., an expanding population caused widespread destruction of salt-marsh vegetation and subsequent dispersal of the nesting birds over a much wider region of the coast. The studies were not detailed enough to detect whether there was evidence of changes in fecundity and survival as a result of the habitat degradation.

Although we have no evidence for the universality of the processes we have described, there is some support for the view that Snow Goose colonies go through some growth phase followed by habitat destruction and a predicted population decline. (We have as yet no evidence in support of this final stage). We imagine that in late stages of the cycle, emigration to new possible nesting areas will occur with subsequent increases in numbers followed by a decline in these new areas. If this is the pattern, two management implications follow. Firstly, coastal salt marshes (with *Puccinellia phryganodes* and *Carex subspathacea* as dominant plant species) are crucial to nesting Snow Geese and their ecology needs to be thoroughly studied. We need to know the ecological processes occurring after destruction of the habitat by geese. Do the same plant species eventually recolonize and do geese eventually return? If so, what is the time interval between destruction and reutilization? It is clear that the coastal marshes, whether close to nesting goose colonies or not, play a vital role in the management of the resource and should be preserved from human degradation. At the moment, the Canadian Arctic coast is sufficiently unexploited for protection by Government agencies to be a realistic possibility. Native communities have little direct interest in these habitats and are sensitive to the need to preserve Arctic wildlife. Top priority should be given to the long-term preservation of the coastal marshes.

The major destroyers of the habitat are undoubtedly the geese themselves, so salt-marsh destroyed by geese can probably recover over a period of several years. Presumably Snow Geese and coastal marsh plants coexisted well before European man become interested in the species. However, salt-marshes are unlikely to recover from human interference, and protection from alternative human uses is essential in the long-term survival of snow and other arctic geese. The destruction of coastal arctic salt-marshes by Snow Geese may itself be a recent phenomenon. As new sources of winter food have become available to the geese through changes in agricultural practices, winter starvation may be much less prevalent now than in the past. If greater food availability in winter has allowed populations to increase, then this may have led to greater destruction of coastal arctic marshes by the geese than that occurring prior to the development of agriculture in North America.

A second management conclusion, if our scenario of colonies which 'boom and bust' is correct, is that colonies must be managed, if at all, on a colony-by-colony basis. If each colony is at a different stage in the cycle, global management would be an insensitive tool. At the colony level one might be able to prolong the cycle by, say, increasing the pressure on birds at the growth phase and relaxing it at the decline phase. There is one problem even here. Although each breeding colony is geographically distinct during the breeding season, some colonies lie within the migration route of others. Much of the documented destruction of salt-marshes occurs early in the season when birds uproot the below-ground vegetation, prior to the growth of new shoots. The geese responsible may comprise migratory as well as resident birds; it is known that large numbers of Snow Geese from the McConnell River colony use the salt-marshes of LPB in the spring and clearly contribute to the destruction.

The above discussion leads up to the question of the purpose of management. If the aim is to preserve nesting Snow Geese in the traditional nesting colonies then individual colony management would be necessary, whereas if the aim is to maximize the global numbers of geese, then preservation of arctic salt-marshes is crucial. If a reduction in numbers of Snow Geese is thought desirable (to prevent degradation of arctic salt-marshes or to reduce crop damage) then restricting food supplies in the wintering grounds may be attempted.

As a guide to further studies concerning the management of Snow Geese, it is clear that regularly collected data of clutch size or brood size, gosling weight, and first-year mortality would be the most useful parameters to assess the demography of individual arctic colonies. If colonies are to be managed on an individual basis, mass ringing immediately prior to fledging would provide much of the required information. This could form the basis for management on a colony-by-colony basis and would assist in the setting of hunting regulations and the detection of future declines in the global population.

We have outlined possible management policies that would be applied in the northern breeding range of the species. Most management policies, however, are applied in the migration or wintering range of the species; the Snow Goose is a widely hunted species and hunting regulations are liberal. Most deaths of adults and around 30 per cent of deaths of fledged young are by the gun. Despite this, the global population in the mid-continent has increased. Natural changes in fecundity and survival appear to be occurring in the LPB population despite this hunter kill, which so far as we can tell is little affected by natural mortality. Management achieved through adjustment of the hunting regulations is probably unnecessary and likely to be ineffective at the present time. There is no evidence that hunting has affected the changing survival rates in the LPB population. It is important to continue to monitor the populations, because in the event of large-scale population declines due to natural mortality, the species could be in jeopardy if hunter kill was added to the other incumbent hazards.

Our current view of the population dynamics of Snow Geese (which is an extrapolation from the facts that we have discovered at LPB) is that Snow Goose breeding colonies are in dynamic equilibrium with their major food supply obtained on the arctic salt-marshes. In newly established colonies food is readily available and production exceeds mortality. As the colony grows, geese begin to degrade their environment and density dependence occurs mainly through increased post-fledging mortality. The habit destruction may eventually lead to the abandonment of the colony and the establishment of new colonies elsewhere. If this is a correct view of Snow Goose dynamics, the logical approach for managers is either not to attempt to regulate numbers at all so long as the global population is maintained or to manage on a colony-by-colony basis. Since birds from different colonies merge during migration and in the wintering areas, it is difficult to see the effectiveness of attempts to control numbers of manipulating mortality rates through the hunting regulations unless global declines in the population are detected. The reduction of food availability on the wintering grounds might be an effective way of reducing population levels. However, this is hardly a practical approach since winter food availability is more likely to be influenced by the needs of agriculture rather than the needs of waterfowl management.

Summary

We describe the demography of a Lesser Snow Goose colony at La Pérouse Bay, Manitoba, which has been the subject of a detailed study since 1973. We outline the major features of the fecundity and viability of the population, draw reasonable biological inferences from the facts and suggest suitable management proposals.

During the early years of the study, fecundity exceeded mortality and the population growth of approximately 8 per cent per annum could be accounted for by local recruitment. Immigration and emigration of females is an infrequent event. Most breeding males in the population are immigrants from other colonies and thus the demography of the LPB colony may be affected by conditions at other colonies.

During the course of the study the colony has increased in size and three fitness components have changed systematically. Clutch size has declined by 0.7 eggs, gosling weight at fledging (and subsequent adult weight) has decreased by 11 per cent and first-year survival has decreased significantly. This has resulted in a decline in the growth rate of the population and is consistent with the operation of density-dependent processes. All these declines could be explained by a decline in the quality and availability of food, and the decline in gosling weight and first-year survival strongly suggests degradation of the arctic salt-marshes where geese rear their broods. Mortality due to hunting is not responsible for the decrease in survival.

Snow Geese nest in large colonies and colony demography is probably driven by local food availability. There may be a natural growth and decline of colonies and as long as new arctic salt-marshes are available and protected, the overall population can remain stable. There is need for basic demographic information on other major arctic colonies. At the present time population levels in Snow Geese are unlikely to be regulated by manipulating the hunting regulations.

Acknowledgements

We are most indebted to all the people who have braved late winter storms, polar bears, and mosquitoes in order to discover the secrets of the Snow Geese of La Pérouse Bay. In particular we would like to thank Rocky Rockwell who shares this research project with us and Dov Lank who is the keeper and manipulator for all our computer-based data. These two and Charles Francis have shared their ideas with us and contributed their time to make this a joint scientific effort.

References

Alisauskas, R. T. (1980). Nutrient reserves of lesser snow geese during winter and spring migration. D.Phil. Thesis, University of Western Ontario.

Ankney, C. D. and MacInnes, C. D. (1978). Nutrient reserves and reproductive performance of female Lesser Snow Geese. *Auk*, **95**, 459–71.

Barry, T. W. (1962). Effect of late seasons on Atlantic Brant reproduction. *Journal of Wildlife Management*, **26**, 19–26.

Boyd, H., Smith, G. E. J., and Cooch, F. G. (1982). The Lesser Snow Geese of the eastern Canadian Arctic. Canadian Wildlife Service Occasional Paper, **46**.

Brownie, C., Anderson, D. R., Burnham, K. P., and Robson, D. S. (1985). Statistical inference from band recovery data—a handbook (2nd edn). United States Fish and Wildlife Service Research Publication, **156**.

Cooch F. G. (1958). The breeding biology and management of the Blue Goose (*Chen caerulescens*). Ph.D. thesis, Cornell University, New York.

Cooch, E. G., Lank, D. B., Rockwell, R. F., and Cooke, F. (1989). Long term decline in fecundity in a Snow Goose population: evidence for density-dependence? *Journal of Animal Ecology*, **58**, 711–26.

Cooke, F. and Findlay, C. S. (1982). Polygenic variation and stabilizing selection in a wild population of lesser snow geese (*Anser caerulescens caerulescens*). *American Naturalist*, **120**, 543–47.

Cooke, F. and Rockwell, R. F. (1988). Reproductive success in a Lesser Snow Goose population. In *Reproductive success* (ed. T. H. Clutton-Brock), pp. 237–50. University of Chicago Press, Chicago.

Cooke, F., MacInnes, C. D., and Prevett, J. P. (1975). Gene flow between breeding populations of Lesser Snow Geese. *Auk*, **92**, 493–510.

Davies, J. C., Rockwell, R. F., and Cooke, F. (1988). Body-size variation and fitness components in Lesser Snow Geese (*Chen caerulescens caerulescens*). *Auk*, **105**, 639–48.

Finney, G. (1975). Reproductive strategies of the lesser snow goose, *Anser caerulescens caerulescens*. D.Phil. Thesis, Queen's University, Kingston, Ontario.

Finney, G. H. and Cooke, F. (1978). Reproductive habitats in the snow goose, the influence of female age. *Condor*, **80**, 147–58.

Fordham, R. A. and Cormack, R. M. (1970). Mortality and population change of Dominican Gulls in Wellington, New Zealand. *Journal of Animal Ecology*, **39**, 13–27.

Healey, R. F. (1985). Estimating population parameters of a nesting snow goose colony using the Jolly-Seber method. M.Sc. Thesis, Queen's University, Kingston, Ontario.

Jefferies, R. L. (1988). Vegetational mosaics, plant-animal interactions and resources for plant growth. In *Plant evolutionary biology* (ed. L. Gottlieb and S. K. Jain), pp. 341–69. Chapman and Hall, New York.

Lakhani, K. H. and Newton, I. (1983). Estimating age-specific bird survival rates from ring recoveries—can it be done? *Journal of Animal Ecology*, **52**, 83–91.

Lank, D. B., Mineau, P., Rockwell, R. F., and Cooke, F. (1989*a*). Intraspecific nest parasitism and extra-pair copulation in lesser snow geese. *Animal Behaviour*, **37**, 74–89.

Lank, D. B., Cooch, E. G., Rockwell, R. F., and Cooke, F. (1989*b*). Environmental and demographic correlates of intraspecific nest parasitism in lesser snow geese *Chen caerulescens caerulescens*. *Journal of Animal Ecology*, **58**, 29–45.

Leslie, P. H. (1945). On the use of matrices in certain population mathematics. *Biometrika*, 33, 183–212.

Newton, I. (1977). Timing and success of breeding in tundra-nesting geese. In *Evolutionary Ecology* (ed. B. Stonehouse and C. M. Perrins), pp. 113–26. MacMillan, London.

Richards, M. H. (1986). A demographic analysis of Lesser Snow Goose band recoveries. M.Sc. Thesis, Queen's University, Kingston, Ontario.

Rockwell, R. F. and Barrowclough, G. F. (1987). Gene flow and the genetic structure of populations. In *Avian genetics* (ed. F. Cooke and P. A. Buckley), pp. 223–55. Academic Press, London.

Rockwell, R. F., Findlay, C. S., and Cooke, F. (1983). Life history studies of the lesser snow goose (*Anser caerulescens caerulescens*). I. The influence of age and time on fecundity. *Oecologia*, 56, 318–22.

Rockwell, R. F., Findlay, C. S., Cooke, F., and Smith, J. A. (1985). Life history studies of the lesser snow goose (*Anser caerulescens caerulescens*). IV. The selective value of plumage polymorphism, net viability, the timing of maturation and breeding propensity. *Evolution*, 39, 178–89.

Rockwell, R. F., Findlay, C. S., and Cooke, F. (1987). Is there an optimal clutch size in Snow Geese? *American Naturalist*, 130, 839–63.

Ryder, J. P. (1970). A possible factor in the evolution of clutch size in Ross' Goose. *Wilson Bulletin*, 82, 5–13.

Sedinger, J. S. and Raveling, D. G. (1986). Timing of nesting by Canada geese in relation to the phenology and availability of their food plants. *Journal of Animal Ecology*, 55, 1083–102.

Seber, G. A. F. (1982). *The estimation of animal abundance*, 2nd edn. MacMillan, London.

Sulzbach, D. S. (1975). A study of the population dynamics of a nesting colony of the Lesser Snow Goose (*Anser caerulescens caerulescens*). M.Sc. Thesis, Queen's University, Kingston, Ontario.

Tanner, J. T. (1978). *Guide to the study of animal populations*. University of Tennessee Press, Knoxville.

9 Constraints on the demographic parameters of bird populations

C. M. PERRINS

Introduction

For a population to remain stable, the breeding birds must produce enough young which will themselves survive to breed—to replace the adults that die. For any population to be in balance, the following must hold true:

No. births + No. immigrants = No. deaths + No. emigrants

These parameters may not necessarily be in balance over short periods, say within a year or even between years, but over a long time such a balance must hold true. In practice few studies have measured all the variables in the above equation, especially since emigration and immigration have frequently proved intractable to study. However, in many cases, such as small populations of endangered species or of species living on islands, questions about such movements are not relevant; there may be no other populations for emigrants to go to or for immigrants to arrive from.

Many more studies have measured the survival rates of adults and looked to see whether sufficient new young adults (ignoring the question of whether these were locally raised or immigrants) are entering the breeding population to replace the adults that die. If the following holds true, then the population is capable of maintaining itself:

No. recruits to breeding population = No. adults dying

Annual adult survival rates have been produced for a wide range of species, though they are still heavily biased towards certain groups of birds and also to species living in north temperate regions. If one knows the adult survival rate, it is possible to deduce the proportion of young that must survive to breed in a stable population, but accurate actual measures of recruitment are still rather few. At equilibrium, the number of young surviving (per pair) to breed must equal twice the annual adult mortality.

Different species of birds have very different demographic patterns; a range of these is shown in Fig. 9.1. At the two extremes lie birds such as Blue Tit *Parus caeruleus* with an annual adult survival rate of perhaps as little as 30 per cent, whereas in some Procellariiforms, such as Sooty

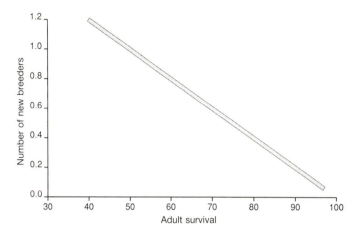

Fig. 9.1. The relationship between annual adult survival rate and the number of young (per pair) which must achieve breeding status in order for a population to maintain its numbers. All populations that maintain their numbers must produce enough young to fall on or above the line.

Albatross *Diomedea fuscea* and Fulmar *Fulmarus glacialis*, as many as 97 per cent or 98 per cent of the adults survive from one year to the next. Not surprisingly, the number of young that have to survive to breeding age in order for such populations to remain stable is very different, ranging from 1.4 young per pair per year in Blue Tits to 0.06 young per pair per year in the Albatross.

Knowledge of such demographic patterns is important. We need to know the survival rates for the different age groups of populations so that, if a population is declining, we can try to pinpoint where the problem lies.

Recent improvements in analytical methods enable us to estimate these figures more accurately, but we still lack sufficient field information about the demographies themselves. In particular, my aim in this paper is to emphasize how little we know about the plasticity of a species' demography, by which I mean the range of survival rates which a species can show and yet still maintain its numbers. I shall discuss this mainly using data from studies of the Mute Swan *Cygnus olor* and then go on to look at ways in which it may be possible to deduce further information about species for which there is only a limited amount of actual data available.

Mute Swan populations

Studies and study areas

In Britain in recent years there have been a number of studies of the Mute Swan, undertaken in part because the species has been declining in large areas of the country. These studies show quite marked differences in demographic characteristics between different areas. The four main studies were carried out in the following study areas:

(1) The Outer Hebrides

Spray (1981) studied the birds in these islands off the west coast of Scotland during the period 1978–82. These birds breed mainly on lowland lochs, some of which are saline; there are marked differences in nesting success on different types of lochs.

(2) Abbotsbury, Dorset

This is the site of the famous Swannery on the Fleet, a stretch of shallow water lying behind the Chesil Bank; here the swans nest colonially and have done so since at least 1393 and almost certainly from well before that date. The population is heavily managed. In particular, the majority of the cygnets that survive are hand-raised in pens, so that we do not have a measure of natural cygnet survival rates (between hatching and September when the hand-raised young are released). However, for the rest of the time the birds seem to live fairly normal lives in a wild situation. The population has been studied since 1968 (Perrins and Ogilvie 1981).

(3) The English Midlands

A long-term study in and around Staffordshire, started in the early 1960s on a wide range of waters from ponds to rivers (e.g. Coleman and Minton 1980).

(4) The River Thames and vicinity

A long-term study from the Edward Grey Institute, mainly by C. M. Perrins, C. M. Reynolds, P. E. Bacon, M. E. Birkhead and E. J. Sears (see e.g. Birkhead and Perrins 1985; Sears 1986). This study covers two different regions. These are (i) the Oxfordshire Study Area, started in the early 1960s; this area covers the Thames through Oxfordshire (from Lechlade to Goring plus the major tributaries, the rivers Windrush, Thame, and Cherwell, the latter only as far north as Upper Heyford); and (ii) the Lower Thames Area, covered since 1979, which includes the main river downstream from Goring to Richmond (few swans occur between this point and London) plus some riparian gravel pits. The swans in the

Oxford area were the subject of an earlier study (Perrins and Reynolds 1967); at that time the population was larger, but later underwent a fairly marked decline. The results of this earlier study are presented alongside the more recent ones.

Results

The main demographic data are given in Table 9.1, which shows clutch size, hatching success, cygnet and adult survival for each of the five study areas. In a number of ways the swans in the Hebrides and at Abbotsbury seem similar, with high adult survival rates. Also, most results from the Oxfordshire Study Area are very similar to those from the Midlands, not surprisingly perhaps, since the areas are not very far apart and the habitats are similar.

The demographic status of these populations is quite different. During the study, both the Abbotsbury and the Hebridean populations were increasing or holding their own, but all the other populations declined. The number of breeding pairs in the Midlands fell steadily from about 70 in the early 1960s to 35 in 1985, a drop of 50 per cent. Similarly, the number of breeding pairs on part of the Lower Thames (Sunbury to Henley) fell by about 76 per cent between 1958 (when 29 pairs with broods were recorded

Table 9.1 Some demographic features of British swan populations

Area	Hebrides	Abbotsbury	Midlands	Oxford I	Oxford II	Lower Thames
Clutch size	6.1	4.8	6.6	6.0	6.9	6.8
% Nests lost	30	38	49		14	6
Brood size						
at hatch[a]			2.6	4.0	3.8	3.4
Brood size						
at fledging	1.8	(1.9)	1.9	2.0	2.8	1.7
Annual survival rate						
to 1 year	1.04	+	.79	1.3	1.03	.65
to 2 years	.78	+	.53	.89	.69	.45
to 3 years	.59	+	.37	.67	.46	.31
to 4 years	.44	+	.28	.43	.31	
Adult survival	.90	.94	.82	.82	.82	.77
Balance						
per annum	0.24	(.82)	−0.08	0.07	−0.05	−0.15

[a] Includes failed nests.

[b] No. of cygnets per pair that survive to June of the following year; the first of these measures is from September so is less than a year.

+ Not available on a per brood basis. Survival of first-year individuals from September to the following June was about 68 per cent; thereafter about 90 per cent per annum.

at swan-upping in late July) and 1985 (when there were only 7 pairs on the same stretch of river).

Declines would have been greater were it not for immigration. Birkhead (1982) and Sears (1986) calculated the Balance Per Annum (BPA: the difference between the number of cygnets per pair surviving to breeding age and the number of adults dying per pair per year) for the populations on the Thames. The BPA on the Lower Thames was −0.15, but this might be an under-estimate since the assumption was made that birds on this stretch most commonly started to breed at 3 years old as opposed to 4 years elsewhere; if an age of four was used, the figure for this stretch of river would have been −0.25. The first study of the population in the Oxford area (Perrins and Reynolds 1967) was started in 1960, before any decline in numbers became evident. This suggested that the population was about in balance, with an adult survival rate of 82 per cent and a BPA of 0.07. The more recent, comparable, figures for the Oxfordshire Study Area are of the order of −0.05. The Midlands population had a similar adult survival rate, but a lower nesting success due to higher vandalism and was not in balance, declining by about 50 per cent in 25 years. A point to note, but one that will not be discussed further here is that, for some time, the declines were probably slowed by the presence of large non-breeding populations. However, the immature birds in these populations probably suffered higher mortality than the breeding pairs with the result that, eventually, they were reduced to such low levels that they could no longer delay the decline.

Hence, while both reproductive output and adult survival rate are key parts to the equation, one can find populations of Mute Swans with very different demographic parameters which are stable. In the cases given here, breeding populations were stable with adult survival rates of anywhere from 94 per cent down to as low as 82 per cent. These approximate to a mean expectation of further life of adults which varies from 16.2 to 5.05 years.

Figure 9.2 shows the relationship between adult survival rate in these populations and the BPA of the population. These figures need to be treated with considerable caution since, in many cases, they are based on small numbers of years and/or pairs; also the BPA has often been calcu-lated from the estimates of cygnet survival and these measures are open to error. Furthermore, the age of first breeding varies between the different populations and though it tends to decline as the population declines, has not been accurately quantified in all studies. As mentioned above, the average age of first breeding is about four, but on the Lower Thames it may be nearer three and at Abbotsbury nearer five; the age of first breeding is not known for the Hebrides.

None the less, the overall pattern shown by Fig. 9.2 is clearly true; Mute

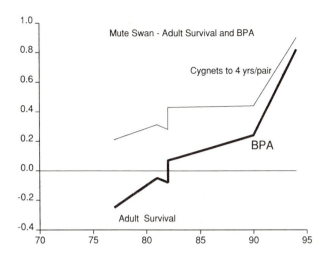

Fig. 9.2. The relationship between the number of cygnets raised per pair and the annual adult survival rate in different studies of the Mute Swan. The upper line shows the number of cygnets raised per pair, the lower one the Balance Per Annum of the population after the number of adults dying had been taken into account; any population falling below the zero base-line 0.0 cannot maintain its numbers. (See text for fuller explanation.)

Swan populations can maintain their numbers in the face of very variable adult losses. In this species, annual adult mortality varies from about 6 per cent to almost 20 per cent before the populations are unable to sustain their numbers. Note also that some populations (those with adult survival rates of about 81–82 per cent) have similar adult survival rates, but marked differences in their BPAs. The much lower productivity of the Midlands population, due to vandalism of nests, results in this population being unable to maintain its numbers with such an adult survival, compared with the other populations with better nesting success.

Both the populations that show very high adult survival have low chick production. In the Hebrides, such cygnet mortality is 'natural', in part coming about because good breeding sites (those on calcareous lochs) are in short supply. Pairs on such lochs raise 2.9 cygnets per year whereas those on the poorer lochs raise only about 1.5. At Abbotsbury, even with human assistance, the number of cygnets raised per nesting pair is usually below 2.0; presumably success would be even lower without such management. Low nesting success may well be a general feature of colonially nesting swans, since in Denmark such birds raise only 0.9 cygnets per pair compared with the 2.6 raised by pairs nesting solitarily (Bloch 1970). Notwithstanding the poor production of cygnets, the overall patterns in the Hebrides and at Abbotsbury is one of populations that are well in balance and are clearly able to withstand occasional years of poor production.

The high survival rate of adult swans in these populations raises the question of whether this is directly related to the low productivity. Are the adults finding survival easier because of reduced competition from smaller numbers of juveniles? A similar situation was reported for Great Tits *Parus major* by Kluyver (1970) on the island of Vlieland; there he removed 60 per cent of the nestlings and noticed that the survival rate of the breeding adults doubled.

Whatever the reason, it is clear that the Mute Swan populations can persist with very different demographies. Judging from the data of Fig. 9.2, the lowest adult survival rate in sustainable populations is of the order of 82 per cent. Not surprisingly, the degree of nesting success has a marked influence on the population at this level of mortality.

Theoretical considerations

In order to be able to manage populations efficiently, we need to be able to predict what will happen in the face of changes. Yet, as mentioned, there are relatively few cases where we have one reliable measure of a species' demography, let alone more; in this respect the Mute Swan data may be unique for birds living in natural or semi-natural conditions. However, there are at least some data from some hunted species, particularly some of the wildfowl and gamebirds, where Maximum Sustainable Yields have been calculated for a number of species and the concomitant survival rates given. These again show how plastic a species' demography can be. For example, the Partridge *Perdix perdix*, may have adult mortality rates ranging from 35 per cent to 72 per cent in stable populations (Potts 1986). In this species, in contrast to the Mute Swan, there does not seem to be a significant difference in survival rates between stable and decreasing populations; the mean survival rates of adults in 21 stable populations was 48 per cent, while that for 15 decreasing populations was 52 per cent. However, this species has a very high clutch size (15 or more) and Potts (loc. cit.) considered the reproductive output to be the most important factor affecting population trends in this species.

In view of the paucity of data, it seems worth asking whether there are any guidelines that can be used to predict aspects of a bird's demography. At one level there is, since some birds can be classed as '*r*' selected and others as '*K*' selected; for example, many passerines raise large numbers of chicks in a year, breed when they are 1-year-old, and have only a short life-span while some of the larger birds of prey and some seabirds lay only a single egg each year, take many years to attain maturity, and have long life-spans. The question raised here is whether, given one demographic para-meter, say the adult survival rate, one can say anything about the rest of a species' demography which can help in management. Can we predict at what point the lowering of the adult survival rates brings a population to the point where it can no longer sustain itself?

We can certainly see that there are limits below which populations are unlikely to be able to sustain themselves. One such example is shown in

Fig. 9.3, which shows, over a range of adult survival rates, the number of young that must survive to breed if the population is to remain stable. A number of assumptions has been made in producing this figure. It is assumed:

(i) the survival rate of juveniles (defined here as birds in their first year of life, from fledging to the following breeding season) is never higher than that of breeding adults;

(ii) 5 per cent of breeding adults (birds that have bred in a previous season) fail to breed in any one year (e.g. after loss of their mate or divorce); and

(iii) 10 per cent of young birds fail to breed at the first age at which breeding occurs in that species.

Figure 9.3 plots, against a range of adult survival rates, such information for species that start breeding at ages of 1 ,2, 3, and 4 years and shows

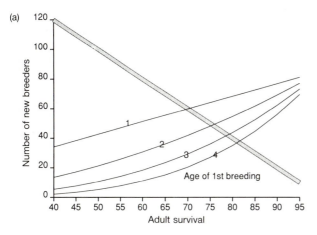

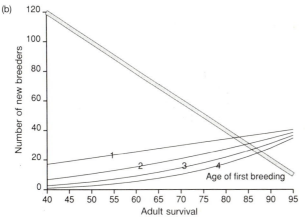

Fig. 9.3. The effect of age of first breeding on the number of young per 100 pairs that must be produced for a population to remain stable. (a) Population where, on average, each pair fledges 1.0 young; (b) Population in which only 0.5 young per pair are raised. The thin lines are the lines of maximum production; any population whose characteristics did not permit increases above the thick line would not be able to maintain sufficient numbers.

these against the number of young that must survive to breed in order to maintain stable numbers.

The levels of the curves for the numbers of surviving young are, obviously, heavily dependent on the number of young that the species concerned raises each year; this number is the product of: clutch size, number of broods, and nesting success. In the two diagrams of Fig. 9.3 it is assumed (Fig. 9.3a) that pairs raise one chick to fledging whereas in Fig. 9.3b only 0.5 chicks are raised per pair. For many of the species that lay clutches of one or two eggs and have only a single brood each year, these curves are probably at approximately the right level. For example, it is probably common for birds to have at least a 10 per cent hatching failure and at least a 10 per cent loss of nestlings; many species have higher losses than this. The losses would give maximum fledging success for birds with a clutch of one of about 0.8 young per pair. For species raising many young, such curves are obviously a misrepresentation of the demographic pattern. Figure 9.4 plots the information in a different way; Fig. 9.4a shows the BPA for birds laying clutches of 1–4 eggs while Fig. 9.4b shows the BPA for species that take from 1 to 4 years to start breeding. From these certain limitations on populations become apparent. The following are barely possible:

1 Even with recruits all breeding at age 1, it is not possible for a species with a <70 per cent adult survival rate to have a clutch of one, nor a species with an annual survival rate of <53 per cent to survive with a clutch size of two. Likewise, species with survival rates of about 43 per cent must produce three or more eggs per year.
2 For species with only a single egg and which do not start breeding until age two, three, or four the minimum survival rates are 75 per cent, 78 per cent and 81 per cent, respectively.

It should be noted that although, in these diagrams, those species that lay one egg and have survival rates in excess of about 75 per cent appear to be able to produce sufficient young 'to meet their needs', several factors bias this figure in their favour.

First, a number of species take even longer than 4 years to start breeding; for example Puffins *Fratercula arctica* often do not start breeding until 5–6 years of age, Manx Shearwaters *Puffinus puffinus* until 6–7, Fulmars 7–9, and the Sooty Albatross, which seems to hold the record, does not start until the age of 13. Figure 9.5 shows the production curves for species, such as the last two mentioned, which start to breed at seven and thirteen repectively; obviously the number of young that reach breeding age is much reduced when the age of first breeding is increased so much.

Second, in many of these longer lived species, it is common to find some members of a cohort not breeding until long after the time at which others

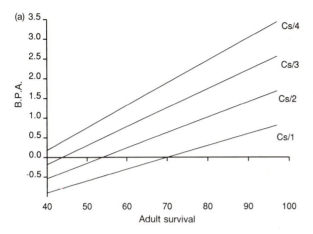

Fig. 9.4. The Balance Per Annum in populations with (a) different clutch sizes (Cs) and (b) different ages of first breeding. (See text for further explanation.)

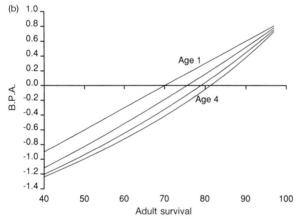

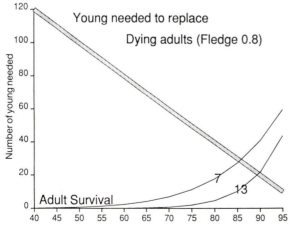

Fig. 9.5. The number of young needed to maintain a population in species with very prolonged periods of immaturity. It is assumed that each pair raises 0.8 chicks per year. (See text.)

have started to do so. It seems likely that many measures of age of first breeding may be too low because observers are more likely to record those individuals that start early, hence underestimating the average age at which breeding starts. The proportion of birds that do not breed as early as the rest of their cohort is poorly known as is the proportion of adults that fail to breed in any year, though in the Mute Swan the latter varies from 20 per cent to 30 per cent. Again, if anything, the figures used seem likely to veer on the conservative side in that it is likely that in some wild populations the proportions of non-breeders are slightly higher than those given here; if this is the case the lines showing the production needed in the figures are too low.

Third, little allowance has been made for juveniles (defined here as birds of less than 1 year of age) or immatures (defined here, in contrast to juveniles, as being birds that have survived to the age of 1 year but still have not started to breed) having lower survival rates than adults. This survival rate has been taken as being 10 per cent lower than adults in the first year, but the same as adults in later years. Both of these assumptions may result in estimates that are too high. For example, using data from species of European birds collated by Saether (pers. comm.), the average survival of first-year birds was only 67.0 per cent ($N = 59$, range 31.6 per cent to 98.4 per cent) of that of adults; a 90 per cent value has been used in the calculations given here. Figure 9.6 shows the relationship between first-year and adult survival rates.

Fourth, in the case of long-lived species which do not start breeding at the age of one, no allowance has been made for the fact that the birds in their immature years probably also have survival rates lower than those of the breeding adults. One might suggest that it is the rule that juveniles and immatures have lower chances of survival than breeding adults, despite the

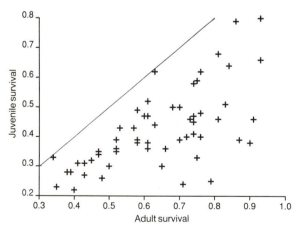

Fig. 9.6. The relationship between annual adult survival rate and juvenile survival in 56 different bird species.

dangers associated with breeding. An exception to this rule is the Red Grouse *Lagopus lagopus*, where a higher proportion of adult than juvenile males are killed by hunters because adults fly singly over the guns while the young birds fly over in flocks and hence are less likely to be killed. Two other cases are unusual; in one study, the young of the Waved Albatross *Diomedea irrorata* had survival rates from fledging to first breeding that were indistinguishable from those of the parents (Harris 1973). In the Herring Gull *Larus argentatus* in a period of population increase the survival rate of immature birds (but not juveniles) was the same as that of adults (Chabrzyk and Coulson 1976). Normally, however, this seems unusual and the survival rate of immatures tends to be lower than that of adults. Even if such survival rates of immature birds were only slightly lower than those of adults, this would lead to a considerable reduction in the number of young surviving to breed in species with many years of immaturity.

Fifth, in the longer-lived birds, individuals breeding for the first time or two often have a breeding success lower than that of the older age-groups. Indeed in the Flamingo *Phoenicopterus ruber*, discussed in this symposium, many birds start to breed at around the ages of five and six, but even at the age of 10 years they have not achieved a breeding success as high as that of older birds (the older marked birds are now 27 years old or older). As yet, the study has not continued for long enough for us to know at what age the breeding success reaches a peak.

As can be seen from the figures presented, birds with low survival rates cannot have a low clutch size. The corollary does not however, seem to hold true (Fig. 9.7). Birds with high survival rates do not necessarily have small clutches; an outstanding example of this is the Barnacle Goose *Branta leucopsis* which has a clutch of 4.5 and a survival rate of 0.96. In

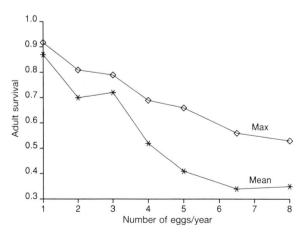

Fig. 9.7. The relationship between the number of eggs laid per year and the annual adult survival rate (for 87 bird species). The lower line shows the mean adult survival rate, the upper one the highest annual adult survival rate within the same species. Birds laying clutches of 6 and 7 eggs are lumped and all birds laying 8 or more eggs are presented with the data for 8 eggs.

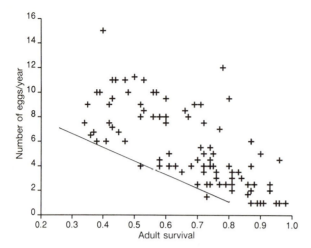

Fig. 9.8. The data from Fig. 9.7 plotted against the theoretical line for the relationship between annual adult survival and number of young that survive in stable populations.

fact, since Barnacle Geese do not breed every year, the annual clutch size will average less than 4.5. Nevertheless, there is, overall, a correlation between clutch size and survival rate as shown in Fig. 9.8.

Comparison of theory and field data

Can data of this sort be used to predict anything about wild populations for which we have insufficiently reliable data? Mortality estimates such as those described here have been used in some studies of raptors to estimate the reproductive output necessary to maintain populations (Henny and Ogden 1970; Newton 1979), but broader generalizations do not seem to have been made. Figures 9.3 and 9.4 are purely theoretical. In order to see whether they bear any relation to nature, I have taken the table of immature and adult survival rates of European birds collated by Saether (pers. comm.) (with a few additions and amendments) and added information on clutch size and number of broods per year. Figure 9.8 shows the number of eggs laid per year (clutch size × number of broods) plotted against adult survival. It must be stressed that these data are based on published sources where the survival rates have been estimated in a wide variety of different ways. Survival rates are, at least at times, difficult to calculate (Clobert and Lebreton, this volume); in particular some calculations based on ringing recoveries have been shown to be lower than those derived from local studies of breeding populations. Hence it would not be too surprising to find that some of the survival rates used here were not very accurate and, probably, were on the low side.

A cut-off line has been drawn along the lower edge of the points (the two points below the line are Snipe *Gallinago gallinago* with 52 per cent survival and Sandwich Tern *Sterna sandwicensis* with 73 per cent survival of adults). These data show a similar cut-off in the low clutch/low survival area as the theoretical model in which birds laying only a single egg per year have survival rates in excess of 85 per cent, those that lay two have survival rates in excess of about 80 per cent, and those that lay three have survival rates of 70 per cent or more. In view of the potential for errors and biases in the data, the agreement between observation and theory is reasonably good. Hence Fig. 9.8 provides some support for the suggestion that there may be limits to the product of clutch size and survival rates.

The chances of any population being able to maintain its numbers is obviously greatly affected by the reproductive success. As can be seen in Table 9.1 and Fig. 9.2, Mute Swans in the Midlands and Oxford areas have similar adult survival rates, but those in the Midlands had a much lower reproductive output, largely as a result of vandalism of nests. Subsequent rates of survival for the remaining young in the two areas seem very similar. Not surprisingly however, the Midlands population was declining faster than the Oxford one. Similarly, in the Great Tit *Parus major*, a bird with a high reproductive output, unusually heavy nesting losses seem to affect the population's ability to increase (McCleery and Perrins this volume).

This raises the question of by how much it is possible for juvenile or immature survival to increase to compensate for increased adult mortality if this should increase. Again we only really have any indication about such figures from gamebird and waterfowl populations and even in these we seldom have much information about unhunted populations.

1 Juvenile or immature survival cannot rise above that of adults *even* when that of adults has decreased. This is because, where man is responsible for mortality, e.g. by shooting, juveniles are usually more susceptible than adults, at least in part, because they are more naive about hunters (though see example of Red Grouse cited above). A possible exception, though I do not know of an example, would be where juveniles or immatures moved, say for winter, to a safer place than the adults.

2 In natural situations, the problem seems likely to be that, if adult mortality increases then so will that of the juveniles and immatures. Hence, there is every likelihood that a population will be put in jeopardy by an increase of mortality amongst adults. Again the most likely exception would be if the juveniles or immatures were for some reason in safer conditions than the adults. Some sort of heavy predation pressure, or disease, in breeding colonies of colonial birds might yield such a situation where the mortality of adults increases greatly, but that of the younger birds does not. This example would not of course be likely to lead to long-term stability with different demographic characters, since high adult

losses in colonies would be associated with very poor reproductive output which would in itself have serious affects in the longer term; the breeding populations would be having catastrophic breeding failures and future cohorts would be small.

Hence it is unlikely that juvenile survival is every likely to increase sufficiently to compensate for more than a small amount of extra adult mortality; indeed it is more likely that if adult survival decreases then that of the juveniles is likely also to decrease, in parallel.

There is a further point concerning very long-lived species, which is that a small reduction in the survival rates of adults requires a proportionately much larger increase in the survival of juveniles in order to maintain stable numbers. For example, if a species with a 95 per cent survival rate encounters conditions that lead to a reduction in survival rate of 1 per cent, the numbers of young that must survive to maintain numbers changes from 0.1 per pair to 0.12, an increase of 20 per cent.

Do endangered species have different demographies?

From the point of view of conservation, it is important to know whether we can use this knowledge of demographic patterns to see where the problems lie with endangered species and, in particular, to try and pinpoint the stages of the life cycle which are the crucial ones for any given species. However, again we have too few data at present to be able to suggest effective conservation strategies.

The demographies of the endangered or declining species discussed by Green and Hirons (1990) in this symposium have been compared with the data for those species on which Fig. 9.8 is based; these are largely species whose populations are holding their own. The results are shown in Table 9.2. The endangered species are significantly longer lived than the main group. This is not unexpected since K-selected species are more slow to replace lost numbers after a disaster. In addition, and associated with longer lives, they tend to lay small clutches. More interestingly, perhaps, there seems to be a tendency for the survival rates of the juveniles of these species to be lower, compared with that of the adults of their species, than is the case in the main group (59 per cent of that of adults compared with 67 per cent for the main group), but this difference is not significant $(P = 0.17)$. It must be emphasized that such comparisons need to be made with extreme caution. Not only are the two groups compared here dissimilar in their positions on the $r-K$ continuum but also, for comparison, populations should be stable and there is every likelihood that some of the endangered species are not. None the less the observed discrepancy might be taken as suggesting that the most 'fragile' part of the life cycle of the endangered species was that

Table 9.2 Comparison of demographic features of endangered species with those of other birds (see text for explanation)

	Endangered species	Others
N	13	59
Mean annual adult survival rate	0.85 $P = \pm 0.001$	0.67
Mean no. eggs per year	2.44 $P = \pm 0.001$	5.60
Mean juvenile survival	0.53 $P = .22$	0.43
Juv. survival/adult survival	0.59 $P = .17$	0.67

of the survival of young birds from fledging to becoming established in the breeding population.

In conclusion, therefore, the published data do begin to provide information from which it may be possible to extrapolate to other species. However, we badly need more detailed demographic studies, on both common and rare species, before we can be certain that there are genuine differences in their demographies.

Summary

Using data from several studies of the Mute Swan, it is shown that the species can maintain stable populations with a wide range of different demographies; in this species populations can remain stable with adult losses varying from 6 per cent to 18 per cent per annum.

Some simple models are produced of the range of demographies that are possible in birds. These are compared with published data of both common and endangered species: the models are quite well supported by the data. However, the samples are too small to allow reliable deductions to be made about characteristics of endangered species, though there may be lower survival of young birds during the first year.

Acknowledgements

I am grateful to B.-E. Saether and R. Green and G. Hirons for permission to use their data and to Nicholas Aebischer, Ian Newton, and Jean Sears for helpful comments on the manuscript.

References

Birkhead, M. E. (1982). Population ecology and lead poisoning in the Mute Swan. D.Phil Thesis. Oxford University.

Birkhead, M. E. & Perrins, C. M. (1985). The breeding biology of the Mute Swan on the River Thames with special reference to lead poisoning. *Biology and Conservation*, **31**, 1–11.

Bloch, D. (1970). [The Mute Swan *Cygnus olor* breeding in a colony in Denmark.] *Dansk. Orn. Foren. Tidssk*, **64**, 152–62. [In Danish with English Summary.]

Chabrzyk, G. and Coulson, J. C. (1976). Survival and recruitment in the Herring Gull *Larus argentatus*. *Journal of Animal Ecology*, **435**, 187–203.

Coleman, A. E. and Minton, C. D. T. (1980). Mortality of Mute Swan progeny in an area of south Staffordshire. *Wildfowl*, **31**, 22–8.

Green, R. and Hirons, G. (1990). The relevance of population studies to the conservation of threatened birds. In *Bird population studies: Relevance to conservation and management* (ed. C. M. Perrins, J.-D. Lebreton, and G. J. M. Hirons), pp. 594 633. Oxford University Press, Oxford.

Harris, M. P. (1973). The biology of the Waved Albatross *Diomedea irrorata* of Hood Island. *Ibis*, **115**, 483–510.

Henny,C. J. and Ogden, J. C. (1970). Estimated status of Osprey populations in the United States. *Journal of Wildlife Management*, **34**, 214–17.

Kluyver, H. H. (1970). Regulation of numbers in populations of Great Tits (*Parus m. major*). *Proc. Adv. Study Inst.* Dynamics numbers of populations (Oosterbeek, 1970), pp. 507–23.

Newton, I. (1979). *Population ecology of raptors*. T. & A. D. Poyser, Berkhampstead.

Perrins, C. M. and Ogilvie, M. A. (1981). A study of Abbotsbury Mute Swans. *Wildfowl*, **32**, 35–47.

Perrins, C. M. and Reynolds, C. M. (1967). A preliminary study of the Mute Swan, *Cygnus olor*. *Wildfowl Trust Annual Report*, **18**, 74–84.

Potts, G. R. (1986). *The Partridge*. Collins, London.

Sears, J. (1986). A study of Mute Swans in relation to lead poisoning. D.Phil Thesis, University of Oxford.

Spray, C. (1981). An isolated population of *Cygnus olor* in Scotland. *Proceedings of the 2nd International Swan Symposium*, Sapporo, Japan. International Wildfowl Research Bureau, Slimbridge, England.

10 Population studies of White Storks (*Ciconia ciconia*) in Europe

FRANZ BAIRLEIN

Introduction

Population studies of the White Storks have been made for a long time. Locally, annual counts were carried out at the turn of the century; more extensive work was begun in the 1930s. In 1934, the first international census was run, covering considerable areas of the European breeding range (Schüz 1936, 1940). This was subsequently followed by further international counts in 1958, 1974, and 1984 (summarized by Boettcher-Streim and Schüz 1989; Sauter and Schüz 1954; Schüz 1980; Schüz and Szijj 1960, 1975). Thus, long-term, concurrent information on breeding pairs is available for many parts of the species' breeding range. These censuses revealed a general decrease in stork numbers in Europe, although the decline was not continuous and rather different changes occurred in various parts of the species' range. In general, storks breeding west of the migratory divide (Fig. 10.1) and migrating to W. African winter quarters sharply decreased or even became extinct, whereas the eastern population (breeding east of the migratory divide, Fig. 10.1) appears to be relatively stable or even locally increasing in size and extending its breeding range. Besides these international censuses comprehensive, long-term annual counts of breeding storks are available from some regions, including ringing data and annual data on breeding success, so allowing a more detailed analysis.

The information presented in this paper comprises

(i) a brief review of the changes in White Stork numbers in the light of the four international censuses; and
(ii) more detailed analyses of changes in fecundity and survival from the most comprehensive long-term regional counts, in order to establish reasons for the observed decline in the European White Storks and to evaluate consequences for management policy.

Material

The sources of data for this review are the summaries of the international stork censuses (see above) and many regional and local reports on stork

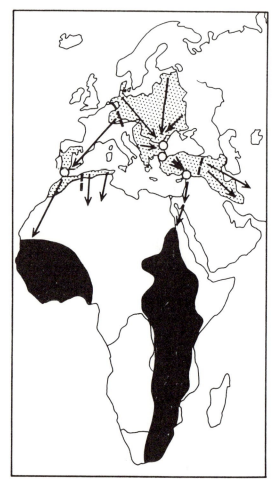

Fig. 10.1. Breeding range (dotted areas), migratory routes (arrows), and wintering grounds (black areas) of White Storks. The broken line in central Europe indicates the migratory divide; the open circles show important migratory bottlenecks (from Hölzinger 1987).

counts which cannot be completely cited here. Most of them are listed in Creutz (1985a), Schüz (1986), and Schüz and Zink (1955). Further unpublished information from spot checks was provided by G. Rheinwald.

These international censuses covered almost all of the breeding distribution and, in many areas, particularly in the first censuses, the basic questionnaires on storks were filled in by non-ornithologists, e.g. school teachers or postmen. Thus, the accuracy of the counts and between-census comparisons may be a bit biased.

Long-term population studies of breeding White Storks carried out intensively in some areas and including data on breeding success, allow a more detailed analysis of population growth and the possible underlying factors. Some of the most comprehensive counts are considered in this review, using both the published information and processing of unpublished recent updates, provided by W. Assfalg, H. R. Henneberg, and R. Schlenker. These long-term counts and their unpublished supplementary data were statistically processed using regression analyses, arithmetic means, and the appropriate statistical tests.

In almost all White Stork censuses the following abbreviations are used: **HPa**: number of pairs occupying a nest; **HPo**: number of pairs without fledglings; **HPm**: number of pairs with at least one fledgling; **JZG**: total number of fledged young in the population; **JZa**: average number of fledglings per pair (JZG/HPa); **JZm**: average number of fledglings per successful pair (JZG/HPm).

Results

Population trends

For areas counted at least three times in the international censuses of 1934, 1958, 1974, and 1984, the changes in number of breeding pairs (HPa) are shown in Fig. 10.2. In most areas, data are available from 1958 onwards. Thus, the data of Fig. 10.2 are shown as changes relative to the pairs counted in 1958.

In many areas the numbers in 1934 were exceptionally high compared with counts in any other period this century, corresponding to a general increase in stork numbers in Europe in the 1930s, though even higher numbers are reported from some areas at the turn of the century (Creutz 1985*a*; Dallinga and Schoenmakers 1987).

If we compare the changes in stork numbers in the three major groups of storks with respect to their different migratory behaviour (cf. Fig. 10.1), fairly different changes are obvious. Within the western population, the numbers of breeding storks declined drastically during the census periods. In the Netherlands (referred to here as belonging to the western population although some birds migrate to the east), in S. W. Germany, and in France the numbers of breeding storks have declined greatly. The same situation is also obvious in the more detailed regional population counts, some of which are shown, as examples, in Figs. 10.3–5. In the Netherlands, Alsace, and Baden-Württemberg the recent increase in stork numbers is related to the release of birds from captivity. In Belgium, Switzerland, Alsace, and parts of S. W. Germany the wild population has become almost extinct during the past decades.

The Iberian White Stork numbers also declined sharply, though less dramatically than those in western central Europe (Bernis 1981; Chozas 1986*a*).

Storks breeding in Denmark (Dybbro 1972) and N., central, and S. E.

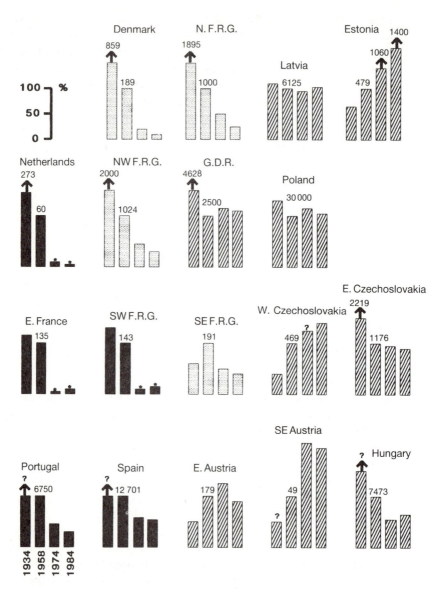

Fig. 10.2. Changes in numbers of breeding pairs (HPa) of White Storks in various parts of Europe, as obtained by the four international counts in 1934, 1958, 1974, and 1984. Variation in numbers is shown as a percentage of the population size in 1958 (number above the 1958 bar = 100%). Numbers above the arrows show the corresponding number of HPa in that particular year. Black bars, western populations (due to the migratory divide, cf. Fig. 10.1); stippled bars, 'mixed' populations; hatched bars, eastern populations. Stars indicate that the figure includes birds released from captivity.

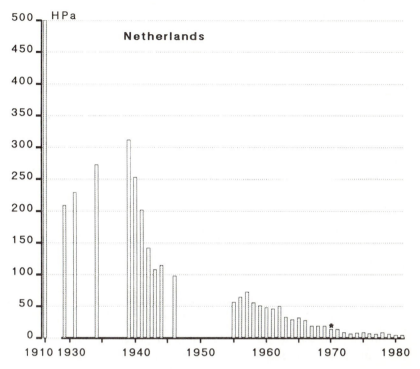

Fig. 10.3. Annual counts of breeding White Storks (HPa) in the Netherlands (after Dallinga and Schoenmakers 1984). The star marks the first year of breeding for storks that were released from captivity.

Germany in the 'mixed population', i.e. at both sides of the migratory divide where breeders may use either route, have also shown a marked and steady decrease in the number of breeding pairs since 1934 or at least 1958. The Swedish population became extinct in the 1950s.

The fluctuation of stork numbers in the Oldenburg area (N. W. Germany), which has been counted almost annually since the 1920s (Fig. 10.6; Tantzen 1962; H. R. Henneberg, personal communication) reflects quite well the general situation in central Europe (Creutz 1985*a*; Dallinga and Schoenmakers 1987). An increase in the 1930s, after a sharp decrease between 1910 and 1928, was followed by a marked decline until the early 1950s. After a period of stability or even increase the stork population in central Europe has been steadily decreasing since the early 1960s.

Farther east, however, in the 'eastern population' the situation is rather different. In the German Democratic Republic a sharp population decline occurred between 1934 and 1958, followed by a period of stability (Creutz

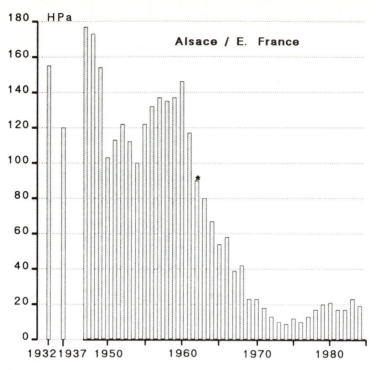

Fig. 10.4. Annual counts of breeding White Storks (HPa) in Alsace/France (after Schierer 1986). The star indicates the first year with storks bred from captivity.

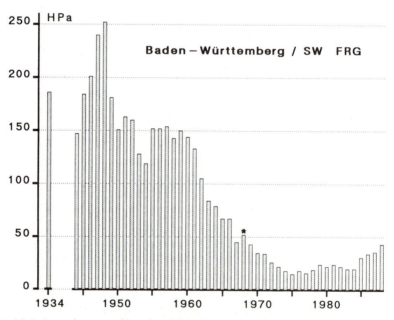

Fig. 10.5. Annual counts of breeding White Storks (HPa) in Baden-Württemberg/S.W. Germany (after Bairlein and Zink 1979; Schlenker 1986; W. Assfalg and R. Schlenker, personal communication). The star indicates the first year with storks from captivity.

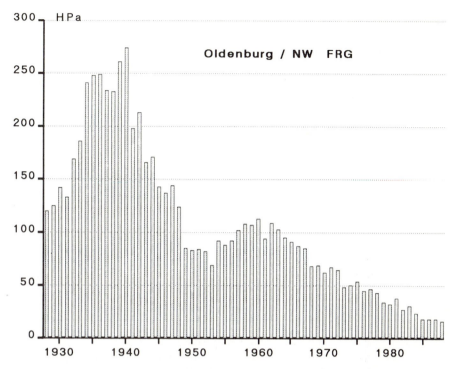

Fig. 10.6. Annual counts of breeding White Storks (HPa) in Oldenburg/N.W. Germany (after Tantzen 1962; Schüz and Szijj 1975; H. R. Henneberg, personal communication).

1985*b*). Regional long-term studies in the GDR show the same general trend (Fig. 10.7; see e.g. Dornbusch 1982; Zuppke 1982; Kaatz and Stachowiak 1987).

Overall, both the Polish and Latvian populations changed only slightly, although there were significant regional variations. In Poland, the southern populations are increasing, whereas the northern populations are declining (Jakubiec 1985). In the Latvian Republic, the populations in the west are decreasing whereas the northern populations are increasing. A pronounced increase has occurred in Estonia, where the entire population has increased about threefold, associated with an expansion of the breeding range to the NE (Veroman 1962, 1987; Veroman and Grempe 1984). Increasing numbers of White Storks are also reported from the western part of Czechoslovakia (Rejman 1986, 1988), and from E. and S.E. Austria (Triebl and Frühstück 1979; Weissert 1986; see Fig. 10.8), although in the Lake Neusiedl area (E. Austria in Fig. 10.2) a recent decrease has taken place. The populations in eastern Czechoslovakia

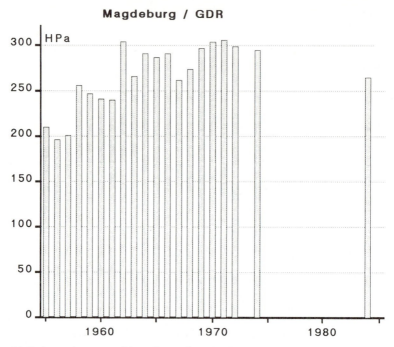

Fig. 10.7. Annual counts of breeding White Storks (HPa) in the area of Magdeburg, East Germany (after Schüz and Szijj 1960, 1975).

(Rejman 1986, 1988) and Hungary (Jakab 1986, 1987) declined after 1934, followed by a stabilization over recent decades. The latter may be, however, a consequence of improved counting efficiency in the most recent censuses.

Scattered information is also available from other areas. The number of breeding White Storks in the southern European USSR is increasing, whereas those of Rumania, Yugoslavia, Greece, and Turkey seem to be decreasing continuously (Boettcher-Streim and Schüz 1989; Fiedler 1986; Hölzinger and Künkele 1986; Klemm 1983; Kumerloeve 1988).

Only a few reports are available from northern Africa. In Morocco, about 24 000 breeding pairs were estimated in 1935, whereas in 1969 and 1974 only 15 000 and 13 500 pairs, respectively, were estimated (Ruthke 1986). The Algerian population increased between 1935 and 1955 from 6500 pairs to 8800 pairs, but declined subsequently to 2000 pairs in 1974 (Ledant *et al*. 1981). The Tunisian population decreased until 1973 at an annual rate of about 15 per cent and appeared to stabilize thereafter (Lauthe 1977).

In summary, the entire White Stork population, both in Europe and northern Africa, shows a declining trend, although there is a pronounced

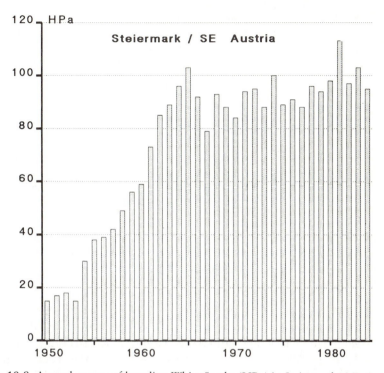

Fig. 10.8. Annual counts of breeding White Storks (HPa) in Steiermark, S.E. Austria (after Weissert 1986).

variation in population changes in various regions, and even an expansion in breeding range and an increase in numbers of breeding pairs in some eastern regions. However, that increase may not compensate for the decrease in stork numbers across many parts of the species' breeding range. The population of western central Europe has almost totally disappeared.

Changes in fecundity, age of first breeding, and mortality

Fecundity

As shown in Table 10.1, periods with declining stork numbers in Oldenburg (1930–49, 1960–88; cf. Fig. 10.6) were associated with a higher proportion of unsuccessful pairs, though not significantly in the latter period, and a corresponding decrease in the average productivity per pair (JZa). During the decline since 1960, the productivity per pair, either JZa or JZm, was very low, although the proportion of HPo changed less. Thus, in addition to a decrease in numbers, there has also been a decrease in fecundity. This is also shown by the correlation coefficients among the

Table 10.1 Breeding success and reproduction of White Storks in Oldenburg/N.W. Germany for different time periods in relation to the pattern of population growth (cf. Fig. 10.1). For the complete set of years (1928–88, last row), rank correlation coefficients (r) are shown, indicating the long-term trends in variation of reproductive success. Statistically significant variation either between subsequent periods or of correlation coefficients are indicated by asterisks.

Period	Annual population change (%)[1]	% Unsuccessful pairs (% HPo)	Mean annual total of fledged young per pair (JZa)	Mean annual total of fledglings per successful pair (JZm)
1928–39	+7.8***	25.8 ± 3.5	2.11 ± 0.15	2.82 ± 0.12
1930–49	−10***	41.3 ± 6.0*	1.63 ± 0.18(*)	2.75 ± 0.04
1950–9	+3.6**	33.2 ± 2.0	2.03 ± 0.10(*)	3.03 ± 0.08**
1960–88	−6.8***	35.1 ± 2.3	1.58 ± 0.08***	2.41 ± 0.07***
1928–88		0.193	−0.406**	−0.505***

[1] derived from a regression of ln (HPa) on time (year).
(*) $0.10 < P > 0.05$; *$P < 0.05$; **$P < 0.01$; ***$P < 0.001$.

complete data set: productivity decreased significantly, whereas the proportion of unsuccessful pairs (HPo) increased slightly and was not significant. Similar relationships were found in the Netherlands (Table 10.2), in Alsace (1947–60 versus 1961–74; Table 10.3), and in Baden-Württemberg (1948–60 versus 1961–75; Table 10.4); and also in Denmark (though not significantly, Dybbro 1972), in two other long-term counts in S. Germany (Rheinland-Pfalz, W. Bavaria; F. Bairlein, unpublished, A. Burnhauser, personal communication), and in Spain (Bernis 1981; Chozas 1986*b*). Thus, there is a clear-cut relationship between average annual fecundity and the annual population changes in various parts of Europe.

Interestingly, the annual productivity of storks in Alsace and Baden-Württemberg is lowest in the last period with a higher proportion of

Table 10.2 Breeding success and reproduction of White Storks in the Netherlands (after Table 3b in Dallinga and Schoenmakers 1984). The variables altered significantly across the time periods

Period	Annual population change (%)	% HPo	JZa	JZm
1931–40	+3.8[1]	20.1 ± 3.4	2.44 ± 0.15	3.03 ± 0.08
1941–50		36.4 ± 4.5	1.87 ± 0.15	2.92 ± 0.07
1955–65	−8.2[2]	34.0 ± 4.0	1.87 ± 0.12	2.87 ± 0.10
1966–80		44.3 ± 5.5	1.39 ± 0.16	2.47 ± 0.15

[1] 1931–9; [2] 1939–72.

Table 10.3 Breeding success and reproduction of White Storks in Alsace/France. For explanation see Table 10.1

Period	Annual population change (%)	% HPo	JZa	JZm
1947–60	+/−	14.4 ± 2.3	2.66 ± 0.09	3.08 ± 0.05
1961–74	−19.8***	19.8 ± 2.3	2.36 ± 0.09*	2.94 ± 0.08
1975–88	+6.9**	46.5 ± 6.1***	1.44 ± 0.19**	2.65 ± 0.11(*)
1947–88		0.621***	−0.654***	−0.476***

+/− = no significant trend.

Table 10.4 Breeding success and reproduction of White Storks in Baden-Württemberg/S.W. Germany. For explanation see Table 10.1

Period	Annual population change (%)	% HPo	JZa	JZm
1948–60	+/−	18.5 ± 1.8	2.46 ± 0.08	3.02 ± 0.05
1961–75	−14.4***	29.4 ± 2.7**	2.03 ± 0.12**	2.85 ± 0.04
1976–88	+6.5*	35.5 ± 4.6	1.64 ± 0.13*	2.54 ± 0.08*
1948–88		0.550***	−0.652***	−0.551***

+/− = no significant trend.

unsuccessful pairs when birds were released from captivity for breeding in the wild (see below).

Data on the productivity of eastern storks are relatively scarce, particularly those breeding in the sharply increasing N.E. populations. Some of the data derived for eastern storks are summarized in Table 10.5. Between different periods, variation in breeding success is less pronounced than in the western populations. Moreover, comparing the fecundity, in particular JZa, of eastern and western storks in recent decades, these data indicate that, on average, the productivity tends to be higher within the increasing eastern populations than in the declining western populations.

Age of first breeding

A further change that may affect the population growth was found in the age of first breeding. In White Storks this differs between various parts of the breeding range, although the reasons are not known. Age of first breeding is considerably higher in northern parts of the range than in S.W. Germany and Alsace where most storks start breeding at an age of 3 (on average 3.6 years compared to 4.9 years farther north; Meybohm and Dahms 1975; Zink 1967).

Table 10.5 Breeding success and reproduction of White Storks in some eastern parts of the European breeding range. For explanation see Table 10.1. + = increase; +/− = no significant trend.

Area	Period	Annual population change (%)	% HPo	JZa	JZm
Magdeburg/GDR[1]	1955–74	+2.1***	22.1 +1.8	2.07 +0.06	2.66 +0.06
Silesia/PL[2]	1928–34	+	15.5	2.71	3.21
	1973–78	+	22.3	2.01	2.59
Estonian SSR[3]	1954–61	I	24.0 +1.9	2.22 +0.10	2.92 +0.06
Burgenland/	1948–58	+	31.4 ± 4.6	1.94 ± 0.20	2.81 ± 0.12
E.Austria[4]	1963–70	+/−	23.0 ± 3.3	2.00 ± 0.11	2.59 ± 0.17
Steiermark/	1950–65	+13.9***	30.9 ± 3.2	1.90 ± 0.11	2.74 ± 0.06
S.E. Austria[5]	1966–84	+/−	29.4 ± 2.2	1.82 ± 0.08	2.62 ± 0.05

[1] Schüz and Zijj (1960, 1975); [2] Profus (1986); [3] Veroman (1962); [4] Sauter and Schüz (1954); Schüz and Szijj (1960), Triebl and Frühstück (1979); [5]Schüz and Szijj (1960), Weissert (1986). ***$P < 0.001$.

In S.W. Germany, the age of first breeding was found to be significantly higher during the period with decreasing numbers (1961–76) compared to the previous period when the population was fairly constant (1955–60; Table 10.6; Bairlein and Zink 1979). The reasons for that change are unknown. However, White Storks, in particular males, are highly philopatric (Zink 1963, 1967) which may reduce the chances of finding a mate in settling first-breeders.

Mortality

As suggested by Lack (1966) and recently further evaluated by Dallinga and Schoenmakers (1984) and Kanyamibwa et al. (1989) the annual output of fledglings seems to have only a minor influence on subsequent changes in the breeding populations. Changes in annual mortality, either of young between fledging and sexual maturity (mainly from their third

Table 10.6 Age of first breeding of White Storks in Baden-Württemberg/ S.W. Germany

Period	Age of first breeding (years)	Proportion of 2- and 3-year-old first breeders (%)
1955–60	3.49 +0.12 ($n = 68$)	60.3
1961–76	3.78 +0.09 ($n = 112$)	44.6
P ÷	0.05	0.05

year of life onwards; see above) or of adults or of both may be more critical.

In some populations, sufficient information from ringed storks is available for the analysis of annual mortality rates, either obtained from recoveries or from resightings of marked birds. As summarized in Table 10.7, average annual mortality rates were higher in periods of declining numbers. The feature is consistent across the three populations considered, although the methods of estimation of mortality rate were different.

Table 10.7 Changes in average annual mortality of first-year birds (1y) and older ones (2y-) in the western population

Area	Period	Annual mortality		Source
		1y	2y-	
S.W. Germany	1950–9	0.60	0.26	Bairlein and Zink (1979)
	1960–76	0.74	0.36	
Alsace	1950–9	0.52	0.25	Lebreton (1981)
	1960–71	0.61	0.40	
The Netherlands	1912–35	0.68	0.31	Dallinga and
	1936–55	0.73	0.50	Schoenmakers (1984)

Furthermore, Kanyamibwa *et al.* (1989) found a relationship between the annual adult survival rates of storks in Alsace and the environmental conditions in the W. African Sahel-zone: the decrease in adult survival could be linked to the severity of the drought in the winter quarters, most probably via the supply of the migratory locusts—one of the most important foods for wintering storks in sub-Saharan Africa which is related to rainfall. Dallinga and Schoenmakers (1987) also mentioned a close relationship between the supply of locusts in Africa and the changes in number of breeding storks in Europe.

Besides these severe effects of the African food resources on stork survival, some other factors are considered to be important (e.g. Riegel and Winkel 1971). Most recoveries of storks record the cause of death as collision with overhead power lines which affect especially the inexperienced young birds (Fiedler and Wissner 1986). In S.W. Germany, more first-year storks were found killed by power lines after 1960, when the stork numbers decreased drastically (52 per cent compared to 36 per cent before 1960; Bairlein and Zink 1979). Furthermore, many recoveries of storks from outside the breeding range can be linked to direct persecution by man. Other factors include intraspecific aggression, predators, collisions with vehicles, falling down factory chimneys, bad weather, and pesticides. According to Zink (1967) mortality rates seem to be similar on both sides of the migratory divide. However, analysis of data from recent years comparing survival of 'eastern' and 'western' storks is still lacking.

Most of the variation in stork numbers in western Europe appears to be related to the reduced survival in recent decades (Lebreton 1978; Kanyamibwa *et al*. 1989), which seems to be particularly influenced by the ecological situation in Africa.

Discussion

Possible causes of decline

In S.W. Germany, the variations in annual mortality, fecundity, and age of first breeding between years of constant stork numbers (1950–60) and years with declining stork numbers (1961–74) explained almost all of the annual loss of breeding pairs (Bairlein and Zink 1979). However, the relative importance of possible factors for the variation in stork numbers and, in particular, for the rapid decline in many areas is not clearly understood. A detailed multivariate analysis, not so far undertaken, may help to disentangle possible combined effects.

As shown above, changes in population size differ in the different parts of the breeding range. However, the reasons for such regional differences are not fully established. Besides some changes in climate since the beginning of the last century (Dallinga and Schoenmakers 1984), there have been various changes in agricultural practice resulting in habitat changes and reduced food supplies within the breeding range; these differences as well as regional factors occurring during migration and in the African wintering grounds are considered to be the major causes of the difference in the population changes between the western and eastern populations (e.g. Dallinga and Schoenmakers 1987; Heckenroth 1986; Lack 1966). The western population arrives in its winter range in W. Africa at the end of the rainy season, whereas the eastern population, wintering in eastern Africa, follows the inter-tropical rain-front, which offers a much better source of food (Dallinga and Schoenmakers 1984).

As shown by Dallinga and Schoenmakers (1987) and Kanyamibwa *et al.* (1989) fluctuations in numbers of breeding pairs in Europe are closely related to locust plagues in Africa. Severe drought and subsequent over-grazing and desertification of the Sahelian wintering sites may have reduced the vegetation and the abundance of insects. As reported by Kanyamibwa *et al*. (1989) the amount of rainfall in the Sahel zone of W. Africa has decreased significantly over the past 35 years. Furthermore, the use of pesticides and, in particular, locusticides may have significant effects on the storks by lowering the food supplies. Direct effects of pesticides on storks are less obvious, although effects may include subsequent deposition of body reserves for migration and breeding, fertility or eggshell thickness and, as a consequence, breeding success (Conrad 1977).

The most important factor in Africa is seen in recent decades in the reduction of food supplies, either by drought, locust control, or habitat destruction by man (including overgrazing). Therefore, it may be suggested that wintering storks are prevented from acquiring sufficient body reserves for return migration and breeding, which affects timing of migration, subsequent breeding success, and survival. For example, delayed return to the breeding grounds in the Oldenburg/N.W. Germany area resulted in an increased proportion of unsuccessful pairs (HPo) and lowered productivity in recent decades (e.g. Dallinga and Schoenmakers 1984; Lack 1966).

However, even if the feeding conditions in Africa are an important factor, it should be emphasized that various breeding populations, though wintering in the same Afrotropical region, are not equally affected (see above). Thus, factors occurring during over-wintering in Africa do not appear to account for all observed variation between the various breeding populations. Within the western population both the N. African and the Iberian storks decreased less than those in the Netherlands, Alsace/France, and S.W. Germany, respectively, though all of them winter in W. Africa. Furthermore, only some of the eastern populations show a decrease, others are still stable or even increase. Consequently, some other reasons must be involved in the recent decline of European storks.

It is obvious that stork numbers decreased particularly in those areas where, irrespective of their wintering grounds, drastic changes and intensification in agriculture occurred in the breeding areas, e.g. Netherlands, Upper Rhine Valley, N. Germany, and Denmark. Thus, a rather important factor may be the intensification of agriculture (e.g. agricultural practice, use of fertilizers and pesticides) and the alteration of feeding habitats for storks in many parts of Western Europe within the past few decades. Drainage of wetlands as well as the lowering of water-tables resulted in loss of feeding grounds and these alterations coincide with the loss of breeding storks (e.g. Löhmer 1974). For example, lowering of water-tables may affect the availability of earthworms, which are considered to be an important food for rearing newly-hatched young (Creutz 1985a). Thus, decreasing food supplies on the breeding grounds can dramatically affect the breeding success, resulting in the observed long-term negative trends in JZm (see Tables 10.1–5). In contrast, therefore, as recently reported by Schneider (1988) for parts of N. Yugoslavia in areas where good alluvial wetlands still exist, one continues to find both maximum densities and good breeding success of White Storks. Moreover, the recent extension of pasture land, especially clover and alfalfa cultures in S. Poland coincides with the increase in stork numbers there (Profus and Mielczarek 1981).

Weather, although sometimes seriously affecting the annual breeding success (e.g. Zink 1967), seems to play a minor role in either the long-term decline of stork numbers (Kanyamibwa *et al*. 1989) or the regional differences in population growth. A further (minor) cause for the decrease of stork numbers may be the extension of overhead power lines since the 1950s (see above), which may have taken place to different extents in various parts of Europe. On the other hand, transmission pylons became an important nesting site for storks in the German Democratic Republic (Creutz 1986), in parts of Czechoslovakia (Rejman 1986), as well as in Poland (Jakubiec 1985) and Hungary (Jakab 1984).

Another factor involved in the continuous decline of stork numbers, though not the primary one and as yet little investigated, may result from the decreasing density of breeding pairs. Small populations run the risk of genetic deterioration which can lead to reduced fertility in adults, increased juvenile mortality and, consequently, increasing risk of extinction (Burgman *et al*. 1988). Furthermore low density and, hence, increased nearest-neighbour distances may affect pairing, resulting in delayed age of first breeding and decreased reproduction (Bairlein and Zink 1979).

Among the confusing puzzle of possible reasons for both the recent decline and the regional differences in changes of numbers of breeding White Storks in Europe and northern Africa, alterations of the feeding conditions in both the breeding and wintering grounds and in stopover sites during migration have to be considered as the primary factors responsible. Thus, the main aim of preservation programmes must be the conservation or even the restoration of adequate suitable feeding habitats and their management (Goos 1977; Heckenroth 1986; Hölzinger 1986; Melchior 1987; Sackl 1987). Schneider (1988) describes characteristics of a good stork habitat for breeding as permanent pastures that are periodically flooded, meadows that are cut only once (after the end of June), and some open water near the nesting site for collecting water for the young. If necessary, artificial nest sites may be offered successfully (e.g. Hornberger 1956). During migration and wintering the storks depend on a variety of large insects and small vertebrates particularly in the northern savannas. Maintaining habitats with that food source may further support many other migrating terrestrial birds, e.g. raptors (Thiollay 1985).

Reintroduction of storks

In 1948, in view of the extinction of the Swiss White Stork population (Fig. 10.9), M. Bloesch started to breed storks in captivity, to keep until maturity, and to release for subsequent breeding in the 'wild' (for summaries on these efforts see Bloesch 1980; Boettcher-Streim 1986). Storks for breeding were primarily taken from Algeria. In 1960, the first free-ranging pair of such storks bred in Switzerland. Tremendous efforts in breeding and releasing of storks, either at the 'mother-station' in Altreu or

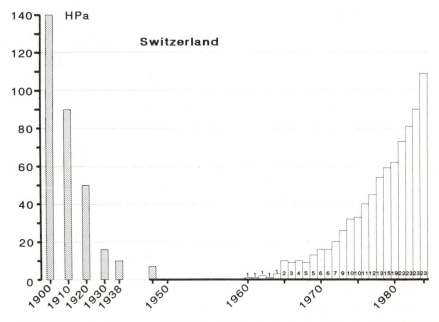

Fig. 10.9. The development of the White Stork population in Switzerland. The wild population (stippled bars) died out in the late 1950s. Increasing efforts at reintroduction resulted in the recent increase in numbers of breeding pairs, released from captivity (open bars). The numbers in the bars show the numbers of breeding stations involved in the reintroduction scheme (after Bloesch 1980; Boettcher-Streim 1986).

at various 'satellite-stations' spread over Switzerland, resulted in a rapid increase in breeding storks in Switzerland (Fig. 10.9). Furthermore, birds from Swiss enclosures settled in Alsace and S.W. Germany. Stimulated by that 'success', conservationists established their own reintroduction schemes in Alsace, S.W. Germany, Belgium, and the Netherlands. Recent stabilization or even an increase in stork numbers in these areas (cf. Figs. 10.4 and 10.5) are due to such released storks. In Alsace and Baden/S.W. Germany, almost all of the White Storks, recently breeding in the wild, are of captive origin (Fig. 10.10).

Although conservationists may be excited by the reintroduction of storks as a 'last hope to stop the decline of the White Stork' (Schulz 1986), one should be concerned objectively about the biological background of such reintroduction schemes, in particular when using introduced birds of completely different origin and with different behaviour, e.g. timing and route of migration. The reproductive success of White Storks breeding in the wild in Switzerland (on average 1.86 JZa; data reported in Bloesch 1980; Boettcher-Streim 1986) is low compared to the values obtained in

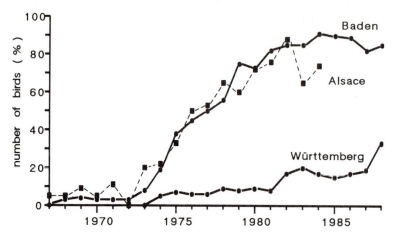

Fig. 10.10. The proportion of White Storks released from enclosures and now breeding in the 'wild' in Alsace and Baden-Württemberg (after Schierer 1986; Schlenker 1986; W. Assfalg and R. Schlenker, personal communication).

Alsace and Baden-Württemberg in periods before the stork numbers declined. Furthermore, in Alsace and Baden-Württemberg, breeding success and reproductive output are lowest in recent years, coinciding with the increasing proportion of birds released from captivity and probably reflecting their low success in dealing with the 'natural' situation. In many cases, these birds do not migrate and may depend on artificial feeding (Bloesch 1980); they are living in a kind of 'semi-captivity'. Therefore, from a biological point of view, all these previous attempts at captive breeding and artificial reintroduction to the wild as a method to stop the recent decline of the White Stork in Europe are of doubtful value. They may even cause new threats for the wild population (for details of discussion see Arbeitsgemeinschaft der Deutschen Vogelschutzwarten 1985; Löhmer and Schulz 1989). Reintroduction carried out for a limited period of time and under a strict scientific programme (Epple and Hölzinger 1986) may be a helpful tool and presents a challenge for conservationists in the education of the public. Promoting the protection and restoration of adequate habitats, either in Europe or Africa, are the only sensible recommendable methods to protect the White Stork in western central Europe.

Summary

For some decades the numbers of White Storks have been declining seriously in many parts of the breeding range in Europe. In particular in western populations that migrate to western Africa via Gibraltar, the numbers of breeding pairs have been reduced dramatically or even exterminated in some areas. Some populations farther east, however, which migrate to eastern and southern Africa via the Bosporus seem to have declined less, and to be stable or even increasing in size. The periods of rapid decline in White Stork numbers in W. and central Europe are associated with changes in annual adult and juvenile survival, age of first breeding, and reproductive success. These changes mainly reflect changes in environmental quality, in the breeding grounds, during migration, or in the African winter quarters. Geographical variations in population growth primarily indicate differences in climate, differences in agricultural use of landscape, and different conditions during migration. Preservation programmes must include habitat conservation and management. The reintroduction of storks by using birds released from captivity, as carried out intensively in some areas of western-central Europe, seems to be an inappropriate method.

Acknowledgements

For providing unpublished data or comments on the manuscript, I wish to thank W. Assfalg, W. Boettcher-Streim, H. R. Henneberg, G. Rheinwald, R. Schlenker, E. Schüz, G. Zink, and an anonymous reviewer. Thanks are also due to C. M. Perrins who checked the English of the text.

References

Arbeitsgemeinschaft der Deutschen Vogelschutzwarten (1985): Stellungnahme der Arbeitsgemeinschaft der Deutschen Vogelschutzwarten zum Problem des Aussetzens von Weißstörchen. *Berichte Deutsche Sektion des Internationalen Rates für Vogelschutz*, **25**, 161–5.

Bairlein, F. and Zink, G. (1979). Der Bestand des Weißstorchs *Ciconia ciconia* in Südwestdeutschland: eine Analyse der Bestandsentwicklung. *Journal für Ornithologie*, **120**, 1–11.

Bernis, F. (1981). *La poblacion de las cigüeñas Españolas*. Facultad de Biologia, Universidad de Madrid.

Bloesch, M. (1980). Drei Jahrzehnte Schweizerischer Storchansiedlungsversuch (*Ciconia ciconia*) in Altreu, 1958–1979. *Ornithologischer Beobachter*, **77**, 167–94.

Boettcher-Streim, W. (1986). Der Wiederansiedlungsversuch des Weißstorchs in Altreu/Schweiz. *Beihefte zu den Veröffentlichungen für Naturschutz und Landschaftspflege in Baden-Württemberg*, **43**, 315–28.

Boettcher-Streim, W. and Schüz, E. (1989). Bericht über die IV. Internationale Betandsaufnahme des Weißstorchs 1984 und Vergleich mit 1974 (b. Übersicht). In *Weißstorch-White Stork. Proceedings 1. International Stork Conservation Symposium* (ed. G. Rheinwald, J. Ogden and H. Schulz), pp. 195–219. Dachverband Deutscher Avifaunisten.

Burgman, M. A., Akcakaya, H. R., and Loew, S. S. (1988). The use of extinction models for species conservation. *Biological Conservation*, 43, 9–25.

Conrad, B. (1977). *Die Giftbelastung der Vogelwelt Deutschlands*. Kilda Verlag, Greven.

Chozas, P. (1986*a*). Status und Verbreitung des Weißstorchs in Spanien 1981 und früher. *Beihefte zu den Veröffentlichungen für Naturschutz und Landschaftspflege in Baden-Württemberg*, 43, 181–8.

Chozas, P. (1986*b*). Fortpflanzungs-Parameter des Weißstorchs (*Ciconia ciconia*) in verschiedenen Zonen Spaniens. *Beihefte zu den Veröffentlichungen für Naturschutz und Landschaftspflege in Baden-Württemberg*, 43, 221–34.

Creutz, G. (1985*a*). Der Weißstorch *Ciconia ciconia*. Ziemsen Verlag, Wittenberg.

Creutz, G. (1985*b*). Die Entwicklung des Storchbestandes in der DDR 1958–1984. *Vogelwelt*, 106, 211–4.

Creutz, G. (1986). Zum Vorkommen des Weißstorchs (*Ciconia ciconia*) in der DDR 1974–1984. *Beihefte zu den Veröffentlichungen für Naturschutz und Landschaftspflege in Baden-Württemberg*, 43, 121–3.

Dallinga, H. and Schoenmakers, M. (1984). Populatieveranderingen bij de ooievaar *Ciconia ciconia ciconia* in de periode 1850–1975. Thesis, University of Groningen.

Dallinga, H. and Schoenmakers, M. (1987). Regional decrease in the number of White Storks (*Ciconia c. ciconia*) in relation to food resources. *Colonial Waterbirds*, 10, 167–77.

Dornbusch, M. (1982). Zur Populationsdynamik des Weißstorches *Ciconia ciconia* (L.). *Berichte der Vogelwarte Hiddensee*, 3, 19–28.

Dybbro, T. (1972). Population studies on the White Stork *Ciconia ciconia* in Denmark. *Ornis Scandinavica*, 3, 91–7.

Epple, W. and Hölzinger, J. (1986). Bestandsstützung und Wiedereinbürgerung des Weißstorchs (*Ciconia ciconia*) in Baden-Württemberg. *Beihefte zu den Veröffentlichungen für Naturschutz und Landschaftspflege in Baden-Württemberg*, 43, 271–82.

Fiedler, G. (1986). Der Brutbestand des Weißstorchs (*Ciconia ciconia*) in einem Abschnitt der Kroatischen Posavina (Jugoslawien) 1974 and 1984. *Larus*, 36–7, 293–6.

Fiedler, G. and Wissner, A. (1986). Freileitungen als tödliche Gefahr für Weißstörche. *Beihefte zu den Veröffentlichungen für Naturschutz und Landschaftspflege in Baden-Württemberg*, 43, 257–70.

Goos, H. (1977). Gesichtspunkte zum Schutz des Weißstorches. *Berichte Deutsche Sektion des Internationalen Rates für Vogelschutz*, 17, 69–72.

Heckenroth, H. (1986). Zur Situation des Weißstorchs (*Ciconia c. ciconia*) in der Bundesrepublik Deutschland, Stand 1984. *Beihefte zu den Veröffentlichungen für Naturschutz und Landschaftspflege in Baden-Württemberg*, 43, 111–20.

Hölzinger, J. (1986). Einführung zum Artenschutzsymposium Weißstorch. *Beihefte zu den Veröffentlichungen für Naturschutz und Landschaftspflege in Baden-Württemberg*, 43, 7–14.

Hölzinger, J. (1987). *Die Vögel Baden-Württemberg*, Vol. 1: Gefährdung und Schutz. E. Ulmer, Stuttgart.

Hokelenger, J. and Künkele, S. (1986). Beiträge zur Verbreitung des Weißstorchs (*Ciconia ciconia*) in Nordgriechenland (Mazedonien, Thrakien). *Veröffentlichungen für Naturschutz und Landschaftspflege in Baden-Württemberg*, 43, 173–9.

Hornberger, F. (1956). Ist der Rückgang des Weißen Storches durch künstliche Nestanlagen aufzuhalten? *Jahreshefte des Vereins für vaterländische Naturkunde Württemberg*, 111, 207–22.

Jakab, B. (1984). A gólya (*Ciconia ciconia*) populációdinamikájának föbb tényezöi. *Puszta*, **2/11**, 89–103.

Jakab, B. (1986). Zur Populationsdynamik des Weißstorchs in Ungarn 1958 bis 1979. *Beihefte zu den Veröffentlichungen für Naturschutz und Landschaftspflege in Baden-Württemberg*, **43**, 329–41.

Jakab, B. (1987). Der Bestand des Weißstorchs in Ungarn von 1958–1984. *Falke*, **34**, 47–50.

Jakubiec, Z. (1985). Populacja bociana bialego *Ciconia ciconia* L. w Polsce. *Polska Academia NAUK*, *Studia Naturae*, *Ser. A.*, **28**, 5–262.

Kaatz, C. and Stachowiak, G. (1987). Untersuchungen zur Reproduktion der Population des Weißstorchs (*Ciconia ciconia*) im Kreis Kalbe/Milde. *Beiträge zur Vogelkunde*, **33**, 205–14.

Kanyamibwa, S., Schierer, A. Pradel, R., and Lebreton J. D. (1989). Changes in adult annual survival rates in a Western European population of the White Stork (*Ciconia ciconia*). *Ibis*, **132**, 27–35.

Klemm, W. (1983). Zur Lage des Weißstorchs (*Ciconia ciconia*) in der S. R. Rumänien. *Ökologie der Vögel*, **5**, 283–93.

Kumerloeve, H. (1988). Zur Brutsituation des Weißstorchs, *Ciconia ciconia*, in Anatolien (Stand 1987). *Bonner Zoologische Beiträge*, **39**, 361–70.

Lack, D. (1966). *Population studies of birds*. Clarendon Press, Oxford.

Lauthe, P. (1977). La cigogne blanche en Tunisie. *L'Oiseau et R.F.O.*, **47**, 223–42.

Lebreton, J. D. (1978). Une modèle probabiliste de la dynamique des populations de Cigogne blanche *Ciconia ciconia* L. en Europe occidentale. In: *Biométrie et Ecologie*, **1**, 277–343 (eds. Legay, J. M. and Tomassone, R.) Société francaise de Biométrie, Paris.

Lebreton, J. D. (1981). Contribution à la dynamique des populations d'oiseaux. Thesis, Université de Lyon.

Ledant, J.-P., Jacob, J.-P., Jacobs, P., Malher, F., Ochando, B., and Roche, J. (1981). Mise à jour de l'avifaune Algérienne. *Gerfaut*, **71**, 295–398.

Löhmer, B. (1974). Zwanzig Jahre Bestandsaufzeichnung und Beringung im Weißstorchforschungskreis Leine—Steinhuder Meer. *Beiträge zur Naturkunde Niedersachsen*, **27**, 92–9.

Löhmer, B. and Schulz, H. (1989). Zucht und Auswilderung—ein Beitrag zur Rettung des Weißstorchs? *Die niedersächsische Gemeinde*, **41/2**.

Melchior, F. (1987). Beschreibung und vergleichende Analyse von Horststandorten des Weißstorchs in Niederbayern und der Oberpfalz. *Jahresberichte Ornithologische Arbeitsgemeinschaft Ostbayern*, **14**, 3–86.

Meybohm, E. and Dahms, G. (1975). Über Altersaufbau, Reifealter und Ansiedlung beim Weißstorch (*C. ciconia*) im Nordsee-Küstenbereich. *Vogelwarte*, **28**, 44–61.

Profus, P. (1986) Fur Brutbiologie und Bioenergetic des Weißstorchs in Polen. *Beihefte in den Veröffentlichungen für Naturschutz und Landschaftspflege in Baden-Württemberg*, **43**, 205–220.

Profus, P. and Mielczarek, P. (1981). Zmiany liczebnosci bociana bialego *Ciconia ciconia* (Lineeaeus, 1758) w poludniowej Polsce. *Acta Zoologica Cracova*, **25**, 139–218.

Rejman, B. (1986). Über die Internationale Bestandsaufnahmen des Weißstorchs in der Tschechoslowakei, besonders den vierten Zensus 1984. *Beihefte zu den Veröffentlichungen für Naturschutz und Landschaftspflege in Baden-Württemberg*, **43**, 153–65.

Rejman, B. (1988). *Ciconia ciconia* 1988 v CSR. [Leaflet, obtainable from the author: 57001 Litomysl, CSSR.]

Riegel, M. and Winkel, W. (1971). Über Todesursachen beim Weißstorch (*C. ciconia*) an Hand von Ringfundangaben. *Vogelwarte*, **26**, 128–35.

Ruthke, P. (1986). Zum Status des Weißstorchs (*Ciconia ciconia*) in Marokko. *Beihefte zu den Veröffentlichungen für Naturschutz und Landschaftspflege in Baden-Württemberg*, **43**, 189–95.

Sackl, P. (1987). Über saisonale und regionale Unterschiede in der Ernährung und Nahrungswahl des Weißstorchs (*Ciconia ciconia*) im Verlauf der Brutperiode. *Egretta*, **30**, 49–80.

Sauter, U. and Schüz, E. (1954). Bestandsveränderungen beim Weißstorch: Dritte Übersicht, 1939–1953. *Vogelwarte*, **17**, 81–100.

Schierer, A. (1986). Vierzig Jahre Weißstorch—Forschung und -Schutz im Elsaß. *Beihefte zu den Veröffentlichungen für Naturschutz und Landschaftspflege in Baden-Württemberg*, **43**, 329–41.

Schlenker, R. (1986). Der Weißstorch-Bestand in Baden-Württemberg 1974–1984. *Beihefte zu den Veröffentlichungen für Naturschutz und Landschaftspflege in Baden-Württemberg*, **43**, 105–9.

Schneider, M. (1988). Periodisch überschwemmtes Dauergrünland ermöglicht optimalen Bruterfolg des Weißstorches (*Ciconia ciconia*) in der Save-Stromaue (Kroatien/Jugoslawien). *Vogelwarte*, **34**, 164–73.

Schüz, E. (1936). Internationale Bestandsaufnahme am Weißstorch 1934. *Ornithologische Monatsberichte*, **44**, 33–41.

Schüz, E. (1940). Bewegungen im Bestand des Weißstorchs seit 1934. *Ornithologische Monatsberichte*, **48**, 1–14.

Schüz, E. (1980). Status und Veränderungen des Weißstorch-Bestandes. *Naturwissenschaftliche Rundschau*, **33**, 102–5.

Schüz, E. (1986). Zur Bibliographie des Weißstorchs (*C. ciconia*). *Beihefte zu den Veröffentlichungen für Naturschutz und Landschaftspflege in Baden-Württemberg*, **43**, 361–72.

Schüz, E. and Szijj, J. (1960). Bestandsveränderungen beim Weißstorch: Vierte Übersicht, 1954–1958. *Vogelwarte*, **20**, 258–72.

Schüz, E. and Szijj, J. (1975). Bestandsveränderungen beim Weißstorch: Fünfte Übersicht, 1959–1972. *Vogelwarte*, **20**, 258–72.

Schüz, E. and Zink, G. (1955). Bibliographie der Weißstorch—Untersuchungen der Vogelwarten Rossitten-Radolfzell und Helgoland—Verzeichniswerk 1955. *Beihefte Vogelwarte*, **18**, 81–5.

Schulz, H. (1986). WWF/ICBP White Stork Project. *Newsletter*, **1**.

Tantzen, R. (1962). Der Weiße Storch im Lande Oldenburg. *Oldenburger Jahrbuch*, **61**, 105–213.

Thiollay, J.-M. (1985). Ecology and status of several European raptors and the White Stork in their winter quarters in West Africa. In *Migratory birds: Problems and prospects in Africa* (eds. A. MacDonald and P. Goriup), pp. 5–8. International Council for Bird Preservation, Cambridge.

Triebl, R. and Frühstück, H. (1979). Erhebungen über den Weißstorch (*Ciconia ciconia*) im Burgenland von 1963–1973. *Natur und Umwelt im Burgenland*, Sonderheft **2**, 1–31.

Veroman, H. (1962). Vom Bestand des Weißstorchs in Estland (Estnische SSR). *Vogelwarte*, **21**, 291–2.

Veroman, H. (1987). Distribution and population changes of the White Stork *Ciconia ciconia* in the northwestern USSR. *Ornis fennica*, **64**, 23.

Veroman, H. and Grempe, G. (1984). Ausbreitung des Weißstorchs im europäischen Nordwesten der Sowjetunion. *Vogelwarte*, **32**, 316–17.

Weissert, B. (1986). Die Besiedlung der Steiermark durch den Weißstorch. *Beihefte zu den Veröffentlichungen für Naturschutz und Landschaftspflege in Baden-Württemberg*, **43**, 147–52.

Zink, G. (1963). Populationsuntersuchungen am Weißen Storch (*Ciconia ciconia*) in SW-Deutschland. *Proceeding XIII. International Ornithological Congress*, pp. 812–18.

Zink, G. (1967). Populationsdynamik des Weißen Storchs, *Ciconia ciconia*, in Mitteleuropa. *Proceedings XIV. International Ornithological Congress*, pp. 191–215.

Zuppke, U. (1982). Der Bestand des Weißstorchs, *Ciconia ciconia*, im Kreis Wittenberg (Bezirk Halle) von 1976 bis 1980. *Beiträge zur Vogelkunde*, **28**, 175–87.

11 Population studies and conservation of *Puffins* Fratercula arctica

M. P. HARRIS and S. WANLESS

Introduction

Seabirds are abundant in virtually all seas and oceans and as such are subject to a wide range of environmental conditions, both natural and influenced by man. The Common Puffin *Fratercula arctica* (L.) is an attractive bird of the islands and cliffs in the North Atlantic. It breeds in burrows and among boulders from Brittany (France) and Maine (USA) northwards to Spitzbergen and north Greenland. Many colonies have declined during this century due either to natural events, such as oceanographic change (Harris 1984), or to human-induced mortality and disturbance (Evans and Nettleship 1985), but total numbers in the early 1980s were estimated at 5–6 million pairs. The species is very popular with members of the general public; thus there is usually concern, and sometimes subsequent funding of research, when its future appears threatened. Here we report on the results of three such projects, in Scotland, Norway, and the USA, and discuss the relevance of the findings to the conservation of seabirds.

The information presented in this paper comes from studies in three widely separated parts of the species range:

(a) our own work of 1972–88 on population changes, breeding parameters, and inter-colony movements of Puffins occurring on the Isle of May (56° 11′ N, 2° 34′W) and east Britain (where conditions for puffins are, or were, very favourable) (Harris 1983, 1984; Harris and Hislop 1978);

(b) a study started by G. Lid in 1970 and developed by T. Anker-Nilssen and co-workers on Hernyken (67° 26′N, 11° 52′E) Røst, Lofoten Islands, Norway, where Puffins have produced few young during 1969–88 (Lid 1981; Anker-Nilssen 1987; Barrett *et al*. 1987; T. Anker-Nilssen and O. W. Røstad, unpublished); and

(c) the successful attempts by S. W. Kress, National Audubon Society and Canadian Wildlife Service to reintroduce Puffins to Eastern Egg Rock (43° 52′N, 69° 23′W) Gulf of Maine, USA (Kress 1979–87; Kress and Nettleship 1988). The relatively few birds and the intensity of observations have produced valuable data on individual birds.

Most data are available for the Isle of May population, which forms the basis for the discussion of population dynamics. However, relevant data are also included from the other studies and from Skomer (51° 45′N, 5° 17′W), Wales (Ashcroft 1979) where the population was increasing slowly after a period of decline.

Methods

Methods have varied somewhat between the studies. Details can be found in the references cited above and only salient points are given here. Population estimates and changes in numbers are based on counts of occupied burrows in fixed sample plots either used directly (Isle of May) or scaled up to whole colony estimates (Røst), or are counts of burrows where food was brought to young (Eastern Egg Rock).

Breeding success is the proportion of young fledging from a known number of eggs laid (Isle of May) or young hatched (Røst), or the proportion of young hatched which were fed for at least 21 consecutive days (Eastern Egg Rock). Ages of first breeding come from birds ringed as chicks and later found with an egg or a chick or carrying fish, expressed as calendar years after hatching (e.g. a bird ringed in 1973 found on an egg in 1978 was classed as 5 years old). Any Puffin with less than two grooves on the outer part of the bill is considered to be 1–3 years old and thus immature (Petersen 1976a; Harris 1984). A Puffin ringed as a chick and seen at another colony 4 or more years later is considered to have emigrated.

Adult annual survival rates come from sightings of colour-ringed birds which were known to have bred previously, although not necessarily in the year in question. The basic survival figures are proportions of birds seen in year n which were known to be alive in year $n + 1$ because they were seen again in either year $n + 1$ or a subsequent year. In addition, estimates of survival were obtained using the Leslie and Chitty (1951) model; no estimate was possible in 1983–4 as the study population was altered somewhat in 1984. Of the 14 years for which we had survival estimates by both methods, six estimates obtained from Leslie models were lower than the known actual survival and two were biologically meaningless, being 100–105 per cent. Therefore, we use the basic survival estimates for calculating recruitment rates and in regressions, and the estimates from Leslie models to compare changes in survival during the study as these latter estimates allow for the fact that some birds not recorded in the most recent years could still be alive.

Survival of young to breeding age (taken as 5 years) is based on large-scale retrapping, mainly using mist-nets, using the calculations described in detail by Rothery (1983). The overall average survival of all cohorts followed was obtained by weighting each estimate by taking into account: (i) the within-year sampling variation, and (ii) the year-to-year variation, over and above that expected from (i). Year-to-year variation in survival estimates was tested using $x^2 = \Sigma Wi(m_i - m_w)^2$, where Wi is the weight attached to each estimate and m_w is the weighted average, and d.f. = number of cohorts minus one.

Recruitment is defined as the accession of mature birds to the breeding colony and is calculated from the number of adult birds required to replace the overwinter mortality of adults plus (where applicable) the number that provide the yearly increase in the colony. The recruitment rate in any year is the number of recruits expressed as a proportion of the total number of breeders in that year. The results for the Isle of May update all previously published data.

Results

Population changes

The colony on the Isle of May increased spectacularly from a few pairs in 1960 to *c.* 10 000 pairs in 1982. The number of burrows in sample plots increased at 19 per cent per annum between 1973 and 1981; the rate of increase then declined and there was little change in numbers during 1985–8 (Fig. 11.1). Four counts of the entire colony made between 1973 and 1988 indicated that this pattern reflected changes in the colony as a whole. In marked contrast, Puffin numbers on Hernyken declined at 12 per cent p.a. from 120 000 burrows in 1979 to 44 000 burrows in 1987.

Puffins were extirpated on Eastern Egg Rock by over-hunting in 1887, although two small colonies persisted elsewhere in the Gulf of Maine. Numbers in these neighbouring colonies increased slowly to *c.* 1100 pairs

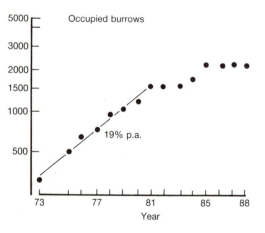

Fig. 11.1. Counts of occupied Puffin burrows in sample areas on the Isle of May, 1973–88.

in the 1950s, but have since remained fairly stable (Nettleship and Evans 1985). In 1973–81, 730 young Puffins from Newfoundland were successfully reared artificially on Eastern Egg Rock. Puffins bred again on Eastern Egg Rock in 1981 and the five pairs included eight introduced birds and two unringed birds from elsewhere. Counts of pairs feeding chicks increased gradually for several years—1982 (14), 1983 (10), 1984 (14), 1985 (20) and then remained stable—1986 (19), 1987 (18) and 1988 (16).

Adult survival

Adult survival on the Isle of May was very high with the observed range from 86.0 to 98.4 per cent (Table 11.1). There was a general decline towards the end of the study but it is unclear whether the apparent very low survival in 1987–8 was real or because many birds were overlooked. The mean values of overwinter survival calculated by the Leslie models were 96.0 per cent (SE = 1.8) in 1972–81 ($n = 9$ years) and 91.4 per cent (SE = 1.8) in 1981–7 ($n = 5$ years) supporting our view that survival was lower during the latter years. There was a positive correlation between the proportional change in numbers of burrows in the sample plots between consecutive years and the minimal overwinter survival of the adults (Fig. 11.2). Adult survival rates for Puffins on Skomer also decreased from

Table 11.1 Adult survival, chick production and proportion of immatures present during the chick rearing period for Isle of May Puffins 1973–88

	Overwinter survival		Chick production		Total birds	
	No. Alive in year n	% Alive in year $n + 1$	Eggs laid	Fledged/pair	Handled	% Immature
1973	130	94.6	58	0.74	No data	
1974	125	98.4	No data		No data	
1975	132	94.7	No data		No data	
1976	141	97.9	No data		676	12
1977	281	95.0	51	0.73	491	15
1978	328	93.6	100	0.87	739	28
1979	286	95.5	139	0.90	277	22
1980	279	95.0	119	0.76	380	37
1981	294	86.1	35	0.89	527	15
1982	232	87.5	124	0.92	753	15
1983	179	86.0	168	0.79	503	22
1984	180	93.9	144	0.88	502	14
1985	166	89.2	166	0.79	101	5
1986	178	84.3[+]	136	0.80	317	10
1987	163	76.1[+]	176	0.93	94	0
1988	—	—	157	0.88	47	0

[+] Possibly too low as some adults could still be alive and not all birds are recorded each year.

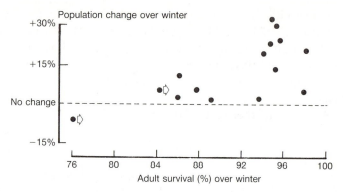

Fig. 11.2. The relationship between population changes and adult survival rate between successive years on the Isle of May, 1973–88. The correlation is significant ($r = 0.65$, d.f. $= 13$, $P < 0.02$). The two points with arrows refer to 1987 and 1988 where the estimates of adult survival may have been too low (see text).

a mean of 95.8 per cent p.a. in 1973–8 to a mean of 83.4 per cent p.a. in 1978–86 (Perrins 1988). Adult survival rates of over 95 per cent p.a. have also been recorded on Great Island (Newfoundland), Ostrova Aynovskiye (Murmansk), and Eastern Egg Rock (Hudson 1985; Kress 1987).

Breeding success

Breeding success was high on the Isle of May throughout the period of study with, on average, $0.84 \pm SE$ 0.02 young fledged per pair laying ($n = 13$ years, Table 11.1) and there was no systematic change during the period (Spearman Rank correlation coefficient $r_s = 0.44$, $n = 13$). Most failures were associated with broken or deserted eggs. Originally it was thought that only about 70 per cent of mature adults bred each year (following Ashcroft 1979). However, this figure is now considered to have been too low. During the 1980s virtually all adults had burrows and probably bred and a figure of 90 per cent of the population breeding in any year is assumed for the period 1973–81.

Puffins on Røst reared normal numbers of chicks in the 1960s. Conditions then changed and between 1969 and 1988 success was high only in 1974, 1983, and 1985 and moderate in 1984; in 5 years no young at all were reared. Breeding also failed in 1986, 1987 and 1988 (T. Anker-Nilssen, personal communication). Anker-Nilssen (1987) estimated a mean annual production of 0.11 young/breeding pair during the period 1969–85 but this may have been an overestimate, being based on the optimistic assumption that the hatching rate was similar to that at productive colonies.

On Eastern Egg Rock, hatching success was not determined but 81 (81 per cent) of 99 chicks hatched 1981–7 were considered to have been reared to fledging age. Total breeding output (young fledged/pair laying) at other east Atlantic colonies in the last 20 years has generally been high, e.g. 0.61–0.66 on Skomer, 0.70 and 0.75 on Fair Isle, Shetland, 0.69–0.84 on St Kilda, and 0.50–0.85 at four Norwegian colonies in 1980–2 when the Puffins on Røst were rearing few young (Ashcroft 1979; Harris 1984; Nettleship 1972; Barrett *et al.* 1987; Fair Isle Bird Observatory unpublished data). A wide range of breeding success (0.28–0.93) was reported from Newfoundland, with the lowest success being attributed to disturbance and predation by gulls (Nettleship 1972).

Recruitment

The proportion of recruits in the breeding population on the Isle of May averaged 19.5 per cent (SE = 1.7) from 1974 to 1988. Recruitment rates varied considerably from year to year ranging from 6.4 per cent recruits in the population in 1979 to 30.1 per cent recruits in 1976. Apart from 2 years of apparent low recruitment in 1977 and 1979, the proportion of recruits to the breeding population tended to decrease during the study (Fig. 11.3).

Age of first breeding

Studies on the Isle of May, St Kilda, and the Gulf of Maine all indicate that Puffins seldom breed at 3 years old; some birds were recorded breeding at 4 years, more at 5 or 6 years, and others not until they were older (Table 11.2). However, the data are biased because some birds may well have bred before they were recorded. Therefore these results do not give an accurate assessment of the distribution of age of first breeding. For later analyses, we assume that birds have chosen a nesting colony when 4 years old and first bred at 5 years.

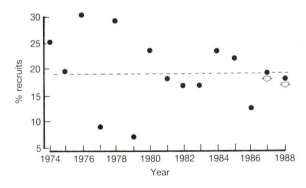

Fig. 11.3. Annual values of recruitment to the Isle of May Puffin population 1974–88. The dotted line indicates the mean level. The points for 1987 and 1988 (arrowed) use estimates of adult survival which may have been too low; if so, the estimate of recruitment will be too high.

Table 11.2 Ages of Puffins at the time first found breeding. Details from Kress and Nettleship (1988) and personal records.

Colony	No. of individuals found breeding at age (years)					Remarks
	3	4	5	6	Older	
Isle of May	1	14	16	28	29	Colony increasing
St Kilda	0	3	7	1	2	Colony stable
Gulf of Maine	0	11	23	12	6	Introduced birds
Total	1	28	46	41	37	

Colony fidelity

Once a Puffin has bred it rarely, if ever, moves to another colony, or even to another part of the colony, even when conditions appear unfavourable (personal observations). Immature birds, however, often visit several colonies before settling down and many of these end up breeding away from their natal colony (Harris 1983; Kress and Nettleship 1988).

Up to 1988, 334 (18 per cent) of 1783 chicks colour-ringed on the Isle of May between 1973 and 1979 were resighted when they were 4 or more years old. Thirty-two (9.6 per cent) of these had emigrated (Fig. 11.4). In a detailed, independent study large numbers of birds which were definitely breeding were checked during three seasons on the Isle of May and the Farne Islands, Northumberland (100 km away). Thirty-nine young that had been ringed on the Isle of May were found breeding, 30 at the natal colony and 9 on the Farne Islands—an emigration rate from the Isle of May to the Farne Islands of 23 per cent. In the wider survey, only 11 (34 per cent) of the 32 birds that had emigrated were seen on the Farne Islands whilst 21 (66 per cent) were seen elsewhere. A more realistic estimate of emigration is given by 26 (i.e. 9 × 32/11) of 56 (i.e. 26 + 30) birds moving to other colonies—an emigration rate of 46 per cent. This figure is a minimum estimate since many British Puffin colonies were not checked. There are no grounds for suspecting that birds from the Isle of May emigrated preferentially to the colonies that were searched since they covered a wide geographic area (Fig. 11.4) and range of colony sizes (Cramp et al. 1974).

The Isle of May also received immigrants from other colonies including chicks ringed in 1973–9 on the Farne Islands (40), Craigleith, Firth of Forth (4), Sule Skerry, Orkney (1), and Fair Isle, Shetland (1). It is difficult to assess the relative numbers of young emigrating out of and immigrating into the Isle of May, but clearly there was a considerable exchange of young birds between colonies in Britain.

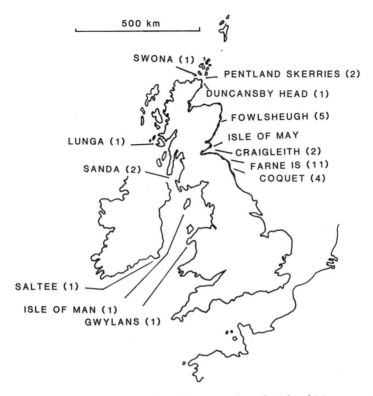

Fig. 11.4. Locations of young Puffins colour-ringed on the Isle of May as young in 1973–9 and seen elsewhere when 4 or more years old. The numbers in brackets refer to the number of individuals seen at the named colonies.

Similar examples of colony interchange were recorded on Eastern Egg Rock. Thirty-six of the young reared artificially in the colony returned to breed whilst 18 were found breeding on other islands in the Gulf of Maine. In addition, up to six birds in a season found breeding on Eastern Egg Rock were unringed so must have been reared elsewhere.

Post-fledging survival

On the Isle of May the survival of chicks from the 1973–8 cohorts was high with up to 25 per cent of some cohorts actually being seen back at the natal colony when aged 5 years or older (Table 11.3). The calculated survival of cohorts varied from 18 per cent to 33 per cent. The 1976 figure of 73 per cent (SE = 42 per cent) is suspect, being based on small samples of birds both ringed and retrapped. The weighted mean survival of the 1973–8 chicks was 26 per cent. Survival of the 1978 cohort was noticeably

Table 11.3 Survival of young Puffins ringed on the Isle of May 1973–8 to breeding age (5 years). Analysis as in Rothery (1983)

Year	No. of young ringed	No. birds seen			% actually known to survive to 5 years	Calculated No. alive	Survival % (SE)
		Year 5 and later	Year 5 and not later	Later year but not in year 5			
1973	169	11	2	27	24	45	27 (4)
1974	177	9	6	26	23	58	33 (7)
1975	162	4	3	20	17	42	26 (8)
1976	80	2	7	11	25	59	73 (42)
1977	225	4	14	12	13	72	32 (13)
1978	269	5	12	9	10	48	18 (6)
Mean 1973–7							28 (3)

lower than other years although the difference was not statistically significant ($x_5^2 = 4.54$; $0.10 > P > 0.05$). A lower survival for 1978 is in accord with the reduced numbers of immature Puffins seen on the Isle of May during the mid-1980s (see below) when the 1978 cohort would have been recruiting into the population. Excluding the 1978 value increases the weighted mean survival to 28 per cent. Given an emigration rate of 46 per cent, then 52 per cent of young should have survived to breed. The comparable 1978 figure would be 33 per cent.

Large-scale retrapping of Puffins on the Isle of May stopped in 1984 so that no data are available for survival of cohorts after 1978. However, the proportion of immature birds present in the colony during chick rearing declined from a mean of 21 per cent of birds handled between 1976 and 1983 to 6 per cent between 1984–8 (Table 11.1). (The 1987 and 1988 figures are based on small samples and visual observations recorded some immature birds present.) This could have been due to a lower immature survival rate or to immatures visiting the colony less often or when older when they would have been indistinguishable from adults.

In the Gulf of Maine, 146 (20 per cent) of the 730 young reared between 1973 and 1981 were seen in later years. Apparent survival of individual cohorts varied dramatically, with from 0 to 56 per cent of young being identified in subsequent years. Very few immatures have been seen ashore at Røst during the last 15 years, e.g. only one of 4140 full-grown birds ringed in 1979–87 had less than two bill grooves and so was definitely immature (G. Lid and T. Anker-Nilssen, personal communications).

Population dynamics

The data presented in the preceding sections can be used to calculate theoretical rates of population change for Puffins on the Isle of May. These can then be compared with the trends shown by the annual counts of occupied burrows.

The study period was divided into two parts; first, the steady increase phase up to 1981 and second, the period of slower increase and stability from 1981 onwards (Fig. 11.1). Two calculations of theoretical population changes were made for each period (Table 11.4). These used the proportion of young reared and calculated to have survived to breeding age based on: (a) just those returning to the Isle of May, and (b) including those that emigrated (assuming that emigration was balanced by equal immigration). Between 1973 and 1981 the numbers of Puffins on the Isle of May could have increased by either (a) 7 per cent p.a. or (b) 16 per cent p.a. The observed increase in the number of burrows from 1973 to 1981 averaged 19 per cent p.a. which suggests that there was net immigration. Between 1982 and 1988, numbers should have either (a) decreased by 1 per cent p.a. or (b) increased at 5 per cent p.a. (Table 11.4). In fact the number of burrows increased between 1982 and 1984 and then stabilized. This could have been due to fewer birds coming from other colonies, but the data for immature survival for this period are limited so it may well be

Table 11.4 Calculations of changes in a population of 1000 burrow holding pairs of Isle of May Puffins for 1973–81 and 1982–8. Difference between (a) and (b) is explained in text.

	1973–81		1982–8	
	(a)	(b)	(a)	(b)
(i) Chick production	0.84		0.84	
(ii) Proportion of pairs laying	0.90		1.00	
(iii) Fledged per 1000 pairs	756		840	
(iv) Survival of young to breeding age	28%	52%	18%	33%
(v) Number of young surviving to breeding age (iii × iv)	212	393	151	277
(vi) Adults dying[a]	80	80	172	172
(vii) Surplus (v − vi)	132	313	−11	105
(viii) Calculated annual increase of population (vii × 100/2000)	+7%	+6%	−1%	+5%

[a] Uses survival estimates obtained using Leslie models (see text).

that survival of birds reared on the Isle of May is lower than suggested by the results from the 1978 cohort.

Food of young

Puffins feed their young on a variety of mid-water shoaling fish but the majority of the diet in most colonies consists of sandeels, mainly *Ammodytes marinus* Raitt, also herring *Clupea harengus* L., sprat *Sprattus sprattus* (L.) and a few species of Gadidae, especially saithe *Pollarchius virens* (L.). In north Norway and Newfoundland, capelin *Mallotus villosus* (Müller) is an important prey. There has been no suggestion of food shortage for chicks on the Isle of May from 1972 to 1988 even though there have been considerable changes in the species of fish fed to young birds. Sandeels were the commonest prey except for the period from 1974 to 1978 when sprats formed 50–86 per cent of the diet (by weight). During the 1980s the proportion of sprats has declined and the importance of herring has increased gradually.

The diet of chicks on Røst has varied considerably but in general, young were fed on saithe at the start of the chick-rearing period with either herring or sandeel becoming more important as the season progressed. As on the Isle of May, the proportion of herring increased during the 1980s following restrictions on the human fishery. In 1983, when Puffins bred successfully, sandeels were clearly the most important component of the diet. In the Gulf of Maine, herring were usually the commonest prey although sandeels and other species were sometimes fed to the chicks.

In winter Puffins appear to eat the same species of fish augmented by pelagic polychaete worms, pelagic crustacea, and a few pelagic molluscs (Harris 1984).

Discussion

Population dynamics

Many seabirds, including the Puffin, have relatively low reproductive rates, delayed maturity and high adult survival, i.e. are classical K-selected species (MacArthur and Wilson 1967) showing characteristics that tend to produce relatively stable populations. Large increases do sometimes occur, but these are the result of prolonged periods of slow growth, e.g. the British Fulmar *Fulmarus glacialis* (L.) population has increased 10- to 20-fold, but this took over a century of constant expansion (Cramp *et al.* 1974). Any decreases can, of course, be extremely sudden if a disaster occurs, because even relatively small changes in adult mortality have important consequences for future population trends. Typically, K-selected seabird populations have intrinsic rates of increase of less than 10 per cent p.a., and where colonies show increases greater than the calculated rates, the excess is usually attributed to net immigration. For example, Nelson (1978) considered that no Gannet *Sula bassana* (L.) colony could sustain an increase of more than 3 per cent p.a. by its own production.

The data from the Isle of May show that many of the Puffin population parameters changed throughout the study. The stabilization of numbers of birds in the colony could be explained by a reduced expectation of life of breeding adults probably coupled with a decline in immature survival, and conceivably less immigration of birds from other colonies. In contrast, both the proportion of adults breeding each year and breeding success either increased slightly or remained constant. In a long-term study of the Kittiwake *Rissa tridactyla* (L.), Coulson and Thomas (1985) similarly showed changes in many population parameters and demonstrated that a decline in the number of breeding birds was due to higher adult mortality and a decline in almost every component of breeding output.

Accurate estimates of adult survival and breeding success are relatively straightforward to obtain whereas survival rates of young birds have proved extremely difficult to quantify because, as shown by several studies (e.g. Chabrzyk and Coulson 1976; Brooke 1978; Ollason and Dunnet 1983) rates of philopatry are much lower than was previously thought. Although many Puffins do return to breed at their natal colony, many others, perhaps even the majority, move elsewhere. Ollason and Dunnet (1983) modelled the numbers of breeding Fulmars at a colony and found that using observed values of survival and recruitment, the population could exhibit a wide range of annual changes varying from −38 per cent to +63 per cent. The main factor influencing population changes was recruitment which varied from 0 to 64 per cent p.a. and averaged 10 per cent. Other long-term studies (e.g. Porter and Coulson 1987; Duncan 1978) have shown the critical importance of recruitment in influencing the numbers of seabirds and how much of this recruitment can come from outside any particular colony. The studies on the Puffin reinforce the plea of Porter and Coulson (1987) for more attention to be paid to the study of recruitment.

The study on Røst did not attempt to document any population parameters except chick survival at the colony and changes in numbers but as extremely few chicks have been produced in recent years, recruitment from the colony's own output must have been negligible. Extensive immigration is unlikely because nearby populations are either relatively small or have also suffered breeding failures (Lid 1981; Barrett *et al*. 1987). Furthermore very few immatures have been recorded in recent years. If, as assumed by T. Anker-Nilssen and O. W. Røstad (unpublished), the 12 per cent p.a. decline equates to the annual adult mortality, then the adult survival rates on Røst, Skomer, and the Isle of May are all currently, at 83–91 per cent p.a., well below the figure of 95 per cent recorded in the 1970s (Ashcroft 1979; Harris 1984). There is at present no way of telling which value, if either, is 'normal'.

Factors influencing population size

The marked increase in the numbers of many seabirds in the North Sea during the 20th century has probably been due to a relaxation of human persecution combined with increases in the numbers of small fish on which seabirds feed (Cramp *et al*. 1974). The reasons behind the changes in fish numbers are poorly understood, but may well be associated with a reduction in numbers of large predatory fish and herring (which eat larval sandeels; Hardy 1924). These changes occurred partly as a result of increased efficiency in human fisheries and also, in the case of the herring, oceanographic change (Anderson and Ursin 1978; Jones 1983; Furness and Monaghan 1987). The halt in the increase of Puffins and Guillemots *Uria aalge* (Pont.) (Harris and Wanless 1988) on the Isle of May has coincided with similar stabilizations or declines in a wide range of seabird species at many other British colonies (Heubeck *et al*. 1986; Mudge 1986; Benn *et al*. 1987). Although population parameters are not available to interpret the changes observed in these colonies, the monitoring programmes do reveal some general features of the current change in seabird fortunes. First, changes appear to have occurred initially in northern Britain (Harris, in press); second, the changes have been most marked among surface feeding species such as the Kittiwake; and third, a reduction in numbers preceded any decline in breeding success suggesting a fairly widespread decrease in adult and/or immature survival.

Lack (1968) considered that the availability of winter food regulated the numbers of seabirds. Outside the breeding season Puffins disperse widely into the open sea and are seldom seen from land. Virtually nothing is known of their winter ecology or behaviour and they are rarely found dead during searches of beaches during the winter. However, numbers of auks, including some Puffins, have apparently died of starvation in several recent winters (e.g. Hudson and Mead 1984; Blake 1984) and (up to 1988) increasing numbers of adult Puffins now return to the Isle of May in spring still showing traces of winter plumage. For instance, in the last week of March 1988, 77 (27 per cent) of 287 birds examined closely still retained some dark features on the face; in contrast, none out of a sample of 687 on a similar date in 1975 and 1976 had any traces of winter plumage. A delay in the pre-breeding moult is consistent with the hypothesis that food availability during the winter has decreased, but would not be expected if the decline in survival rate was due to increased pollution (either oil or chemical or fishing nets); these are factors that have previously been considered to have an adverse effect on survival (Nettleship and Evans 1985). The diet of Puffins in the winter is largely unknown but sandeels, their main prey during the summer, will be less available since they are buried in the sand at this time (Winslade 1974); also the numbers of some other species of small fish are now declining after increasing in the 1960s and 1970s (Corten 1986; Anon. 1988).

Ringing recoveries indicate that adult Puffins from the Isle of May winter mainly in the North Sea with a small minority getting as far south as the Bay of Biscay. In contrast, Puffins from Skomer move further south to Iberia and some individuals enter the Mediterranean. Thus there is relatively little overlap in the wintering range (Harris 1984). The decline in adult survival of Puffins on both the Isle of May and Skomer suggests that the factor(s) responsible might be operating over a very wide area.

During chick rearing, Puffins usually feed within about 30 km of the colony (Bradstreet and Brown 1985). They make several feeding trips each day, and since they are inefficient fliers and dive extensively (Wanless *et al*. 1988) foraging costs are likely to be high. Any reduction in food availability would, therefore, be expected to have serious consequences. This hypotheses is supported by the evidence from Røst where Puffins have usually reared chicks only in years when numbers of small herring have been high (Lid 1981; Anker-Nilssen 1987). Similarly in 1987 and 1988, Puffins in Shetland brought in very small loads of small fish and many chicks died (Martin 1989). Breeding success on Eastern Egg Rock was reduced in 1983, a year when the numbers of herring in the area was very low (Kress 1983).

Conservation

Much of our knowledge about the Puffin and other seabirds relates to the stage of the life cycle when they are at their breeding colonies, and perceived threats to birds at this time can be countered. For instance, we can protect nesting areas from erosion and ensure that ground predators are not introduced. In some parts of their range large numbers of Puffins are still taken for human consumption, e.g. 150 000–200 000 birds are killed each year in Iceland (Petersen 1982). However, Petersen (1976*b*) found that only 7 per cent of the 100 000 Puffins killed annually in the Westmann Islands were breeding birds. It is difficult to assess the importance of this 'fowling' on the Puffin population of the Westmann Islands because of (i) the lack of adequate counts to assess whether the population is increasing or declining, and (ii) the paucity of relevant biological data.

If the total population of Iceland is 3 million pairs of which one-third nest on the Westmann Islands (Harris 1984; Nettleship and Evans 1985), and we assume that the population parameters collected on the Isle of May apply also to this population, the annual killing of 100 000 immatures could be removing anywhere from one-quarter to two-thirds of the birds expected to recruit into the population. The low figure for fowling assumes high survival of immatures and adults and much immigration, the high figure the reverse. Further speculation is futile and more data relevant to these exploited populations are needed before we can add anything to the comments of Nettleship and Evans (1985) that this level of human predation 'may be significant'. Kress's work in the Gulf of Maine has shown that, as a last resort, at least some species of seabird can be re-introduced into areas from which they have been exterminated, but such reintroductions are very costly in terms of time and money.

The greatest threats to seabirds, however, are found in the marine environment. In 1986, 1987, and 1988, Puffins, Kittiwakes, and Arctic Terns *Sterna paradisaea* Pont. in the main part of the Shetland Islands had difficulty in feeding their young and in 1988 all three species reared very few young (Heubeck 1989; Monaghan and Uttley 1989). There is good circumstantial evidence that adults were unable to obtain the younger classes of sandeels with which they usually feed their chicks (Monaghan *et al*. 1989). Puffins at Hermaness, the northern tip of Shetland, fed their young mainly on small rockling *Ciliata/Gaidropsarus* spp. and saithe and suffered a severe breeding failure (Martin 1989). In contrast young Puffins on Fair Isle, 150 km to the south, which are usually fed on sandeels, received 60 per cent of whiting *Merlangius merlangus* (L.) in their diet but 75 per cent of pairs still managed to rear a chick (Harris and Riddiford 1989). The reason for this food shortage is far from clear. The spawning stock of sandeels in the area remained moderately high and it seems likely that there has been some environmental change influencing either the numbers or the behaviour of young sandeels (Kunzlik 1989). Overfishing of the stock, for fishmeal, is unlikely to be the ultimate cause.

As yet we do not understand the factors that affect the ability of seabirds to feed their chicks at a time when birds are forced to forage fairly close to the colonies. During the winter seabirds are dispersed over vast areas of the world's oceans. We know virtually nothing of their ecology when they are away from the breeding colonies but individuals are potentially at risk of oiling, chemical contamination, predation, being caught in fishing nets, driven ashore by storms, and food shortage [for reviews of threats to seabirds see Bourne (1976) and Croxall *et al*. (1984)]. This lack of knowledge hinders our ability to put forward an effective conservation strategy. All we can do is recommend the obvious, that is reduce pollution in the sea, seek sensible ways of managing fish stocks, and lobby for fundamental, multi-species programmes of research to improve our understanding of predator-prey interactions in marine communities.

Summary

Breeding success and adult and immature survival of Puffins are usually high, but the last is difficult to determine as a high proportion of young breed away from their natal colony. Populations can increase rapidly, e.g. 19 per cent p.a. on the Isle of May 1973–81. The calculated intrinsic rate over this period was only 16 per cent which suggests that there was net immigration. Later, adult and immature survival declined and the calculated increase was 5 per cent p.a. The population had by then stabilized and immature survival was probably lower than the single estimate suggested; immature survival and immigration have critical effects on population size. Conditions in the North Sea, possibly winter food, appear to have changed. This change may be widespread as the annual survival rates for adults on Skomer and the Isle of May were significantly correlated and the birds from these colonies winter in different areas. Little can be done to conserve most species of seabird except to manage fish stocks sensibly and control pollution.

Acknowledgements

We thank many people for help with the field work over the years: S. W. Kress and T. Anker-Nilssen for unpublished data and for improving the manuscript, P. Rothery and S. Buckland for help with determining the survival rate of Puffins, and the Nature Conservancy Council for allowing us to work on the Isle of May NNR. The work was funded partly by the commissioned research programme of the Nature Conservancy Council.

References

Anon. (1988). Report of the working group on industrial fisheries. International Committee for the Exploration of the Sea, CM/198 Assess: 15.

Anderson, K. P. and Ursin, E. (1978). A multispecies analysis of the effects of variations of effort upon stock composition of 11 North Sea fish species. *Rapports et Proces-Verbaux des Reunions, Conseil Permanent International pour l'Exploration de la Mer*, **172**, 286–91.

Anker-Nilssen, T. (1987). The breeding performance of Puffins *Fratercula arctica* on Røst, northern Norway in 1979–85. *Fauna norvegicus Series C, Cinclus*, **10**, 21–38.

Ashcroft R. E. (1979). Survival rates and breeding biology of Puffins on Skomer Island, Wales. *Ornis Scandinavica*, **10**, 100–10.

Barrett, R. T., Anker-Nilssen, T., Rikardsen, F., Valde, K., Røv, N., and Vader, W. (1987). The food, growth and fledging success of Norwegian Puffin chicks *Fratercula arctica* in 1980–1983. *Ornis Scandinavica*, **18**, 78–83.

Benn, S., Tasker, M. L., and Reid, A. (1987). Changes in numbers of cliff-nesting seabirds in Orkney 1976–85. *Seabird*, **10**, 51–7.

Blake, B. F. (1984). Diet and fish stock availability as possible factors in the mass death of Auks in the North Sea. *Journal of Experimental Marine Biology and Ecology*, 76, 89–103.

Bourne, W. R. P. (1976). Seabirds and pollution. In *Marine pollution* (ed. R. Johnston), pp. 403–502. Academic Press, London.

Bradstreet, M. S. W. and Brown, R. G. B. (1985). Feeding ecology of the Atlantic Alcidae. In *The Atlantic Alcidae* (ed. D. N. Nettleship and T. R. Birkhead), pp. 263–318. Academic Press, London.

Brooke, M. de L.(1978). The dispersal of female Manx Shearwaters *Puffinus puffinus*. *Ibis*, 120, 545–51.

Chabrzyk, G. and Coulson, J. C. (1976). Survival and recruitment in the herring gull *Larus argentatus*. *Journal of Animal Ecology*, 45, 187–203.

Corten, A. (1986). On the causes of the recruitment failure of herring in the central and northern North Sea in the years 1972–78. *Journal du Conseil International Exploration de la Mer*, 42, 281–94.

Coulson, J. C. and Thomas, C. S. (1985). Changes in the biology of the kittiwake *Rissa tridactyla*: a 31-year study of a breeding colony. *Journal of Animal Ecology*, 54, 9–26.

Cramp, S., Bourne, W. R. P., and Saunders, D. (1974). *The seabirds of Britain and Ireland*. Collins, London.

Croxall, J. P., Evans, P. G. H., and Schreiber, R. W. (ed.) (1984). *Status and conservation of the world's seabirds*. ICBP Technical Publication, No. 2, Cambridge.

Duncan, N. (1978) The effects of culling Herring Gulls (*Larus argentatus*) on recruitment and population of dynamics. *Journal of Applied Ecology*, 15, 697–713.

Evans, P. G. H. and Nettleship, D. N. (1985). Conservation of the Atlantic Alcidae. In *The Atlantic Alcidae* (ed. D. N. Nettleship and T. R. Birkhead), pp. 428–87. Academic Press, London.

Furness, R. W. and Monaghan, P. (1987). *Seabird ecology*. Blackie, Glasgow.

Hardy, A. C. (1924). The herring in relation to its animate environment, Part I. The food and feeding habits of the herring. *Fishery Investigations*. Ser. II, 7, 1–39.

Harris, M. P. (1983). Biology and survival of the immature Puffin, *Fratercula arctica*. *Ibis*, 125, 56–73.

Harris, M. P. (1984). *The Puffin*. Poyser, Calton.

Harris, M. P. (in press). Population changes in British Common Murres and Common Puffins. In *Population biology and conservation of marine birds* (ed. W. A. Montevecchi and A. J. Gaston), Canadian Wildlife Service, Ottawa.

Harris, M. P. and Hislop, J. R. G. (1978). The food of young Puffins. *Journal of Zoology, London*, 185, 213–36.

Harris, M. P. and Riddiford, N. J. (1989). The food of some young seabirds on Fair Isle in 1986–88. *Scottish Birds*, in press.

Harris, M. P. and Wanless, S. (1988). The breeding biology of Guillemots *Uria aalge* on the Isle of May over a six-year period. *Ibis*, 130, 172–92.

Heubeck, M. (1989). Breeding success of Shetland's seabirds—Arctic Skua, Kittiwake, Guillemot, Razorbill and Puffin. In *Seabirds and sandeels*. Proceedings of a symposium held in Lerwick, Shetland, 15th–16th October 1988 (ed. M. Heubeck), pp. 11–18. Shetland Bird Club, Lerwick.

Heubeck, M., Richardson, M. G., and Dore, C. P. (1986). Monitoring numbers of Kittiwakes *Rissa tridactyla* in Shetland. *Seabird*, 9, 32–42.

Hudson, P. (1985). Population parameters for the Atlantic Alcidae. In: *The Atlantic Alcidae* (ed. D. N. Nettleship and T. R. Birkhead), pp. 233–65. Academic Press, London.

Hudson, R. and Mead, C. J. (1984). Origins and ages of auks wrecked in eastern Britain in February–March 1983. *Bird Study*, **31**, 89–94.

Jones, R. (1983). The decline in herring and mackerel and the associated increase in other species in the North Sea. *FAO Fisheries Report*, **291**, 507–20.

Kress, S. W. (1979–88). *Egg Rock Up-date*. National Audubon Society, Ithaca.

Kress, S. W. and Nettleship, D. N. (1988). Re-establishment of Atlantic Puffins (*Fratercula arctica*) at a former breeding site in the Gulf of Maine. *Journal of Field Ornithology*, **59**, 161–70.

Kunzlik, P. A. (1989). Small fish around Shetland. In *Seabirds and sandeels*. Proceedings of a symposium held in Lerwick, Shetland, 15th–16th October 1988 (ed. M. Heubeck), pp. 38–47. Shetland Bird Club, Lerwick.

Lack, D. (1968). *Ecological adaptations for breeding in birds*. Methuen, London.

Leslie, P. H. and Chitty, D. (1951). The estimation of population parameters from data obtained by means of the capture–recapture method. I: The maximum likelihood equations for estimating the death rate. *Biometrika*, **38**, 269–92.

Lid, G. (1981). Reproduction of the Puffin on Røst in the Lofoten Islands in 1964–80. *Fauna norvegicus Series C, Cinclus*, **4**, 30–9.

MacArthur, R. H. and Wilson, E. O. (1967). *The Theory of Island Biogeography*. Princeton University Press.

Martin, A. R. (1989). The diet of Atlantic Puffin *Fratercula arctica* and Northern Gannet *Sula bassana* chicks at a Shetland colony during a period of changing prey availability. *Bird Study*, **36**, 170–80.

Monaghan, P. and Uttley, J. (1989). Breeding success of Shetland's seabirds: Arctic Tern and Common Tern. In *Seabirds and sandeels*, Proceedings of a symposium held in Lerwick, Shetland, 15th–16th October 1988 (ed. M. Heubeck), pp. 3–5. Shetland Bird Club, Lerwick.

Monaghan, P., Uttley, J. D., Burns, M. D., Thaine, C., and Blackwood, J. (1989). The relationship between food supply, reproductive effort and breeding success in arctic terns *Sterna paradisaea*. *Journal of Animal Ecology*, **58**, 261–74.

Mudge, G. P. (1986). Trends of population change at colonies of cliff-nesting seabirds in the Moray Firth. *Proceedings of the Royal Society of Edinburgh*, **91B**, 73–80.

Nelson, J. B. (1978). *The Gannet*. Poyser, Berkhampstead.

Nettleship, D. N. (1972). Breeding success of the Common Puffin *Fratercula arctica* on different habitats of Great Island, Newfoundland. *Ecological Monographs*, **42**, 239–68.

Nettleship, D. N. and Evans, P. G. H. (1985). Distribution and status of the Atlantic Alcidae. In *The Atlantic Alcidae* (ed. D. N. Nettleship and T. R. Birkhead), pp. 54–155. Academic Press, London.

Ollason, J. C. and Dunnet, G. M. (1983). Modelling annual changes in numbers of breeding Fulmars, *Fulmaris glacialis*, at a colony in Orkney. *Journal of Animal Ecology*, **52**, 185–98.

Perrins, C. M. (1988). Skomer Island 1987: seabird survival figures. Unpublished report to Nature Conservancy Council.

Petersen, A. (1976a). Size variables in Puffins *Fratercula arctica* from Iceland and bill features as criteria of age. *Ornis Scandinavica*, **7**, 185–92.

Petersen, A. (1976b). Age of first breeding in Puffin *Fratercula arctica*. *Astarte*, **9**, 43–50.

Petersen, A. (1982). Sjofuglar. In *Fuglar* (ed. A. Gardarsson), pp. 15–60. Rit Landverndar, Reykjavik.

Porter, J. M. and Coulson, J. C. (1987). Long-term changes in recruitment to the breeding group, and the quality of recruits at a Kittiwake *Rissa tridactyla* colony. *Journal of Animal Ecology*, **56**, 675–89.

Rothery, P. (1983). Estimation of survival to breeding age in young Puffins. *Ibis*, **125**, 71–3.

Wanless, S., Harris, M. P., and Morris, J. A. (1988). Diving behaviour of guillemot *Uria aalge*, puffin *Fratercula arctica* and razorbill *Alca torda* as shown by radio-telemetry. *Journal of Zoology, London*, **216**, 73–81.

Winslade, P. (1974). Behavioural studies on the lesser sandeel *Ammodytes marinus* (Raitt). The effect of temperature on activity and the environmental control of the annual cycle of activity. *Journal of Fish Biology*, **6**, 587–99.

12 Survival rates of Greater Flamingos in the west Mediterranean region

A. R. JOHNSON, R. E. GREEN, and G. J. M. HIRONS

Introduction

Greater Flamingos *Phoenicopterus ruber roseus* are large, gregarious birds that feed by filtering invertebrates from the mud and water of brackish marshes and saline lagoons (Jenkin 1957). The world population has been roughly estimated at about 800 000 individuals, about 10 per cent of which are found in southern Europe and N. Africa west of 12°E (Kahl 1975). We will refer to this as the western Mediterranean population. The Flamingos of this region are not completely isolated from those in other areas, as has been shown by ring recoveries. However, compared with the number of movements recorded within the region, movements involving the Middle Eastern/eastern Mediterranean and West African regions are few (Johnson 1989).

Flamingos lay their eggs on the ground or in raised mud nests and breed in large, dense colonies. In spite of the numbers of adult birds present at colonies they are ineffective in deterring predators from taking eggs or young. Even birds much smaller than Flamingos, such as gulls, are able to take eggs with little difficulty. Therefore, successful breeding depends on the inaccessibility of colonies. Consequently, these are usually on islands and set in large desolate areas rarely penetrated by mammalian predators (Cramp and Simmons 1977).

In the past 40 years Flamingos have bred in four western Mediterranean countries; France, Spain, Tunisia, and Morocco. Breeding last occurred in Morocco in 1968 and in Tunisia in 1976. In France breeding has taken place in the Camargue in every year, except for the period 1964–8. Breeding was infrequent in Spain until 1977, but has occurred in all but 3 years since then.

Before attaining breeding age, and in winter, Flamingos from the few breeding colonies around the western Mediterranean are distributed over a vast area. The distances and routes travelled vary considerably between individuals and hence they are exposed to widely differing ecological conditions (Green *et al*. 1989). In this paper we examine the consequences of this distribution pattern for survival rates and the dynamics of the population.

Methods

Counts

In the Camargue, southern France, studies of Flamingos began in the late 1940s. The numbers of birds that bred and the numbers of chicks fledged were monitored and, in the period 1947–51, 5669 young were marked with numbered metal rings. Gradually the counting programme was extended to cover breeding colonies and midwinter populations throughout the western Mediterranean region (Johnson 1983). Numbers of breeding pairs have been estimated by counting incubating birds, either directly or from aerial photographs. Numbers of young fledged were taken from direct counts or photographs of creches of chicks that were, on average, more than half-grown. Co-ordinated midwinter counts of Flamingos in western Mediterranean wetlands were made in conjunction with January wildfowl counts organized by the International Waterfowl Research Bureau. The counts were made by a large number of people by various methods chosen to suit conditions at different sites which varied enormously in size and accessibility. We are therefore unable to assess their accuracy and have accepted the data at face value provided that all wetlands with resident Flamingos were covered.

Movements and survival

Since 1977, 8–33 per cent of the 2000–8590 young Flamingos reared each year in the Camargue have been marked individually with plastic rings inscribed with permutations of three or four characters (Johnson 1989). Up to 1985, 6021 birds had been marked. Rings can be read through a telescope from a distance of up to 400 m. Observations of ringed birds have been made in southern France by Tour du Valat staff since ringing began. In the breeding season (late March to July) most resightings resulted from a programme of continuous observation of the Camargue breeding colony from a tower hide 70 m away. Resightings outside France came from the observations of amateur ornithologists and the staff of local research and conservation bodies, supplemented by occasional expeditions from Tour du Valat. Coverage of these regions varied according to the distribution of effort and enthusiasm. The development of a similar programme in Spain, also involving observations at the breeding colony at Fuente de Piedra, has resulted in an increase in resightings there in recent years. Over 70 000 resightings of ringed birds have now been collected.

Resightings were classified according to time and location. Time periods of 1 year were used for survival analyses and ran from 1 August to 31 July in the following calendar year. This period was chosen because the young

Flamingos are ringed in late July and most leave the area of the breeding colony permanently during August. Age-specific survival rates were calculated separately for each cohort of ringed young using Cormack's method (Cormack 1964). Survival rates were constrained not to exceed unity by the method of Buckland (1980).

For some analyses we wished to examine survival rates separately for birds wintering in different areas. Resightings were classified according to three wintering zones (Fig. 12.1), which we called France, Spain, and Tunisia although the last two include parts of adjacent political land divisions. Many ringed Flamingos were also seen in Sardinia, but since this is known from counts and ring resightings to be an important staging area for birds moving between France and Tunisia (Johnson 1983; Green *et al*. 1989), birds seen only in Sardinia were not used in analyses concerning wintering zones.

Many more Flamingos spend the spring and summer in France than winter there. Regular counts have shown that birds leave France for other wintering zones in August–October. Many birds return to France in March and April (Johnson 1983, 1989). We therefore considered a bird to have located itself outside France during the non-breeding season if it was seen in Spain or Tunisia between 1 August and 29 February in the

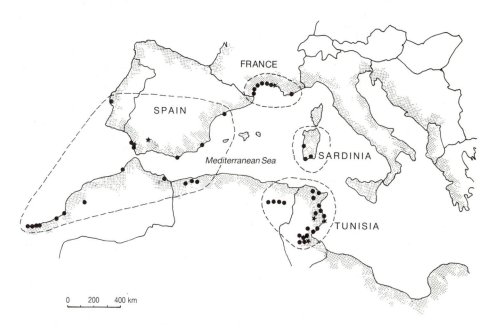

Fig. 12.1. Principal western Mediterranean wetlands (circles) used by Greater Flamingos. Breeding colonies used since 1969 are marked with stars. The dashed lines divide the range into four zones (see text).

following year. A bird was considered to have spent the winter in France if it was seen there between 1 November and 29 February.

We noticed that birds seen in a particular zone in one winter tended to occur there again in subsequent winters (see below). This tendency was exploited for the purpose of estimating survival rates specific to birds, wintering in each of the three zones. We combined two approaches in carrying out this analysis. Our first method was to treat the first record of a bird in a wintering zone during the non-breeding season as being equivalent to ringing it there. Subsequent survival and resighting probabilities were then estimated, using all resightings regardless of location and time of year, by Cormack's method. The minority of birds that were observed to change wintering zone were considered to winter in their first zone until they were seen in winter elsewhere, at which time they were withdrawn from the dataset for the first zone and reallocated as if newly ringed in the second zone. Using this method alone has the disadvantage that estimates of resighting probabilities were based on small samples for the first 1 or 2 years after ringing until the numbers of birds allocated to specific wintering areas has built up. Therefore we also estimated resighting probabilities by the Manly–Parr method (Manly and Parr 1968), allocating data for an individual in any particular year to the wintering zone in which it was seen in the first subsequent year of resighting during the non-breeding season. This approach substantially increased samples for the years immediately after ringing. The final estimates of resighting probabilities were taken to be the weighted means of the estimates obtained by these two methods, using the reciprocals of their variances as weights. Survival rates were then recalculated using these composite resighting probability estimates.

The resighting probabilities obtained by the above method were estimates of the chance of a bird, categorized according to its wintering location, being seen during subsequent years in any location at any time of year, given that it was alive. We also estimated, using the same method, the probabilities of birds allocated to a particular wintering zone being resighted in that zone during the non-breeding season. We used these estimates and wintering zone-specific survival rates to estimate the overall probability that a bird settling in a particular wintering zone in its first winter would be recorded as wintering there in at least one non-breeding season before dying. Using this value, the number of marked birds from each cohort settling in each of the three wintering areas in their first year was estimated. This calculation assumes that all birds remain faithful to their first wintering zone; although this is not the case we would expect the small proportion of birds that change wintering areas to have a small effect on the results.

The relationship was investigated of survival from one year to the next of Flamingos wintering in France with weather in the winter of the first year of the pair and to January population size in that year from counts. Survival probability was assumed to be related to these independent variables by a logistic function. The parameters of these functions were estimated by non-linear least squares. It should be noted that significance tests on regression of point estimates of survival on other variables need to be interpreted with caution if successive estimates from the same marked group of birds are used, because the estimates are not statistically independent. However, in our analyses this problem only applies to the data for adults. For first- and second-year birds the survival estimates made for different cohorts were independent of one another.

Results

Survival rate in relation to time, age, and cohort

Survivorship curves for each cohort of plastic-ringed birds, using all resightings irrespective of where they were made, are shown in Fig. 12.2. Mortality was high in the first and second years of life with only 50–60 per cent of the numbers ringed remaining by the third year. Thereafter the annual survival rate was high and relatively constant (about 95 per cent) with the obvious exception of the unusually low survival for all cohorts from 1984 to 1985.

In some ways this analysis is unsatisfactory. Members of each of these cohorts spent time in different countries where the effort made to read their rings varied widely and where their chances of surviving might have differed.

Monthly counts of Flamingos in southern France indicate that the number of birds present in winter is about 30–50 per cent of the numbers present in spring and summer (Johnson 1983). Counts, recoveries of metal-ringed birds, and resightings of plastic-ringed birds show that the Flamingos leaving the Camargue do so along two main routes; one to the S.S.E. to Sardinia and Tunisia and the other to the S.W., mainly to Spain and Morocco (see Fig. 2 in Johnson 1989).

Marked individuals have exhibited a wide range of variation of movement patterns. Some immatures and breeding adults make regular movements from breeding colonies to the same wintering sites in successive years; some immatures remain in areas remote from breeding colonies for several years, while others make brief visits to colonies from their first summer onwards. Some birds have been seen regularly in France throughout their lives.

We examined the faithfulness of Flamingos to the three wintering zones shown in Fig. 12.1. Multiple resightings within the same winter indicated

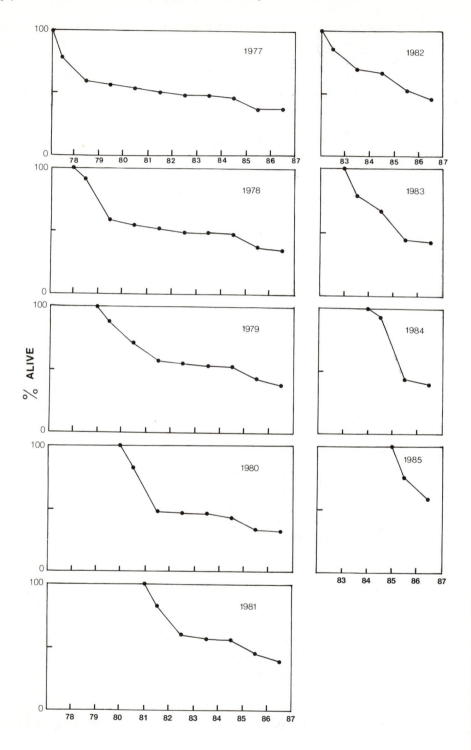

that birds rarely move between Tunisia and Spain within the same non-breeding season (Green *et al*. 1989). It was also the case that birds seen in a wintering zone in one year also tended to be seen there in subsequent years (Fig. 12.3). The number of birds recorded as changing wintering grounds from Spain or Tunisia to France overestimates the true proportion that did so because a greater effort was made to read rings in France than elsewhere. In another paper we show that there was a significant tendency for birds to remain faithful to the wintering area adopted in the first year until at least the seventh year of life, even when only those birds known to have moved between separate wintering and summering areas were considered (Green *et al*. 1989).

The above result was used to examine survival rates separately for birds identified as wintering in each of the three zones (see above for method; Fig. 12.4). Birds wintering in France generally survived well, even in the first and second years of life and there was little variation in survival between years compared with birds wintering in Tunisia and Spain. The obvious exception is the survival rate from 1984 to 1985. In that year survival was much lower than average, particularly for young birds. For all three wintering zones, survival showed no consistent trend with age. However, in years where there were large differences in survival between age classes, the survival of first and/or second year birds tended to be lower than that of older birds. The survival of birds wintering in France from 1984 to 1985 showed a tendency to increase with age even for 3- to 8-year-old birds (Fig. 12.5).

Birds wintering in Tunisia showed no evidence of reduced survival between 1984 and 1985 (Fig. 12.4). In Spain survival was low during that period for first years, but average for older birds.

Survival in relation to temperature in the wintering zones

There was an obvious cause of the unusual mortality in 1984–5 of Flamingos wintering in France: during a cold spell in January 1985 the

---

Fig. 12.2. Proportions of Flamingos, ringed as chicks in the Camargue, France, estimated to be alive within the western Mediterranean region in successive ecological years (1 August–31 July) after ringing, by Cormack's method. Each graph represents the results for one cohort. The number of chicks ringed in each year is shown in the list below. Arithmetic means of estimates (± 1 S.E.) of annual survival rate, combining cohorts for birds at least 3 years old, are: from ringing to the first-year, 0.837 ± 0.021; first- to second-year, 0.715 ± 0.041; second- to third-year, 0.888 ± 0.040; adult (3 years or older), 0.931 ± 0.022.

1977, $n = 557$	1980, $n = 761$	1983, $n = 720$
1978, $n = 650$	1981, $n = 697$	1984, $n = 781$
1979, $n = 651$	1982, $n = 652$	1985, $n = 552$

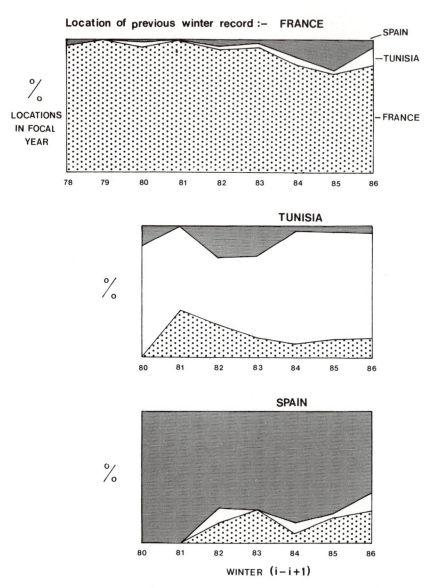

Fig. 12.3. Proportions of ringed Flamingos previously seen in a particular wintering zone, that were seen in each of three wintering zones in the focal winter. The three diagrams show proportions of birds in the three zones in relation to ecological year (1978–86) for birds seen previously in France (upper diagram), Tunisia (middle), and Spain (lower). Sample sizes (numbers of birds allocated to a wintering zone in each of the years) are shown opposite. Fewer than five birds with known wintering history were seen in Tunisia and Spain in 1978 and 1979 so data for these years were omitted from the diagram.

daily minimum temperature in the Camargue remained below freezing point for 15 days in succession. The wetlands in which most Flamingos spend the winter were completely covered by ice. Birds were only able to feed in the sea or where rice grain was artificially provided in areas maintained free from ice. About 3000 dead Flamingos were collected.

We compared the number of birds marked with plastic rings and found dead in France (141) during the 1984–5 winter with the number of marked birds wintering in France that were estimated from the survival analysis to have disappeared between 1984 and 1985. The number found dead was less than half of the number estimated to have disappeared (Fig. 12.6). The mortality rate of the older age classes for an entire year is usually about 5 per cent (Fig. 12.2), so it seems unlikely that much of this discrepancy was due to deaths outside the winter period. It seems likely that many more birds died in the cold spell or soon afterwards than were collected.

If we assume that the proportion collected of all birds that died whilst wintering in France in the 1984–5 winter was the same as the proportion collected of marked birds estimated to have disappeared, then we estimate that 6646 birds disappeared rather than the 3000 that were found dead. An independent, but approximate, estimate of the numbers of birds that died can also be obtained by comparing counts of Flamingos wintering in France before the 1984–5 winter, in January 1984, and immediately after the cold spell in late January 1985. Unfortunately there was no count during the 1984–5 winter before the cold spell. The difference in numbers, 10 000, supports our suggestion that many more birds died than were collected. There was no evidence from similar counts in Spain, Tunisia, and Sardinia that the missing birds had moved elsewhere and only a small proportion of ringed birds changed their wintering zone after that year (Fig. 12.3).

The high mortality rate in January 1985 seemed to be attributable to the ice preventing Flamingos from feeding. Periods with low temperatures are rarely long enough to cause extensive freezing of wetlands on the Mediterranean coast of France. We looked for evidence for effects of less

France	Tunisia	Spain
1978, $n = 30$	1980, $n = 6$	1980, $n = 8$
1979, $n = 59$	1981, $n = 20$	1981, $n = 9$
1980, $n = 92$	1982, $n = 25$	1982, $n = 30$
1981, $n = 119$	1983, $n = 26$	1983, $n = 37$
1982, $n = 188$	1984, $n = 49$	1984, $n = 85$
1983, $n = 179$	1985, $n = 54$	1985, $n = 104$
1984, $n = 223$	1986, $n = 88$	1986, $n = 130$
1985, $n = 95$		
1986, $n = 144$		

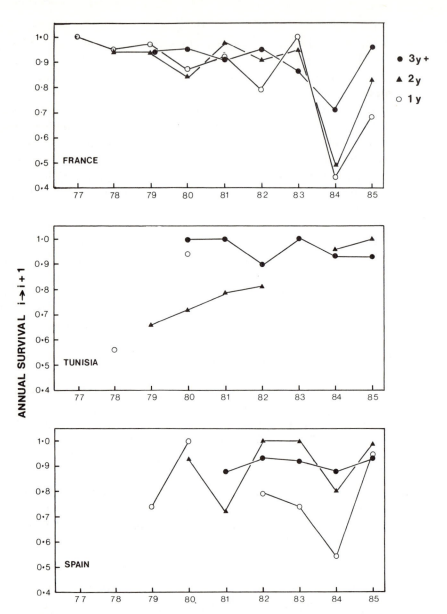

Fig. 12.4. Annual survival rates according to year for Flamingos identified as wintering in France, Tunisia, and Spain (see text). Separate survival rates are shown for first-years (open circles) and second-years (triangles). The mean survival rate of birds at least 3 years old is also shown (filled circles). Points representing survival from year i to year $i+1$ are plotted at year i. Survival rate estimates for which the estimated number of marked birds alive in year i was less than 20 have been omitted. Note that the survival rate given for first-years is from the first winter to the following ecological year, rather than from fledging (ecological year is 1 August–31 July).

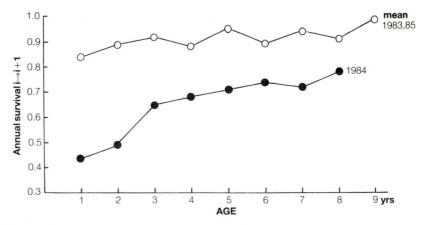

Fig. 12.5. Annual survival rate in relation to age for Flamingos that winter in France in a winter in which a severe cold spell occurred (filled circles) and the mean of survival rates for the previous and subsequent winters (open circles).

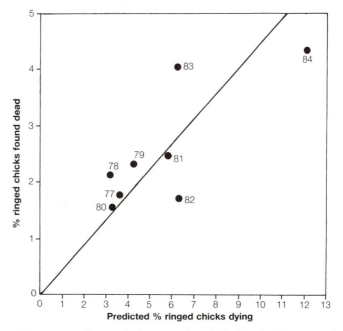

Fig. 12.6. The number of marked Flamingos found dead in the Camargue during the cold spell of January 1985 in relation to the number of birds that wintered in France and were estimated to have disappeared between 1984 and 1985 from the survival analysis. Each point represents birds from one cohort. The number of birds both found dead and estimated to have disappeared are expressed as a percentage of the number of chicks of that cohort originally marked. The line is a linear regression of proportion disappeared on proportion found dead, fitted by least squares, and constrained to pass through the origin.

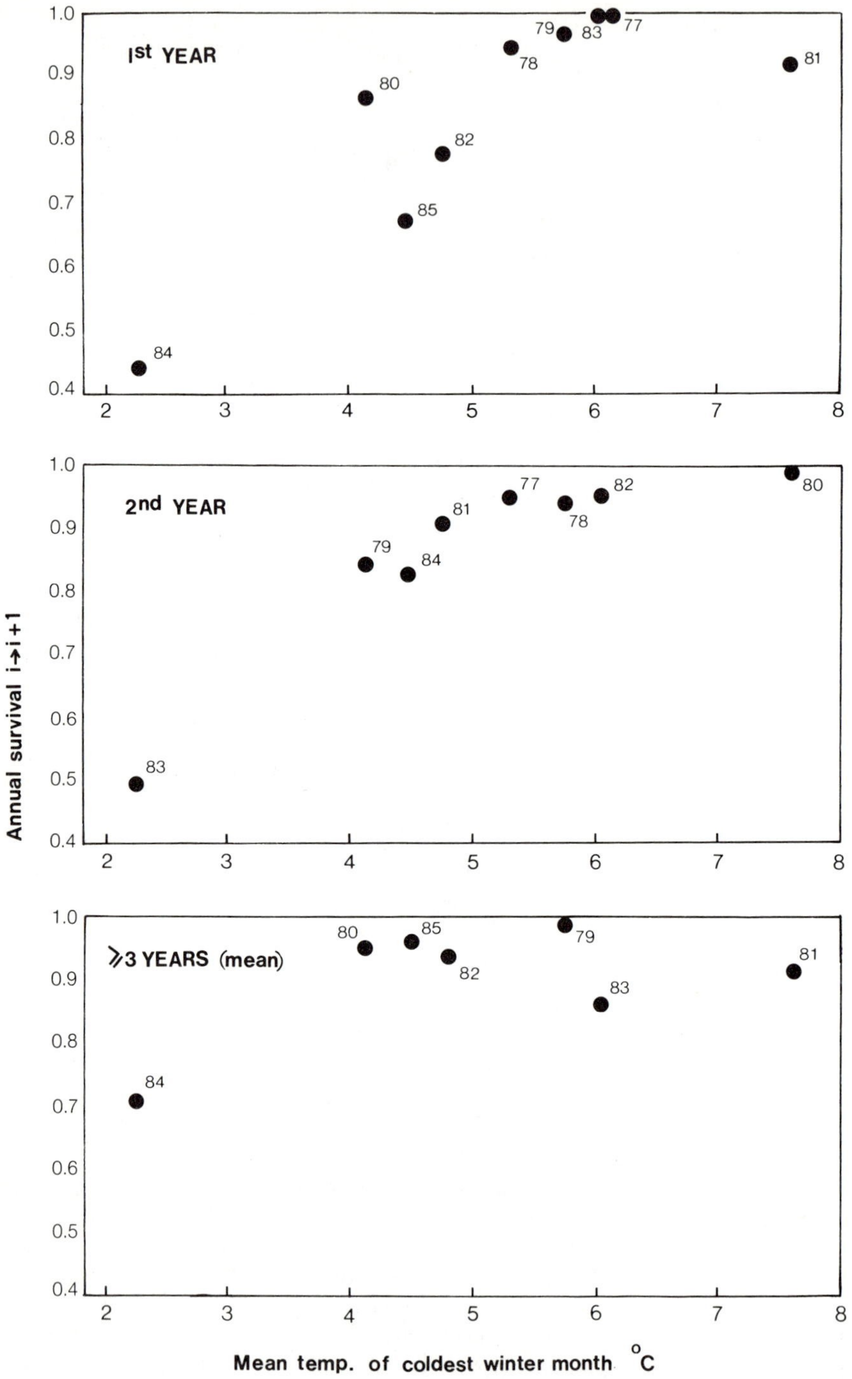
1st YEAR
2nd YEAR
≥3 YEARS (mean)
Annual survival i→i+1
Mean temp. of coldest winter month °C
1.0
0.9
0.8
0.7
0.6
0.5
0.4
2
3
4
5
6
7
8
77
78
79
80
81
82
83
84
85

severe conditions on the survival of Flamingos wintering in France by plotting annual estimates of survival against the mean temperature of the coldest winter month in the Camargue (Fig. 12.7). The survival of first-year birds was depressed in years in which the temperatures of the coldest winter month was less than 5 °C. A smaller effect was also evident for second-year birds, while in older birds reduced survival associated with cold weather was only evident in the year with the extreme cold spell (1984–5). We thought it likely that some of the unexplained variation in survival rates might be attributable to variation in the number of Flamingos wintering in France. Intraspecific competition for good feeding areas might cause a higher proportion of birds to die in years in which the wintering population was high. However, when the size of the population in January was included with temperature in a multiple regression model, there was no significant effect of population size on survival.

There was considerable annual variation in the survival rate of Flamingos wintering in Tunisia and Spain, but there was no correlation of survival with the temperature of the coldest winter month in those regions. The lack of evidence for reduced survival for all age classes between 1984 and 1985 for Flamingos wintering in Tunisia and Spain (Fig. 12.3) may be due to the higher winter temperatures experienced there than in France. Although annual variations in the mean temperature of the coldest winter month were correlated across the three wintering zones, the temperatures in the coldest winters in Tunisia and southern Spain were much higher than those that produced depression of survival for any age class in France (Fig. 12.8 cf. Fig. 12.7).

Effects of geographical variation in survival rates on the dynamics of the Flamingo population

The effect on the whole western Mediterranean Flamingo population of differences in the pattern of annual variation in survival rates observed between components of the population wintering in different areas depends on the relative numbers of birds utilizing the three main wintering

---

Fig.12 7. Annual survival rates of Flamingos wintering in France from ecological year i to year $i+1$ in relation to the mean temperature of the coldest winter month in year i. For first-years and second-years the numbers next to the points indicate the cohort. For older birds a mean survival rate was calculated for all available age classes and the number identifies the winter of year i. Logistic functions fitted to the data indicated a significant effect of winter temperature on survival for first-years ($t_7 = 5.79$, $P = 0.001$), second years ($t_6 = 22.05$, $P \neq 0.001$), and older birds ($t_5 = 2.69$, $P \neq 0.05$). When the effect of January population size was incorporated into the model there was no significant effect of this variable ($P \neq 0.05$) for any age class.

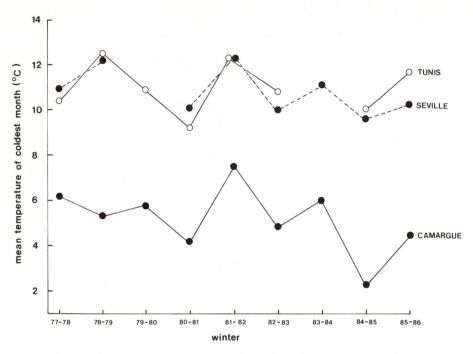

Fig. 12.8. The mean temperatures of the coldest winter month at stations chosen to represent conditions in the three Flamingo wintering zones, southern France (Salin-de-Giraud, Camargue), Tunisia (Tunis), and Spain (Seville).

zones. The distribution of Flamingos in winter has changed in recent times. The number of birds that winter in France increased from a few hundreds in the 1960s to a maximum of 22 414 in January 1982. During the same period the number of birds wintering in Tunisia has declined (Johnson 1983) and the number wintering in Spain has fluctuated erratically (Fig. 12.9). There has been no obvious trend in the size of the wintering population of the western Mediterranean region as a whole. Therefore the proportion of the Flamingo population wintering in France has increased substantially, even over the short period covered by the co-ordinated counts shown in Fig. 12.9. In 1977–8, 17 per cent of the western Mediterranean population was wintering in France compared with 31 per cent in 1982–4, the last co-ordinated count before the severe winter of 1984–5. This change in distribution has resulted in a larger proportion of the population being exposed to the risk of cold winters and therefore increased the impact of the cold spell in France in January 1985 on the Flamingo population as a whole.

The preference of individual Flamingos for particular wintering areas becomes established early in life (Green *et al*. 1989) and so the factors that

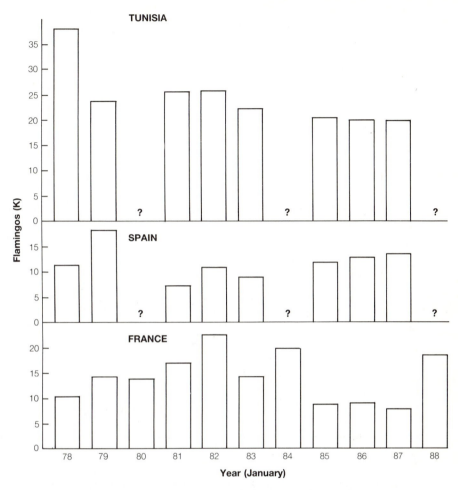

Fig. 12.9. Co-ordinated January counts of Flamingos (thousands) in Tunisia, Spain, and France. Note that the count made in the year 1984–5 was made immediately after the cold spell in January 1985.

determine the choice of wintering area must also operate at this time. The proportion of young that were ringed and remained in France in their first winter varied from year to year and was not related to the total number of Flamingos wintering in France. Most first-year Flamingos leaving France do so in September and October. When there was little rainfall in these months a lower proportion of the ringed young tended to stay over winter (Fig. 12.10). This might be because rainfall increases the extent of temporarily flooded areas suitable for feeding. However, the proportion of young birds leaving France in their first autumn also tended to increase

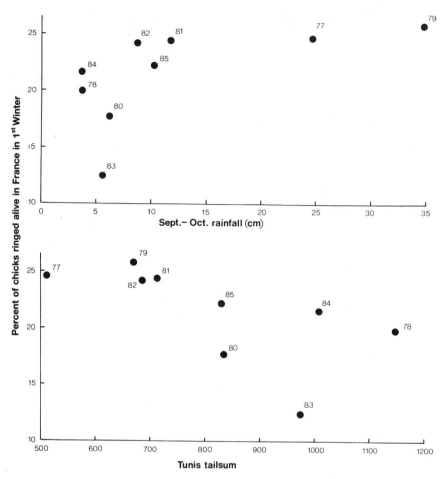

Fig. 12.10. The proportion of Flamingos, ringed as chicks in the Camargue, which were estimated to settle in France for their first winter (see Methods) in relation to September–October rainfall in their first autumn ($r_7 = 0.615$, $t_7 = 2.06$, $P = 0.08$) and a score representing the prevalence of following winds in the direction of Tunisia in their first autumn ($r_7 = -0.627$, $t_7 = 2.13$, $P = 0.07$; see Green *et al*. 1989). Each point represents one cohort.

with the occurrence of following winds in the direction of Sardinia and Tunisia in September and October (Fig. 12.10). We have shown elsewhere that when conditions are favourable for a S.S.E. crossing of the Mediterranean a higher proportion of the young leaving France spend the winter in Tunisia rather than Spain (Green *et al*. 1989). Because following winds towards Tunisia tend to be prevalent in dry autumns it is not possible to disentangle these two effects.

Effects of abnormal winter mortality on the size of the breeding population

Although it affected total population size, the increase in mortality of Flamingos due to the cold spell in France in January 1985 had no discernible effect on the numbers of Flamingos breeding around the western Mediterranean in the subsequent summer (Fig. 12.11). The number breeding in France actually increased. Moreover, there was no indication of a positive overall relationship between the size of the breeding population of Flamingos in the western Mediterranean region and the total population size from counts in the preceding January, even though there was considerable variation in both (Fig. 12.12). This suggests that the proportion of birds that breed may, on average, decline with increasing population size. However, a formal test of this is not possible because the accuracy and precision of the midwinter counts have not been evaluated. Furthermore, the number of incubating birds or nests counted underestimates the number of pairs that attempt to breed. The extent of this bias depends on the proportion of breeding attempts that fail during incubation

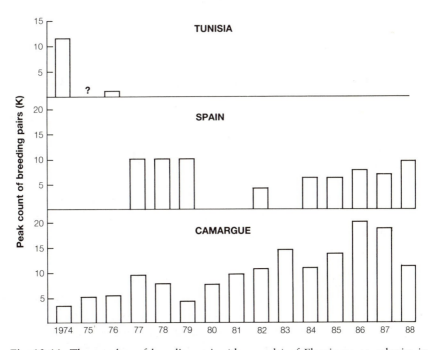

Fig. 12.11. The number of breeding pairs (thousands) of Flamingos at colonies in Tunisia, Spain, and France, estimated by counts of peak numbers of incubating birds or nests.

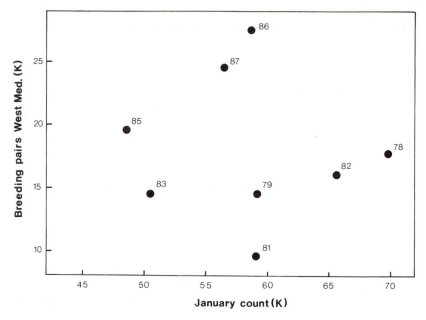

Fig. 12.12. The number of breeding pairs of Flamingos in the western Mediterranean region in relation to the size of the total population in the previous January (thousands).

and this is known to vary from year to year (Green and Hirons 1988). The lack of correlation between breeding population and total population may be due in part to variation in the proportion of birds below breeding age. Since these cannot be distinguished in the counts, the correlation between the sizes of the breeding population and populations of breeding age cannot be examined.

Several lines of evidence suggest that the size of the breeding population in a given year may be close to a limit set by the availability of safe nesting sites. The whole of the breeding colony at Fuente de Piedra, Spain and most of that in the Camargue are on islands that are densely packed with nests. Nests that fail during incubation are taken over almost immediately by new pairs.

Annual variation in the number of breeding pairs of Flamingos in the western Mediterranean region arises in two ways. In the 20 years 1969–88 the number of breeding colonies occupied in one year has varied between one and three (see Fig. 12.1). Occupation of the two breeding sites in Spain appears to depend on water levels in the marshes within the feeding range of the colonies, since breeding did not occur in springs following dry winters in the Marismas del Guadalquivir (Fig. 12.13). The second source

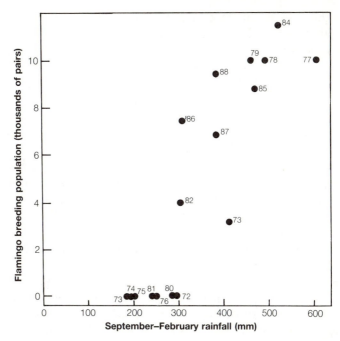

Fig. 12.13. The number of pairs of Flamingos breeding in Spain (thousands) in rela-tion to rainfall in the previous winter (September–February) at Seville, near the main feeding areas of breeding birds in the Marismas del Guadalquivir. There is a significant correlation ($r_{15} = 0.877$, $t_{15} = 7.07$, $P \neq 0.001$).

of variation in the size of the total breeding population occurs because part of the colony in the Camargue lies on a dyke between two lagoons near the island that holds most of the nests. In some years the dyke is not occupied by breeding pairs at all or is abandoned early in the season. Predators have access to the colony along the dyke and disturbance by them may be the immediate cause of abandonment or failure to establish. However, shortage of suitable feeding areas may predispose this part of the colony to failure in some years (unpublished data).

Trends in the western Mediterranean Flamingo population

We estimated whether a Flamingo population with the observed annual survival rates would be expected to be increasing or decreasing. We used a model in which survival rates were age-specific but constant over time, with the values given in Fig. 12.2. The number of young produced per year in the whole region is known since 1976. In the period 1976–88 the number fledged has ranged between 3500 and 12 600 (mean 8107). In eight winters during this period a co-ordinated count was made of Flamingos in all their major haunts in the region. Population estimates ranged between 48 027 and 69 869 birds (mean 58 566). It was assumed that the number of young produced per year was in constant proportion (0.138 = 8107/58 566) to the size of the winter population. We ran a model incorporating these values until a stable age distribution was achieved. The population multiplication rate was then 1.019 [95% confidence limits, 0.946–1.092, estimated by the method of Lande (1988)]. The broad confidence limits (C.L.) are mainly due to marked annual variation in the number of young fledged in relation to the total population; 72 per cent of the sampling variance of the population multiplication rate was due to this component. Population multiplication rate was also estimated from a linear regression analysis of the logarithms of the eight winter population estimates on year and gave an estimate of 0.979 (95% CL., 0.950–1.010). The estimates of population multiplication rate obtained by the two methods were not significantly different (Fisher-Behrens test, d = 1.04, P > 0.30). Hence both methods indicate that the Flamingo population has been approximately stable over the past 10–15 years. The sensitivity of population multiplication rate to changes in demographic parameters was examined using a Leslie matrix model. Age-specific survival rates were the average rates given in Fig. 12.2, the age of first breeding was 8 years (Green and Hirons 1985), and the fecundity rate of birds aged 8 years or more was 0.1856 young per breeding adult per year. This fecundity rate was chosen to obtain a stable population, given the other parameter values. Sensitivity coefficients, calculated by the method of Goodman (1971), indicate that population multiplication rate would be most sensitive to alteration in the adult survival rate (Table 12.1).

Table 12.1 Sensitivity coefficients indicating the effect of unit change in survival rate, fecundity, or age at first breeding on the population multiplication rate. Demographic parameters were assumed not to vary with time or population density. Age at first breeding was considered to be 8 years. Survival rates were set at the mean values given in Fig. 12.2 and production of young was set at the level that would give a stable population. The coefficients were calculated by the method of Goodman (1971)

Parameter	Sensitivity coefficient
Young fledged per breeding bird per year	0.251
Chick to first-year survival	0.056
First- to second-year survival	0.065
Second- to third-year survival	0.052
Adult survival	0.924
Age at first breeding	0.003

In the foregoing analysis we assumed that no density-dependent processes were operating. We were unable to find evidence of density-dependent mortality for Flamingos wintering in France. However, we suspect that the number of birds that can breed successfully may be limited by the availability of suitable nesting sites. If so, this would make the adult fecundity rate dependent on adult population density. We explored the effects of such limitation by adapting the model described above. We assumed that when the number of birds of breeding age exceeded some threshold the excess birds were unable to breed. This resulted in population regulation at a stable equilibrium provided that the production and survival of young was sufficient at least to replace adult mortality. We examined the effects on equilibrium population size of varying adult survival rate and the rate of production of fledged young per breeding adult.

In such a model total population size at equilibrium was directly proportional to the ceiling population of breeding birds. The ratio of total population to breeding population increased with increasing adult survival rate and increasing rate of production of young per breeding adult, provided that these rates exceeded thresholds for population persistence (Fig. 12.14). If this model accurately represents Flamingo population dynamics then it would be expected that a reduction in the number or size of breeding sites or the quality and extent of feeding places within range of breeding sites would eventually affect total population size in direct proportion. Increases in survival or breeding success would increase the total population size by increasing the number of non-breeding birds. However, a relatively small decline in average survival and/or birth rates from the present mean levels might precipitate a rapid population decline.

Conclusions

This study demonstrates that the demography of a single population of birds can be affected in complex ways by environmental factors if the geographical range is large. Flamingos originating in the same colony, but wintering in different parts of the western Mediterranean region, were found to have differing patterns of survival. A detailed analysis of survival rates of birds wintering in France showed that they were correlated with winter temperature, but not with local population density, and that susceptibility to cold winters declined with age. Hence, the effects of environmental factors on survival measured over the whole population are modified by its age structure and geographical distribution. Because Flamingos are long-lived and develop affinities to particular wintering areas early in life, changes in the selection of wintering areas can have consequences for overall survival rates for many subsequent years.

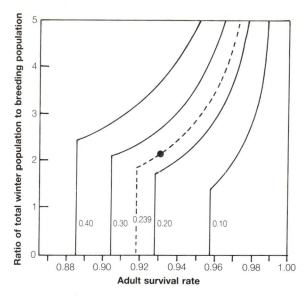

Fig. 12.14. Midwinter total population size in relation to adult annual survival rate, the production of fledged young per breeding bird per year, and the maximum number of birds that can breed, from a model of the western Mediterranean Flamingo population. Survival rates are assumed to be age-specific with values equal to the means given in Fig. 12.2. Survival is assumed not to vary with time or population density. The number of birds that can breed is assumed to be limited by the availability of suitable breeding places. Total population size was found to be directly proportional to the ceiling breeding population size so the ordinate shows the ratio of total population: breeding population. Each curve shows the relationship between total population: breeding population and adult survival for a specified level of production of young per breeding adult. The vertical line at the left of each curve shows the level of adult survival below which the population cannot be sustained at a given level of productivity. The dashed line shows predicted equilibrium populations for the observed mean rate of production of young per breeding adult for the period 1976–88 (0.239). The circle indicates the observed mean adult survival rate.

The large geographical range of the western Mediterranean Flamingo population also leads to complexity in the pattern of reproduction. In the past 30 years the distribution of birds between breeding sites has changed considerably. Flamingos have ceased to breed in Tunisia, but colonies in France and Spain have increased in size or frequency of use. In recent years no more than three colonies have been used in one year and only two regularly produce large numbers of young. We suggest that the availability of undisturbed sites inaccessible to predators, and within range of large areas of brackish marshland suitable for feeding, is likely to impose an upper limit on population size, though this is modified by the average levels of breeding success and survival. If this suggestion is correct then the western Mediterranean Flamingo population is extremely vulnerable to changes in its breeding habitat caused by human activities. Wetlands throughout Europe are subject to drainage, pollution, and exploitation. Pollution and changes in the hydrology of the few wetlands used by breeding Flamingos could reduce the food supply, and increased use of wetlands for recreation could deny the birds access to undisturbed nesting sites. The future of the western Mediterranean Flamingo population depends on the implementation of measures to protect the surviving wetlands from overexploitation.

References

Buckland, S. T. (1980). A modified analysis of the Jolly–Seber capture-recapture model. *Biometrics*, **36**, 419–35.

Cormack, R. M. (1964). Estimates of survival from the sighting of marked animals. *Biometrika*, **51**, 429–38.

Cramp, S. and Simmons, K. E. L. (1977). *Handbook of the birds of Europe, the Middle East and North Africa: Birds of the western Palearctic*, Vol. I. Oxford University Press.

Goodman, L. A. (1971). On the sensitivity of the intrinsic growth rate to changes in the age specific birth and death rates. *Theoretical Population Biology*, **2**, 339–54.

Green, R. E. and Hirons, G. J. M. (1985). The population dynamics of Camargue flamingos. Unpublished report, Station Biologique de la Tour du Valat.

Green, R. E. and Hirons, G. J. M. (1988). Effects of nest failure and spread of laying on counts of breeding birds. *Ornis Scandinavica*, **19**, 76–8.

Green, R. E., Hirons, G. J. M., and Johnson, A. R. (1989). The origin of long-term cohort differences in the distribution of greater flamingos *Phoenicopterus ruber roseus* in winter. *Journal of Animal Ecology*, **58**, 543–55.

Jenkin, P. M. (1957). The filter-feeding and food of flamingos (*Phoenicopteri*). *Philosophical Transactions of the Royal Society, London [B]*, **240**, 410–93.

Johnson, A. R. (1983). Etho-ecologie du flamant rose (*Phoenicopterus ruber roseus* Pallas) en camargue et dans l'ouest Palearctique. Unpublished Thesis, Université Paul Sabatier de Toulouse.

Johnson, A. R. (1985). Les effets de la vague de froid de janvier 1985 sur la population de flamants roses hivernant en France. ICBP-IWRB Flamingo Research Group Special Report No. 2.

Johnson, A. R. (1989). Movements of greater flamingos in the western Palearctic. *Terre et Vie*, **44**, 75–94.

Kahl, M. P. (1975). Distribution and numbers—a survey. In *Flamingos* (eds. J. Kear and N. Duplaix-Hall), pp. 93–102. Poyser, Berkhamsted.

Lande, R. (1988). Demographic models of the northern spotted owl (*Strix occidentalis caurina*). *Oecologia (Berlin)*, **75**, 601–7.

Manly, B. F. J. and Parr, M. J. (1968). A new method of estimating population size, survivorship and birth rate from capture-recapture data. *Trans. Soc. Br. Ent*, **18**, 81–9.

13 Population regulation of seabirds: implications of their demography for conservation

J. P. CROXALL and P. ROTHERY

Introduction

Despite the paucity of pertinent information there has been considerable discussion on the subject of population regulation in seabirds (e.g. Ashmole 1963, 1971; Lack 1954, 1966; Birkhead and Furness 1985; Furness and Monaghan 1987). This is perhaps because seabirds include several groups of species showing the lowest reproductive rates (smallest clutch size, longest deferred sexual maturity) and highest adult survival rates (Table 13.1). They are therefore a convenient paradigm of extreme K-selected species. In the context of conservation and management, the characteristics and habitats of seabirds give rise to two important considerations. First, many species, because of their long generation times are potentially rather vulnerable to adversity. Second, because changes in parameters of seabird populations will tend to reflect integration of environmental effects over long temporal and large spatial scales, they can be regarded as potentially important indicators of widespread chronic and possibly local acute effects (see e.g. Croxall *et al.* 1988).

The above demographic characteristics, however, when combined with a highly pelagic existence, make seabirds much less suitable for actual studies of population regulation. In particular, long periods of sexual immaturity, during much of which the juveniles remain at sea, makes assessment of mortality rates and any possible causes very difficult. Collecting pertinent information on the nature and significance of effects of environmental variables on seabirds is not easy and in very few cases are there reliable data, for instance on food availability, even during the breeding season. The low reproductive rates mean that by the time significant adverse changes in breeding, population size, and adult survival etc. have been detected, the factors underlying these changes must have operated for many years. Consequently, effective conservation measures may not be readily apparent, since it will often be difficult first to determine the causes of changes and second to redress the damage. With typically low recruitment rates, even if remedial action is taken

Table 13.1 Some mean demographic and biological characteristics of the main families of seabirds

Group	Age (years) at first breeding	Adult annual survival rate (%)	Clutch size	Chick-rearing period (days)
Sphenisciformes				
Spheniscidae (penguins)	4–8	75–85 (–95)[a]	1–2	50–80 (–350)[b]
Procellariiformes				
Diomedeidae (albatrosses)	7–13	92–97	1	116–150 (–280)[c]
Procellariidae (petrels)	4–10	90–96	1	42–120
Hydrobatidae (storm petrels)	c.4–5	c.90+	1	55–70
Pelecanoididae (diving petrels)	2–3	75–80	1	45–55
Pelecaniformes				
Pelecanidae (pelicans)	3–4	c.85	2–3	55–60
Sulidae (gannets, boobies)	3–5	90–95	1–2 (–3)[d]	90–120 (–170)[c]
Phaethontidae (tropic birds)	?	?	1	60–90
Fregatidae (frigate birds)	?c.9–10	?	1	140–170+
Phalacrocoracidae (cormorants)	4–5	85–90	2–3	60–90
Charadriiformes				
Alcidae (auks)	2–5	80–93	1–2	15–20
Lariiformes				
Laridae (gulls)	2–5	80–85	2–3	25–50
Sternidae (terns)	2–5	77–90	1–3	25–40
Stercorariidae (skuas)	3–8	c.93	2	25–40

[a]Emperor Penguin *Aptenodytes forsteri*; [b]Emperor Penguin 170 days; King Penguin *A. patagonicus* 350 days; [c]Great albatrosses; [d]Boobies; [e]Red-footed Booby *Sula sula* to 140 days, Abbott's Booby *S. abbotti* to 170+ days.

immediately, the population is still likely to decline for a number of years.

In this paper we have three aims. First, to review the main current ideas on how seabird populations might be regulated. Second, briefly to examine the relative importance of adult and juvenile survival, breeding frequency (and the problems of emigration/immigration) as factors influencing the population change. Finally, to consider the kinds of information required to detect adverse change in seabird populations within sufficiently short time-spans for effective management/conservation action to be taken. We confine our attention mainly to well-studied seabird species which often show more extreme demography e.g. albatrosses, petrels, kittiwakes.

Population regulation

There seems to be a general consensus in the ornithological literature that most bird populations are regulated, usually in a density-dependent fashion, and that important regulatory factors are food, breeding space, parasites, and predation. Evidence for these assertions that is other than correlative, inferential, or circumstantial is very scarce—and particularly so for seabirds.

Density-independent effects

Although environmental conditions are known to influence seabirds in many ways (e.g. breeding success, juvenile survival) there have been few suggestions that such events could regulate populations. However, the severity of the 1982/3 El Nino Southern Oscillation (ENSO) and its spectacular and widespread effects on seabird populations (e.g. Schreiber and Schreiber 1984; Bayer 1986; Hays 1986; La Cock 1986; Valle and Coulter 1987) rekindled interest in this topic. Some indication of the magnitude of possible effects comes from data on Galapagos Fur Seals *Arctocephalus galapagoensis*, whose ages of first breeding (4–5 years) and annual survival (*c.* 80–85 per cent) are similar to many seabird species. Trillmich and Dellinger (in press) show that breeding success was greatly reduced, in that all pups born in 1982 died within 5 months of birth and in 1983 pup production was only 11 per cent of 'normal'; also juvenile and adult survival was drastically affected (Table 13.2), with most potential recruits, all territorial males, and half the adult females disappearing entirely.

Table 13.2 Mortality of Galapagos fur seals after the 1982–3 ENSO event (data from Trillmich and Dellinger, 1990)

Element of population	Proportion died (%)
Pups	100
1-year-olds	100
2-year-olds	100
3-year-olds	70
Breeding males	100
Breeding females	50

For seabirds, detailed demographic data are lacking. We know only that breeding populations of Galapagos Penguins *Spheniscus mendiculus*, Flightless Cormorant *Nannopterum harrisi*, and Blue-footed and Masked Boobies *Sula nebouxii* and *S. dactylatra* were reduced by 77, 49, 31 and 61 per cent respectively (Valle and Coulter 1987; Gibbs *et al.* 1987). Except for the Galapagos Penguin, numbers of these species had largely recovered by 1984.

ENSOs of such severity are very rare, but the occurrence about every four years of events that are even much less severe is likely to have a demonstrable effect on populations of resident marine vertebrates. Whether this effect could act to keep populations below levels at which other regulatory factors (e.g. space, food) might operate merits investigation, at least in the case of tropical Pacific seabirds. In view of the potential conservation implications, there would seem to be a particular scope for simulation studies to model the demographic consequences of the regular imposition of various levels of density-independent mortality on Galapagos seabirds (including one species each of the main types of seabird there—penguin, cormorant, booby, petrel).

Predation

Large flying seabirds have few predators other than man. Individual seals, patrolling close to a particular colony, may take large numbers of penguins (e.g. Bonner and Hunter 1982), but the effects are insignificant at a population level. Skuas and gulls take many small petrels and shearwaters at their breeding colonies, but again at the population level such mortality is of limited significance. However, the nocturnal habits of small petrels and their reduced attendance at colonies on moonlit nights (e.g. Watanuki 1986) demonstrate that predation is an important selective force.

In contrast, effects on seabirds of predation by man and alien introduced predators can be very significant. This topic has been extensively reviewed (e.g. many papers in Croxall *et al*. 1984; Moors 1985) and is largely outside the scope of this paper. It is surprising, however, that despite adequate knowledge of the demography of, for example, auks and small petrels, the effects of various levels of predation by humans (as in the harvesting of Brunnich's Guillemot *Uria lomvia* in Greenland and Newfoundland for example) or by cats (e.g. predation on burrowing petrels at Marion Island) have not been evaluated in terms of potential effects on future prey population trends which could then be translated into appropriate management and conservation proposals or action (but see Hudson 1985). The scope for simulation studies is clear.

Parasitism

The only detailed study of seabird-parasite interactions involves the role of the argasid tick *Ornithodorus amblus* in causing desertion of nests by Peruvian guano birds (Duffy 1983*a*). This showed that high tick densities were correlated with high proportions (30–60 per cent) of birds deserting. Excluding ENSO years, similarly high levels of desertion have occurred in 30 per cent of years between 1940 and 1978 (Duffy 1983*a*). While the level and frequency of reduction in breeding success may not necessarily be significant in terms of population regulation, the effects of parasites might be more important at sites where the regular collection does not occur of guano deposits, including all nest material and substrate-dwelling parasites. Colony desertions following tick outbreaks have been reported for other seabirds (Feare 1976; King *et al*. 1977), and the potential role of parasites (including viral infections such as puffinosis) in influencing seabird populations needs critical assessment.

Breeding space

Most seabirds breed colonially and defend small territories around the actual nest site. Even so breeding numbers may be limited by available space (but not necessarily overall population size) as indicated by substantial increases following the provision of additional habitat (artificial islands) for guano birds in Namibia and Peru (Crawford and Shelton 1978; Duffy 1983*b*). In contrast, many surveys, particularly in temperate and polar regions, have indicated that large expanses of potentially suitable habitat exist (based on comparisons of vegetation type, slope, aspect, etc. with occupied sites) (see e.g. Dunnet *et al*. 1979; Croxall and Prince 1980; Hunter *et al*. 1982).

Regulation of population by limited available breeding space requires that all suitable sites are occupied and a pool of potential immediate recruits of both sexes exist (i.e. birds capable of breeding but not doing so). In practice, in addition to the pool of recruits, we would expect to observe (a) variation in nest site quality, (b) that birds at better nests have higher breeding success and/or higher chances of survival, and (c) that use of these sites varies in a density-dependent fashion. Two studies, in northeast England, provide important evidence that such conditions can be fulfilled, at least on a local scale.

For the European Shag *Phalacrocorax aristotelis* population on the Farne Islands, Potts *et al*. (1980) classified nest sites in terms of objective quality criteria and showed that pairs at better sites fledged significantly more chicks (in fact over twice as many) than those at poorer sites. Nest quality accounted for 84 per cent of the variance in breeding success, the only other identified factor of any significance being previous breeding experience. Because the Farne Island shag population was in a phase of consistent expansion at the time, the use of poorer sites increased and mean site quality and average chick production decreased (Fig. 13.1).

In the Black-legged Kittiwake *Rissa tridactyla* at North Shields, Porter and Coulson (1987) showed that:

1 As size of the breeding population increased, recruitment rate decreased; recruitment rate and adult mortality rate were correlated (Fig. 13.2);

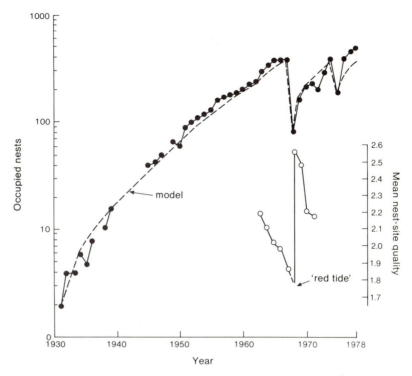

Fig. 13.1. Breeding population of the European Shag on the Farne Islands 1931–78 (●) in comparison with changes in mean quality of nest site, 1962–71 (○), showing the improvement following the 'red tide' of 1968 coinciding with the population crash (from Potts *et al*. 1980).

2 Of the nest sites (window ledges) at this warehouse colony, central sites were used preferentially, with birds breeding at these sites showing a higher reproductive success (Coulson 1968);

3 Provision of additional central sites led to 80 per cent of these being used, all by pairs that included at least one recruit; i.e. established pairs did not change nesting sites;

4 Current (in 1986) overall use of sites was only abut 65 per cent of that available.

This latter example illustrates the dilemma of recruitment/immigration facing birds existing in a pool of potential recruits. They can (a) breed now at North Shields, but at a low quality site with probable low lifetime reproductive success; or (b) delay breeding until a high quality site becomes available at North Shields; or (c) go and breed at another colony on the assumption that any such colony might offer, on average, a better

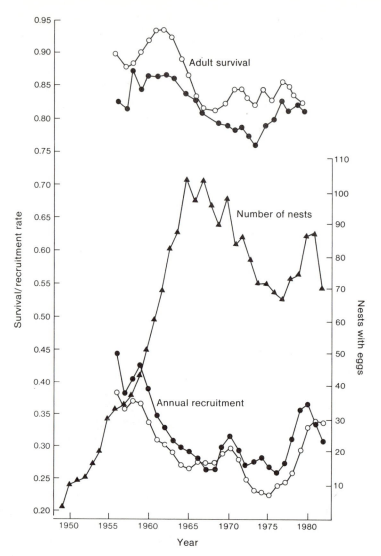

Fig. 13.2. Breeding populations of the Black-legged Kittiwake at North Shields 1952–82 (▲), in comparison with survival and recruitment rates (5-year running means) of adult males (●) and females (○) (from Coulson and Thomas 1985; Porter and Coulson 1987).

chance of a good quality site than presently offered at North Shields. The proportion of birds actually making any one of the above three choices is not clear. However (i) few birds appear to adopt the first alternative, otherwise the colony would use more than 65 per cent of sites; (ii) when recruitment rates increased (because adult mortality rates increased which presumably increased availability of better quality sites) this was accompanied by a reduction in age of first breeding, suggesting that more birds were prepared to recruit under these circumstances. However, changes in the age of first breeding have been observed in other seabird species and populations where there is no indication of any restriction in nest site availability, differences in nest site quality, or competition for nests (e.g. Coulson *et al*. 1982; Weimerskirch and Jouventin 1987; Croxall *et al*. 1990) so demographic changes of this type do not in themselves provide any evidence for the regulating effects of breeding space.

Therefore, the regulation of seabird populations through limitations on breeding space will depend crucially on the processes of recruitment, emigration, and immigration. It is unfortunate that we know so little about the factors influencing the dynamics of these processes in seabirds (but see Brooke 1978; Harris 1983). In these circumstances the more general role of breeding space in limiting seabird populations, other than at the local scale of certain breeding colonies, remains uncertain. Other evidence often adduced as potentially correlated with restrictions on nest site availability (e.g. Birkhead and Furness 1985) includes the occurrence of floaters (non-breeding sexually mature birds) and overt competition for nest sites. In neither case is the evidence very convincing. Populations of many, if not most, seabird species will contain floaters, because recruitment typically takes place over several years and individuals are usually sexually mature one or more years before recruitment takes place (Hector *et al*. 1986*a*, *b*). There are few, if any, adequate quantitative data on the extent and significance of intraspecific competition for nest sites among seabirds. Even Duffy's (1983*b*) study of intraspecific competition for nesting space showed that the two species, Guanay Cormorant *Phalacrocorax bougainvilli* and Peruvian Booby *Sula variegata*, whose habitat overlapped most showed little intraspecific aggression and were unable to displace each other from already acquired sites. Competition was chiefly, therefore, through a scramble to occupy space at the beginning of each breeding season.

Food availability

There are numerous sources of information indicating that food availability influences aspects of seabird biology. One line of evidence comes from many studies showing that changes in some aspect of the ability of adults to supply food to offspring (e.g. in any one, or combination of, delivery frequency, mass, and quality) causes changes in the growth and/or survival of offspring. Table 13.3 shows an example of particularly detailed data, from a study of Grey-headed Albatross *Diomedea chrysostoma* at South Georgia. These studies, which include twinning and other manipulation

Table 13.3 Comparison of various elements of reproductive performances (mean values ± standard deviation) in Grey-headed Albatrosses at Bird Island, South Georgia in an average year (1981–2) and a year of reduced food availability (1983–4). Differences between years significant at: *$P < 0.05$, **$P < 0.01$, ***$P < 0.001$; other differences not significant. For further details see Croxall *et al.* (1988) and Prince *et al.* (submitted).

Parameter	1981–2		1983–4
Breeding population size (pairs)	444		394
Meal mass (g)	670 ± 259		687 ± 180
Foraging trip duration (h)	53 ± 37	***	94 + 75
Time budget at sea:			
% flight	72 ± 9	*	60 ± 17
% night on sea (= feeding)	50 ± 14	**	76 ± 7
% day on sea (= resting)	50 ± 9	**	6 ± 3
Chick growth:			
rate (g d^{-1})	44 ± 6	***	25 ± 5
peak mass (g)	4642 ± 727	***	3023 ± 553
fledge mass (g)	3561 ± 428	***	2454 ± 382
fledge period (d)	141 ± 5	***	149 ± 8
Breeding success (%)	53		28
Breeding population size (pairs)			
next year	333		235
change (%)	−25		−40
change from predicted (%)	+23 (271)		−55 (424)

experiments, are not reviewed here, but it should be noted that none takes account of the potential effect of changes in the costs of breeding on adult survival.

A second line of evidence comes from correlations between changes in population parameters of seabirds and changes in prey stocks (Fig. 13.3). Relevant data come from comparisons of:

(1) seabird stocks with prey stocks, e.g. Namibian guano birds versus biomass of Cape Pilchard *Sardinops ocellata* (Crawford and Shelton 1978) or Arctic Terns *Sterna paradisaea* populations in Scotland versus Herring *Clupea harengus* stock estimates (Monaghan and Zonfrillo 1986);

(2) seabird stocks with prey harvest, e.g. Peruvian guano bird numbers with Anchoveta *Engraulis ringens* catch (Santander 1980; Duffy and Siegfried 1987);

(3) seabird harvests with prey harvests, e.g. biomass of squid in the diet of Northern Gannet *Sula bassana* in Newfoundland versus local squid catches (Montevecchi *et al.* 1988);

(4) elements of predator reproductive performance (breeding success, chick growth rate, etc.) versus prey stock size, e.g. Arctic Tern chick

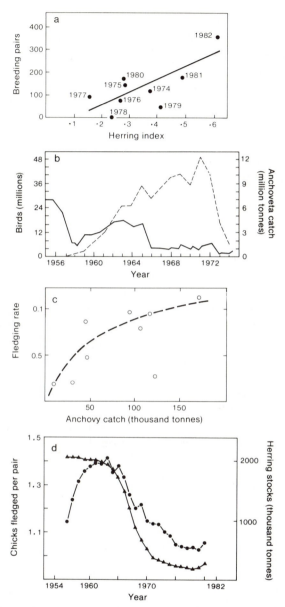

Fig. 13.3. Relationships between seabird populations and prey stocks. (a) Numbers of breeding terns in Firth of Clyde, Scotland in relation to an index of North Sea Herring stocks (Monaghan and Zonfrillo 1986). (b) Numbers of Peruvian guano birds in relation to annual catches of Anchoveta (Santander 1980). (c) Brown Pelican production (fledging rate) versus anchovy catch in the Southern Californian Bight 1971–9 (Anderson *et al.* 1980). (d) Chick production (5-year running mean) of Black-legged Kittiwakes (●) versus North Sea Herring stocks (▲) (Coulson and Thomas 1985).

growth rates and breeding success in Shetland (Monaghan *et al.* 1989), Black-legged Kittiwake breeding success at North Shields (Coulson and Thomas 1985), and Atlantic Puffin *Fratercula arctica* breeding success in Norway (Lid 1981) versus Herring stock estimates; also Brown Pelican *Pelecanus occidentalis* breeding success in Southern California versus Northern Anchovy *Engraulis mordax* stock estimates (Anderson *et al.* 1980, 1982).

There are two confounding problems with this approach. First, the evidence relates changes in the seabird population to changes in prey abundance or prey availability to fishing boats, rather than prey availability to the bird. Second, for many of the relationships, correlations are best at the extremes (i.e. when resources are particularly high or low there is a good match with the seabird index) but the association is weak at intermediate levels.

A third and rather different approach is based on evidence, or rather inferences, for intra-specific competition for food in seabirds as a potential key element in density-dependent population regulation. The basic argument for potential density-dependent processes relating to food were discussed by Lack (1954, 1966) in relation to regulation in the non-breeding season and by Ashmole (1963, 1971) in relation to regulation in the breeding season; these were well-summarized by Birkhead and Furness (1985). Basically, Lack argued that in the non-breeding season, food supplies were scarcest, seabird numbers were highest, and therefore potential competition was strongest. Ashmole argued (see Fig. 13.4) that because seabird densities were so many times greater in the breeding season than the non-breeding season (because the foraging range of the

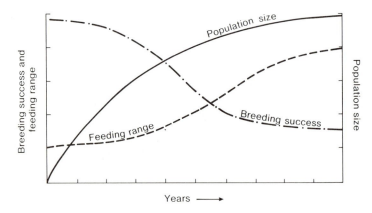

Fig. 13.4. Relationships between colony size, foraging range, and breeding success suggested by Ashmole's model of seabird regulation (from Furness and monaghan 1987).

breeding adults was constrained by the need regularly to feed offspring), and that the apparently superabundant prey in the breeding season did not necessarily imply superabundant prey availability, the most likely regulatory outcome was density-dependent reproductive success. Lack also argued that mortality in most seabird species mainly occurs outside the breeding season. However, this is not necessarily so for adults and, even if true, could not be considered unequivocal support for his view because deaths outside the breeding season might result from e.g. low fledging weight or poor adult condition as a result of effects during the breeding season.

Some support for Ashmole's view comes from indications that seabirds might significantly deplete local resources during the breeding season. Various studies of the food requirements of breeding seabirds have suggested that they may consume a significant proportion of local prey stocks. However, the numerous uncertainties inherent in the assumption on which the calculations are founded (reviewed in Croxall 1987) suggest caution in any inferences derived therefrom. The only deliberate test of Ashmole's hypothesis involved measuring fish population densities within and without the foraging range of Double-crested Cormorants *Phalacrocorax auritus* (Birt *et al*. 1987). This survey, done late in the chick rearing period, showing significantly lower fish densities in the vicinity of the colony than those recorded more than 40 km away, well outside the foraging range of breeding birds. Unfortunately, without a similar survey of fish densities prevailing before the breeding season these results cannot be regarded as conclusive.

Additional support comes from evidence that, within species, colonies of different sizes (i.e. producing different foraging densities at sea) might show differences in reproductive performance. For Brunnich's Guillemots Gaston *et al*. (1983) and Furness and Barrett (1985) found significant correlations between colony size and chick fledging weight. In a more comprehensive study, Hunt *et al*. (1986) compared five seabird species (two kittiwakes, two guillemots, and Red-faced Cormorant *Phalacrocorax urile*) having populations of very different size on two islands in the Pribilof archipelago, Alaska (Table 13.4). They found a consistently significant inverse relationship between population size and (a) chick growth rate for all species except Red-legged Kittiwake *Rissa brevirostris*, (b) chick weight for both guillemot species, but (c) no consistent relationship with breeding success in any species. Further analyses by Birkhead and Furness (1985) were consistent with intra- rather than inter-specific effects being the prime determinant. In addition there are indications that colonies of different size might be distributed, within species, to avoid excessive local densities of seabirds at sea during the breeding season. The predicted inverse relationships between colony size and numbers at other

Table 13.4 Reproductive performance in relation to population size of seabirds at St George (G) and St Paul (P) Islands, Alaska. Data from Fig. 1 of Hunt *et al*. (1986) averaged over 1976–8. For sample sizes see Hunt *et al*. (1986), Table 2

Species	Population size (pairs)		Chick growth rate (g d⁻¹)		Chick fledging weight (g)		Breeding success (chicks/nest)	
	G	P	G	P	G	P	G	P
Red-faced Cormorant	5000	2500	52	62	—	—	1.5	1.1
Black-legged Kittiwake	72 000	31 000	13	14	440	445	0.4	0.5
Red-legged Kittiwake	220 000	2200	12	13	390	400	0.4	0.4
Common Guillemot	190 000	39 000	8	9	175	200	0.4	0.5
Brunnich's Guillemot	1 500 000	110 000	8	13	160	215	0.5	0.5

colonies within specified distances were revealed in analyses of British populations of Northern Gannets, Atlantic Puffins, European Shags, and Black-legged Kittiwakes (Furness and Birkhead 1984). However, the distances at which significant effects were detected did not correlate well with the normal foraging ranges of the species in the breeding season, particularly in the cases of the Puffin and Shag. Because so many factors influence the location and size of seabird colonies, it is unlikely that failure to have found the expected relationship would have been taken as serious evidence against Ashmole's hypothesis; overall the evidence is inconclusive.

In general there is good evidence that changes in food availability may affect the reproductive output of seabirds and correlate with overall changes in population size. There are also strong inferences that intraspecific density-dependent competition for food has the potential to act as a regulating mechanism, perhaps especially during the breeding season (but this largely reflects the absence of relevant data from other times of year).

Conclusions

1. Density-independent environmental effects (frequency of catastrophies) and parasites may be more important influences, at least on seabirds in tropical areas, than was hitherto recognized and are deserving of further investigation.

2. The availability of nest sites can act as an important constraint on the

breeding population in many seabird species, especially in a proximate fashion and at a local level.

3. Of any single factor, food availability is probably the one most likely consistently to provide an ultimate limit to overall numbers of seabirds and may frequently be significant as a proximate factor at the level of the breeding population.

4. So far we have discussed the main potential factors affecting population levels as if each factor operated in isolation. In reality all will be operating at some level and may be acting in concert or opposition. The Black-legged Kittiwake is a useful example because it is clear that the size of the breeding population at North Shields was strongly influenced by nest site availability and quality. However, the relationship between Herring stocks, demographic rates, and breeding success (Fig. 13.3d) indicates that food availability may have played the dominant role in influencing the main parameters (adult and juvenile survival and breeding success) which were responsible for the observed changes in population size at North Shields and probably over much larger areas.

5. The relative contribution of various factors to overall population regulation will, of course, also vary with the temporal and spatial scales being considered. Definition of the time-span, population, or area being considered is not always explicit in discussions of these topics and thus may significantly bias interpretation.

6. Few if any of the data discussed here comes from natural systems (in the sense of unperturbed by man). Some combination of man-induced artificial influences (e.g. direct human predations, mortality through introduced animals, habitat destruction, pollution, competition with seabirds for common prey resources, etc.) may have changed the relative importance of some or all of the factors already discussed as being potential regulators of seabird populations. This distinction is of considerable relevance when considering the relationship between species' theoretical sensitivity and vulnerability in practice.

Demography and conservation

Effective conservation of seabird populations requires the ability to detect adverse changes as quickly as possible, to assess the significance of these, and to determine the possible cause. An attempt to distinguish between responses due to natural events and those caused by human activity may also be important.

Conventional expression of population change is: Population Change = (Recruitment + Immigration) − (Mortality + Emigration). In practical terms, adequately detailed demographic studies usually try to estimate annually the size of the breeding population and its breeding success.

Within the breeding population the main elements distinguished are recruits (often separating immigrants from birds reared within the study population) and previous breeders; and within the latter, the number and proportion that died or emigrated since the last sampling occasion and those that are still alive, but not breeding. The collection of these data enables estimation of the key variables, i.e. adult and juvenile survival, breeding frequency and also, often with rather varying degrees of accuracy, of emigration and immigration rates.

Changes in breeding populations are affected to different extents by changes in each of these main demographic parameters and the balance between them differs for different species. This theoretical sensitivity depends on the species' population structure and generation time, i.e. on the precise values of the main demographic parameters themselves. Identifying the most sensitive elements of the demography of a particular species is obviously of considerable relevance to conservation, particularly when matched with good estimates of the species' practical vulnerability, i.e. the magnitude of the existing impacts on each parameter.

We illustrate this approach using first a simple model involving adult survival, juvenile survival, and breeding success. We then develop the method to discuss the specific case history of the Wandering Albatross at South Georgia.

Seabird demography: a simple model

For the model, let N_i denote the number of females in the breeding population in year i. If the average adult survival is denoted by s_A, the average juvenile survival to breeding status by s_j, and the mean number of female fledglings per breeding pair by f, then the expected population size in year $(i + 1)$ is given by

$$N_{(i+1)} = s_A N_i + f s_j N_{(i+1-k)}$$

where k denotes the age at which the birds join the breeding population. Simple models of this type for bird populations were introduced by Capildeo and Haldane (1954).

For losses due to adult mortality to be balanced by the new recruits we must have, on average, $(1 - s_A) = f s_j$. Figure 13.5 shows values of s_j and f required to balance model populations for a range of adult survival values which covers that of most seabirds. For illustration, two hypothetical species are shown. Species P has a relatively high adult survival rate (95 per cent) and low recruitment, typical of the Atlantic Puffin, Northern Fulmar *Fulmarus glacialis*, and many other auks and Procellariiformes. By contrast, species S has a lower adult survival (80 per cent) and higher recruitment, more typical of shags, gulls, penguins, and pelicans.

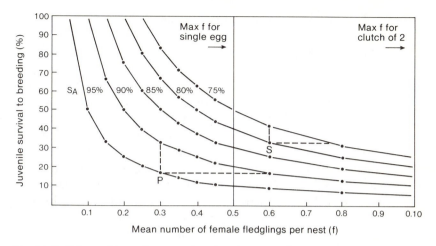

Fig. 13.5. Values of juvenile survival to breeding and the number of female chicks fledged per female required to maintain model populations with different adult survival rate (s_A). P illustrates a species with high survival and low recruitment; S a species with relatively low survival and higher recruitment. See text for details of the model.

In both P and S, the changes in juvenile survival or breeding success required to offset a 5 per cent decrease in adult survival are indicated. For species S, the required increases are 0.09 for juvenile survival and 0.16 for the mean number of fledged female chicks; for species P, the corresponding increases are 0.17 and 0.30. Note that the required increase in breeding success would not be possible from a single clutch of eggs. In this sense, species P is relatively more sensitive to changes in adult survival.

The same point can be examined by looking at changes in size of the breeding population for specified changes in the values of the demographic parameters. The data of Table 13.5 are for a reduction of 0.05 in any parameter and with the assumption that species S breeds at 3 years old and species P at 5 years. This shows that declines in breeding population size are more sensitive to changes in adult survival than in juvenile survival or breeding success. The effect is more pronounced in species P with relatively high adult survival and low reproductive rate.

Table 13.5 Long-term population declines associated with a reduction of 0.05 in any demographic parameter

Parameter	Species P	Species S
Adult survival (s_A)	4.2%	3.6%
Juvenile survival (s_J)	1.5%	2.4%
Breeding success (f)	0.8%	1.3%

The changes imposed on the model in the above analysis are rather arbitrary and, furthermore, differential effects of changes in survival and recruitment may be affected by regulatory factors that are not included in the model. However, the analysis accords fairly well with the assessment of the relationship between the nature of threats and the severity of the consequences they represent for different types of seabirds. Thus the species of *Pterodroma* petrels (highly K-selected seabirds) exposed to threats to adult survival from cats are at much greater risk than those species facing threats to eggs and juveniles from rats; this fits well with reviews of their population status (see various authors in Croxall *et al.* 1984; and Moors 1985). At the other extreme, persistent disturbance to breeding colonies (whether through direct disturbance or habitat loss) or persistent chronic exploiting of eggs are observed to have more disastrous effects on tern and booby populations (Feare 1984; de Korte 1984, personal communication) than actual predation on adults.

The simple model makes no allowance for intermittent breeding. In most annually breeding seabirds a population of birds that have already bred may not do so in any particular year. These may be individuals who lost their partner before the start of a breeding season, but particularly in species with naturally low reproductive rates and high longevity, a signifi-cant proportion of the existing breeding population may not breed in some years. Changes in the proportion of these intermittent breeders are a very important element of year-to-year population fluctuations in the Black-browed Albatross *Diomedea melanophrys* (Prince *et al.* submitted). Fluctuations of similar magnitude in the breeding population of the Northern Fulmar, which were mainly attributed to changes in the immigration and emigration rate (Ollason and Dunnet 1983) may be influ-enced by the same phenomenon. A number of seabird studies have used field methods which precluded distinguishing, from year to year, changes in rates of recruitment, emigration, and intermittent breeding. When new demographic studies of K-selected seabirds are being established, it is crucial to allow for the ability to make these distinctions.

Wandering albatross demography at South Georgia

The above analysis enables us to make an assessment of the theoretical sensitivity of seabird populations to changes in the demographic para-meters. This assessment, however, needs to be matched with the observed values of these demographic parameters in actual populations. To illustrate the comparison between theoretical sensitivity and practical vulnerability we use our data for the Wandering Albatross *Diomedea exulans* at Bird Island, South Georgia (Croxall *et al.* 1990), where the breeding population has declined by 22 per cent since 1961.

In general, comparison is easiest for species with low or constant immigration/emigration rates. At Bird Island, fledgling Wandering Albatrosses show 99 per cent philopatry. Furthermore, birds that breed on the island have never been recorded breeding elsewhere.

Wandering Albatrosses broadly follow a biennial breeding pattern. Birds that breed successfully generally return 2 years later to breed again, but some defer their next breeding attempt by 3 or 4 years. Birds that fail usually try again the following year, but some of these birds then defer their breeding by 2 or 3 years. This pattern of breeding periodicity has been built into the mathematical model, given below, as follows. For failed birds we assume that, subject to their surviving, a proportion P_{f1} return the following year whereas P_{f2} and P_{f3} defer their next breeding attempt by 2 and 3 years respectively. For successful birds, the corresponding pattern is denoted by P_{s2}, P_{s3}, and P_{s4}.

Adult survival and breeding success are related because if either parent dies the chick usually dies. Thus, if adult survival is denoted by s_A and b is the proportion of females that fledge a chick, then the average proportion of females that fail but survive to the next breeding season is $s_A - b$. The number of females in the breeding population in year $(i + 1)$ is made up of: (a) failed birds surviving from years i, $(i - 1)$, and $(i - 2)$; (b) successful bird surviving from years $(i - 1)$, $(i - 2)$, and $(i - 3)$; and (c) the fledglings from previous years which survive to breed for their first time in year $(i + 1)$. Thus, if N_i denotes the size of the breeding population in year i then a simple model for the average population changes is as follows:

$$N_{(i+1)} = P_{f1}(s_A - b)N_i + P_{f2}s_A(s_A - b)N_{(i-1)} + P_{f3}s_A^2(s_A - b)N_{(i-2)}$$
$$+ P_{s2}s_AbN_{(i-1)} + P_{s3}s_A^2bN_{(i-2)} + P_{s4}s_A^3bN_{(i-3)}$$
$$+ s_jb\Sigma a_kN_{(i+1-k)}/2$$

In the last term, s_j denotes overall survival of fledglings to breed and a_k is the proportion of birds surviving to breed at age k.

Analysis of data from captures of marked birds (Croxall *et al.* 1990) yielded the following estimates for the demographic parameters. Adult survival 94 per cent; juvenile survival 32 per cent (range 25–35 per cent); breeding success 64 per cent (range 52–73 per cent). Breeding frequency: $P_{s2} = 0.75$, $P_{s3} = 0.15$, $P_{s4} = 0.10$, $P_{f1} = 0.80$, $P_{f2} = 0.15$, $_{f3} = 0.05$. Age at first breeding: $a_8 = 0.036$; $a_9 = 0.071$, $a_{10} = 0.250$, $a_{11} = 0.286$, $a_{12} = 0.179$, $a_{13} = 0.071$, $a_{14} = 0.071$, $a_{15} = 0.036$. Using the estimates in the above model produces a population decline of 0.6 per cent per annum. This is close to the observed decline of 1 per cent per annum based on the population counts (Croxall *et al.* 1990).

A diagrammatic representation of the elements of the demographic balance sheet (Fig. 13.6) illustrates the equivalence, in terms of the size of

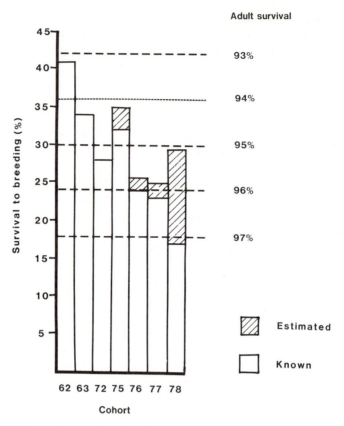

Fig. 13.6. Values of survival to breeding required to maintain model populations of Wandering Albatross with different adult survival rates and breeding success of 64 per cent. Histograms show the observed survival rates to breeding for seven cohorts of chicks at Bird Island. See text for details of the model.

the breeding population, between changes in juvenile survival to breeding and changes in adult survival. In particular, a reduction of 1 per cent in adult survival must be offset by an increase of 6 per cent in juvenile survival to prevent a decline. It also shows that, given our estimate of a current adult survival rate of 94 per cent, recruitment from recent cohorts is insufficient to maintain breeding population size.

The current population decline of Wandering Albatrosses at South Georgia is thus consistent with our estimates of adult survival and recruitment and suggests that survival is now lower than in the 1960s, when it was 96 per cent for adults (Tickell 1968) and 34–41 per cent for juveniles (Croxall *et al.* 1990). The most likely cause of this incidental mortality of albatrosses is due to fishing operations, especially long line fishing for Southern Blue-fin Tuna *Thunnus maccoyii* (Brothers 1990; Croxall *et al*. 1990). About four times as many juvenile as adult albatrosses are caught at the fishery (Brothers 1990) so the practical vulnerability in this species is very close to the converse of the theoretical sensitivity. Depending on the accuracy with which the age of the fishery-killed birds is assessed, the annual losses to the Bird Island population represent 2–3 per cent of the adults and 14–26 per cent of the juveniles. These levels of mortality are broadly consistent (though juvenile mortality is currently higher than is recorded) with those required to produce the observed reductions in breeding population size.

Monitoring population parameters

The use of seabirds as 'monitors' of the marine environment is becoming increasingly widespread. For most studies the monitoring involves measurement of breeding population size and breeding success and sometimes of indices of reproductive performance. Monitoring other population parameters (e.g. survival rates) is usually regarded as part of the background information necessary for interpreting changes. As is made clear here, detection of change in demographic parameters is a crucial element of any overall monitoring programme, but the relative importance of the various elements will depend to some extent on the life-style of the species concerned.

Obviously any monitoring study or system requires long-term commitments, and such studies should usually combine annual or periodic measurements of demographic parameters with, usually annual, measurement of breeding population size and indices of reproductive performance. In theory such a system should provide data capable of assessing changes at a variety of temporal scales. In this paper we are mainly concerned with the demographic parameters, Croxall *et al*. (1988) having reviewed the practical and theoretical advantages and disadvantages of many indices of reproductive performance in seabirds. In the demographic context, two sets of different constraints may be relevant when comparisons need to be made between what is desirable and what is actually possible (i.e. for financial and/or logistic reasons). First, the priority of different parts of the monitoring system differs for different types of species. Thus for extreme *K*-selected species the order might be adult survival, juvenile survival, breeding success, whereas for an extreme *r*-selected species the sequence might be breeding success, juvenile survival, adult survival. Second, the different population parameters integrate environmental (and other) effects over different time-spans; there are different lag times before the effects can be measured and different accuracies with which effects are likely to be detected (for detailed discussion see Croxall *et al*. 1988).

For adult survival, very large samples are required to detect significant change. For instance, in the Wandering Albatross study described above, for a sample of 1000 birds the standard error of an annual survival estimate, assuming a 100 per cent recapture rate, is 0.8 per cent; for such long-lived birds a consistent reduction in survival of 1–2 per cent per annum can be highly significant. Also accurate estimates of survival in one year cannot be made until a further 2 to 3 years have elapsed, because of the extent to which birds are known to show intermittent breeding.

Estimation of juvenile survival also needs large samples of birds and in species with long deferred sexual maturity there will be a considerable delay in obtaining results. This may be short-circuited, for some species, by obtaining estimates of survival from the birds that attend the colony in the years prior to that of the first breeding attempt.

Estimation of breeding success is usually very straightforward, requiring only estimates of the breeding population size, mean clutch size, and the number of chicks fledging. However, as indicated by Hunt *et al*. (1986), breeding success is not necessarily a very sensitive indicator of reproductive performance. The results of several recent studies support this (see e.g. Table 13.3) and suggest that other indices of reproductive performance, especially those relating to provisioning rate, might be preferable.

Within any monitoring programme the final balance between the parameters selected for study will reflect some compromise between what is desirable and what is possible. To achieve the best balance, it is essential to make an assessment of the species' population structure and dynamics and to complement this with an evaluation of the factors most likely currently to be limiting (or significantly influencing) population size.

Summary

We briefly review the main current ideas on how seabird populations may be regulated, considering evidence for the role of density-independent (catastrophe) effects and density-dependent factors, such as predation, parasitism, breeding space, and food. We then consider the potential relative importance of adult and juvenile survival and breeding frequency for influencing population change in different types of seabird. As a particular example we consider the Wandering Albatross at South Georgia and compare its theoretical sensitivity to those factors with current data on breeding success, recruitment, and adult survival which illustrate the actual present vulnerability of this species. Finally we consider some of the practical constraints relevant to incorporating demographic parameters into routine monitoring studies using seabirds.

Acknowledgements

We thank Dirk Briggs for drawing the diagrams, Julie Thomson for typing the manuscript, Peter Prince and Mary Taylor for help with the preparation of this report, Nigel Brothers for allowing us to use his data, and Mike Harris and Peter Hudson for critical review of the manuscript. We thank also the organizers, hosts, and other participants of the symposium for creating a stimulating forum for discussion.

References

Anderson, D. W., Gress, F., Mais, K. F., and Kelly, P. R. (1980). Brown Pelicans as anchovy stock indicators and their relationships to commercial fishing. *CalCOFI Reports*, **21**, 54–61.

Anderson, D. W., Gress, F., and Mais, K. F. (1982). Brown Pelicans: influence of food supply on reproduction. *Oikos*, **39**, 23–31.

Ashmole, N. P. (1963). The regulation of numbers of tropical oceanic birds. *Ibis*, **103**, 458–73.

Ashmole, N. P. (1971). Seabird ecology and the marine environment. In *Avian biology*, Vol. 1 (eds. D. S. Farner, J. R. King, and K. C. Parkes), pp. 224–86. Academic Press, New York.

Bayer, R. D. (1986). Breeding success of seabirds along the mid-Oregon coast concurrent with the 1983 El Nino. *Murrelet*, **67**, 23–6.

Birkhead, T. R. and Furness, R. W. (1985). Regulation of seabird populations. In *Behavioural ecology* (eds. R. M. Sibley and R. H. Smith), pp. 145–67. Blackwell, Oxford.

Birt, V. L., Birt, T. P., Goulet, D., Cairns, D. K., and Montevecchi, W. A. (1987). Ashmole's halo: Direct evidence for prey depletion by a seabird. *Marine Ecology Progress Series*, **40**, 205–8.

Bonner, W. N. and Hunter, S. (1982). Predatory interactions between Antarctic Fur Seals, Macaroni Penguins and Giant Petrels. *British Antarctic Survey Bulletin*, **56**, 75–9.

Brooke, M. de L. (1978). The dispersal of female Manx Shearwaters *Puffinus puffinus*. *Ibis*, **120**, 545–51.

Brothers, N. B. (1990). Approaches to reducing albatross mortality and associated bait loss in the Japanese longline fishery. Biological Conservation.

Capildeo, R. and Haldane, J. B. S. (1954). The mathematics of bird population growth and decline. *Journal of Animal Ecology*, **23**, 215–23.

Coulson, J. C. (1968). Differences in the quality of birds nesting in the centre and on the edges of a colony. *Nature*, **217**, 478–9.

Coulson, J. C. and Thomas, C. (1985). Changes in the biology of the Kittiwake *Rissa tridactyla*: a 31-year study of a breeding colony. *Journal of Animal Ecology*, **54**, 9–26.

Coulson, J. C., Duncan, N., and Thomas, C. (1982). Changes in the breeding biology of the Herring Gull *Larus argentatus* induced by reduction in the size and density of the colony. *Journal of Animal Ecology*, **51**, 739–56.

Crawford, R. J. M. and Shelton, P. A. (1978). Pelagic fish and seabird interrelationships off the coasts of South West and South Africa. *Biological Conservation*, **14**, 85–109.

Croxall, J. P. (ed.) (1987). *Seabirds: feeding ecology and role in marine ecosystems*. Cambridge University Press.

Croxall, J. P. and Prince, P. A. (1980). Food, feeding ecology and ecological segregation of seabirds at South Georgia. *Biological Journal of the Linnean Society*, **14**, 103–31.

Croxall, J. P., Evans, P. G. H., and Schreiber, R. W. (ed.) (1984). *Status and conservation of the world's seabirds*. ICBP, Cambridge.

Croxall, J. P., McCann, T. S., Prince, P. A., and Rothery, P. (1988). Reproductive performance of seabirds and seals at South Georgia and Signy Island, South Orkney Islands, 1976–87: Implications for Southern Ocean monitoring studies. In *Antarctic Ocean and Resources Variability* (ed. D. Sahrhage), pp. 261–85. Springer-Verlag, Berlin.

Croxall, J. P., Rothery, P., Pickering, S. P. C., and Prince, P. A. (1990). Reproductive performance, recruitment and survival of Wandering Albatrosses *Diomedea exulans* at Bird Island, South Georgia. *Journal of Animal Ecology*, **59**, 775–96.

Duffy, D. C. (1983*a*). The ecology of tick parasitism on densely nesting Peruvian seabirds. *Ecology*, **64**, 110–19.

Duffy, D. C. (1983*b*). Competition for nesting space among Peruvian guano birds. *Auk*, **100**, 680–8.

Duffy, D. C. and Siegfried, R. W. (1987). Historical variations in food consumption by breeding seabirds of the Humboldt and Benguela upwelling regions. In *Status and conservation of the world's seabirds* (eds. J. P. Croxall, P. G. H. Evans, and R. W. Schreiber), pp. 327–46. ICBP, Cambridge.

Dunnet, G. M., Ollason, J. C., and Anderson, A. (1979). A 28-year study of breeding Fulmars *Fulmarus glacialis* (L.) in Orkney. *Ibis*, **121**, 293–300.

Feare, C. J. (1976). Desertion and abnormal development in a colony of Sooty Terns *Sterna fuscata* infested by virus-infected ticks. *Ibis*, **118**, 112–15.

Feare, C. J. (1984). Human exploitation. In *Status and conservation of the world's seabirds* (ed. J. P. Croxall, P. G. H. Evans, and R. W. Schreiber), pp. 691–9. ICBP, Cambridge.

Furness, R. W. and Barrett, R. T. (1985). The food requirements and ecological relationships of a seabird community in North Norway. *Ornis Scandinavica*, **16**, 305–13.

Furness, R. W. and Birkhead, T. R. (1984). Seabird colony distributions suggest competition for food supplies during the breeding season. *Nature*, **311**, 655–6.

Furness, R. W. and Monaghan, P. (1987). *Seabird ecology*. Blackie, Glasgow.

Gaston, A. J., Chapdelaine, G., and Noble, D. G. (1983). The growth of Thick-billed Murre chicks at colonies in Hudson Strait: inter- and intra-colony variation. *Canadian Journal of Zoology*, **61**, 2465–75.

Gibbs, H. L., Latta, S. C., and Gibbs, J. P. (1987). Effects of the 1982–83 El Niño event on Blue-footed and Masked Booby populations on Isla Daphne Major, Galapagos. *Condor*, **89**, 440–2.

Harris, M. P. (1983). Biology and survival of the immature Puffin *Fratercula arctica*. *Ibis*, **125**, 56–73.

Hays, C. (1986). Effects of the 1982–83 El Niño on Humboldt Penguin colonies in Peru. *Biological Conservation*, **36**, 169–80.

Hector, J. A. L., Croxall, J. P., and Follett, B. K. (1986*a*). Reproductive endocrinology of the Wandering Albatross *Diomedea exulans* in relation to biennial breeding and deferred sexual maturity. *Ibis*, **128**, 9–22.

Hector, J. A. L., Follet, B. K., and Prince, P. A. (1986*b*). Reproductive endocrinology of the Black-browned Albatross *Diomedea melanophris* and the Grey-headed Albatross *D. chrysostoma*. *Journal of Zoology*, **208**, 237–54.

Hudson, P. J. (1985). Population parameters for the Atlantic Alcidae. In *The Atlantic Alcidae* (ed. D. N. Nettleship and T. R. Birkhead), pp. 233–61. Academic Press, London.

Hunt, G. L., Eppley, Z. A., and Schneider, D. C. (1986). Reproductive performance of seabirds: the importance of population and colony size. *Auk*, **103**, 306–17.

Hunter, I., Croxall, J. P., and Prince, P. A. (1982). The distribution and abundance of burrowing seabirds (Procellariiformes) at Bird Island, South Georgia. I. Introduction and methods. *British Antarctic Survey Bulletin*, **56**, 49–67.

King, K. A., Keith, J. O., and Mitchell, G. A. (1977). Ticks as a factor in nest desertion of California Brown Pelicans. *Condor*, **79**, 507–9.

de Korte, J. (1984). Status and conservation of seabird colonies in Indonesia. In *Status and conservation of the world's seabirds* (ed. J. P. Croxall, P. G. H. Evans, and R. W. Schreiber), pp. 527–45. ICBP, Cambridge.

La Cock, G. D. (1986). The Southern Oscillation, environmental anomalies, and mortality of two southern African seabirds. *Climatic Change*, **8**, 173–84.

Lack, D. (1954). *The natural regulation of animal numbers*. Oxford University Press.

Lack, D. (1966). *Population studies of birds*. Clarendon Press, Oxford.

Lid, G. (1981). Reproduction of the Puffin on Rost in the Lofoten Islands in 1964–80. *Cinclus*, **4**, 30–9.

Monaghan, P. and Zonfrillo, B. (1986). Population dynamics of seabirds in the Firth of Clyde. *Proceedings of the Royal Society of Edinburgh*, **90B**, 363–75.

Monaghan, P., Uttley, J. D., Burns, M. D., Thaine, C., and Blackwood, J. (1989). The relationship between food supply, reproductive effort and breeding success in Arctic Terns *Sterna paradisaea*. *Journal of Animal Ecology*, **58**, 261–74.

Montevecchi, W. A., Birt, V. L., and Cairns, D. K. (1988). Dietary changes of seabirds associated with local fisheries failures. *Biological Oceanography*, **5**, 153–61.

Moors, P. J. (ed.) (1985). *Conservation of island birds. Case studies for the management of the threatened island species*. ICBP, Cambridge.

Ollason, J. C. and Dunnet, G. M. (1983). Modelling annual changes in numbers of breeding Fulmars *Fulmarus glacialis* at a colony in Orkney. *Journal of Animal Ecology*, **52**, 185–98.

Porter, J. M. and Coulson, J. C. (1987). Long-term changes in recruitment to the breeding group, and the quality of recruits at a Kittiwake *Rissa tridactyla* colony. *Journal of Animal Ecology*, **56**, 675–89.

Potts, G. R., Coulson, J. C., and Deans, I. R. (1980). Population dynamics and the breeding success of the Shag *Phalacrocorax aristotelis* on the Farne Islands, Northumberland. *Journal of Animal Ecology*, **49**, 465–84.

Prince, P. A., Rothery, P., and Croxall, J. P. (submitted). Demography of Black-browed and Grey-headed Albatrosses at Bird Island, South Georgia. *Journal of Animal Ecology*.

Santander, H. (1980). *The Peru Current system 2: Biological aspects*. UNESCO, Paris.

Schreiber, R. W. and Schreiber, E. A. (1984). Central Pacific seabirds and the El Niño Southern Oscillation: 1982 and 1983 perspectives. *Science*, **225**, 713–16.

Tickell, W. L. N. (1968). The biology of the great albatrosses, *Diomedea exulans* and *Diomedea epomophora*. *Antarctic Research Series*, **12**, 1–55.

Trillmich, F. and Dillinger, T. (in press). The effects of El Niño on Galapagos pinnipeds. In *Effects of El Nino on pinnipeds in the eastern Pacific* (ed. F. Trillmich). University of California Press.

Valle, C. A. and Coulter, M. C. (1987). Present status of the Flightless Cormorant,

Galapagos Penguin and Greater Flamingo populations in the Galapagos Islands, Ecuador, after the 1982–83 El Niño. *Condor*, **89** 276–81.

Watanuki, Y. (1986). Moonlight avoidance behaviour in Leach's Storm Petrels as a defense against Slaty-backed Gulls. *Auk*, **103**, 14–22.

Weimerskirch, H. and Jouventin, P. (1987). Population dynamics of the Wandering Albatross *Diomedea exulans* of the Crozet Islands: causes and consequences of the population decline. *Oikos*, **49**, 315–22.

14 Changes in the population size and demography of southern seabirds: management implications

PIERRE JOUVENTIN and HENRI WEIMERSKIRCH

Introduction

Oceans occupy some 70 per cent of the world surface. The greater part of the northern hemisphere consists of land, while oceans represent the majority of the southern hemisphere, in a contiguous body of water known collectively as the Austral of Antarctic Ocean in its circumpolar reaches. Here the seabird communities are the richest in the world in terms of both actual numbers and numbers of species. Despite the impact of sealing and whaling on subantarctic ecosystems, many of the islands and their avifaunas remain relatively undamaged. However, the situation is changing rapidly. With the increase of human activities, it is essential to monitor the seabird species likely to be affected, in order to identify the causes of demographic changes. When trying to explain the changes in population sizes, it is important to detect the environmental factors that could cause changes and to identify past and present human influences on seabird populations.

This paper gives an account of the long-term demographic studies that are being carried out at four localities in very different climatic zones in the southern ocean and representing the range of ecological situations experienced by seabirds in the southern hemisphere. For each of the major natural- and human-induced factors that are likely to cause changes in seabird numbers, we provide here an example of our recent findings.

Study area and methods

Long-term monitoring studies have been conducted over the last 30 years on four groups of islands in the Austral Ocean:

1 In the Antarctic zone, on the Pointe Géologie Archipelago (66° 40′S, 140° 01′E), Adélie Land.
2 At the confluence of sub-Antarctic and Antarctic waters, on the Kerguelen archipelago (49°S 70°E).

3 In the sub-Antarctic zone, on Possession Island (46° 25′S, 51° 45′E) in the Crozet archipelago.
4 In the subtropical zone, on Amsterdam Is. (37° 50′S, 77° 30′E).

A total of 44 species of seabird breed in these four localities (Jouventin *et al*. 1984; Weimerskirch *et al*. 1989). Twenty-three of these species are now monitored on a regular basis (Table 14.1). Some 120 000 individuals have been ringed. Field records are stored in a data bank and analysed for estimation of demographic parameters, population fluctuations, and modelling.

Only ground-nesting populations are censused on a regular basis: direct

Table 14.1 Study site, first year of ringing, frequency and first year of census, and start of demographic studies on the 23 seabird (and a sheathbill) species monitored by French programmes

Species	Study site[1]	Frequency (years) of census (first year of census	Start of demographic studies
King Penguin (*Aptenodytes patagonicus*)	C	1 (1962)	1975
King Penguin (*Aptenodytes patagonicus*)	K	5 (1963)	—
Emperor Penguin (*A. forsteri*)	T	1 (1954)	1965
Gentoo Penguin (*Pygoscelis papua*)	C	1 (1970)	1984
Sub-Antarctic Rockhopper P. (*Eudyptes c. chrysocome*)	K	4 (1987)	1987
Subtropical Rockhopper P. (*E. c. moseleyi*)	A	5 (1982)	1987
Macaroni Penguin (*E. chrysolophus*)	C	5 (1981)	—
Macaroni Penguin (*E. chrysolophus*)	K	5 (1963)	—
Wandering Albatross (*Diomedea exulans*)	C	1 (1960)	1966
Wandering Albatross (*Diomedea exulans*)	K	1 (1971)	—
Amsterdam Albatross (*D. amsterdamensis*)	A	1 (1982)	1983
Black-browed Albatross (*D. melanophrys*)	K	1 (1978)	1978
Yellow-nosed Albatross (*D. chlororynchos*)	A	5 (1982)	1978
Light-mantled Sooty Albatross (*Phoebetria palpebrata*)	C	5 (1981)	1968
Sooty Albatross (*P. fusca*)	C	5 (1981)	1968
Southern Giant Petrel (*Macronectes giganteus*)	C	1 (1981)	—
Southern Giant Petrel (*Macronectes giganteus*)	T	1 (1984)	1963
Northern Giant Petrel (*M. halli*)	C	5 (1981)	—
Antarctic Fulmar (*Fulmarus glacialoïdes*)	T	1 (1963)	1963
Snow Petrel (*Pagodroma nivea*)	T	5 (1983)	1963
Belcher Prion (*Pachyptila belcheri*)	K	—	1985
Blue Petrel (*Halobaena caerulea*)	K	—	1985
White-chinned Petrel (*Procellaria aequinoctialis*)	C	5 (1982)	1984
Grey Petrel (*P. cinerea*)	K	—	1986
White-headed Petrel (*Pterodroma lessoni*)	K	—	1986
Common Diving Petrel (*Pelecanoïdes urinatrix*)	K	—	1985
Antarctic Skua (*Catharacta maccormicki*)	T	1 (1967)	1963
Lesser Sheathbill (*Chionis minor*)	C	—	1978
Lesser Sheathbill (*Chionis minor*)	K	—	1986

[1]A = Amsterdam; C = Crozet, K = Kerguelen; T = Adelie Land.

counts of the breeders are generally performed with the exception of larger colonies of penguins (King, Macaroni) where photographs are taken from helicopters. Numbers given always relate to the annual breeding population (in pairs) even in biennial breeding species. Annual survival of adult birds was estimated using the Maximum Likelihood Method (Cormack 1964; Seber 1973) and the survival of immatures using binomial statistics. Life expectancy corresponds to 2-m/2m where m is the average mortality (l-survival). Fledging frequency is defined as the average time taken for a pair to fledge one chick.

Some islands are completely undisturbed by man, while others shelter various associations of mammals introduced at different times (ranging from centuries to only 10 years ago).

Results

Demographic parameters

Table 14.2 shows average demographic parameters estimated for 14 species over periods of 7 to 25 years. The annual survival of adults, breeding success, and the rate of recruitment are all highly variable from year to year for each species. Averaged data should consequently be treated with caution, particularly for populations that have been through adverse or favourable periods while also having undergone periods of stability. For example, the mean survival of Wandering Albatross and Emperor Penguin corresponds only to periods of stability for the two populations. Although the two large penguins *Aptenodytes* spp. are of comparable age when they first breed, they have significantly different reproductive and survival rates (Table 14.2).

Albatrosses generally have a deferred sexual maturity, but it appears that the age of first breeding varies greatly according to the species. Survival is high in the two largest species and the two sooty albatrosses which breed in alternate years and so have a high fledging frequency. Yellow-nosed and Black-browed Albatrosses breed annually and have a relatively low survival. In the latter species, low survival is associated with a decrease of the monitored population. The three fulmarine petrels breed each year and have similar ages of sexual maturity and similar survival rates.

Changes in the size of some populations

During the last 25 years, the monitored populations of Adélie, Macaroni, and King Penguins have increased, while the population of Emperor Penguin has decreased in Adélie Land (Fig. 14.1). The populations of

Table 14.2 Demographic parameters estimated for 14 bird species breeding in French Antarctic and sub-Antarctic territories

Species	Age at first breeding (years)		Mean annual survival (%)	Mean life expectancy (years)	Breeding frequency (years)	Fledging frequency (years)	References
	Min.	Mean					
Gentoo Penguin	2	3.5	86.5	7	1	1.0	C. A. Bost, unpublished data
King Penguin	3	6.8	95.5	16.2–24.5	2/3	3.2	Unpublished data
Emperor Penguin	3	5.3	91.0	10.6	1	1.6	Unpublished data
Wandering Albatross	7	11.3	96.8	30.8	2	2.7	Weimerskirch and Jouventin (1987)
Amsterdam Albatross	—	—	97.7	43.0	2	2.4	Jouventin et al (1990), Unpublished data
Black-browed Albatross	6	9.8	89.0	10.6	1	2.1	Weimerskirch et al. (1987)
Yellow-nosed Albatross	6	8.9	91.2	10.9	1	1.9	Jouventin et al. (1983), Weimerskirch et al. (1987)
Sooty Albatross	9	12.2	95.1	19.9	2	3.9	Jouventin and Weimerskirch (1984), Weimerskirch et al. (1987), Weimerskirch (1982)
Light-mantled Sooty Albatross	8	12.0	97.3	36.5	2	5.0	Weimerskirch et al. (1987), Jouventin and Weimerskirch (1990a)
Southern Giant Petrel	6	8	91.0	10.6	1	1.6	Jouventin and Weimerskirch (1985), Unpublished data
Antarctic Fulmar	7	8.5	92.4	12.7	1	1.7	Jouventin and Weimerskirch (1985), Unpublished data
Snow Petrel	5	8.5	93	13.8	1	2.1	Guillotin and Jouventin (1980), Jouventin and Weimerskirch (1985), Unpublished data
Antarctic Skua	3	5.2	91.2	10.9	1	1.1	Jouventin and Weimerskirch (1979)
Lesser Sheathbill	3	4	89.0	8.6	1	0.9	Jouventin and Weimerskirch (1985), Unpublished data

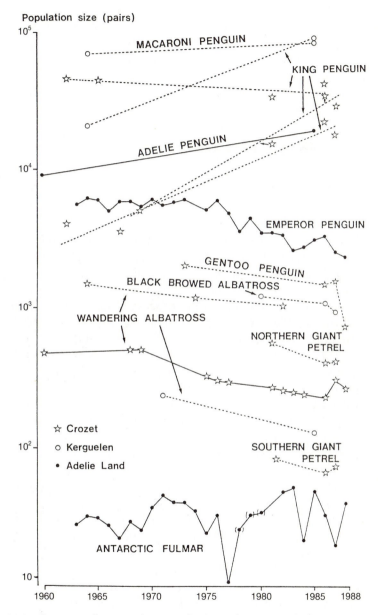

Fig. 14.1. Changes in the sizes of some seabird populations (partly from Jouventin and Weimerskirch 1990*b*).

Wandering and Black-browed Albatrosses and of both Giant Petrels have all decreased (Fig. 14.1). The only population of small Procellariiformes that has been monitored regularly is the Antarctic Fulmar in Adélie Land, which has fluctuated greatly during the period considered without any overall trend. On the sub-Antarctic islands, populations of small Procellariiformes that nest in burrows have not been censused regularly, but on the Kerguelen mainland are all progressively disappearing from the lowlands.

Natural causes of demographic change

Climate has a profound influence on the breeding success of seabirds nesting along the coast of the Antarctic continent and particularly of the Emperor Penguin that breeds on the sea-ice during the long Antarctic winter. This is consequently an interesting locality in which to study the influence of climatic conditions on the breeding of seabirds.

The number of pairs of Antarctic Fulmars laying has varied from 9 to 50 during the last 25 years (Fig. 14.1). Except for 1976/7, which will be considered separately, the number of breeders depended mostly on the amount of snow cover from the colony in spring, with frozen snow preventing the birds from reoccupying their nests. Adverse climatic conditions accounted for only 20–30 per cent of breeding failures and the overall breeding success varied from 52 per cent to 80 per cent for 20 nesting seasons, with two particularly bad seasons of 32 per cent and 37 per cent (Fig. 14.2). The breeding success of Snow Petrel was, on average, significantly lower than that of the Antarctic Fulmar and has varied from 18 per cent to 80 per cent during the last 25 years (Fig. 14.2). This species nests mostly in crevices and is precluded from breeding, or suffers a very high losses of eggs, when nests are flooded by melting snow which freezes at night and then remains over the nest. In Antarctic Fulmar and Snow Petrel, annual adult survival averaged 92.4 per cent and 93.0 per cent, respectively, during the last 25 years. A higher variability of survival in the Fulmar (Fig. 14.2) might be due at least in part to the smaller size of the sample used for the estimation ($n = 267$), compared with Snow Petrel ($n = 824$).

The number of eggs laid each year by the Emperor Penguin was not affected by climatic conditions. However, the breeding success and survival of adult birds are closely connected with the width of sea-ice between the colony and open seawater, and with the solidity of the ice holding the colony, i.e. with temperature, weather, and sea conditions (Jouventin and Weimerskirch 1990b). An abnormal extension of the sea-ice had several deleterious effects on the Emperor Penguin. The length of foraging trips of adults was increased, and consequently the feeding and

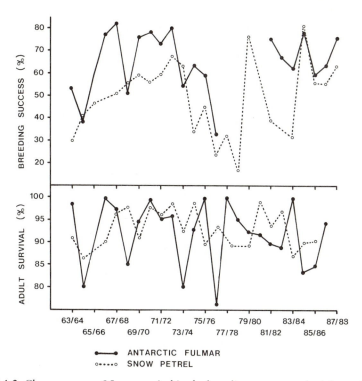

Fig. 14.2. Changes over a 25 year period in the breeding success and adult survival of Antarctic Fulmar and Snow Petrel in Adélie Land.

survival of chicks decreased as also did the survival of males that have to return to the sea after 75 days of incubation fasting (Fig. 14.3). Conversely, an early break up of the sea-ice decreases fledging success by drowning the eggs and chicks. The dramatic decrease of the Emperor Penguin colony in Adelie Land that started in 1976–7 (−75 per cent per year 1987) resulted from several factors:

(1) The extensive sea ice of 1972 resulted in the death through starvation of almost all chicks from the colony (Fig. 14.3, centre) with consequently very little recruitment into the population 4–6 years later.

(2) The overall decrease in breeding success since 1972 has consistently decreased the number of recruits (Fig. 14.3, centre).

(3) Adult survival was lower between 1972 and 1977 than after 1978 (Fig. 14.3, bottom) and probably contributed to the first steep population decline.

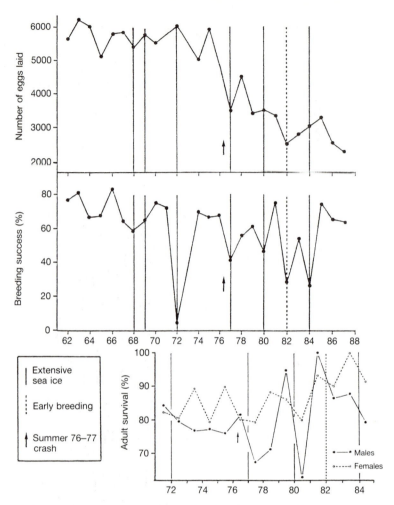

Fig. 14.3. Changes in population size, breeding success, and survival of male and female Emperor Penguins in Adélie Land (redrawn from Jouventin and Weimerskirch 1990*b*).

The 1976–7 summer season was very unfavourable for species breeding in Adélie Land (Fig. 14.2), with unprecedentedly low numbers of Antarctic Fulmars, Adélie Penguins, and Snow Petrels returned to their breeding grounds (Fig. 14.1, unpublished data). The factor that caused this general breeding failure might also explain the decreased numbers of breeding Emperor Penguin during the winter of 1977 (Fig. 14.3).

Impact of bases

In Antarctica, almost every snow-free rocky area close to the sea has been colonized by krill-eating penguins. Such areas are increasingly in demand for scientific bases, roads and, now, airstrips and these constructions inevitably limit space for breeding birds. This form of direct influence on the size of populations is 'negative' for timid species such as the Southern Giant Petrel, which has decreased to 14 per cent of the original population since the building of the base in the Pointe Geologie Archipelago (Thomas 1986) but 'positive' for scavengers such as the Antarctic Skua which has increased about two-fold in the same place (Jouventin and Guillotin 1979; Thomas 1986).

Introduced mammals

Introduced mammals, especially predators, are the main threat to birds on sub-antarctic islands. The difficulties of landing and settlement have led to a patchwork of mammals of up to 10 of such species (Jouventin *et al.* 1984). Rats (*Rattus rattus* and *R. norvegicus*) were accidental introductions and have proved almost impossible to eradicate. Although ship rats *Rattus rattus* are considered to have a limited impact on seabird populations (Moors and Atkinson 1984), a comparison of burrowing petrels on Possession Island where rats occur and Est Island where they are absent (both in the Crozet archipelago) reveals a tremendous difference in numbers of individuals and of species (Table 14.3). On the same archipelago, cats that were introduced during the last century to the Cochons Islands have eradicated several species of petrels, notably species larger than those eradicated by rats (Table 14.3). On the Kerguelen mainland, cats were introduced in the 1950s (Derenne 1976) and their spread is still continuing (Weimerskirch *et al.* 1989). Near the base where they were introduced their diet is changing: burrowing petrels have almost disappeared there and the cats now survive by feeding on rabbits (Weimerskirch *et al.* 1989).

Amsterdam Is. is the most damaged of the four groups of islands. Forest originally occupied 27 per cent of the island but covered only 5 per cent of the area in 1875 (Decante *et al.* 1987). Other types of vegetation are also disappearing due to erosion over most of the island. Bird habitat has been degraded by a combination of several fires and the expansion of a feral cattle population introduced in 1871.

The introduction of brown rats, cats, and pigs has played a major part in depleting the local seabird populations. Cats that were introduced to Amsterdam Is. at least 50 years ago, now eat less than 3 per cent birds but about 85 per cent rodents in the areas from which birds have virtually

Table 14.3 Comparison of the composition of communities of burrowing petrels (numbers of species and breeding pairs) on three islands of the Crozet group in relation to the presence or absence of introduced predators

Species	Possession Island (ship rats)	Est Island (no introduced predator)	Cochons Island (cats)
Salvin's Prion	10s of 1000s	Millions	4 000 000
Fairy Prion		10s of 1000s	10s of 1000s
Blue Petrel		10s of 1000s	
Great-winged Petrel	10s	10s of 1000s	
White-headed Petrel	100s		
Kerguelen Petrel	1000s	10s of 1000s	
Soft-plumaged Petrel	100s	10s of 1000s	
White-chinned Petrel	10s of 1000s	10s of 1000s	
Grey Petrel	100s	1000s	
Wilson's Storm Petrel	10s	10s of 1000s	
Black-bellied Storm Petrel		1000s	
Grey-backed Storm Petrel		100s	
South Georgia Diving Petrel	10s of 1000s	Millions	1 000 000
Kerguelen Diving Petrel		Millions	

disappeared; however birds still represent 36 per cent of the diet of cats in areas where they are present (Furet 1989). A preliminary analysis of subfossil bones shows that the avifauna of Amsterdam Island comprised 22 species (Table 14.4). Only seven species of birds breed there today, of which four are extremely rare, including about 15 breeding pairs of the endemic Amsterdam Albatross (see Roux *et al.* 1983; Jouventin *et al.* 1990). Among the species that have disappeared, four were probably endemic and remain undescribed: a flightless duck (Martinez 1987), a storm petrel, and probably two subspecies of the genera *Pterodroma* and *Procellaria*.

Considering the difference between the past and present breeding distributions of Amsterdam Albatross (Jouventin *et al.* 1990) and the recent colonization of the plateau by cattle, it was imperative to control the cattle population so as to halt the degradation of the vegetation and particularly the draining of peat bog on the high plateau (Decante *et al.* 1987). A fence cutting the island into two parts was constructed in 1987. In the protected part, the cattle have been eradicated and in the other part the remaining population will be limited. In this way it is hoped to protect at least the breeding grounds of the relict population of Amsterdam Albatrosses.

Table 14.4 Past and present avifauna of the Amsterdam and Saint Paul Islands

	Endemic form	Past records (+ subfossil deposits)	Present status (A = abundant, R = rare)	
			Amsterdam	St Paul
Northern Rockhopper Penguin		X	A	A
Amsterdam Albatross	O	X	R	
Yellow-nosed albatross			A	R
Sooty Albatross		X	A	R
MacGillivray's Petrel	O	X		R
Fairy Prion				R
Procellaria sp.	?	X		
Grey Petrel		X	R	
Flesh-footed Shearwater		X		A
Little Shearwater		X		?
Great Winged Petrel		X		
Soft Plumaged Petrel		X	?	
Pterodroma sp. 1		X		
Pterodroma sp. 2	?	X		
White-bellied Storm Petrel				R
White-faced Storm Petrel		X		
Storm Petrel sp. 1		X		
Storm Petrel sp. 2	O	X		
Diving Petrel sp.		X		
Sub-Antarctic Skua		X	R	
Sub-Antarctic Tern			R	R
Flightless Duck	O	X		
Past avifauna	4	18	7	8
	22 species			

Recovery from past exploitation

During the past 30 years the numbers of Antarctic penguin have increased considerably, probably due to increased food supply made available by the reduction of whale and seal stocks (Jouventin and Weimerskirch 1990*b*). However, a large number of penguins, were also slaughtered during the last century, particularly on the sub-Antarctic islands, and increases could be due to the recovery of the populations from past harvesting. Increases in Adélie and Chinstrap Penguin populations along the Antarctic continent and in the Antarctic Peninsula probably result from an increased food supply because these populations were not exploited. On the other hand, King Penguins on the Kerguelen and Crozet islands were extensively slaughtered during the last century and their populations are now increasing at the high rates of 7 per cent and 3.1 per cent per year, respectively (Weimerskirch *et al.* 1989; unpublished data).

On Possession Island (Crozet archipelago) four of the five King Penguin colonies have increased at extremely high rates (7 per cent to 14 per cent/ year), although year-to-year variation can be important with many fewer birds coming ashore to breed in some years, e.g. 1987 (Fig. 14.4). The overall annual rate of increase of Possession Island population is about 3 per cent, rather higher than anticipated given the low fecundity of the species (one egg/clutch, laid every 2 years or twice every three years) and the poor breeding success on Crozet (0.25–0.48 chick fledged/year according to year and colony). Estimates of annual survival give a minimum rate during the first year at sea of 0.31, of 0.89–0.92 during the immature stage (1–6 years), and a survival of at least 0.941 for breeding adults.

Emigration from the Possession Island colonies is estimated to be almost non-existent for adult breeding birds, but could reach up to 20–30 per cent for juveniles during their first year at sea. A simulation using the Leslie (1945) method indicates that the rate of increase of 3 per cent/year observed for the Possession Island population can only result from the maximum values for fledging frequency and survival presented above, with an adult survival of 97 per cent. This adult survival rate appears to be

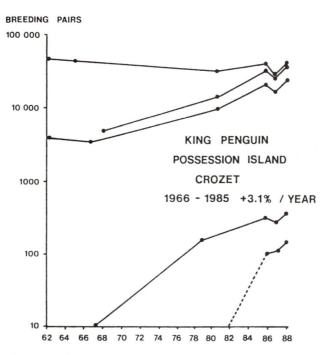

Fig. 14.4. Changes in the sizes of five colonies of King Penguin on Possession Island of the Crozet archipelago.

extremely high compared to any other Penguin or even seabird species. In the absence of natural predators on adults except Killer Whales *Orcinus orca* and Leopard Seal *Hydrurga leptonyx*, which are occasional predators of King Penguins at Crozet and Kerguelen, a survival of 97 per cent suggests that conditions must be presently very favourable. The observed rates of increase of the King Penguin population could alternatively be explained through simulation by a juvenile emigration rate of 45 per cent/year, although this figure is not supported by our data.

It is difficult to know for Crozet and Kerguelen whether colonies are still below or exceed their past levels.

Interactions with fisheries

With the increasing development of fisheries in the Austral Ocean since the 1960s, there is a need to assess the interactions between seabird populations and commercial fisheries. This assessment requires long-term studies of the birds on the breeding grounds, together with a good knowledge of the ecology at sea of the different species and of the fisheries.

Numbers of Wandering Albatross have decreased at all the breeding localities where censuses have been carried out (Weimerskirch and Jouventin 1987). A decrease was first documented on South Georgia where numbers declined by 19 per cent over 15 years (Croxall 1979). At that time the decrease was suspected to result from less favourable environmental conditions in the feeding zones of immatures and of the non-breeding half population off Australia (Croxall 1979). Later it was realized that the Crozet population had halved during the 1970s on three islands of the group. The decrease was suspected to be caused by the fisheries industry because of increasing numbers of ringing recoveries from accidental killing on fishing lines (Jouventin *et al.* 1984). A study on Possession Island showed that high adult mortality was a major cause of the population decline and that females had a significantly lower annual survival rate than males (89.9 per cent and 93.8 per cent respectively between 1968 and 1983; Weimerskirch and Jouventin 1987). The immature survival rate was low, but progressively increased in the cohorts born from 1966 to 1978, which was paralleled by a decrease in the maximum and mean ages at first breeding of the recruit (Fig. 14.5). Adult males foraged mainly in Antarctic waters while females and immatures went to sub-Antarctic and subtropical waters (Weimerskirch and Jouventin 1987; Weimerskirch *et al.* 1990).

Long-line fisheries, which are the major cause of mortality were concentrated in southern subtropical waters, and adult females and immatures of both sexes are consequently exposed to a potentially heavier mortality than males (Weimerskirch and Jouventin 1987). The differences

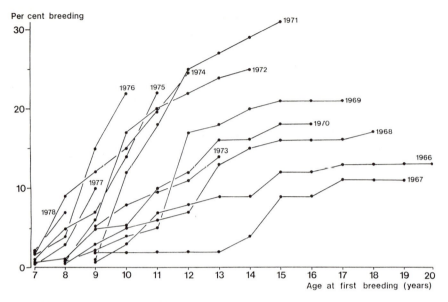

Fig. 14.5. Accumulative total by each age of birds breeding for the first time in each cohort of fledgling Wandering Albatrosses born between 1966 and 1976 (from Weimerskirch and Jouventin 1987).

in foraging zones between the two sexes and between adults and immatures probably explain the differential mortality observed within the population. The mortality due to fisheries affects the Crozet population both during and outside the breeding because the fishing zones overlap with the foraging range of both breeding and non-breeding Wandering Albatrosses from Crozet (Weimerskirch *et al*. 1985).

At Kerguelen, between 1978 and 1988, the population of Black-browed Albatrosses has been reduced by 30 per cent over the last 10 years (Fig. 14.1). The population winters off the southern coasts of Australia (Weimerskirch *et al*. 1985), where there is a long-line fishery which has reported the killing of Kerguelen Albatrosses. During the breeding season the species forages solely over the Kerguelen shelf (Weimerskirch *et al*. 1988) where a Russian fleet trawls for fish. The fishery provides considerable additional food resources to Black-browed Albatrosses through trawler offal, but on the other hand birds are killed when feeding at the trawls. To measure the interactions between fisheries and the albatross population, adult survival and breeding success have been compared with fisheries statistics (Weimerskirch, in preparation).

A preliminary analysis showed that the demographic parameters of the albatrosses were not correlated with total catches of fish, trawling effort, or number of trawlers on the shelf. Annual breeding success and adult survival appears to be strongly correlated to the 'catch per unit effort' of the fleet. It is easy to understand that the breeding success of albatrosses is related to the density of fish on the shelf. The survival of adults might be lower when densities of fish are low and consequently they try to catch the fish in the trawls. The above demonstrates the necessity for studies dealing not only with demographic mechanisms, but also with information on the ecology of birds at sea and on the fisheries.

Conclusions

Because the ecosystems of the Southern Ocean have remained largely unmodified until recently, monitoring of their seabird populations over the last 25 years has provided a unique opportunity to link population changes with human activities and so to distinguish natural changes from those that are induced by humans. Although the effects of natural factors such as climatic conditions on the breeding performance of Antarctic seabirds can be readily measured, the effects of complex factors affecting the marine environment are much more difficult to comprehend, requiring a multi-disciplinary approach and yearly measures of marine productivity, which are not always available.

For some species, such as Gentoo and Emperor Penguins or the Antarctic Fulmar, changes seem to be naturally induced. Numbers fluctuate greatly from year to year, but we need a better knowledge of the key environmental factors: the marine biology around the breeding ground for Gentoos, and quantitative data concerning sea-ice and snow for Emperors and Antarctic Fulmars. For several krill-eaters and also King Penguins, recovery from past exploitation seems to explain the present increase.

The decrease of seabird populations in the southern hemisphere, and particularly of Procellariiformes, is caused mainly by introduced predators (burrowing petrels) and by mortality due to fisheries (larger species). Several points are worth emphasizing:

1 Soils, plants, and animals are linked in these insular ecosystems; humans and even non-predatory mammals such as cows and rabbits have a significant impact on seabirds.
2 Increases in mortality probably occur in a greater number of Procellarii-form species than has previously been suspected.
3 The whole of the Austral Ocean is threatened; it must be managed through the auspices of international programmes (e.g. CCMALR). Seabirds have to be considered as predators of marine resources, since their study on land is easily and inexpensively carried out.

4 The potential damage to bird populations is not incidental, as first estimates indicate that tens of thousands of albatrosses are killed every year, but our understanding of these changes is presently hindered by the limitation of data concerning the impact of fisheries.
5 Populations of Procellariiformes are most sensitive to human disturbance because of the nature of their demographic strategy.

These practical considerations are founded on a theoretical understanding of population dynamics. The ultimate cause of negative correlations between the sizes of Procellariiformes populations and the amount of human activity is found in the sophisticated adaptations of this group to a pelagic life. For example, albatrosses have relatively small populations which are naturally limited by sparsely distributed resources and not by predators. The sizes of their populations have only a limited potential for increase, which is provided by a higher recruitment from the large non-breeding population to compensate for a small increase in adult mortality. Having the slowest population turn-over of all birds, they are not adapted to the disturbances induced by human activities throughout the world and we cannot be optimistic about the future of their populations.

Summary

For the last 30 years, the population of 23 species of seabirds have been monitored at the following four localities in the Southern Ocean, ranging from Antarctic to tropical waters: Pointe Geologie archipelago on Adélie land, the Kerguelen and Crozet archipelagos, as well as Amsterdam Island in the Indian Ocean. For this network of observatories, firstly, we provide estimates of the demographic parameters of 14 species and changes in the population sizes of 10 species (several penguins are becoming more numerous while several petrels and albatrosses are declining in numbers). Secondly, we analyse natural causes of demographic change for three Antarctic seabirds: Antarctic Fulmar *Fulmarus glacialoides*, Snow Petrel *Pagodroma nivea*, and Emperor Penguin *Aptenodytes forsteri*. Thirdly, human causes of changes are reviewed, such as the impact of scientific bases and the influence of introduced mammals on sub-Antarctic islands (e.g. cats and rats which eradicate burrowing petrels). Recovery from past harvesting explains some observed changes, such as the presently increasing King Penguin *Aptenodytes patagonica* populations. The development of fisheries in the Southern Ocean induces a significant accidental mortality among large Procellariiformes such as albatrosses. We conclude that at the present time the long-lived petrels and albatrosses are particularly vulnerable to human activities, and their future prospects appear particularly bleak.

Acknowledgements

The administration of 'Terres Australes et Antarctiques Françaises' provided financial and logistical support to the monitoring programmes. We acknowledge the assistance of all those involved in the landing, recovering, and monitoring operations on the four southern localities. We are grateful to D. Besson for typing the manuscript and aid in the processing of data, L. Ruchon for drawing the figures, Dr C. P. Doncaster for helpful grammatical corrections, and Dr M. P. Harris for his extensive help in the improvement of the manuscript.

References

Cormack, R. M. (1964). Estimates of survival from the sighting of marked animals. *Biometrica*, **51**, 429–38.

Croxall, J. P. (1979). Distribution and population changes in the Wandering Albatross *Diomedea exulans* at South Georgia. *Ardea*, **67**, 15–21.

Decante, F., Jouventin, P., Roux, J. P., and Weimerskirch, H. (1987). Projet d'aménagement de l'île Amsterdam. Stretie-Taaf-Cebas.

Derenne, P. (1976). Note sur biologie du chat haret de kerguelen. *Mammalia*, **40**, 531–5.

Furet, L. (1989). Régime alimentaire et distribution du Chat haret (*Felis catus*) sur l'île Amsterdam. *Terre et Vie*, **44**, 31–43.

Guillotin, M. and Jouventin, P. (1980). Le Pétrel des neiges à Pointe Gélogie. *Le Gerfaut*, **70**, 51–72.

Jouventin, P. (1975). Mortality parameters in Emperor Penguins *Aptenodytes forsteri*. In *The Biology of Penguins* (ed. B. Stonehouse), pp. 435–46. MacMillan, London.

Jouventin, P. and Guillotin, M. (1979). Socioécologie du Skua antarctique à Pointe Géologie. *Terre et Vie*, **38**, 109–27.

Jouventin, P. and Weimerskirch, H. (1984). L'Albatros fuligineux (*Phoebetria fusca*), exemple de stratégie d'adaptation extrême à la vie pélagique. *Terre et Vie*, **39**, 401–29.

Jouventin, P. and Weimerskirch, H. (1985). Population dynamics and monitoring of seabirds in french antarctic and subantarctic islands. In *Population and monitoring studies of seabirds* (ed. M. L. Tasker), pp. 6–7. Proceedings of the 2nd International Conference of the Seabird Group.

Jouventin, P. and Weimerskirch, H. (1990*a*). Demographic strategies in southern albatrosses. *Proceedings of the XIXth International Ornithological Congress*, pp. 857–65. Ottawa, 1986.

Jouventin, P. and Weimerskirch, H. (1990*b*). Long term changes in seabird and seal populations at four localities in relation to their demography. In *Monitoring ecological changes as a tool for science and conservation* (ed. K. R. Kerry). Springer Verlag, Berlin.

Jouventin, P., Roux, J. P., Stahl, J. C., and Weimerskirch, H. (1983). La biologie et la fréquence de reproduction de *Diomedea chlororynchos*. *Le Gerfaut*, **73**, 161–71.

Jouventin, P., Stahl, J. C., Weimerskirch, H. and Mougin, J. L. (1984). The seabirds of the French Sub-Antarctic islands and Adélieland. In *Status and Conservation of the*

World's Seabirds (ed. J. P. Croxall, P. G. H. Evans and R. W. Schreiber). International Council for Bird Preservation, Technical Publication, **2**, 603–25.

Jouventin, P., Martinez, J., and Roux, J. P. (1990). Breeding biology and current status of the Amsterdam Island Albatross. *Ibis*, **131**, 171–89.

Leslie, P. H. (1945). On the use of matrices in certain population mathematics. *Biometrica*, **33**, 183–212.

Martinez, J. (1987). Un nouveau cas probable d'endémisme insulaire: le Canard de l'île Amsterdam. *Geobios*, **10**,211–47.

Moors, P. J. and Atkinson, I. A. E. (1984). Predation on seabirds by introduced animals and factors affecting its severity. In *Status and conservation of world's seabirds* (eds. J. P. Croxall, P. G. H. Evans, and R. W. Schreiber), pp. 667–90. I.C.B.P. Technical Publication, **2**.

Roux, J. P., Jouventin, P., Mougin, J. L., Stahl, J. C., and Weimerskirch, H. (1983). Un nouvel Albatros (*Diomedea amsterdamensis nova species*) découvert sur l'île Amsterdam (37° 50′S, 77° 35′E). *L'Oiseau et la R.F.O.*, **53**, 1–11.

Seber, G. A. F. (1973). *The estimation of animal abundance and related parameters*. Griffin, London.

Thomas, T. (1986). L'effectif des oiseaux nicheurs de l'archipel de Pointe Géologie (Terre Adélie) et son évolution au cours des trente dernières années. *L'Oiseau et la R.F.O.*, **56**, 349–68.

Weimerskirch, H. (1982). La stratégie de reproduction de l'Albatros fuligineux à dos sombre. In *Les Ecosystèmes Subantarctiques* (eds. P. Jouventin, L. Massé, and P. Tréhen, P.), **51**, 437–48. C.N.F.R.A., Paris.

Weimerskirch, H. and Jouventin, P. (1987). Population dynamics of the Wandering Albatross (*Diomedea exulans*) of the Crozet Islands: causes and consequences of the population decline. *Oïkos*, **49**, 315–22.

Weimerskirch, H., Clobert, J., and Jouventin, P. (1987). Survival in five southern albatrosses and its relations with their life history. *Journal of Animal Ecology*, **56**, 1043–56.

Weimerskirch, H., Jouventin, P., Mougin, J. L., Stahl, J. C., and Van Beveren, M. (1985). Banding recoveries and the dispersion of seabirds breeding in french austral and antarctic territories. *Emu*, **85**, 22–33.

Weimerskirch, H., Bartle, J. A., Jouventin, P., and Stahl, J. C. (1988). Foraging ranges and partitioning of feeding zones in three species of southern albatrosses. *Condor*, **90**, 214–9.

Weimerskirch, H., Zotier, R., and Jouventin, P. (1989). The avifauna of the Kerguelen Islands. *Emu*, **89**, 15–29.

Weimerskirch, H., Lequette, B., and Jouventin, P. (1990*a*). Development and maturation of plumage in the Wandering albatross *Diomedea exulans*. *Journal of Zoology* (London), **219**, 411–21.

Weimerskirch, H., Zotier, R., and Jouventin, P. (1990*b*). The avifauna of the Kerguelen islands. *Emu*.

Further issues

15 Distribution of birds amongst habitats: theory and relevance to conservation

CARLOS BERNSTEIN, JOHN R. KREBS, and
ALEX KACELNIK

Introduction

In this article we will review certain models of the distribution of animals between adjacent habitats or patches. The introduction explains how the models discussed in the later part of the paper might be relevant to the conservation and management of bird populations.

The pivotal importance of patchiness in the ecology of individuals, populations, and communities is now widely recognized (e.g. May and Southwood 1990). In the particular context of conservation, the general question of distribution of animals between neighbouring habitats or patches arises from the fact that conservationists are often able to set aside only a small patch within a larger framework of different kinds of habitat, whilst animals may be able to move between habitat patches. For example in the Camargue a small proportion of freshwater marshes are set aside as protected areas for wildfowl, whilst the majority of marshes and other habitat types are open to exploitation by hunters. In order to understand the consequences of this conservation strategy for the species one is trying to conserve, it is crucial to know how populations in patches of conserved habitat interact with those from neighbouring, non-protected areas.

The example of freshwater marshes in the Camargue can be used to illustrate a range of possibilities. First, populations in the protected marsh and in adjacent patches of habitat might be independent entities, so that the conserved populations can be maintained without regard to the rest of the environment round about. Another possibility is that the populations in the conserved marsh are sustained only by immigration from surrounding areas, in which case preserving a single marsh surrounded by marshes in which hunting occurs will not be a successful long-term strategy. A third possibility is that all birds move freely and reciprocally between habitats, in which case the strategy of conserving one marsh may again be unsuccessful in the long run if all individuals sooner or later go to the marshes where they are shot.

This is by no means an exhaustive list of the potential relationships between adjacent habitats, nor is it necessarily intended to imply that we have identified the most likely ones for Camargue marshes. The example serves simply to illustrate the fact that understanding of factors governing the distribution of animals between habitats is important for conservation. In the next section we describe three examples which develop this theme in more detail.

Examples of the problem of populations in adjacent habitats

Birds in woodland and hedgerow

A typical habitat in lowland Britain and many other parts of Western Europe consists of a mosaic (whose elements are often in the range of a few hectares in size) of woodland, hedgerow, cereal fields, and pastures. Many plants and animals occur in more than one of these habitat types: common farmland birds such as tits *Parus* spp., starlings *Sturnus vulgaris*, and chaffinches *Fringilla coelebs* occur in woods, hedges, orchards, and gardens. What is the relationship between populations of these species in the different kinds of habitat? Would it be feasible, for example, to conserve the population of a species in, say, woodland, without regard to the population in surrounding hedges? This is a pertinent question in England, where 600 000 miles of hedgerow that existed in 1945 were disappearing at a rate of 6000 miles per year in 1970 (Hooper 1970).

One of the pieces of evidence that is sometimes cited in this context is the work of Krebs (1971), who showed that in southern England, great tits living in hedgerows have a lower reproductive success and occur at a lower density than in adjacent small areas of mixed deciduous woodland. Furthermore, many great tits are excluded from breeding in the good habitat (woodland) as a result of territorial interactions in the spring. The excluded individuals breed or attempt to breed in suboptimal hedgerow habitat, but if the opportunity arises to move into woodland (opportunities created by either natural mortality of the residents or, as in the case of Krebs' study, by experimental removal of residents), they set up territories and breed in the woods.

In discussions about the role of hedgerows for woodland birds, two diametrically opposed interpretations of Krebs' results could be suggested. On the one hand it could be argued that hedgerows are populated by surplus birds which are evicted from woodland and condemned to breed in a habitat where there are poor chances of success, so that hedges cannot be viewed as important resources from the point of view of conserving woodland birds. The opposite line of argument stems from the suggestion that the hedgerow populations act as a pool from which come immigrants to the woodland population following any decline in the woodland population (Brown 1969; Fretwell 1972). In this way the woodland population is buffered against dramatic fluctuations in density, so that the stability of populations in the good habitat depends on the presence of a population in the adjacent suboptimal habitat.

In fact the evidence supports this second view. Populations of great tits in optimal habitats such as woodland fluctuate less than those in suboptimal areas such as hedgerow and, at least in part, this comes about because young birds, regardless of where they were born, settle in optimal habitats if space is available, and only occupy suboptimal areas when the best habitat is fully occupied by territorial pairs (Kluijver and Tinbergen 1953; Krebs and Perrins 1977). In many years, about 50 per cent of all first-year breeding great tits in Wytham Wood near Oxford are birds that were born outside the wood, many of them presumably in the suboptimal hedgerow habitat surrounding the wood. Thus, because of the way birds distribute themselves between habitats and given that the agricultural environment consists of a mosaic of patches of woodland separated by fields and hedges, the conservation of populations in patches of woodland depends critically on the presence of a population in adjacent hedgerow. To generalize from this example: if, as is often the case, one kind of habitat is optimal for the species in question in terms of the density of birds present or their breeding success, it may be important not only to try to conserve the optimal habitat but also suboptimal habitats, especially if the latter are common relative to the former.

Wildfowl refuges

A similar analysis can be applied to Owen's (1979) discussion of the idea of establishing refuges for wintering geese as a means of combining conservation of these birds with a reduction in the damage they do to agricultural crops. The principle behind this strategy is that by making the refuges more attractive than the farmland, it should be possible to lure birds away from places where they can damage crops. As Owen recognizes, the effectiveness of this policy depends on how birds distribute themselves amongst habitats: if the refuges are designed to constitute an optimally attractive habitat, at what density in this habitat will the birds spill over into the less desirable, farmland habitats? Will the effect of creating refuges simply be to increase overall goose population size to the point at which they spill out of the refuges into farmland? As in the case of hedges and woods, the outcome of alternative conservation policies cannot be predicted without an understanding of distribution amongst habitats.

Setaside and field margins

A topical issue in conservation and land use in Britain is the question of what to do with surplus agricultural land arising from the planned reduction in cereal and dairy production under the Common Agricultural Policy. One currently (1989) fashionable scheme is to decrease the area of farmed land by increasing the width of field margins. Presumably these margins will act as refuges for native wildlife. The ecological questions raised by this policy include the following:

(a) Are long thin strips of habitat surrounded by hostile areas, as represented by field margins adjacent to intensively farmed cereal fields, appropriate for conservation?

(b) Are field margins likely to act as refuges from which pest species might invade adjacent fields?

(c) Alternatively will they act as reservoirs for natural enemies of pest species and hence reduce the likelihood of pest outbreaks in crops?

The answers to these questions depend, among other things, on the way in which animals or plants distribute themselves between adjacent habitats. Also critical to an understanding of these problems is the question of scale, which will determine whether the populations in neighbouring patches form part of a single larger population or whether they are more or less autonomous. Most of the discussion that follows is based on the assumption that, for birds at least, there is the potential for extensive interchange between populations in adjacent habitats.

In each of these examples we have posed questions relevant to conservation to which the answers depend on an understanding of habitat distribution. In the next section we turn to the theoretical analysis of this problem.

Models of habitat distribution

In the example referred to above of great tit populations in southern England, the distribution of birds in woodland and farmland habitats is determined by territorial exclusion: the good habitats are occupied first and at a higher average density, and later settlers are forced into poorer habitats (measured by breeding success) where the density of birds is lower. This kind of distribution is often referred to as a 'despotic' distribution (Fretwell and Lucas 1970; Fretwell 1972). In a seminal paper, Fretwell and Lucas (1970) were the first to attempt a formal analysis of habitat distribution, characterizing the assumptions and predictions of different models. Brown (1969), Orians (1969), Parker (1970), and Royama (1970) developed some of the same ideas independently, but did not couch them in terms of a general theory of habitat distribution.

Sutherland (1983), Sutherland and Parker (1985), and Parker and Sutherland (1986) made important modifications to Fretwell and Lucas' original models.

Fretwell and Lucas based their models on the following three premises.

1 Habitats can be ranked in terms of their suitability, ultimately defined in terms of fitness of the individuals living in the habitat, but in practice measured as some presumed correlate of fitness such as foraging success, breeding success, or survival.

2 Suitability is related to population density in each habitat: in the simplest case the relationship is negative, reflecting competition for limited resources, and all animals are assumed to be equal competitors. We shall use the term zero-density suitability to describe the relative ranking of habitats when no competitors are present.

3 Animals settle first in the most suitable habitats.

Two general classes of habitat distribution that might arise from these three premises were identified: 'Ideal free distributions' and 'Despotic' distributions.

The ideal free distribution

An ideal free distribution arises when all individuals are able to select the most suitable habitat available at the time of settling. This implies that each animal has complete information about the current suitability of alternative habitats (in this sense the animals are 'ideal') and that it is free to settle in any habitat (there is no territorial exclusion). Given these properties, the predicted pattern of settling is shown in Fig. 15.1a. The first settlers occupy the best habitat, which reduces its suitability, until eventually the second best habitat has a suitability (its zero-density suitability) equal to that of the best area, at which point new settlers will settle equally in the two areas, so that the suitability remains equal for all individuals; or in other words so that no individual could improve its fitness by moving to a new habitat.

The three predictions of the simplest version of the ideal free model are as follows:

(a) The lower ranking habitats, in terms of zero-density suitability, will be occupied only after specifiable densities of competitors have settled in the higher ranking habitats (the densities could be specified in any particular case if the curves in Fig. 15.1a were known);

(b) at equilibrium, the suitability of all habitats will be the same, so that individuals, regardless of habitats, will have equal fitness, or equal success as measured by a proximate correlate of fitness;

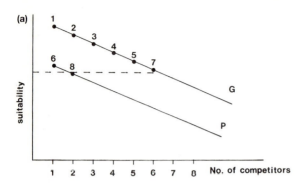

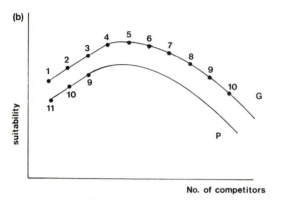

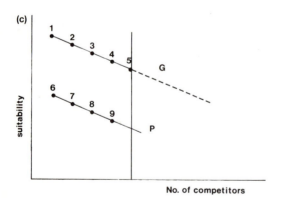

Fig. 15.1. (a) The Fretwell and Lucas model of the ideal free distribution. G = good habitat, P = poor habitat. Numbers refer to settling sequence of competitors. The horizontal broken line shows the equilibrium distribution with 8 settlers. Each new arrival decreases the habitat suitability for previous residents. (b) The ideal free distribution when there is an advantage of group living. The suitability increases at first and then decreases due to competition as more settlers arrive. Individuals 1–10 settle in habitat G, 11 settles in P, at which point it pays one individual and then another to shift to P and the density in G drops. Although the two individuals are labelled '9' and '10', they do not have to be the last arrivals. (c) The despotic distribution. Individuals 1–5 settle in G and later settlers are excluded. If they had been allowed to settle in G, fitnesses of an occupant of G would have been shown by the broken line.

(c) equilibrium density in each habitat will be correlated with the zero-density suitability.

The ratio of equilibrium densities can be predicted quantitatively, but to do so it is necessary to distinguish between two different kinds of resource. If the rate-limiting step in consumption per predator is not only the input of resources to a patch, but includes other factors such as handling time and interference, as might be the case of, say, oystercatchers feeding on a mussel bed, the equilibrium densities in each habitat are given by:

Proportion of competitors in site i = K proportion of prey in site $i^{1/m}$

where K is a constant and m is the 'interference' coefficient, the decrease in exploitation efficiency per additional competitor in a habitat (the slope of the lines in Fig. 15.1a, if both axes have a log scale, Sutherland 1983). If, however, the resources are consumed as soon as they appear (i.e. input of resources is the only rate-limiting step) and the total intake of resources in a patch is independent of the number of consumers, the equilibrium distribution of competitors is when:

Proportion of competitors in site i = C relative rate of input of resources in
site i

where C is a constant (usually it has the value 1). Thus if, for example, there are two habitats, one with twice the rate of input of resources of the other, there should be twice as many competitors in the better habitat at equilbrium. This is referred to as the 'input matching rule' (Parker 1978). The latter case is the one that has been most extensively studied in the laboratory, although its ecological relevance is perhaps less obvious that the former (Bernstein, Kacelnik, and Krebs, in preparation).

Fretwell and Lucas (1970) extended their analysis beyond the simple case illustrated in Fig. 15.1a to encompass examples in which the suitability of a habitat first increases with increasing density and then decreases (Fig. 15.1b). This could arise, for example, because of one of the many benefits of group-living (Pulliam and Caraco 1984) which could lead to an advantage from the presence of other individuals at low densities, but is eventually offset by competition at higher densities. This may lead to equilibrium distributions in which the densities in good and bad habitats shift suddenly as a function of overall population size (Fig. 15.1b). Notice also from this example, that the distribution achieved under ideal free conditions is not always the same as the distribution that would maximize average fitness of individuals. In Fig. 15.1b, individuals would experience a higher mean suitability (fitness) with a distribution of 6 individuals in the good habitat and 5 in the poor than they experienced at the ideal free distribution of 8 and 3.

Despotic and related distributions

For many animals it is unrealistic to assume that competitors are free to select any habitat at the time of settling. The situation described for great tits, in which early settlers exclude later settlers from the best areas (areas with the highest suitability) is probably common in territorial birds, and was incorporated into Fretwell and Lucas's framework as the despotic distribution (Fig. 15.1c). If animals obey a despotic distribution, the following predictions will hold:

(a) later settlers will be excluded from the best habitats;
(b) fitness will be lower in habitats with lower zero-density suitability;
(c) density may or may not be higher in the best habitats.

Sutherland and Parker (1985) and Parker and Sutherland (1986) present a model of despotic habitat distribution in which there are differences in competitive ability within a species but there is no territoriality (they refer to this as the 'interference' model). Here, each individual is free to move between habitats but some individuals are more severely affected by competition than others, which can be represented as different slopes of lines in Fig. 15.2. As density in the best habitat, G, increases, the adverse effect on poor competitors is greater than on good competitors, so that with increasing density, it pays the former to shift to the next habitat (P) sooner than is advantageous to the latter. This is essentially the same model as that developed by Rozenzweig (1979, 1985) for interspecific competition between a dominant and subordinate species (see below). The model predicts three regions of habitat use for the subordinate individuals. At low densities of dominants they use the good habitat, as density of dominants increases they use both, and as density increases still further they use exclusively the poorer habitat (Fig. 15.2).

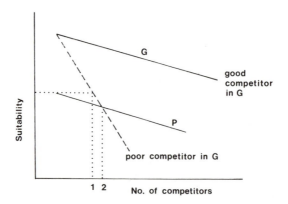

Fig. 15.2. As the density of competitors in the good habitat G increases, the poor competitors suffer a steeper decline in suitability than do the good competitors. At density 1 in the good habitat it pays poor competitors to use both habitats G and P. At density 2 it pays poor competitors to use habitat 2 exclusively. The graph illustrates a special case, chosen for simplicity, in which the density of competitors is such that only poor competitors use P. The effect of competitor density on poor competitors is less severe in P because all the competitors here are poor.

At equilibrium the average fitness will be lower in the poorer habitat. The equilibrium densities in the two habitats depend on the relative competitive abilities of strong and weak competitors: if poor competitors are very much more affected by interference than are good competitors, the equilibrium density in poorer patches is likely to be higher than in good patches.

Mixture of ideal free and despotic distributions

It may well be that a mixture of the two classes of model described above is the best description of many natural situations. Individuals are usually not of equal competitive ability, but given this constraint, each competitor may select habitats in which its fitness is highest. An example is Whitham's (1980) study of *Pemphigus* aphids settling on leaves of the narrow leaf cottonwood *Populus angustifolia*. The larger leaves have a higher zero-density suitability for aphids because they yield more food and therefore give a higher reproductive success, and within a leaf size class, suitability declines with density of aphids. At equilibrium, the average fitness of aphids is the same for all leaf sizes and all densities, supporting the ideal free model. However, within a leaf, sites near the base are more suitable than distal sites, and here differences in competitive ability determine who gets the best site: aphids fight for places at the leaf base. Thus within a leaf the aphids are despotically distributed, but as a whole the population shows ideal free distribution. This probably comes about because, after the good competitors have taken the best sites within a leaf, the less successful competitors adjust their distribution between leaves until no individual could do better by moving to another leaf.

Sutherland and Parker (1985) and Parker and Sutherland (1986) discuss the ideal free distribution with unequal competitors and conclude that when there are renewing resources, the equilibrium distribution is such that the distribution of 'competitive units' matches the ratio of input of resources to each habitat. This is equivalent to the input matching (see above) rule for equal competitors, but takes into account the fact that some individuals are better able than others to exploit resources. A competitive unit is defined in terms of an individual's relative ability to exploit resources. If, for example, the resource is food and A is capable of eating at twice the rate of B, A is worth two competitive units and B is worth 1.

Figure 15.3 illustrates the equilibrium distributions of twelve competitors between two habitats, one of which (A) has twice the resource input of the other (B), both for the case when all individuals are identical in terms of competitive units and when six of the individuals are twice as good at competing as the other six. An important point to note is that, whereas for the former case there is only one equilibrium distribution, in the latter

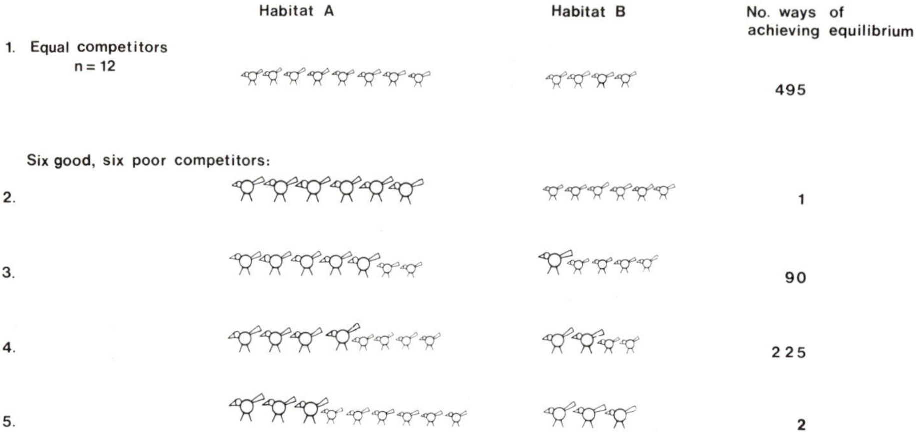

Fig. 15.3. Continuous input model of habitat distribution. Habitat A is twice as good as habitat B. When all competitors are equal, the ideal free distribution is 8 competitors in A and 4 in B (row 1). When 6 of the competitors are twice as good as the other six at consuming resources, distributions 2–5 all conform to the modified ideal free distribution in which the ratio of 'competitive units' matches the ratio of inputs. The final column gives the number of ways in which each distribution can be achieved.

(unequal competitive abilities) there are four distributions which result in the equilibrium of twice as many competitive units in habitat A. Furthermore, if one calculates the number of combinations of individuals which would lead to each distribution, it is apparent that the third possible distribution, in which there are eight animals in the good patch (four good competitors and four poor competitors) and four animals in the bad patch (two of each kind of competitor) can be achieved in more different ways in terms of the location of individual consumers than any of the other distributions. This distribution happens to mimic the ideal free distribution with equal competitors (i.e. it fits the simplest version of the input matching rule); Parker and Sutherland suggest that in some experiments where the results appear to support the equal competitor model, in fact the unequal competitor model is supported, but by chance the distribution of individuals predicted by the two models will often be the same. See Houston and McNamara (1988) for further discussion of this point, including a demonstration that it no longer holds for large numbers of competitors.

Evidence for ideal free and despotic distributions

Parker and Sutherland (1986) review seventeen studies of habitat distribution and reach the following conclusions.

(a) There are several studies that present good evidence for despotic distributions. Parker and Sutherland refer to despotic distributions based on interference and differential competitive abilities rather than on resource defence, although the effect of the latter on habitat distribution is also well-established (review in Patterson 1980). The evidence cited by Parker and Sutherland for a despotic distribution in most studies is simply that success rate (e.g. rate of food intake) differs between habitats. This result could arise for a number of reasons (e.g. a non-equilibrium ideal free system) and more direct evidence of interference should be sought. The studies of Goss Custard and colleagues (Goss Custard *et al*. 1982, 1984) provide one of the best studies to date of a despotic habitat distribution based on interference. These authors show that wintering juvenile oystercatchers feeding on mussel beds are more susceptible to interference than adults (rate of food intake declines with density more sharply for juveniles). They also show that juveniles utilize the most profitable mussel beds in the summer before the wintering adults arrive, but that they move to less profitable beds as adult density increases through the autumn.

(b) Parker and Sutherland conclude that no studies support the simple ideal free model with equal competitive abilities. One possible example is the work of Pleszczynska and Hansell (1980) on settlement of polygamous female lark buntings on male territories. Pleszczynska and Hansell claim to show that territories differ in quality (measured as breeding success) and that suitability is density dependent. By observing the exact sequence of settling by females on territories of different initial qualities and with different densities of females, Pleszczynska and Hansell concluded that each female occupied the territory with the highest suitability at the time of settling. Given that this is the only example of this ideal free distribution, it would seem desirable to replicate it.

(c) Seven studies cited by Parker and Sutherland support the mixture of an ideal free and despotic distribution illustrated in Fig. 15.3 (Courtenay and Parker 1985; Davies and Halliday 1979; Godin and Keenleyside 1984; Harper 1982; Milinski 1979, 1984; Parker 1970, 1978; Whitham 1980). In all cases, the evidence for an ideal free distribution came from the observation of equal payoffs between habitats, whilst within habitats, better competitors obtained more resources. All the studies were of 'continuous input' systems. Other studies not cited by Parker and Sutherland include the work of Kluijver and Tinbergen (1953) on settling densities of tits in mixed and pine woods and the experimental studies of Recer *et al*. (1987), Inman (1990), and Cezilly (1989) on flocks of mallard ducks, starlings' and herring gulls respectively.

Thus it seems likely that two of the commonest patterns of habitat distribution are those represented by the despotic and mixed models.

Interspecific competition and habitat distribution

The models described in Figs. 15.1 and 15.2 refer to habitat distribution by populations of a single species at different densities. Rosenzweig (1979, 1981, 1985) recognized that essentially similar models could be applied to the effects of competition between species on habitat distribution. He presents graphical models for three different cases:

(a) two competing species that differ in their ability to utilize different habitats;
(b) two competing species, one of which is a superior competitor to the other, with the same preferred habitat; and
(c) two competing species both with the same preferred habitat and with equal competitive abilities.

First consider Rosenzweig's model for two competing species A and B with different abilities to exploit two habitats a and b. A is better at exploiting habitat a and B is better in b and each species prefers, given a choice, to exploit that habitat in which it does better. In the absence of a competing species, either species utilizes its own preferred habitat when population density is low, but suitability of the optimal habitat declines with increasing density until it pays to use both the optimal habitat (where density is high) and the suboptimal (where density is low). Thus each species, in the absence of competition, would switch from a specialist to a generalist use of habitats as density of intraspecific competitors increases (Fig. 15.4a) (Rosenzweig calls the switching lines 'isolegs'). Now the effect of interspecific competition can be added in: the effect of species B is to decrease the suitability of habitat b to species A, and vice versa. Imagine a few individuals of B reducing the value of habitat b to species A. This would cause species A to remain a specialist in habitat a at a higher density of A (since the value of the alternative is lower). This effect becomes greater as the number of B individuals increases, which can be represented as a clockwise rotation of the isoleg in Fig. 15.4a dividing the specialist and generalist regions for species A. A similar argument can be applied to species B: its switch point to generalizing is postponed by the presence of individuals of species A in habitat a. Consequently the isoleg dividing the specialist and generalist regions of the graph for species B is rotated anticlockwise. The overall result is summarized in Fig. 15.4b. There are three regions of habitat use: in region S both species specialize on their preferred habitat; in region SG, species A specializes and B is a generalist; and in GS, species A is a generalist and B specializes.

Rosenzweig also modified the model to include a difference in competitive ability between two species, D (dominant) and U (subordinate). Both

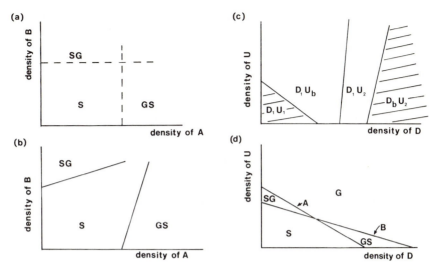

Fig. 15.4. The Rosenzweig isoleg model of habitat selection (after Rosenzweig 1985). (a) Two species, which do not compete, utilizing two different preferred habitats according to the ideal free distribution. When the density of either species increases above a certain point, the species utilizes both habitats. These thresholds are indicated by horizontal and vertical broken lines. (b) The distribution in (a) is now modified to incorporate interspecific competition. This reduces the suitability of each species' alternative habitat and therefore rotates the isolegs (see text). The two isolegs divide the figure into three regions. In region SG, species A specializes on its preferred habitat and species B is a generalist (i.e. uses both habitats). In region S, both species specialize and in region GS, species A is a generalist and species B specializes. See text for further details. (c) Habitat selection isolegs for dominant and subordinate competing species with same preferred habitat. D = dominant; U = subordinate; 1 = uses habitat 1 (better habitat) exclusively; b = uses both; 2 = uses habitat 2. (d) Habitat selection isolegs for two species with the same preferred habitat and two different secondary habitats. The line A refers to the isoleg for the dominant species (D) and the line B for the subordinate species (U). Regions of the graph are defined as for (b).

prefer the same optimal habitat, but D is a superior competitor. This model is similar to that shown in Fig. 15.2 for a single species in which individuals differ in competitive ability. Consider first the subordinate species, U. At low densities of both species, it utilizes only the preferred habitat. The isoleg bounding this region of specialization has a negative slope (in contrast to those of Fig. 15.4b), because an increase in density of either species reduces the suitability of the habitat for U: the more individuals there are of D, the fewer subordinate competitors it takes to push the species U across the isoleg and similarly, the more subordinates there are, the fewer individuals of D it takes (Fig. 15.4c). To the right of this isoleg, the species U generalizes. As the density of D increases further, there comes a point at which it pays U to avoid the good habitat altogether and to specialize in the poor habitat. This second isoleg has a positive slope, because, the greater the density of U in the poorer habitat, the more dominants are required to force a shift to exclusive use of the poorer habitat by subordinates.

For the dominant species, there are only two patterns of habitat use: at low densities of D, this species specializes. D is completely dominant to U in the preferred habitat so D's use of this habitat depends only on its own density. As the density of D increases, suitability of the preferred habitat is reduced to the level where it pays to use both habitats. If the subordinate species has no competitive effect on the dominant, this isoleg would be a vertical line, but if U slightly reduces the suitability of the poor habitat for D, the isoleg has a positive slope (Fig. 15.4c).

Pimm *et al.* (1985) and Rosenzweig (1986) applied this model to competition between three species of hummingbirds competing at artificial feeders: the dominant Blue-throated Hummingbird *Lampornis clemenciae* and two subordinate species, Rivoli's Hummingbird *Eugenes fulgens* and the Black-chinned Hummingbird *Archilochus alexandri*. There were two feeders, representing a good 'habitat' (1–2 M sucrose) and a bad 'habitat' (0.3 M sucrose), respectively. Certain aspects of the isoleg model were supported by the data. The Black-chinned showed three regions of habitat use, switching from good only, to both, and then to bad only as the density of Blue-throats increase. Further, the two isolegs dividing the regions for the Black-chinned had different slopes: the first being negative and the second positive in slope, as predicted by the isoleg model. The Rivoli's Hummingbird also tended to feed exclusively in the good habitat at low Blue-throat densities and then switched to using either both habitats or exclusively the poor habitat at higher densities of Blue-throats. The latter two patterns of habitat use were not separable by an isoleg.

The third version of the isoleg model refers to two competing species D and U which are equal competitors in the same preferred habitat, c, at low densities, but have different alternative habitats at high density. When D is common habitat a is included as preferred habitat and when U gets common habitat b is included. This results in isolegs of negative slope for both species (Fig. 15.4d). For either species, an increase in the density of itself or its competitor decreases the suitability of habitat c and causes a switch to include a or b. In general, where two species prefer the same habitat, the isolegs have negative slopes (Rosenzweig 1985).

How useful is Rosenzweig's approach to generating new insights about habitat use by competing species? Some of the conclusions are not new. Other formulations of interspecific competition, such as niche theory, predict changes in habitat use by subordinates in the presence of dominant competitors (e.g. Werner and Hall 1976) (as in the second version of the model) or increasing niche width (change from specializing to generalizing) with increasing density within a species (Holmes 1961), as predicted by the first version of the model. Predictions that are novel to the isoleg approach include the fact that switch points tend to have negative slopes when habitat preference is shared and positive slopes when they are not, and the fact that the degree of intraspecific competition influences the positive slope of the isolegs for non-shared habitat preferences. These predictions are used by Rosenzweig to examine interspecific competition, but they could also be applied to the distribution of different classes of individuals within a species (e.g. adults versus juveniles). A further interest of the isoleg approach is that it would in principle be possible to make quantitative predictions about the position of isolegs in a two-species system.

From the point of view of conservation, perhaps the major reason for appreciating the role of interspecific competition is this: simply observing the occurrence of a species in a habitat may not be adequate for finding out whether or not the habitat is a preferred one. In certain regions of Fig. 15.4c, for example, the subordinate species might occur exclusively in its suboptimal habitat. A naïve conservation policy might end up trying to protect a species in its suboptimal habitat.

Individual decisions and the ideal free distribution

In this section we ask how decisions by individual competitors could lead to an ideal free distribution. The equilibrium distribution is a property of the populations, so it might be thought that the distribution has to be reached by some kind of group decision. However, because selection acts on individuals more effectively than it does on populations, it is more plausible to postulate that there is a mechanism based on moment-to-moment decision rules of individuals that produces the population distribution. The decision rules must also affect the population's return to equilibrium following a perturbation. Furthermore, to be realistic, we cannot allow individuals to be 'ideal' in Fretwell's sense: they cannot have complete information about the current suitability of all patches. Because the optimal choice of habitat changes with density of competitors and depletion of resources, it seems likely that an appropriate behavioural rule for individual predators might include learning—the ability to modify behaviour as a result of experience.

Learning

A widely used, simple descriptive model of animal learning is the so-called linear operator (Bush and Mosteller 1955), which is a function expressing allocation of behaviour between alternatives as a weighted average of past and present experiences. The linear operator is a discrete time model with the following general structure:

$$P_i(t+1) = \alpha\lambda(t) + (1-\alpha)P_i(t), \quad O < P < 1, 0 < \alpha < 1$$

Where P is a variable associated with option i (P can be thought of as the predators' estimate of rate of capture in patch i), λ is a measure of present reward rate, normally the number of rewards at time t, and α is the memory factor, which determines the relative weight of past and current experience. This model is the optimal learning rule under specific forms of environmental stochasticity (McNamara and Houston 1987); it is simple in the sense of requiring only short-term memory of a continuously updated value and not the recollection of specific past experiences, and has been used successfully to approximate the behaviour of different animals in the laboratory (Kacelnik and Krebs 1985).

Two studies that have used models related to the linear operator to model the distribution of competitors between two patches are those of Lester (1984) and Regelmann (1984). Lester studied the distribution of goldfish between two patches without resource renewal, and Regelmann modelled Milinski's (1984) data on the distribution of sticklebacks between two patches with continuous resource input and unequal competitors.

Bernstein *et al.* (1988) investigated whether or not learning behaviour in predators, acting on an individual basis, would approach the ideal free distribution in a more complicated environment of many patches of a non-renewable resource. The model simulated the behaviour and distribution of 900 predators searching for prey in an environment of 49 patches containing a total of about 4900 prey. The major features of the model were as follows:

(a) Prey were distributed at random between patches.
(b) Within patches, capture rate was determined by a type II functional response (Holling 1959);

$$N_{ai} = a'N_iT/(1 + a'T_hN_i)$$

where N_{ai} and N_i are respectively the number of prey present and the number of prey attacked per predator in patch i, T is the total time available to predators, T_h is the time spent pursuing and handling each prey and a' is a measure of the predators' search efficiency.

(c) Predators were assumed to interfere with each other following the Hassell and Varley (1969) model:

$$a' = Q'P_i^{-m}$$

where Q' is the 'quest constant' (Hassell and Varley 1969) and m is a measure of interference. For computational convenience, Bernstein *et al.* (1988) defined a measure of instantaneous relative profitability q, the fraction of the maximum attack rate (T/T_h) realized by each individual.

(d) Each generation was divided into a number of discrete time units. At each time step each predator left its present patch if its relative capture rate, q fell below γ, which was its estimate of the mean value of q in all the patches of the environment. This rule is an approximation of the marginal value theorem (Charnov 1976), which gives the rate-maximizing departure rule for patches with depletion. Predators were assumed to migrate at random.

(e) Each predator estimated γ individually and according to its particular experience. This learning process was modelled by a linear operator:

$$\gamma_{t+1} = \alpha q + (1 - \alpha)\gamma_t$$

As we noted earlier, the ideal free distribution of animals in an environment with depletion and interference is given by:

Proportion of predators in site $i = K$ proportion of prey in site $i^{1/m}$

This gives rise to three predictions, if animal distributions follow the ideal free distribution:

(a) for $0 < m < 1$, there will be density-dependent mortality of prey (proportionally more predators per prey in high density patches);
(b) capture rate per individual is independent of prey density; and
(c) the actual capture rate per individual and the number of predators in a patch, as a function of the number of prey in it, are predicted by Sutherland's (1983) model.

Any of these different predictions could be used to test the convergence to the ideal free distribution. Here we use the distribution of predators in relation to the distribution of prey to illustrate our results.

The results of Bernstein *et al.* (1988) model are highly dependent on the predators' capture rate. When capture rate is low, so that prey density changes slowly, the predator's 'personal' estimate of the mean capture rate tracks to the true value and animals approach the ideal free distribution (Fig. 15.5a). This result demonstrates that in an environment with many patches and in which predators follow individual rate-maximizing foraging decisions based on learning, the population can reach an ideal free distribution.

In contrast, if capture rate is high, predators fail to converge to the ideal free distribution and the system behaves erratically (Fig. 15.5b). This is because changes in prey distribution occur faster than the speed at which

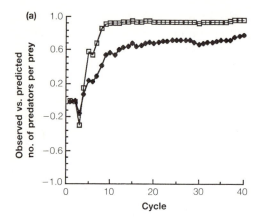

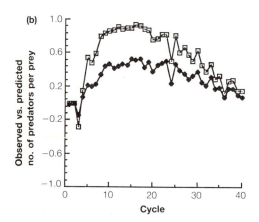

Fig. 15.5. The results of Bernstein *et al.*'s simulation model. (a) With slow depletion, the fit to the ideal free distribution is good. (b) With rapid depletion, the fit is poorer. In each graph, the correlation coefficient (open symbols) and regression coefficients (closed symbols) between observed and predicted (by the ideal free model) number of predators per prey are plotted as a function of the number of cycles of the model. Correlation and regression coefficients of 1 indicate a perfect fit to the ideal free distribution.

learning predators can track changes in the environment (Recer *et al.* 1987). The range of relative payoffs experienced by predators across patches also appears to have a very strong influence in determining whether or not the model will converge to the ideal free distribution: the larger this range, the closer the convergence. Although these results apply to an environment with prey depletion, the model also converges on the ideal free distribution when prey are not depleted (continuous input) (Bernstein *et al.* 1988).

Cost of migration

The simulations of Bernstein *et al.* (1988) assume that patches are not far apart and that there are therefore no costs (in time, metabolism, risk of predation, etc.) associated with moving from patch to patch. This may be true for some experimental systems, but generally in the field there is likely to be a cost of migration. The model of Bernstein *et al*. (1988) can be used to analyse the influence of these costs, measured in terms of time or energy, on the distribution of predators. In order to investigate this, we assumed that migration between patches takes a given number of time steps in the simulation. During this time predators do not feed ($q = 0$) and as a consequence, if migration takes w time steps, then when the animal reaches a new patch, its value of γ has been updated as

$$\gamma_{t+w} = (1 - \alpha)^w \, \gamma_t \qquad (w \geqslant 0)$$

We used the same departure rule as in Bernstein *et al*. (1988): migrate whenever $\gamma > q$. The influence of migration time on the distribution of predators is shown in Fig. 15.6a (see over). As migration becomes more costly (w increases) predators become more reluctant to leave their current patch, because the cost of migration has to be set against the possible gains from future patches, and the model stabilizes further and further away from the ideal free distribution.

To summarize, the model of Bernstein *et al*. shows that competitors acting on an individual basis can reach the ideal free distribution. Whether or not this distribution is reached at equilibrium seems to depend on at least two factors: (i) the relative speed of depletion and learning, and (ii) the cost of migration. The relevance to conservation is that, when distribution patterns of competitors are far from the ideal free distribution, differences in density of competitors between habitats may be a poor indicator of difference in resource density (see also below).

The distribution of prey mortality: scale and the spatial correlation between predators and prey

When predator distributions follow the ideal free distribution, predator density is positively correlated with the density of resources. Under some conditions (e.g. low interference) of ideal free distribution the predators exert spatially density-dependent mortality on the prey. Heterogeneity of predation between patches including spatial density dependence can affect the dynamics and stability of predator prey interactions (Hassell and May 1973; Murdoch and Oaten 1975; May 1978; Hassell 1984; Chesson and Murdoch 1986; Pacala, Hassell and May 1990) and thus be of long-term importance in conservation of populations. The most interesting short-term consequence of a positive correlation between consumer and resource distributions is that variations between habitats in the density of prey or other resources can often be inferred from the distribution of consumers. Conclusions can therefore be drawn about the value of different habitats to

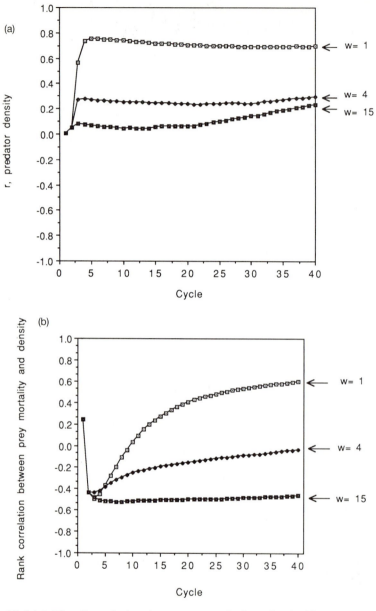

Fig. 15.6. (a) The effect of migration cost (w) on the fit to the ideal free distribution in the model of Bernstein *et al*., (1988). The higher the cost, the poorer the fit. The Y axis shows the correlation between observed and predicted numbers of predators in relation to prey density. The cost of migration is measured in cycles of the model. The data are means of 25 runs of the model. (b) The effect of cost of migration on density-dependent mortality. As the cost increases, mortality changes from positive to negatively density dependent.

consumers and hence about the priorities for habitat conservation. The studies referred to earlier in our brief review of the literature demonstrated that animals often exhibit a mixture of ideal free and ideal despotic distributions, which will often, but not invariably, lead to positive correlations between resource density and consumer density (Parker and Sutherland 1986).

Do results of field studies show a disproportionately high density of consumers in patches of high resource density? Although the relevant literature on birds is rather small, there are many studies of insect parasitoids in which the spatial distribution of parasitoids in relation to their hosts has been documented. These have recently been reviewed by Lessells (1985), Stiling (1987) and Walde and Murdoch (1988), who find that host mortality in natural conditions can be density dependent, inversely density dependent, density independent, or even 'domed' (i.e. having a maximum of mortality for intermediate densities). For instance, out of 49 cases reviewed by Lessells (1985), host mortality was density independent in 13 cases, inversely or directly density dependent in 15 and 17 cases respectively, and 'domed' in 2. How can we account for this variation? One view is that it can be explained in terms of behavioural processes that either do or do not result in parasitoids approaching the ideal free distribution (Royama 1970; Lessells 1985). This point can be illustrated with reference to the model of Bernstein *et al.* (1988).

In Bernstein *et al.*'s (1988) simulation, spatial patterns of mortality were affected both by the speed of learning in relation to depletion and by the cost of migration. First, recall the effects of learning and depletion, when prey capture was low and the predator population converged to the ideal free distribution (Fig. 15.5a). Accumulated prey mortality was density dependent (Fig. 15.7a, see over), while in conditions of fast prey depletion, when predators did not reach the ideal free distribution (Fig. 15.5b), prey mortality became inversely density dependent (Fig. 15.7b). Intermediate capture rates resulted in density-independent prey mortality. Inverse density dependence, as in Fig. 15.7b, is the consequence of predator aggregation not being strong enough to offset the diminishing returns of the functional response.

Turning to the effects of migration, as indicated in Fig. 15.6a, the fit to the ideal free distribution decreases with increasing cost of migration. As the fit decreases, prey mortality passes from density dependence, to independence and finally to inverse density dependence (Fig. 15.6b). A more detailed analysis (Figs. 15.8a and b) shows that the cases classified as density independent can be 'domed' or genuinely density independent. In fact, repeated runs of the model and fitting the results to a parabola show that the basic response is 'domed' density dependence (Bernstein, Kacelnik, and Krebs, in preparation), which was a pattern reported in

Lessell's (1985) review. An intuitive explanation of this surprising result is as follows: at intermediate travel times it is still worth a predator's while to leave the poorest patches and move to patches of intermediate or high prey density. Prey mortality therefore becomes density dependent in the low-intermediate range of densities. However, the cost of migration means that it does not pay individuals to move once they are in a patch of intermediate density and this will lead to inverse density dependence between patches of intermediate and high density. In other words, the cost of migration reduces the likelihood that predators will switch patches. Small costs only affect patches in the medium to high range (producing the domed relationship) but very high costs affect all patches, producing the inverse relationship.

We are suggesting that the features of the individual searching behaviour of predators and parasitoids such as learning and cost of migra-

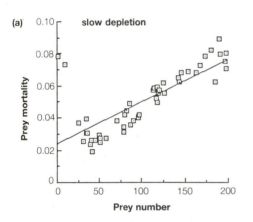

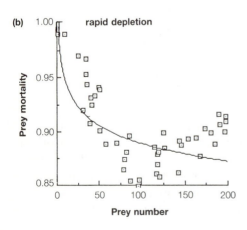

Fig. 15.7. Depending on the speed of depletion, the model of Bernstein *et al*. (1988) produces (a) positive or (b) negative density dependence. These two figures correspond in their parameter values to Fig. 15.5(a) and (b).

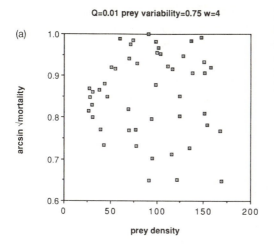

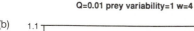

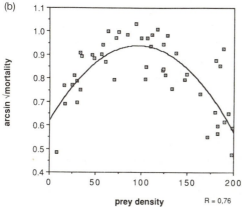

Fig. 15.8. (a) Density independent and (b) domed density dependent relations arising at intermediate costs of migration in the model of Bernstein *et al*. (1988).

tion may explain the different patterns in spatial density dependence. However, this is not the only possible explanation: Walde and Murdoch (1988) present a different view. They note, as does Stiling (1987), that mortality caused by insect parasitoids tends to be positively density dependent on a large scale of patchiness (say whole trees) but negative on a small scale (say leaves or branches within trees). They argue that exactly the opposite result would be expected from the hypothesis that positive spatial density dependence arises through behavioural responses: if behavioural responses are going to have an effect on any scale, it should, they suggest, be on a small scale where predators can readily move between patches. They propose an alternative hypothesis, namely that positive spatial density dependence arises on a large scale because each patch of prey supports a semi-autonomous population and that (for reasons not fully explained by Walde and Murdoch) this gives rise to disproportionately large parasitoid populations in richer host patches.

There are, however a number of problems with Walde's and Murdoch's analysis. First, as already implied, they do not explicitly show how their hypothesis of 'semi-autonomous populations' produces positive density dependence. Second, their analysis of scale of patchiness does not discriminate between two aspects: size of patches and distance between them. If, as we suggest, migration costs can be an important contributor to spatial patterns of host mortality, distance between patches is a crucial variable. Third, they lump together active and mobile parasitoids such as *Nemeritis* with much less mobile groups such as the scale insects (Hassell, personal communication). It could be that the variation in spatial density dependence is in part related to differences in mobility of parasitoids.

Let us return to the bird literature. Perhaps the only group of birds for which good data exist on the spatial distribution of birds in relation to their prey over a range of scales are seabirds such as alcids and larids (Schneider and Piatt 1986; Hunt and Schneider 1987; Piatt 1990). These studies do not measure density dependence, in fact it is likely that seabirds have a negligible effect on their prey, but they report spatial correlations between prey and predator. Interestingly enough, the pattern that emerges is similar to that reported by Walde and Murdoch (1988), namely that there is a good correlation between bird and prey distribution on a large scale but not on a small scale. For example, Piatt (1990) found an increase in the strength of the correlation between density of foraging murres *Uria aalge* and puffins *Fratercula arctica* and their prey (*Mallotus villosus*) when the scale of sampling increased between 1 km^2 and 8 km^2. Since the same individual birds exploit resources on both a large and small scale, the result cannot arise from the mechanism proposed by Walde and Murdoch: it must be due to differences in the behavioural response of seabirds to different scales of patchiness. First, the small-scale patches may be too ephemeral for the birds to keep track of them (Hunt and Schneider 1987) (like the rapid depletion case of Bernstein *et al*.'s model); second, the birds may have difficulty in discriminating the small-scale patches (Abrahams 1986); third, it may actually pay the birds to ignore small patches and travel only to rich areas on a larger scale (Arditi and Dacorogna 1988; Kacelnik and Bernstein 1988; Piatt 1990).

Discussion

Why should biologists concerned with habitats or populations be interested in the question of habitat distribution? In the 'Introduction' we started with the general argument that any conservation strategy should take into account the way in which individuals are distributed between habitats. This is not only because many ecological processes are strongly influenced by patchiness, but also because of the particular problem that conservationists face of being able to protect only small patches of habitat in a much larger mosaic. In the following discussion we consider three questions that conservationists might ask and show how, at least in part, the answers depend on an understanding of the concepts reviewed in this chapter.

1. What is the effect of habitat loss on a population?

Many of today's conservation issues, especially in highly developed regions such as Western Europe, revolve around habitat loss. Estuaries, ancient woodland, heathland and moorland are among many habitats that have been drastically eroded in the last 50 years. To understand the consequences of such habitat loss for particular species, it is necessary to have a model of how individuals are distributed between habitat patches. Two examples of analyses that illustrate this point are those of Goss Custard and Durrell (1989) on wader populations and Dobson (1986) on spotted owls. Dobson's (1986) report of the impact of logging on spotted owl populations has, as one of its central points, the fact that breeding populations of owls generally follow a despotic distribution. This means that at low densities only good habitats are occupied but, as density increases, poor areas in which individuals have low breeding success, are occupied as a result of territorial exclusion from the good areas. The consequences of this is that the individuals in good habitat patches contribute a disproportionately large amount to the total population fecundity and therefore loss of this habitat has a dramatic effect on population growth. Goss Custard and Durrell (1989) show that in wintering populations of oystercatchers, juveniles are the first to be excluded from the best habitat patches and that loss of these good habitats would therefore have an especially severe density-dependent effect on juvenile survival. This in turn could have a large effect on equilibrium population size. In these two studies accurate assessment of the impact of habitat loss could not have been made without an understanding of the appropriate model for distribution of individuals between habitat patches. The same is probably true for many other populations.

2. Estimating the value of habitats

Conservationists often talk of the 'habitat requirements' of a species. By this they mean the habitats in which the species can breed and survive, the implication being that conservation policy should be directed towards retaining these habitat types. One frequently used way of estimating habitat requirements is by survey work, recording where individuals are most often detected breeding or feeding. In other words, the distribution of consumers is taken as an indicator of the distribution of habitat values. As we have seen, models of habitat distribution show circumstances in which this assumption can be misleading. Intra- or interspecific competition may lead to the exclusion of most or all individuals of a species from its preferred habitat (the one with the highest zero-density suitability) (Figs. 15.1 and 15.4). Similarly, as we have shown (Figs. 15.7 and 15.8), factors such as the cost of migration and the speed of prey depletion can drastically affect the extent to which the distribution of consumers is correlated with the distribution of resources. Without some knowledge of these processes, estimates of habitat value from the surveys of distribution of consumers could be misleading.

3. What is the best mixture of habitats and scale for conservation?
The choice of habitat mixtures and the scale of patches to be conserved may be influenced by many factors, but here we emphasize that behavioural-ecological processes of distribution of individuals between habitats should be taken into consideration. Thus, as pointed out in the 'Introduction', suboptimal (low zero-density suitability) habitats may provide a 'buffer zone' for populations in optimal habitats. The question of scale of habitat patches for conservation includes both the size of patches and their separation. The importance of distance between patches is emphasized by our analysis of the cost of dispersal or migration between patches. When dispersal costs are high (widely separated patches) high-quality habitat patches may be underexploited simply because they are never colonized from poorer patches (Fig. 15.8).

In this chapter, we have considered one aspect of the problem of habitat distribution, namely how the costs and benefits of occupying habitats might be used to predict equilibrium distributions. Further work is needed to investigate the consequences of these patterns of habitat distribution for population dynamics, genetic separation of local populations, and for the colonization of new habitat patches.

Acknowledgements

We thank AFRC, CNRS, King's College, and The Royal Society for financial support. Colleen Kelly and Andy Hurly made useful comments on the manuscript. George Hunt and John Piatt provided pre-prints.

References

Abrahams, M. V. (1986). Patch choice under perceptual constraints: a cause for departures from the ideal free distribution. *Behavioral Ecology and Sociobiology*, **19**, 409–15.

Arditi, R. and Dacorogna, B. (1988). Optimal foraging on arbitrary food distributions and the definition of habitat patches. *American Naturalist*, **131**, 837–46.

Bernstein, C., Kacelnik, A., and Krebs, J. R. (1988). Individual decisions and the distribution of predators in a patchy environment. *Journal of Animal Ecology*, 57, 1007–26.

Bush, R. R. and Mosteller, F. (1955). *Stochastic models of learning*, Wiley, New York.

Brown, J. L. (1969). The buffer effect and productivity in tit populations. *American Naturalist*, **103**, 347–54.

Cezilly, F. (1989). Contribution à l'étude fonctionelle de l'exploration collective des ressources alimentaires chez les oiseaux. Unpublished Ph.D. Thesis, Université de Provence.

Charnov, E. L. (1976). Optimal foraging: the marginal value theorem. *Theoretical Population Biology*, **9**, 129–36.

Chesson, P. L. and Murdoch, W. W. (1986). Aggregation of risk relationship among host-parasitoid models. *American Naturalist*, **127**, 696–715.

Courtenay S. P. and Parker, G. A. (1985). Mating behaviour of the tiger blue butterfly (*Tarucus theophrastus*): competitive mate searching when not all females are captured. *Behavioural Ecology and Sociobiology*, **17**, 213–21.

Davies, N. B. and Halliday, T. R. (1979). Competitive mate searching in common toads, *Bufo bufo*. *Animal Behaviour*, **27**, 1253–67.

Dobson, A. (1986). Spotted owl guidelines: some criticisms of the environmental impact statement for an amendment to the Pacific Regional Guide. Report for the Wilderness Society.

Fretwell, S. D. (1972). *Populations in a seasonal environment*. Princeton University Press, Princeton.

Fretwell, S. D. and Lucas, H. L., Jr. (1970). On territorial behaviour and other factors influencing habitat distribution in birds. *Acta biotheoretica*, **19**, 16–36.

Godin, J. G. J. and Keenleyside, M. H. A. (1984). Foraging on patchily distributed food by a cichlid fish (Teleosti, Cichlidae): a test of ideal free distribution theory. *Animal Behaviour*, **32**, 120–31.

Goss Custard, J. D. (1985). Foraging behaviour of wading birds and the carrying capacity of estuaries. In *Behavioural ecology* (eds. R. M. Sibly and R. H. Smith), pp. 169–88. Blackwell Scientific Publications, Oxford.

Goss Custard, J. D. and Durell, S. E. A. Le V. dit (1989). Bird behaviour and environmental planning: approaches in the study of wader populations. Paper presented at B.O.U. Symposium 1989.

Goss Custard, J. D. , Durell, S. E. A. Le V. dit, McGrorty, S., and Reading, C. J. (1982). Use of mussel *Mytilus edulis* beds by oystercatchers *Haemotopus ostralegus* according to age and population size. *Journal of Animal Ecology*, **51**, 543–54.

Goss Custard, J. D., Clarke, R. T., and Durell, S. E. A. Le V. dit (1984). Rates of food intake and aggression of oystercatchers *Haematopus ostralegus* on the most and least preferred mussel *Mytilus edulis* beds of the Exe estuary. *Journal of Animal Ecology*, **53**, 233–45.

Harper, D. G. C. (1982). Competitive foraging in mallards: 'ideal free' ducks. *Animal Behaviour*, **30**, 575–84.

Hassell, M. P. (1984). Parasitism in patchy environments: inverse density dependence can be stabilizing. *Journal of Mathematics Applied in Medicine and Biology*, **1**, 123–33.

Hassell, M. P. and May, R. M. (1973). Stability in insect host-parasite models. *Journal of Animal Ecology*, **42**, 693–726.

Hassell, M. P. and Varley, C. G. (1969). New inductive population model for insect parasites and its bearing on biological control. *Nature*, **223**, 1133–6.

Holling, C. S. (1959). Some characteristics of simple types of predation and parasitism. *Canadian Entomologist*, **91**, 385–98.

Holmes, J. C. (1961). Effects of concurrent infections on *Hymenolepsis diminta* (Cestoda) and *Moliniformis dubius* (Acanthocephala) 1. General effects and comparison with crowding. *Journal of Parasitology*, **47**, 209–16.

Hooper, M. (1970). The botanical importance of our hedgerows. In *The flora of a changing Britain*, Botanical Society of the British Isles, Report 11, pp. 58–62.

Houston, A. I. and McNamara, J. M. (1988). The ideal free distribution when competitive abilities differ: an approach based on statistical mechanics. *Animal Behaviour*, **36**, 166–74.

Hunt, G. L. and Schneider, D. C. (1987). Scale-dependent processes in the physical and biological environment of marine birds. In *Seabirds* (ed. J. P. Croxall), pp. 7–41. Cambridge University Press.

Inman, A. J. (1990). Group foraging in starlings: ideal free distributions of unequal competitors. *Animal Behaviour*.

Kacelnik, A. and Bernstein, C. (1988). Optimal foraging and arbitrary food distributions: patch models gain a lease of life. *Trends in Ecology and Evolution*, **3**, 251–3.

Kacelnik, A. and Krebs, J. R. (1985). Learning to exploit patchily distributed food. In *Behavioural ecology* (eds. R. M. Sibly and R. H. Smith), pp. 189–205. Blackwell Scientific Publications, Oxford.

Kluijver, H. N. and Tinbergen, L. (1953). Territoriality and the regulation of density in titmice. *Archives Neerlandaises de Zoologie*, **10**, 265–89.

Krebs, J. R. (1971). Territory and breeding density in the great tit (*Parus major* L.) *Ecology*, **52**, 2–22.

Krebs, J. R. and Perrins, C. M. (1977). Behaviour and population regulation in the great tit (*Parus major*). In *Population control by social behaviour* (eds. F. J. Ebling and D. M. Stoddart), **99**, 23–47. Institute of Biology, London.

Lessells, C. M. (1985). Parasitoid foraging: should parasitism be density dependent? *Journal of Animal Ecology*, **54**, 27–41.

Lester, N. P. (1984). The feed-feed decision: how goldfish solve the patch depletion problem. *Behaviour*, **85**, 175–99.

McNamara, J. M. and Houston, A. I. (1987). Memory and the efficient use of information. *Journal of Theoretical Biology*, **125**, 385–95.

May, R. M. (1978). Host-parasitoids systems in patchy environments: a phenomenological model. *Journal of Animal Ecology*, **47**, 833–44.

May, R. M. and Southwood, T. R. E. (1990). Living in a patchy environment: on arrival. In *Living in a patchy environment* (eds. B. Shorrocks and I. Swingland). Oxford University Press.

Milinski, M. (1979). An evolutionarily stable feeding strategy in sticklebacks. *Zeitschrift für Tierpsychologie*, **57**, 36–40.

Milinski, M. (1984). Competitive resource sharing: an experimental test of a learning rule for ESS's. *Animal Behaviour*, **32**, 233–42.

Murdoch, W. W. and Oaten, A. (1975). Predation and population stability. *Advances in Ecological Research*, **9**, 2–131.

Orians, G. H. (1969). On the evolution of mating systems in birds and mammals. *American Naturalist*, **103**, 589–603.

Owen, M. (1979). The role of refuges in wildfowl management. In *Bird problems in agriculture* (eds. E. N. Wright, I. R. Inglis, and C. J. Feare), pp. 144–56. B.C.P.C. London.

Pacala, S. W., Hassell, M. P., and May, R. M. (1990). Host parasitoid associations in patchy environments. *Nature*, **334**, 150–3.

Parker, G. A. (1970). The reproductive behaviour and the nature of sexual selection in *Scatophaga stercoraria* L. (Diptera: Scatophagidae). VII, The origin and evolution of the passive phase. *Evolution*, **24**, 774–88.

Parker, G. A. (1978). Searching for mates. In *Behavioural ecology: an evolutionary approach* (1st edn.) (eds. J. R. Krebs and N. B. Davies), pp. 214–44. Blackwell Scientific Publications, Oxford.

Parker, G. A. and Sutherland, W. J. (1986). Ideal free distributions when individuals differ in competitive ability: phenotype-limited ideal free models. *Animal Behaviour*, **34**, 1222–42.

Patterson, I. J. (1980). Territorial behaviour and the limitation of population density. *Ardea*, **68**, 53–62.

Piatt, J. L. (1990). The aggregative response of common murres and atlantic puffins to schools of capelin. *Studies in Avian Biology*.

Pimm, S. L., Rosenzweig, M. L., and Mitchell, W. A. (1985). Competition and food selection: field tests of a theory. *Ecology*, **66**, 798–807.

Pleszczynska, W. and Hansell, R. I. C. (1980). Polygyny and decision theory: testing of a model in lark buntings *Calamospiza melanocorys*. *American Naturalist*, **116**, 821–30.

Pulliam, H. R. and Caraco, T. (1984). Living in groups: is there an optimal group size? In *Behavioural ecology* (2nd edn.) (eds. J. R. Krebs and N. B. Davies), pp. 122–47. Blackwell Scientific Publications, Oxford.

Recer, G. M., Blanckenhorn, W. U., Newman, J. A., Tuttle, E. M., Witham, M. L., and Caraco, T. (1987). Temporal resource availability and the habitat-matching rule. *Evolutionary Ecology*, **1**, 363–78.

Regelmann, K. (1984). Competitive resource sharing: a simulation model. *Animal Behaviour*, **32**, 226–332.

Rosenzweig, M. L. (1979). Optimal habitat selection in two species competitive systems. *Fortschritte Zoologie*, **25**, 283–93.

Rosenzweig, M. L. (1981). A theory of habitat selection. *Ecology*, **62**, 327–35.

Rosenzweig, M. L. (1985). Some theoretical aspects of habitat selection. In *Habitat selection in birds* (ed. M. L. Cody), pp. 517–40. Academic Press, New York.

Rosenzweig, M. L. (1986). Hummingbird isolegs in an experimental system. *Behavioral Ecology and Sociobiology*, **19**, 313–22.

Royama, T. (1970). Evolutionary significance of predators' response to local differences in prey density: a theoretical study. *Proceedings Advanced Study Institute: Dynamics and Numbers of Populations* (*Oosterbeek*), pp. 344–57.

Schneider, D. C. and Piatt, J. F. (1986). Scale-dependent correlation of seabirds with schooling fish in a coastal ecosystem. *Marine Ecology—Progress Series*, **32**, 237–46.

Stiling, P. D. (1987). The frequency of density dependence in host-parasitoid systems. *Ecology*, **68**, 854–6.

Sutherland, W. J. (1983). Aggregation and the 'ideal free' distribution. *Journal of Animal Ecology*, **52**, 821–28.

Sutherland, W. J. and Parker, S. A. (1985). Distribution of unequal competitors. *Behavioural ecology: Ecological consequences of adaptive behaviour* (ed. R. M. Sibly and R. H. Smith), pp. 255–74. Blackwell Scientific Publications, Oxford.

Walde, S. J. and Murdoch, W. W. (1988). Spatial density dependence in insect parasitoids. *Annual Review of Entomology*, **33**, 441–66.

Werner, E. E. (1976). Niche shift in sunfishes: experimental evidence and significance. *Science*, **191**, 404–6.

Whitham, T. (1980). The theory of habitat selection examined and extended using Pemphigus aphids. *American Naturalist*, **115**, 449–66.

16 Seasonal and annual patterns of mortality in migratory shorebirds: some conservation implications

P. R. EVANS

Introduction

Population dynamics of shorebirds formed the subject of a research review by Evans and Pienkowski (1984), and the role of migration in population processes has been further discussed by Pienkowski and Evans (1984, 1985). The present chapter updates these reviews and focuses particularly on causes of mortality in shorebird populations, since these have important implications for conservation of the species concerned.

Most shorebirds (waders) are birds of open habitats throughout their annual cycles. The migratory species nest chiefly on Arctic tundra, peatlands, montane plateaux, and lowland wet grasslands. Most of the Arctic and northern temperate breeding species belong to the long-billed sandpiper group (*Scolopacidae*), but a few are short-billed plovers (*Charadriidae*). In winter, all species leave the Arctic, some travelling as far south as Argentina, Australia, and the Republic of South Africa. The total distances travelled between breeding and non-breeding grounds may exceed 12 000 km, probably the longest migrations of any land birds. Different species making the same journey may break the flight into stages of different lengths and use different staging posts (refuelling sites). The reasons for this are not clear; some ideas have been discussed by Piersma (1987) and Evans and Davidson (1990).

On the breeding areas, waders typically nest at low densities over large areas of superficially homogeneous habitat. During migration especially, but also in winter, they aggregate at high densities (often several thousand birds/km²) on coastal sand- and mudflats and on rocky shores, as well as in shallow inland wetlands at more tropical latitudes. The northern and western limits of winter distributions of shorebirds correlate fairly closely with the limit of ice-free intertidal and coastal habitats. Because many species are restricted to open coastal sites, total population counts are possible, using an international network of observers (Smit 1984).

Most migratory northern temperate and Arctic breeding waders lay a single clutch of four eggs, which may be replaced if lost, particularly in temperate areas. Clutch size usually does not vary with date of laying. Adults normally leave the breeding area before juveniles, and adult females of several species leave before males. This difference in timing may be detectable at staging posts some considerable distances from the nesting areas, e.g. in the southern Baltic for birds breeding in western Siberia.

Because many species nest at low densities in rather inhospitable and inaccessible areas, few long-term studies have been made of breeding populations. Indeed, individuals of some species that nest in transient and unstable habitats may not remain faithful to their nesting sites from year to year, so that in these species measurement of annual survival of adults is impossible. Even for species in which faithfulness to the breeding site is the rule, adults from a particular nesting area may spread over a wide variety of wintering sites, in many of which they mix with others of the same species from nesting areas many hundreds of miles away. During migration, populations of the same species from as far afield as arctic Canada and western Siberia may use identical staging areas. It is thus difficult to study population processes in many shorebirds in ways that are closely analogous to those traditionally applied to many resident birds. In particular, within most shorebird species, any density-dependent effects act upon different combinations of individuals at different times during the annual cycle.

In this chapter the life-history strategies of migratory shorebirds are considered first, with high adult survival identified as a key feature. I then consider how this is achieved, what threats man's activities pose to this, and how they might be counteracted.

Breeding success of Arctic waders

In most parts of the Arctic, shorebirds are restricted for breeding to the months of June–August. Because the period from egg-laying, through fledging, to completion of preparation for migration in juveniles is about 2 months, even in the smallest species, egg-laying after the beginning of July is most unlikely to produce any young that will survive. It is not surprising, therefore, that such late nesting has rarely been described.

Conditions across the Arctic are not uniform, because of the pattern of atmospheric circulation. In western Alaska, the lowland tundra is very wet and waders nest primarily on adjacent higher ground. On Ellesmere Island (Arctic Canada) and in north-east Greenland, conditions are more desert-like, but with heavier snowfall in winter in the more extensive mountainous regions of Greenland. In contrast, parts of Siberia, particularly the Taimyr peninsular, lie within the 'Polar desert' region, with relatively little winter and spring snowfall. In Alaska and Ellesmere Island, waders time their nesting so that the precocial young hatch when food, chiefly small dipterans, is most abundant (Holmes 1966). In north-east Greenland, however, marked differences in the timing of breeding occur amongst sites only a few kilometres apart, associated with differences in the date of snow-melt, arising from differences in snow depth which in turn are dependent upon the wind directions during precipitation. In some years snow-melt is so late that in July some areas are still covered and birds do not breed there. Thus the proportion of the population able to breed varies from year to year in Greenland. In Alaska and Ellesmere Island, waders often re-lay if they lose clutches early in the nesting season. This appears to be less common in north-east Greenland and depends critically upon the date within June of any late snowfalls which cause abandonment of nests. In some years, snowfalls in late June may cause total breeding failure (Green *et al*. 1977).

In the Taimyr Peninsula, breeding success of Brent Geese *Branta b. bernicla* and waders is closely correlated with the 3-year cycle of abundance of lemmings. Numbers of three species of shorebirds from this breeding area that reach South Africa in the northern autumn are much lower in the year after the peak of lemming abundance than in the peak year itself. Summers and Underhill (1987) have argued that this results from prey switching by Arctic foxes which, they suggest, prey on ground-nesting birds when lemmings become scarce, although this interpretation has been challenged (de Boer and Drent 1989).

The conclusion from all these studies is that breeding success of waders in different parts of the Arctic is highly variable from year to year, depending on the timing and severity of the weather and of predation. For species to persist, therefore, the survival of adults must be high, so that at least some manage to produce enough recruits during their lifetime to replace those adults that die.

Breeding success of migratory species in the northern temperate zone

At lower latitudes, climatic influences on the outcome of breeding attempts are much less important, but predation may be equally, if not more, important than in the Arctic. Indeed, Pienkowski (1984) suggested that the intensity of predation may set the southern limit to the distribution of the Ringed Plover *Charadrius hiaticula* in Europe and provide an explanation for the reduction in variety of breeding shorebird species at lower latitudes. The latter had previously been attributed by Jarvinen and Vaisanen (1978) to a reduction in habitat diversity. Attacks by mammalian predators may lead to local extinctions of ground-nesting waders, through repeated breeding failures, unless there is immigration from successful nesting areas further away. Alternatively, predation by foxes, for example, may vary from year to year, in parallel with variations in abundance of preferred prey, such as rabbits and voles. Indeed, predators may turn to alternative prey during the course of the nesting season itself, so that heavy predation on eggs is not necessarily followed by similarly heavy losses of chicks. The result of these large local, seasonal, and annual fluctuations in predation, in addition to other unpredictable losses (e.g. to flooding by exceptionally high tides on coastal salt-marshes or on meadows) is that the breeding success of waders in the northern temperate zone is also highly variable from year to year, though not to the same extent as for Arctic nesting species.

Components of the high annual survival of adult shorebirds

Year-to-year survival rates of adult shorebirds breeding in Arctic regions have been determined chiefly on the wintering areas, by observations of the return rates of individually colour-marked birds, rather than on the breeding grounds where few long-term studies have been undertaken. By noting at the non-breeding site whether each marked individual was present until its normal date of departure in spring, it has been possible to subdivide the year-to-year disappearance rates (equivalent to maximum annual mortality) into two components: (i) an estimate of the mortality on the wintering grounds, and (ii) an estimate of the sum of mortality on the breeding area and of losses on migration to and from the breeding area. This approach assumes that individual birds are faithful to a non-breeding site. For many species this assumption seems valid (see review by Evans and Pienkowski 1984), and provides minimum values of survival rates.

For species or populations breeding in the northern temperate zone, several studies have been made on breeding sites, though with a possible bias towards places where breeding densities are particularly high (since these have contained adequate numbers of birds for population studies within a few square kilometres). Again, it has been possible in some studies to determine two components of annual mortality: (i) losses on the breeding grounds themselves, and (ii) the sum of losses in winter and of losses on migration to and from the wintering grounds.

Both these approaches confirm that the annual survival of adult shorebirds is high. Most mortality of Arctic-nesting species occurs on the wintering grounds (where many species spend more than 8 months of the year) and, for temperate zone species also, the majority of the annual mortality occurs away from the breeding grounds. Examples are given in Table 16.1.

Table 16.1 Components of mortality of adult migratory shorebirds

	On wintering grounds(%)	On migration and on breeding grounds (%)	Total annual mortality (%)
Sanderling[1]	10	7	17
Turnstone[1]	8.5	7	15.5
Curlew[2]	13.4	3.4	16.8
	On wintering grounds and on migration %	On breeding grounds %	Total annual mortality %
Redshank[3]	16	8	24
Dunlin[3]	13	4	17

Sources: [1] Evans and Pienkowski (1984); [2]Evans, in preparation; [3] Jackson (1988).

Factors permitting high survival of adult migrant shorebirds on, and on route to the Arctic breeding grounds

On Ellesmere Island, Morrison and Davidson (1989) have studied the arrivals of Knot *Calidris canutus* and Turnstone *Arenaria interpres* in two successive springs. By catching birds over a period of 2 weeks in late May and early June, they have shown that both species arrived on Ellesmere, after a flight of at least 3000 km, with substantial reserves of body fat. Furthermore, on arrival, their flight muscles were larger than at certain other times of year. Because snow and wind conditions often prevented access to food, or foraging itself, birds fed for only short periods on most days before snow-melt began. Knot foraged less than Turnstone and steadily lost weight, particularly body fat; Turnstone also lost weight, but to a lesser degree, until feeding habitats became exposed. It thus appears that the birds survived chiefly because they carried reserves of fat and protein to the breeding grounds from the final staging post used during migration, where they had prepared for the flight by storage of more than twice as much fat as needed for the flight itself. Occasional heavy mortality of adult shorebirds after arrival on Arctic breeding grounds, as a result of late snowfalls, has been recorded (e.g. Morrison 1975), and similar effects have been noted in upland areas of the northern temperate zone, but only rarely.

The implication is that feeding conditions on the final staging post are normally good enough for most birds to deposit appropriate quantities of fat and muscle before departure. It is not known for certain whether the timing of their departure is regulated strictly by date, by the achievement of certain levels of body reserves, or by a combination of both. The few studies that have been made so far suggest a joint influence, with the last birds to leave being those that have the lowest body masses (e.g. Knot in north Norway; Evans and Davidson 1990), but there was very little difference in mean date of departure from year to year. Survival of birds in the Arctic may be critically dependent upon the level of body reserves at departure from the staging post. Of 20 Knots caught in Iceland in late May, just before departure for Greenland, and which survived the severe weather on the breeding grounds in summer 1974, 18 were heavier than the average in Iceland (Wilson 1988).

It is not only at the final staging post between the wintering and breeding grounds that migrating waders sometimes deposit more fat than is necessary for the flight stage that immediately follows. Knot arrive in north Norway from southern North Sea coasts with fat levels of above 10 per cent of total body mass, and with larger flight muscles than in winter (Evans and Davidson 1990). Sanderling *Calidris alba* migrating to breeding grounds in Greenland leave a staging post in north-east England, aiming for a final staging area in Iceland, with more than 30 per cent fat, which is much more than is needed for the flight (Davidson and Evans 1989). The reasons for this could be twofold: (a) to guard against food shortage or bad weather at the next staging post, which might depress the rate of fat deposition and muscle hypertrophy and ultimately reduce the levels that could be carried to the breeding grounds and (b) to enable the migrants to fly economically at a faster speed, since the maximum range speed (Pennycuick 1975) increases with total body mass. This may be important if waders are likely to encounter head winds during a particular flight stage.

There is little information on the effect of feeding conditions on bird behaviour at staging posts. The density of polychaete worms has been shown to affect food intake rate and the duration of stay of Dunlin *Calidris alpina* at a staging post in Morocco (Piersma 1987), but very cold weather in north Norway in May 1985 affected neither the duration of stay nor the rate and extent of gain of body reserves by Knot feeding on bivalves (Evans and Davidson 1990). Delays at any of the staging posts may be compounded to cause late arrival at the breeding grounds, or to force birds to arrive without adequate reserves. This could be particularly important in the case of Arctic-breeding species that have only a very restricted period in which to attempt to nest.

The conservation implications are clear: the environmental quality of staging posts must be maintained so that prey densities remain adequate. In particular, industrial pollution should be prevented. The other major threat to staging areas, many of which are intertidal flats, is reclamation of land for human activity. Loss of habitat must eventually lead, via increased density of birds using the remaining areas, to interference between feeding birds and thus a reduction in food intake by some. This must eventually cause an increase in adult mortality.

Competitive effects need not be restricted to interactions between shore birds of the same species, as Recher (1966) was the first to discuss. Circumstantial evidence of interspecific competition is provided by changes in the foraging areas used by Grey Plovers *Pluvialis squatarola* after the spring departure of Bar-tailed Godwits *Limosa lapponica* from a wintering site in north-east England (Pienkowski 1982). A more fully documented example of avoidance of intraspecific competition is provided by the segregation in migration timing, and feeding areas used, in the Wadden Sea staging area by Knots of the Siberian and Greenland breeding populations (Prokosch 1988). In one North American site, shorebirds removed 60 per cent of prey during one migration period (Schneider and Harrington 1981), although it was not established whether this affected food intake rates of the later migrants that used the area. Because the range of common prey species in intertidal areas is so restricted, there is often considerable overlap in prey taken by different shorebirds (Pienkowski *et al*. 1984). Loss of part of a feeding site is thus likely to heighten inter-specific competition and so affect different bird species to different extents (Evans and Pienkowski 1983).

Factors permitting high survival of adult shorebirds on the non-breeding areas

Although most mortality occurs during the winter months, shorebirds survive well even near the northern limit of their non-breeding range in northern Britain. This is aided by a number of adaptations. Foremost amongst these is storage and, if necessary, use of fat reserves. These show a distinct seasonal pattern, usually reaching a peak in December or January (Evans and Smith 1975; Davidson 1981). Different species store different maximum levels of fat, but in almost all cases these are less than the levels accumulated before migration. Fat serves as an insurance against periods of inadequate food intake, which become less likely as winter progresses. The reduction in fat reserves after January is regulated (Dugan *et al*. 1981) and the levels maintained at any particular date are presumably a compromise between avoiding death either through starvation or through predation, to which heavier birds are presumed to be more susceptible (Lima 1986).

Shorebirds' food intake rates may fall during both cold and windy weather. Coasts and estuaries are extremely exposed habitats and feeding becomes difficult, particularly for long-legged shorebird species, in strong winds. Under these conditions, birds may stay all day on sheltered roosts without attempting to feed (Evans and Smith 1975; Davidson 1981).

Some Grey Plovers defend territories centred around creeks on the mudflats of the Tees estuary; under windy conditions they are able to feed in the sheltered creeks whilst their non-territorial conspecifics abandon attempts to feed (Townshend *et al.* 1984). It is likely that individuals of other species, such as Redshank *Tringa totanus*, may benefit from holding territories partly for the same reason, although other advantages may also be involved.

In cold weather, shorebirds feeding upon polychaetes and on burrowing crustacea and bivalves may be faced with a reduction in prey accessibility (because they may burrow deeper) and in prey availability (because they become less active and hence provide less frequently clues to their location). The latter feature is particularly important to shorebirds foraging visually and relying upon movement of buried prey towards the surface of the substratum, i.e. the plovers (Pienkowski 1983; Evans 1989). The combined effect of these two features is that the density of *available* prey is reduced. Thus, the birds' rate of food intake drops in cold weather, unless foraging is concentrated in patches of particularly high prey density, as observed for some territorial Grey Plovers (Dugan 1982). To compensate for lower intake rates, many shorebird species feed at night as well as by day in winter. The prey spectrum available at night may be greater than by day, at least at certain times of year (Dugan 1981; Pienkowski 1983). A further adaptation for survival in cold weather is movement away from frozen areas. This is seen chiefly in plovers and particularly in those that feed inland, the Lapwing *Vanellus vanellus* and Golden Plover *Pluvialis apricaria*; but it also occurs in some Grey Plovers (Townshend 1982).

The outcome of these several adaptations is that overwinter survival of adults of most shorebird species is not markedly reduced in severe winters (see Evans and Pienkowski 1984). Exceptionally, severe weather may cause changes in behaviour of other bird taxa may be caused by severe weather which then affect shorebirds. The lower survival of Sanderlings in north-east England in the very cold winter of 1978/9 may have been caused in part by influxes of gulls onto the Sanderling's main foraging beaches, from which the shorebirds were then displaced.

Predation is much less important as a contributor to mortality of adult shorebirds than it is to juvenile mortality in winter (see later).

From a conservation standpoint, if it becomes necessary to manage estuarine habitats to enhance survival of shorebirds overwinter, possible methods would seem to be: (a) the provision of sheltered, yet open, feeding areas, e.g. by excavating channels through mudflats; and (b) the discharge of (clean) cooling water from industrial processes over the flats to raise water and mud temperatures slightly, in order to increase the proportion of prey that are available.

Creation of peripheral wetlands may also be important (see below).

Mortality during late summer and autumn migrations

Adults

After breeding, many shorebirds migrate to moulting grounds which are often not their final winter destinations. For many species, these moulting areas are particularly large expanses of intertidal flats, e.g. the Waddenzee/Wattenmeer/Vadehavet in Western Europe, where birds are relatively safe from mammalian predators during the period when their flight feathers are being replaced. In some autumns, parties of adult waders, still in breeding plumage, pause during migration on estuaries where the species is present during the winter but where they do not moult. Such birds stay for only a week or two, presumably until they have refuelled sufficiently to continue migration to the moulting grounds. The fat levels and muscle masses of a flock of Knot, which interrupted their migration in north-east England in late July, were the lowest levels recorded during the annual cycle. If the site at which they refuelled had been lost to land reclamation it seems doubtful whether the birds could have continued to their intended destination. However, loss of a traditional moulting ground could have far more serious effects on survival, since so few sites seem to provide both good food resources and protection against predation.

Juveniles

No estimates are yet possible of the extent of losses of juveniles during their first migration, from their natal areas to overwintering grounds. Nevertheless, Gromadzka (1983) obtained a recovery rate in September which was four times as high for juvenile Dunlin as for adults, from birds ringed on the Polish coast of the Baltic Sea; this is more than half-way between their arctic Russian breeding grounds and their wintering areas around the North Sea coasts, a primarily overland flight. On long-distance migrations over the sea, losses may be even higher. Many small groups of juvenile Knots from Greenland occur along the west coast of Norway in late summer, even though all adults travel through Iceland. How many juveniles die without reaching the Norwegian coast is not known. Juvenile waders of several species have reached Mauritainia in autumn in an emaciated condition, from which some did not recover (Dick and Pienkowski 1979). Possible reasons for higher mortality of juveniles versus adults leaving the Arctic breeding grounds could be that juveniles, departing later, have a greater chance of encountering bad weather over the open sea.

Survival of juveniles on the non-breeding areas

Once established on wintering grounds in north-east England, survival of juvenile Sanderling was as high as that of adults (Pienkowski and Evans 1985). Avian predators are scarce at that site, a finding that is probably not applicable to Sanderlings generally. Nevertheless, adaptations must exist to enable juveniles of all species to cope with normal winter feeding conditions. Juveniles carry higher fat reserves than adults of the same species wintering in the same area, as shown for Dunlin and Knot by Davidson (1981). Another is that juveniles feed for a longer period each day than adults, if necessary. This requirement may lead some birds to move to feeding areas around brackish and freshwater pools adjacent to an estuary, once the main intertidal feeding areas are covered by the tide. Such peripheral wetlands can be of crucial importance to the survival of adults and, more particularly, of juveniles of the smaller shorebird species such as Dunlin on estuaries where land claim has removed the upper tidal zones, so that feeding time has been curtailed. (A general finding is that the normal period of feeding within each tidal cycle is shorter for larger species, so these are less affected by loss of the upper tidal zones.) Evidence that birds using supplementary feeding sites are in poorer condition is provided for Dunlin (Table 16.2). Clearly, the provision of such supplementary feeding sites may be an important management tool to restore the potential of degraded estuaries to support shorebirds.

Feeding habitat in estuaries has also been lost by downshore spread of the cord-grass *Spartina anglica*. Again, small species such as Dunlin and Redshank are most affected as they tend to feed throughout much of the tidal cycle in winter and therefore need to make use of upper tidal zones. Areas colonized by *Spartina* hold reduced densities of prey (Millard and Evans 1984), as well as being unfavourable feeding sites for birds of open

Table 16.2 Body condition of Dunlin in relation to use of supplementary feeding areas on the Tees estuary, north-east England (adapted from Davidson and Evans 1986).

	Feeding during high water		Roosting during high water	
Fat (% of body mass)				
Adults	9.5	(.86)	12.1	(1.02)
Juveniles	10.7	(2.12)	13.4	(1.27)
Muscle index				
Adults	0.263	(.007)	0.274	(.005)
Juveniles	0.247	(.005)	0.274	(.005)

Values are of mean (standard error).
Sample sizes: Feeding—adults 21, juveniles 7; roosting—adults 14, juveniles 12.

habitats. Indeed the populations of Dunlin wintering on British estuaries invaded by *Spartina* have decreased in recent years (Goss-Custard and Moser 1987), which suggests an overall limitation on suitable wintering habitat for the species. Removal of *Spartina* with the herbicide 'Dalapon' has been used as a management tool, with some success; cleared areas have been used by foraging shorebirds, particularly where invertebrate densities have increased above those found in the grass sward itself (Evans 1986).

Along rocky shores, juvenile shorebirds of several species have been shown to be disproportionately at risk to surprise attacks by avian predators such as Sparrowhawks *Accipiter nisus*. On open coasts and estuaries, predation by Merlins *Falco columbarius* can also be significant to smaller wader species (Whitfield 1985), and again juveniles may be more at risk than adults (Table 16.3). Whilst mortality caused by birds of prey can scarcely be considered a conservation problem, shorebird deaths from mammalian predators may be, in some areas. Roosting birds seem most at risk and, because juveniles often roost in atypical sites when they first arrive on the wintering grounds, they are often taken in greater numbers than adults by e.g. foxes (Townshend 1984). Provision of safe roosting sites could be an alternative to predator control.

Table 16.3 Predation on shorebirds by birds of prey and mammals in northern Britain

Shorebird species	Habitat	% or juveniles amongst	
		those present	those taken
Taken by birds of prey			
Redshank	Rocky shore	40	85
Redshank	Estuary	42	85
Turnstone	Rocky shore	18	58
Dunlin	Estuary	35	73
Taken by mammals			
Grey Plover	Estuary	30	67

Conclusion

The preceding sections have illustrated that shorebirds are adapted to erratic breeding performance by maintaining high adult survival rates. These arise from physiological adaptations on breeding and wintering grounds, and on migration staging areas, whereby birds carry reserves of fat and muscle protein as an insurance against poor feeding conditions. The extent of these reserves is a compromise between longer survival with higher reserves and greater risk of being taken by a predator.

A major conservation priority is safeguarding the quality of habitats used by shorebirds during their annual cycle, so that adequate densities of food are available. So also is safeguarding the extent of habitat, particularly of intertidal land at the upper tidal zones, which may be particularly important to provide sufficient feeding time. For high-Arctic nesting species, in particular, preservation of a network of staging areas is vital, as the time-programming of their use of such areas is rigorous and delays caused by poor conditions at a site may be compounded along the migration route.

A further threat to the future survival of waders breeding in north temperate areas is agricultural change. Lowering of the water table of wet meadows, changes in timing of their use by cattle, and earlier cutting of grass, e.g. for silage, have all reduced breeding productivity, even before the erratic influences of predation and unseasonable weather are considered. Some species may not be able to compensate for this and their future survival is uncertain in areas such as the Netherlands (Beintema and Muskens 1987). Management of water levels to prolong the period of good feeding has been suggested as a means of raising breeding success of Snipe *Gallinago gallinago* (Green 1988).

In the longer term, climatic warming threatens to modify the Arctic breeding areas, perhaps rendering them more accessible to predators; and the temperate wintering areas may be affected by raising of the sea-level to cover, permanently, many presently intertidal habitats. The extent to which new habitats will be formed to replace these, as a result of distribution of sediments, is a matter of debate.

References

Beintema, A. J. and Muskens, G. J. D. M. (1987). Nesting success of birds breeding in Dutch agricultural grasslands. *Journal of Applied Ecology*, **24**, 743–58.

Davidson, N. C. (1981). Survival of shorebirds during severe weather: the role of nutritional reserves. In *Feeding and survival strategies of estuarine organisms* (eds. N. V. Jones and W. J. Wolff), pp. 231–49. Plenum, New York.

Davidson, N. C. and Evans, P. R. (1986). The role and potential of man-made and man-modified wetlands in the enhancement of the survival of overwintering shorebirds. *Colonial Waterbirds*, **9**, 176–88.

Davidson, N. C. and Evans, P. R. (1989). Prebreeding accumulation of fat and muscle protein by arctic-breeding shorebirds. Proceedings 19th International Ornithology Congress, Ottawa, pp. 342–52.

de Boer, W. F. and Drent, R. H. (1989). A matter of eating or being eaten? The breeding performance of arctic geese and its implications for waders. *Wader Study Group Bulletin*, **55**, 11–17.

Dick, W. J. A. and Pienkowski, M. W. (1979). Autumn and early winter weights of waders in north-west Africa. *Ornis Scandinavica*, **10**, 117–23.

Dugan, P. J. (1981). The importance of nocturnal foraging in shorebirds: a consequence of increased invertebrate prey activity. In *Feeding and survival strategies of estuarine organisms* (eds. N. V. Jones and W. J. Wolff), pp. 251–60. Plenum, New York.

Dugan, P. J. (1982). Seasonal changes in patch use by a territorial Grey Plover: weather-dependent adjustments in foraging behaviour. *Journal of Animal Ecology*, **51**, 849–57.

Dugan, P. J., Evans, P. R., Goodyer, L. R., and Davidson, N. C. (1981). Winter fat reserves in shorebirds: disturbance of regulated levels by severe weather. *Ibis*, **123**, 359–63.

Evans, P. R. (1986). Use of the herbicide 'Dalapon' for control of *Spartina* encroaching on intertidal mudflats: beneficial effects on shorebirds. *Colonial Waterbirds*, **9**, 171–5.

Evans, P. R. (1989). Predation of intertidal fauna by shorebirds in relation to time of day, tide and year. In *Behavioural adaptation to intertidal life* (eds. M. Vannini and G. Chelazzi). NATO ARW Series, **151**, 65–78. Plenum, New York.

Evans, P. R. and Davidson, N. C. (1990). Migration strategies of waders breeding in arctic and north temperate latitudes. In *Bird migration—Physiology and eco-physiology* (ed. E. Gwinner), pp. 387–98. Springer-Verlag, Berlin.

Evans, P. R. and Pienkowski, M. W. (1983). Implications for coastal engineering projects of studies at the Tees estuary on the effects of reclamation of intertidal land on shorebird populations. *Water Science and Technology*, **16**, 347–54.

Evans, P. R. and Pienkowski, M. W. (1984). Population dynamics of shorebirds. In *Shorebirds: Breeding behaviour and populations* (eds. J. Burger and B. L. Olla), pp. 83–123. Plenum, New York.

Evans, P. R. and Smith, P. C. (1975). Fat and pectoral muscle as indicators of body condition in the Bar-tailed Godwit. *Wildfowl*, **26**, 64–76.

Goss-Custard, J. D. and Moser, M. (1987). Rates of change in the numbers of Dunlin *Calidris alpina* wintering in British estuaries in relation to the spread of *Spartina anglica*. *Journal of Applied Ecology*, **25**. 95–109.

Green, R. E. (1988). Effects of environmental factors on the timing and success of breeding of Common Snipe *Gallinago gallinago*. *Journal of Applied Ecology*, **25**, 79–93.

Green, G. H., Greenwood, J. J. D., and Lloyd, C. S. (1977). The influence of snow conditions on the date of breeding of wading birds in north-east Greenland. *Journal of Zoology, London*, **183**, 311–28.

Gromadzka, J. (1983). Results of bird-ringing in Poland: Migrations of Dunlin *Calidris alpina*. *Acta Ornithologica, Warszaw*, **19**, 113–36.

Holmes, R. T. (1966). Feeding ecology of the Red-backed sandpiper (*Calidris alpina*) in arctic Alaska. *Ecology*, **47**, 32–45.

Jackson, D. B. (1988). Habitat selection and breeding ecology of three species of waders in the Western Isles of Scotland. Unpublished Ph.D. thesis, University of Durham, UK.

Järvinen, O. and Vaisanen, R. A. (1978). Ecological zoogeography of North European waders, or why do so many waders breed in the North? *Oikos*, **30**, 495–507.

Lima, S. L. (1986). Predation risk and unpredictable feeding conditions; determinants of body mass in birds. *Ecology*, **67**, 377–85.

Millard, A. V. and Evans, P. R. (1984). Colonization of mudflats by *Spartina anglica*: some effects on invertebrate and shorebird populations at Lindisfarne. In *Spartina anglica in Great Britain* (ed. P. Doody), pp. 41–8. Nature Conservancy Council, Peterborough.

Morrison, R. I. G. (1975). Migration and morphometrics of European Knot and Turnstone on Ellesmere Island, Canada. *Bird Banding*, **46**, 290–301.

Morrison, R. I. G. and Davidson, N. C. (1989). Migration, body condition and behaviour of shorebirds at Alert, Ellesmere island, NWT. *Syllogeus* (National Museum Natural History, Ottawa).

Pennycuick, C. J. (1975). Mechanics of flight. *Avian Biology*, 5, 1–53.

Pienkowski, M. W. (1982). Diet and energy intake of Grey and Ringed Plovers *Pluvialis squatarola* and *Charadrius hiaticula* in the non-breeding season. *Journal of Zoology, London*, 197, 511–49.

Pienkowski, M. W. (1983). Surface activity of some intertidal invertebrates in relation to temperature and the foraging behaviour of their shorebird predators. *Marine Ecology Progress Series*, 11, 141–50.

Pienkowski, M. W. (1984). Breeding biology and population dynamics of Ringed Plovers *Charadrius hiaticula* in Britain and Greenland: Nest predation as a possible factor limiting distribution and timing of breeding. *Journal of Zoology, London*, 202, 83–114.

Pienkowski, M. W. and Evans, P. R. (1985). Migratory behaviour of shorebirds in the western Palaearctic. In *Shorebirds: migratory and foraging behaviour* (eds. J. Burger and B. L. Olla), pp. 73–123. Plenum, New York.

Pienkowski, M. W. and Evans, P. R. (1984). The role of migration in the population dynamics of birds. *Behavioural ecology: the ecological consequences of adaptive behaviour* (eds. R. M. Sibly and R. H. Smith), pp. 331–52. Symposium British Ecology Society, 25. Blackwell Scientific Publications, Oxford.

Pienkowski, M. W., Ferns, P. N., Davidson, N. C., and Worrall, D. H. (1984). Balancing the budget: problems in measuring the energy intake and requirements of shorebirds in the field. In *Coastal waders and wildfowl in winter* (eds. P. R. Evans, J. D. Goss-Custard, and W. G. Hale), pp. 29–56. Cambridge University Press.

Piersma, T. (1987). Hink, stap of strong? Reisebeperkingen van arctische steltlopers. *Limosa*, 60, 185–94.

Prokosch, P. (1988). Arktische Watvogel im Wattenmeer. *Corax*, 12, 273–442.

Recher, H. F. (1966). Some aspects of the ecology of migrant shorebirds. *Ecology*, 47, 393–407.

Schneider, D. C. and Harrington, B. A. (1981). Timing of shorebird migration in relation to prey depletion. *Auk*, 98, 801–11.

Smit, C. J. (1984). Identification of important sites for waders by co-ordinated counts. In *Shorebirds and large waterbirds conservation* (eds. P. R. Evans, H. Hafner, and P. L'Hermite), pp. 43–51. Commission of the European Community, Brussels.

Summers, R. W. and Underhill, L. (1987). Factors related to breeding production of Brent Geese and waders (*Charadrii*) on the Taimyr Peninsula. *Bird Study*, 34, 161–71.

Townshend, D. J. (1982). The Lazarus syndrome in Grey Plovers. *Wader Study Group Bulletin*, 34, 11–12.

Townshend, D. J. (1984). The effects of predators upon shorebird populations in the non-breeding season. *Wader Study Group Bulletin*, 40, 51–4.

Townshend, D. J., Dugan, P. J., and Pienkowski, M. W. (1984). The unsociable plover—use of intertidal areas by Grey Plovers. In *Coastal waders and wildfowl in winter* (eds. P. R. Evans, J. D. Goss-Custard, and W. G. Hale), pp. 140–59. Cambridge University Press.

Whitfield, D. P. (1985). Raptor predation on wintering waders in south-east Scotland. *Ibis*, 127, 544–58.

Wilson, J. R. (1988). The migratory system of Knots in Iceland. *Wader Study Group Bulletin*, 54, 8–9.

17 The importance of migration mortality in non-passerine birds

MYRFYN OWEN and JEFFREY M. BLACK

Introduction

The migratory flight itself has long been regarded as one of the most hazardous stages of the life cycle of migratory birds. In their review of the importance of migration in shorebird populations, Pienkowski and Evans (1985) recognized that mass deaths probably occurred frequently, though there were great difficulties in quantifying such mortality. They also recognized that these most often related to young birds, and pointed out that this could be for two reasons: failure to grow and lay down sufficient fat reserves to complete migration in the short Arctic summer and possible competition with adults on the staging areas.

The length of a single migratory flight clearly depends on the availability of suitable staging and wintering habitat, and the provision and protection of this habitat is the most important consideration in the conservation of migratory birds. This has been recognized for wetlands in the formulation of the Convention for the Conservation of Wetlands of International Importance Especially as Waterfowl Habitat (the Ramsar Convention) in 1971—the first international convention concerned with protecting habitat. In view of this, the estimation of flight ranges and capabilities of migratory species is vital to their conservation.

This paper reviews the impact of migration mortality in non-passerine birds and summarizes some direct evidence of losses on migration in geese.

Extent and reasons for migration losses

For shorebirds, there are several lines of evidence, summarized by Pienkowski and Evans (1985), to suggest that it is advantageous for birds to winter as close as possible to their breeding areas. In several studies, the survival of shorebirds migrating short distances is higher than that of long distance migrants. Young birds are forced by competitive exclusion to range further and suffer higher mortality rates than adults. Pienkowski and Evans predict that 'major losses of young' shorebirds occur during their first migratory flights. Once on the wintering areas both adults and young of several species have comparatively high survival rates.

First year survival rates in ducks are considerably lower than those of adults, but this is generally regarded as reflecting differential vulnerability to hunting mortality (see, for example, Boyd 1962). However, survival rates refer to the period before the birds reach their terminal wintering grounds, and this could also reflect the longer travelling distances of the young making them more accessible to hunters; in Europe, since hunting pressure is heavier in the south (Tamisier 1985), the longest migrants are at greatest risk. If long flights are in themselves hazardous, hunting mortality may be replacing what would be a natural loss through exhaustion, predation or starvation. In a more natural situation, where there might be density dependent increases in mortality (see, for example, Hill 1984), losses on migration would be expected to assume greater significance.

In many duck species there is differential migration of the sexes, such that females migrate further and leap-frog over the males that occupy more northerly wintering sites (Perdeck and Clason 1983). In addition, there is evidence for the Canvasback *Aythya vallisneria* (Nichols and Haramis 1980) and for the Goldeneye *Bucephala clangula* (Sayler and Afton 1981) that males occupy the most favourable habitats and competitively exclude females. There is no direct evidence that this leads to higher mortality rates in females, though for several species of ducks the overall population sex ratio is markedly skewed in favour of males. For example, in the west European population of Wigeon *Anas penelope* there are estimated to be in excess of 1.3 males per female (Campredon 1983, Owen and Dix 1986). In the same population, males survive longer than females; the time between ringing and death was 771 days for males and 567 days for females (Owen and Mitchell 1988).

Nowadays, most geese and swans migrate over open areas where they can readily stop and refuel frequently. However, before arable agriculture deforested vast tracts, much of the habitat between wintering and breeding areas was unsuitable for feeding. Many species have taken advantage of the new staging areas and interrupted their traditional migrations. For example, the Lesser Snow Goose *Anser caerulescens* traditionally migrated 3300 km direct from James' Bay to the Gulf Coast, but made use of the central United States, especially if its departure body mass was insufficient to make the non-stop flight (Cooch 1958). Some species migrate considerable distances over the sea, where feeding is impossible for these vegetarian birds.

Direct evidence from geese

Since 1973 individual marking has been used to estimate mortality rates in a population of Barnacle Geese *Branta leucopsis* which breeds in the Svalbard (Spitsbergen) archipelago and winters exclusively on the Solway

Firth in northern Britain, some 3200 km away (see Owen 1982 for methods). On autumn migration most of the geese use the small Bear Island, 250 km south of the breeding grounds, as a staging area, but their flight from there (3000 km) is direct (Owen and Gullestad 1984). Since 1970, the number of geese in the population, which is effectively closed (0.1 per cent annual emigration rate and no recorded immigration), has increased from 3000 to more than 11 000. The number of marked birds has varied between 350 and 3500 and the annual resighting rate has been around 95 per cent (93–99 per cent) (Owen 1982, 1984). Ring loss has remained small, around 0.1 per cent annually in the first five years and 0.6 per cent annually in the first ten years. Between 1970 and 1978, numbers grew exponentially following a decline in the mortality rate of birds more than a year old (from about 25 per cent to 10–12 per cent annually (Owen 1982)). Since then, however, density dependent limitations on recruitment have resulted in a levelling out of the growth rate (Owen and Black 1989).

During the period between 1978 and 1981, when observers were on the breeding and wintering grounds as well as on the migration staging areas, the pattern of losses was examined at different times of year. The pattern of last sighting dates of 356 of some 1800 geese marked on a single stretch of coastline is shown in Fig. 17.1, plotted as the observed last sighting in

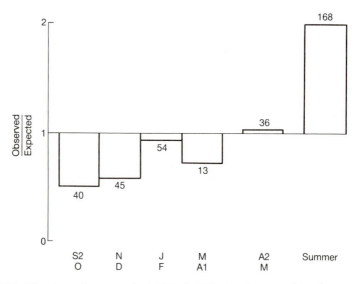

Fig. 17.1. The time when a total of 356 of 1800 ringed geese (older than one year) from one breeding area disappeared from the population. The index is expressed as the observed/expected number of losses based on the probability of sighting geese in different periods (from Owen 1982). S2 is the second half of September, A1 and A2 are the first and second halves of April. Gaps between blocks represent periods during which the geese were migrating.

relation to that expected given the distribution of all sightings. Clearly, there is a much higher chance of birds disappearing in the summer months, including both northerly and southerly migrations, than at any other time. Indeed, nearly half the recorded disappearances (=deaths, see Owen 1982) were during this period. Clearly these losses include deaths during breeding and moult as well as on autumn migration, but losses of adults in summer are small (Owen 1982, Prop *et al*. 1984).

Losses of young

Figure 17.2 shows the proportion of juveniles in flocks of geese in 1983 during the Bear Island staging period and during the time of arrival on the Solway. The young (which are in stable families at this time) remain later in the breeding areas; indeed, in some years we know that they stay on the autumn fattening slopes until driven out by complete snow cover.

As a result of the analysis summarized above, it was suggested that the most important source of natural mortality in this population was losses on autumn migration (Owen 1982). In adults, losses are likely to be the result of birds being unable to lay down sufficient fat reserves, after the stresses of breeding, to complete the journey. Thus it is breeding-related stresses that are manifest through lower migration survival. Because they emerge from incubation in poorer condition than males, females were predicted to suffer higher mortality rates on migration. It was also suggested that young birds would suffer especially in late seasons and at high bird densities, when competition for food would affect the fattening rate (Owen 1982).

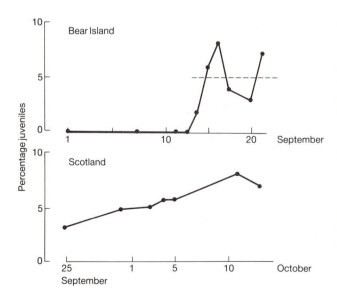

Fig. 17.2. The percentage of young birds in flocks on a staging area in Bear Island (Svalbard) and following arrival in Scotland in 1983. In late summer the geese move 250–500 km south to Bear Island, the migration, in late September, is 3000 km long and takes about 40 hours (see Owen and Gullestad 1984). The dashed line indicates the mean from 13–22 September.

Since the chance of a ringed bird being alive but not identified for two consecutive years was only 0.1 per cent (Owen 1982), and since 80 per cent of returning geese were seen on our reserve within two to three months of arrival, the birds seen in Britain provide an accurate index of survival. Survival estimates for ringed goslings in the early 1980s varied between 79 and 88 per cent, whereas adult survival (between Bear Island and Scotland) was 97–98 per cent in favourable years. Large sample sizes are available from 1986; the results are presented in detail in Owen and Black (1989) and summarised in Table 17.1. There is no significant difference between adults and yearlings (16 months old), but juveniles suffered substantial losses.

Young birds caught in 1986 were aged to the nearest week by plumage characteristics, and weighed. Older and heavier goslings had a significantly higher chance of survival; there was a significant effect of body mass on survival even between goslings of a similar age (Table 17.2). There were also differences between rearing areas on survival and this was not entirely explained by differences in the hatching dates of the young. These differences were consistent with the hypothesis that competition for food in the rearing period determined survival, since losses were related to available food resources and bird density in the different areas (Owen and

Table 17.1 Survival of different classes of Barnacle Geese in 1986

Age class	Seen 1986/1987	Not seen	% Survival
Juveniles	286	153	65.1
Yearlings	148	5	96.7
Adults	1520	74	95.4

Table 17.2 The effect of body mass on survival within the three most common age classes

	3–4 Weeks	4–5 Weeks	5–6 Weeks
Birds that returned Mean mass ± SE	759.9 ± 10.6g $N = 68$	826.8 ± 19.0g $N = 51$	969.7 ± 26.2g $N = 32$
Birds not seen Mean mass ± SE	689.4 ± 16.4g $N = 47$	810.3 ± 26.6g $N = 38$	839.6 ± 46.7g $N = 14$
P value	<0.001	<0.001	<0.02

Black 1989). In the population of Barnacle Geese recently established in the Baltic, there are significant differences between the growth rates of young reared in different grazing localities (K. Larsson, personal communication).

Losses of adults

Of 115 females which had brood patches when caught, indicating that they had nested, 105 (91 per cent) survived to reach Scotland whereas all but one of the 47 females (98 per cent) without brood patches survived; the difference was, however, not significant. We have, as yet, found no significant difference between migration mortality between the sexes, but the trend in most years suggests that fewer females survive. In the very late season of 1979, the survival rate of 1012 females was significantly lower than that of a similar sample of males. In the adjacent seasons 1978 and 1980, both favourable for breeding, no such difference was found. This does affect the average expectation of life of the two sexes; in a sample of 37 females and 41 males hatched in 1972 and followed until 1987, the median lifespan of females was eight years compared with ten for males (Owen and Black 1990).

To obtain an index of migration mortality in adults we examined the proportion of birds lost between February and arrival in Scotland the following autumn. This is after the end of the shooting season and most losses during this period are likely to be incurred on migration (see Fig. 17.1). The migration mortality index is related to population size in Fig. 17.3. Over the same period, there was no significant correlation between the total annual mortality rate (the proportion of ringed birds that disappeared between one October and the next), and population size for either sex. It varied between 6 and 18 per cent in females and between 4 and 15 per cent in males in the ringed sample. The mean mortality rates for years when population size was respectively below and above the median were 10.5 per cent and 11.5 per cent for females and 9.0 per cent and 11.6 per cent for males. This suggests that the density dependent effect on mortality in adults operates between March and September, and (since it is known that losses between ringing and departure are few) most probably on autumn migration.

Evidence from other species

There are a number of seabirds in which survival between fledging and wintering or return to a breeding colony was checked in relation to the distance of the first migration. In the short distance migrants, the Herring Gull *Larus argentatus* (Parsons *et al*. 1976) and the Puffin *Fratercula arctica* in the North Sea (Harris and Rothery 1985), subsequent survival

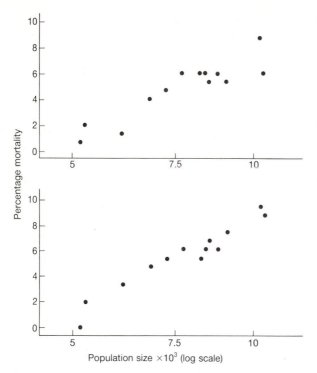

Fig. 17.3. The mortality rate between February and October (most of deaths on autumn migration) in 13 years between 1973 and 1985 in relation to the number of geese in the population in the previous winter, obtained by direct and accurate total counts (Owen 1984). Coefficients of determination (r-squared) 0.919 (P < 0.001) for females (upper figure) and 0.819 for males (lower) (P < 0.001).

was not related to fledging age or body mass. In two long-distance migrants, however, the Manx Shearwater *Puffinus puffinus* (Perrins 1966, Perrins *et al*. 1973) and the Cape Gannet *Sula capensis* (Jarvis 1974), survival after fleding was greater for older and/or heavier young.

Table 17.3 lists the species for which evidence is available on mass and/or age of premigratory young birds and their subsequent arrival. In most of the studies the interval between fledging and the time at which survival was estimated was long; in seabirds, for example, survival cannot be estimated until the birds return to the colony some years later. However, the losses are probably incurred in the immediate post-fledging period just before or during the first migration. The evidence from these studies is consistent with the suggestion that autumn migration mortality is an important factor limiting recruitment into the population of long-distant migrants. In the Brent Goose *Branta bernicla*, which migrates from the Arctic islands of northern Canada to its first staging area in Iceland, the survival rate of young in both years where data are available was around 67 per cent (O'Briain 1987)—very similar to that we have reported for Barnacle Geese. In the Brent, however, the survival rate of yearlings, at around 80 per cent, is considerably lower than the 96 per cent for Barnacle Geese in Table 17.1.

Table 17.3 Some of the studies of non-passerine birds which measured survival in relation to biometrics, age and date of departure of fledglings

Species	Parameters	Effect on	Explanation	Authors
Barnacle Geese	fledging mass and age	Yes	long migration over sea	this chapter and Owen and Black 1989
Brent Geese	age and class	Yes	long migration over sea	O'Briain 1987
Cape Gannets	fledging mass and date	Yes	migrant time at sea	Jarvis 1974
Guillemots	fledging mass and date	No	extended parental feeding on migration	Hedgren 1981
Herring gulls	fledging age and date	No	short migration	Nisbet and Drury 1972 Parsons et al. 1976
Manx Shearwaters	Fledging mass, age, and date	Yes	long migrations	Perrins 1966 Perrins et al. 1973
Puffins	fledging mass and age	No	Short migration	Harris and Rothery 1985
Razorbills	fledging mass and age	No	little competition and short migration	Lloyd 1979
Sparrowhawks	fledging mass	No	Little/no competition	Newton and Moss 1986

Modelling migration range

Clearly the maximum length of a non-stop flight of which a bird is capable is a critical factor in its survival when large distances have to be covered over terrain unsuitable for feeding. This is particularly so for arctic-breeding shorebirds, for which the presence of a few protected staging areas may be vital to the survival of whole populations (Myers 1983, Pienkowski and Evans 1985).

Waterfowl such as geese are under the same constraints, and we list in Table 17.4 the components of a model which would predict flight range in Barnacle Geese (very similar rules would apply to any migrating bird). The clearance of snow from the breeding area is a crucial determinant of laying date and in some years whole populations of arctic birds fail to breed (see, for example, Owen 1984). The timing of autumn migration is determined by the onset of winter, which usually intervenes before the shortening daylength curtails the feeding time of diurnal feeders. The direction and speed of the wind are also of crucial importance to the amount of energy needed for flight. Birds normally wait for following winds before setting off; Barnacles leaving Bear Island did so with a mean following wind speed of 27 km/h (Owen and Gullestad 1984). However, on a long flight, there is no guarantee that these conditions will continue.

Given these conditions, the speed at which birds are able to gain energy in the premigration period (and on staging areas where these exist) is

Table 17.4 Components of a juvenile flight range model for barnacle geese

Variables	Whether known		Factors affecting
Laying date	Yes	25 May–15 June	Spring migration date
			Spring weather
Daily energy gain	No	50–150 kcal/day	Habitat quality
			Daylength
			Density
			Social Status
Autumn migration time	Yes	15 Sept–10 Oct	Autumn weather
			Daylength
Wind speed/direction	Yes	−40 to +40 km/h	Prevalent conditions
		(mean + 27 km/h)	Selection of departure
			conditions by birds
Assumed constants			
Nest stage	Yes	30 days (5 days laying and 25 days inc.)	
Fledging period	Yes	45 days hatch–fly	
Energy consumption	No	62 kcal/h at 64 km/h air speed	

crucial. This depends mainly on habitat quality on the rearing and fattening areas, where competition for resources induces density-dependent effects. Which individuals survive under these circumstances depends largely on social status. Also in Table 17.4 are those factors which are assumed to be constant. The incubation and fledging periods usually vary only slightly, though the effect of density on fledging period is not known. The energy consumed during flight has not been determined empirically for geese but we have assumed a requirement of around 12 times basal metabolic rate during constant flapping flight (Berger and Hart 1974).

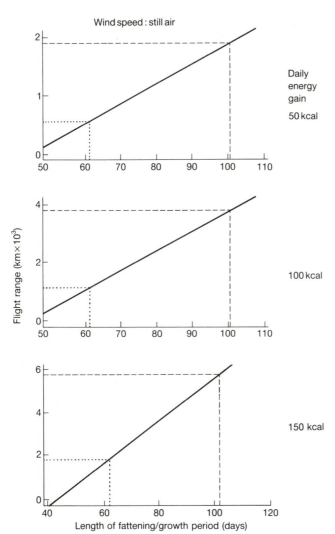

Fig. 17.4. The relationship between flight range and the number of days available for fattening of young post-fledging (days from arrival to departure minus the time needed for laying, incubation and the fledging period). The relationship is given for three different rates of energy accumulation. The dashed line represents the longest period between the date of hatching and arrival of winter so far recorded (103 days) and the dotted line the shortest (62 days). Note that the vertical scale is different in each figure.

A number of these parameters are still unknown, but Fig. 17.4 shows an example of how a single factor, the daily energy accumulated as reserves, affects flight range. In still air and assuming the constants in Table 17.4, a daily energy gain of more than 50 kcal/day is required even in the longest summer on record. In the shortest summer, energy gains of more than 150 kcal/day are required throughout the fattening period. When more of the unknowns have been determined, these values can be fitted into a multivariate model to predict flight range in summers of different lengths.

Importance for conservation

It is extremely difficult to collect evidence on the survival of birds during long migrations but, especially for arctic breeding species, losses of young on the journey may be significant. Where refuelling stops are infrequent, their protection and the maintenance of habitat quality in them is vital to the conservation of such species. This is particularly true of shorebirds and some waterfowl which have specialized habitat requirements on wetlands which may be few and far between. As has been pointed out (for example, Myers 1983), the importance of networks of these wetlands has been recognized and many are now protected under the Ramsar Convention. Because food supplies may be limited at these times, competition may lead to higher mortality on subsequent flights and limit recruitment to the population. Thus the loss of any staging area may bring about a reduction in the total size of a population. If losses of staging areas result in an increase in the necessary flight distance to the next available site, migration losses will similarly increase. If models such as that proposed here can be developed and reliably predict flight range under different conditions, we can provide a better basis for arguing the case for the protection of premigratory fattening areas and migration stopover places.

Acknowledgements

Many people and organizations helped to make the work on Barnacle Geese possible, in particularly Earthwatch (USA), Norsk Polarinstitutt and The Wildfowl and Wetlands Trust. The long term study of these geese which provided the data on mortality rates has depended on a large number of professional and volunteer observers whose contribution is gratefully acknowledged.

Summary

The paper reviews evidence about the survival of non-passerine birds on (chiefly autumn) migration, and presents recent direct evidence from a Barnacle Goose population.

In some years, young Barnacle Geese suffer a mortality rate of up to 35 per cent between late summer and arrival in Scotland after a 3000 km journey from their breeding grounds in Svalbard. Losses are related to the hatching date of the young and, independently, to body mass. It is suggested that there is a density dependent effect operating through competition for food during the rearing and fattening period.

Annual mortality rates of adults over 13 years also show a clear positive relationship to the number of geese in the population in the previous winter. There is evidence from a number of other studies that survival postfledging is related to the length of the migration and this seems to be a consistently important factor in those species whose young have to travel large distances without an opportunity for feeding.

The components of flight range model are presented to predict the maximum distance young birds can fly. The most important components are the length of the summer, wind speed and direction during the flight, and feeding conditions during the rearing and fattening periods.

A network of staging areas for feeding during migration is crucial to the survival of populations, particularly of wetland birds. Understanding the needs of the birds and the factors that influence the minimum desirable distance between such areas can make a crucial contribution to their conservation.

References

Berger, M. and Hart, J. S. (1974). Physiology and energetics of flight. In *Avian Biology* Vol. 4. (ed. D. S. Farner and J. R. King), pp. 415–17. Academic Press, New York.

Boyd, H. (1962). Population dynamics and the exploitation of ducks and geese. In *The exploitation of natural animal populations*. (ed. E. D. LeCren and N. W. Holdgate), pp. 85–94. Blackwells, Oxford.

Campredon, P. (1983). Sex et age ratios chez le canard siffleur *Anas penelope* en periode hivernale en Europe de l'ouest. *Revue Ecologique (Terre et vie)*, 37, 117–28.

Cooch, F. G. (1958). The breeding biology and management of the Blue Goose *Chen caerulescens*, Unpublished PhD thesis. Cornell University.

Harris, M. P. and Rothery, P. (1985). The post-fledging survival of young puffins *Fratercula arctica* in relation to hatching date and growth. *Ibis*, 127, 243–50.

Hedgren, S. (1981). Effects of fledging weight and time of fledging on survival of guillemot chicks. *Ornis Scandinavica*, 12, 51–4.

Hill, D. A. (1984). Population regulation in the mallard *Anas platyrhynchos*. *Journal of Animal Ecology*, 53, 191–202.

Jarvis, M. J. F. (1974). The ecological significance of clutch size in the South African Gannet (*Sula capensis* (Lichtenstein)). *Journal of Animal Ecology*, 43, 1–17.

Lloyd, C. S. (1979). Factors affecting breeding of Razorbills *Alca torda* on Skokholm. Ibis, 121, 165–76.

Myers, J. P. (1983). Conservation of migration shorebirds: Staging areas, geographic bottlenecks, and regional movements. *American Birds*, 37, 23–5.

Newton, I. and Moss, D. (1986). Post-fledging survival of sparrowhawks *Accipiter nisus* in relation to mass, brood size and brood composition at fledging. *Ibis*, **128**, 73–80.

Nichols, J. D. and Haramis, G. M. (1980). Sex specific differences in winter distribution patterns of Canvasbacks. *Condor*, **82**, 406–16.

Nisbet, I. C. T. and Drury, W. H. (1972). Post-fledging survival of Herring Gulls in relation to brood size and date of hatching. *Bird Banding*, **43**, 161–72.

O'Briain, M. (1987). Families and other social groups of Brent geese in winter. Unpublished report. University College, Dublin.

Owen, M. (1982). Population dynamics of Svalbard Barnacle Geese, 1970–1980. The rate, pattern and causes of mortality as determined by individual marking. *Aquila*, **89**, 229–47.

Owen, M. (1984). Dynamics and age structure of an increasing goose population—the Svalbard Barnacle Goose. *Norsk Polarinstitutt Skrifter*, **181**, 37–47.

Owen, M. and Black, J. M. (1989). Factors affecting the survival of Barnacle Geese on migration from the breeding grounds. *Journal of Animal Ecology*, **57**, 603–17.

Owen, M. and Black, J. M. (1990). The Barnacle Goose. In *Lifetime reproductive success in birds* (ed. I. Newton), pp. 349–62. Blackwell, Oxford.

Owen, M. and Dix, M. (1986). Sex ratios in some common British wintering ducks. *Wildfowl*, **37**, 104–12.

Owen, M. and Gullestad, N. (1984). Migration routes of Svalbard Barnacle Geese *Branta leucopsis* with a preliminary report on the importance of the Bjornoya staging area. *Norsk Polarinstitutt Skrifter*, **181**, 67–77.

Owen, M. and Mitchell, C. R. (1988). Movements and migrations of Wigeon *Anas penelope* wintering in Britain and Ireland. *Bird Study*, **35**, 47–59.

Parsons, J., Chabrzyk, G., and Duncan, N. (1976). Effects of hatching date on post-fledging survival in Herring Gulls. *Journal of Animal Ecology*, **45**, 667–75.

Perdeck, A. C. and Clason, C. (1983). Sexual differences in migration and wintering of ducks ringed in the Netherlands. *Wildfowl*, **36**, 95–103.

Perrins, C. M. (1966). Survival of young Manx Shearwaters *Puffinus puffinus* in relation to their presumed date of hatching. *Ibis*, **108**, 132–35.

Perrins, C. M., Harris, M. P., and Britton, C. K. (1973). The survival of Manx Shearwaters *Puffinus puffinus*. *Ibis*, **115**, 535–48.

Pienkowski, M. W. and Evans, P. R. (1985). The role of migration in the population dynamics of birds. *Behavioural ecology: ecological consequences of adaptive behaviour* (ed. R. M. Sibly and R. H. Smith), pp. 331–52. Blackwell Scientific Publications, Oxford.

Pienkowski, M. W., Knight, P. J., and Minton, C. D. T. (1979). Seasonal and migrational weight changes in Dunlins. *Bird Study*, **26**, 124–48.

Prop, J., Van Earden, M. R., and Drent, R. H. (1984). Reproductive success of the Barnacle Goose *Branta leucopsis* in relation to food exploitation on the breeding grounds, western Spitsbergen. *Norsk Polarinstitutt Skrifter*, **181**, 87–117.

Sayler, R. D. and Afton, A. D. (1981). Ecological aspects of Common Goldeneyes *Bucephala clangula* wintering on the Upper Mississippi River, USA. *Ornis Scandinavica*, **12**, 99–108.

Tamisier, A. (1985). Hunting as a key environmental parameter for the Western Palearctic duck populations. *Wildfowl*, **36**, 95–103.

18 Modelling the population dynamics of the Grey Partridge: conservation and management

G. R. POTTS and N. J. AEBISCHER

Introduction

Computers first became available for use in ecological research in the late 1950s and by the mid-1960s a number of population simulation models had been published. These population models were mainly for harvested commercial fish or for insect pests. It was not until 1969 that they were first used on a bird, the Great Tit *Parus major* (Pennycuick 1969).

Even today, when desk-top computers have become ubiquitous and when computers could solve many of the questions that ecologists ask about bird populations, few ornithologists undertake population modelling. The purpose of this paper is therefore to show how, in the case of the Grey Partridge *Perdix perdix*, simple modelling has been an invaluable aid to the more orthodox observational and experimental methods used to study population dynamics. We argue that many practical questions, for example the calculation of sustainable yields, compensatory survival, and the comparison of management options, can best be answered by population simulation modelling combined with data obtained by experimentation.

Methods

Outline of the natural history of the partridge

The partridge is a bird of open landscapes, having originated from the temperate steppes. Breeding begins after the first winter; the pairs usually form in February and then search for permanent grassy nesting cover, preferably on a dry bank. The clutch size is the largest of any species of bird, averaging fifteen or sixteen eggs, but predation on the hen and eggs is high. The chicks hatch in mid- to late-June and, during the first 2 weeks of life, feed mainly on insects in cereals or in other tall grasses. By late summer, adults and young form into 'coveys' (groups of about twelve birds), feeding on unharvested grain and on seeds of stubble-weeds. In winter and spring most partridges feed by grazing on growing cereals or on pastures, especially clover and grass leys. The species is remarkably sedentary and the home range may involve only two or three fields.

Sussex study

The study was conducted on an area of the South Downs, Sussex, comprising five farms and totalling 29 km². Data on Grey Partridges were collected annually from 1955; the main study began in March 1968 and is continuing. Partridge numbers, breeding success, and losses were recorded each year, using the same methods. Routine monitoring of the chick food supply began with the 1970 season. The farms are privately owned and because we were monitoring, we sought no influence over any aspect of farm or game management. Full details of the study area have been given elsewhere (Potts and Vickerman 1974; Potts 1980, 1986). A key feature was the choice of the outer boundaries of the study area: the areas beyond the boundaries were relatively unsuitable for Grey Partridges, which reduced the effects of immigration and emigration.

Having the advantage of knowledge gained from previous intensive studies of the ecology and breeding biology of the partridge, this study concentrated on three different but complementary approaches. First the long-term *monitoring* was used to generate data which defined the overall problem (that of too few partridges) in terms of the life-history dynamics. *Modelling* was then used to simulate the population processes and to design experimental tests of solutions in terms of *management* of the partridge population.

Monitoring, essentially, was a census of partridges on stubble fields in August or early September in which birds were aged and sexed. The methods of converting sex-ratio of adults, young:old ratio, and brood sizes to measures of survival are given in Potts (1986). The precision of the monitoring in terms of measuring population size and survival rates has been estimated elsewhere, most recently by Aebischer and Potts (1990). They concluded that sampling errors were not hindering the analysis and interpretation of the monitoring data. For instance, the accuracy of the prediction of year-to-year changes in partridge density, using mortality rates at different stages of the life cycle estimated from the Sussex sampling, was inversely related to the number of partridges recorded. At the present average density of 5 pairs/km², approximately 25 km² needed to be sampled to achieve 75 per cent accuracy of prediction, which was nearly the whole of the Sussex study area. In the 1950s the same level of accuracy could be achieved over 3 km², simply because densities were five times higher. Likewise, in recent years the number of broods available for estimating chick survival rates has been too low for satisfactory precision on some of the individual farms, although not in the study area as a whole. This was an inevitable consequence of the general decline of partridges throughout the study period.

The mortality rates calculated from the post-breeding censuses were expressed as k-factors, i.e. as log(initial number)−log(number surviving). The k-factors were k_1: loss of eggs when the hen survived, k_2: loss of eggs caused by loss of the hen, k_3: chick losses to age 6 weeks, k_4: shooting losses, and k_5: losses between shooting and nesting. In this paper k_1 and k_2 were combined where necessary to make the data comparable with those obtained in other studies. There was no k value for failure to hatch when fully incubated, since this was rare and there was little variation in clutch size (Potts 1980). Losses of eggs, in the above analysis, were therefore losses of potential chicks.

Our method of estimating chick survival rates was based on:

(1) extensive surveys in Eastern England in the 5 years from 1933 to 1937;
(2) an intensive study on an area of 14 km² near Fordingbridge in the 11 years from 1949 to 1959, involving substantial numbers of individually marked partridges; and
(3) an intensive three-season study of individually marked partridges by David Jenkins in Central Hampshire.

Details of all three studies are given in Potts (1986). The survival rates of chicks from hatching to 6 weeks of age was estimated from the geometric mean brood size (see Appendix 18.1). Provided that the number of broods was greater than ten, the precision of the geometric mean was high (ratio of brood size standard error to brood size mean less than 0.1). The relationship used to estimate chick survival rates from the geometric mean brood size was verified by two independent studies of groups of radio-tracked broods: there was an almost exact match to the regression lines obtained in both studies from a regression of observed within-brood chick survival rates against brood size (Potts 1986, p. 47, Fig. 3.6). Reitz *et al.* (1988) recently proposed a new method which, when applied to a study where mean chick survival rates were directly determined, yielded estimates that were up to double those of the observed rates (see Appendix 18.1). Moreover on average, the new method overlooked five out of every six hens which were observed to have lost all their chicks. For these reasons, we rejected the use of the Reitz *et al.* method.

Chick survival rates were transformed to probits to normalize their frequency distribution, to stabilize the variance, and to ensure that predicted survival rates remained within the range 0–100 per cent. This is normal in work that seeks to quantify adverse effects of pesticides—a major part of our study.

Sussex model

The construction of the simulation model, essentially a minor variant of one used since 1975, was described in detail in Potts (1980), with a complete printout published in Potts (1986) and again here as Appendix 2. This model was considered validated because the outputs accurately represented the population dynamics on the Sussex study area. However, although apparently correct, the predictions needed verification by experiments elsewhere.

From the first runs of the model, it became obvious that the resemblance of the model output to reality, and therefore the interpretation of the model, were crucially dependent on two environmental factors and their perceived relationship with partridge survival rates at different stages of the life cycle:

1. Insect food

(a) Chick survival was dependent on the supply of suitable insect food; and
(b) the supply of suitable insect food was reduced by the use of pesticides.

2. Nest predators

(c) Nest predation was density-dependent; but
(d) predator control removed this density dependence.

So the first result of the modelling was the production of hypotheses ranked according to their importance, thus enabling experimental research to be directed at the crucial points in the system.

The methods used to study these two factors, insect food and nest predators, are summarized below together with the approximate year in which they were first used systematically given in brackets (from Potts 1986).

1. Insect food

(i) Annual, field-by-field surveys of the broods at age 6 weeks to give mean brood size, breeding success, and later to estimate chick survival rates (1903).
(ii) Identification and quantification of the insect food found in chicks (1933).
(iii) Feeding trials to investigate the importance of insect food and prey preferences (1961).
(iv) Field-by-field measure of the abundance of insect food, annually and consecutively (1970).
(v) Intensive studies of effects of pesticides on insect food supply (1973).

(vi) Direct measurement of survival rates of chicks in radio-tracked broods with simultaneous analysis of diet based on faeces collected at their roost sites (1981). The insect fragments in faeces were identified by the method of Moreby (1988).

(vii) Extensive studies with experiments of varying pesticide use in cereal crops, in tandem with (v), examining the effect upon chick survival rates (1982).

In the results section of this paper we present evidence showing that partridge chick survival depends on the supply of certain insects rather than on the supply of insects in general. We investigate whether the supply of these insects is reduced by the normal use of pesticides, whether chick survival rates are thereby reduced, and, conversely, whether lower, selective, pesticide use results in improved chick survival rates.

2. Nest predators

(i) Annual field-by-field surveys of adult male and female partridges in autumn coveys to give sex ratios and later to estimate nest and hen losses (1903).

(ii) Identification of nest predators and quantification of the nest and hen losses caused by predators, by following known nests through incubation (1911).

(iii) Identification and quantification of prey consumed by predators (1926).

(iv) Predator removal experiments (1959).

(v) Comparison of nest and hen losses on keepered and unkeepered areas (1970).

Much of the evidence gathered along the above lines was potentially flawed, as regards causation, for example because some of the differences between keepered and unkeepered areas may have been partly attributable to site rather than management differences; this even applied to the pioneering experiment carried out by Frank (1970).

An experiment on Salisbury Plain, which started in 1985, was designed to overcome such difficulties. In particular two of the points mentioned above were tested: (i) nest predation was density-dependent; and (ii) predator control removed the density dependence. The basic method of Marcström *et al*. (1988) was used: annual predator removal on one area, none on a second, control area, and a swapping of the two areas after a number of years. Our project, designed to last 8 years and now (1988) in its sixth year, is described in detail each year in the Annual Review of The Game Conservancy (e.g. Tapper *et al*. 1988). The full results will not be published until the end of the study, but we present some data and preliminary conclusions here.

Results

Insect food

The relationship between the annual chick survival rate (probits) on five farms in the Sussex study area, and an index of the density of preferred insects in cereals in the same years and locations is shown in Fig. 18.1. The insect abundance index was obtained by summing the densities of five insect taxa, chosen and weighted by their coefficients from a forward stepwise regression on probit chick survival. These five taxa were, in descending order of importance, (1) small diurnal ground beetles (Carabidae); (2) sawfly and other caterpillars (Symphyta and Lepidoptera); (3) leaf beetles (Chrysomelidae) and weevils (Curculionidae); (4) plant bugs (Heteroptera) and leaf hoppers (Cicadellidae); and (5) aphids (Aphididae). No further insect taxa were accepted as statistically significant ($P < 0.05$) in the regression against chick survival rates. The densities of these preferred insects explained 52 per cent of the variation in chick survival rates (probits).

This correlative work was much strengthened by the results of monitor-

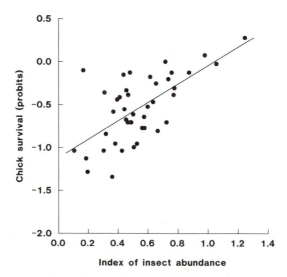

Fig. 18.1. Grey Partridge, Sussex. Annual chick survival to 6 weeks (probits) in relation to an index of the abundance of preferred insects (see text) at the time of hatching in years 1970–87. The index (I) was calculated according to the equation $I = 0.141x_1 + 0.120x_2 + 0.083x_3 + 0.006x_4 + 0.0004x_5$ where x_1, x_2, x_3, x_4, x_5 represent the densities of ground beetles, caterpillars, leaf beetles and weevils, plant bugs and leaf hoppers, and aphids respectively. Survival estimates based on broods less than 10 have been omitted.

ing the diet and survival of chicks in 17 radio-tracked broods in two independent studies, one in Norfolk and one in Hampshire (Fig. 18.2). Chick survival rates were significantly correlated ($P < 0.02$) with the proportion of chick insect diet made up of only taxon 2: caterpillars (Symphyta and Lepidoptera), and taxon 3: leaf beetles (Chrysomelidae) and weevils (Curculionidae). Thus two independent approaches, one long-term and extensive (Fig. 18.1) and the other extremely intensive (Fig. 18.2) drew us to the same conclusion, that amongst the insects serving as a food resource some groups were especially important in determining the survival of partridge chicks.

The overall outcome of the Sussex work on the insect/chick relationship suggested, on the one hand, that the complete loss of insects would reduce chick survival rates to a level below that necessary to replace adult losses, even when these were at their lowest observed values. On the other hand, it was predicted that doubling the insect numbers would restore chick survival rates and, thus, the population density equilibrium levels, to those of the pre-pesticide era.

Preferred insects in the chick diet are those that were selected as statistically significant in a stepwise multiple regression analysis against chick survival rates, and which usually occur more than twice as frequently in the diet as in vacuum net samples (e.g. Green 1984, Table 2). Together they accounted for 56 per cent of the individual insects in the diet of 29 chicks collected in Sussex (Vickerman and O'Bryan 1979) but only 18 per cent—even excluding Collembola—of the arthropod community in the cereal crops of the same area during the period of chick collection.

In order to verify the effects of insect food supply on chick losses, an

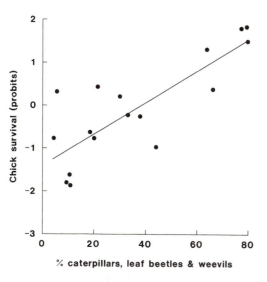

Fig. 18.2. Grey Partridge chick survival (probits) in 17 radio-tracked broods in relation to the proportion of the diet (dry weight), as determined from the faeces, made up of sawfly and other caterpillars (Symphyta, Lepidoptera), leaf beetles (Chrysomelidae) and weevils (Curculionidae). Data from Green (1984, extended *in litt.*) and Rands (1985, extended *in litt.*).

experiment was begun at Manydown, Hampshire, in 1982. Cereal head-lands were sprayed selectively in order to improve the chick food supply: herbicides for broad-leaved weeds, insecticides, and the insecticidal fungicide pyrazophos were excluded. 'Control' headlands were sprayed non-selectively, consistent with normal agricultural practice. The densities of four of the five taxa of preferred insects, i.e. those that could be measured, were raised by a factor of 2.5 (Rands 1985). The average chick survival rate was raised by a factor of 1.8, based on the annual trials at Manydown over 6 years and on associated experiments on eight farms over 5 years (Fig. 18.3).

Nest predators

Direct observations show that the main causes of nest loss in any one study area are either predation or mowing (Potts 1980). Mowing is however a problem only in areas where hedgerow nesting cover is sparse or non-existent and where grass is grown intensively for hay. For example in the 1940s and 1950s, mowing was the main cause of nest loss of Grey Partridges in North America. Elsewhere and more recently, predation has been the main cause of nest loss, even where predators are controlled (for details see Potts 1980).

Although Grey Partridges are now often scarce, they in fact occur at near the average density for all species of bird on farmland (Aebischer and Potts 1990). Moreover they are larger than many birds, passerines especially, and therefore relatively more attractive to most predators. Most importantly they nest only on the ground and, since their preferred nesting

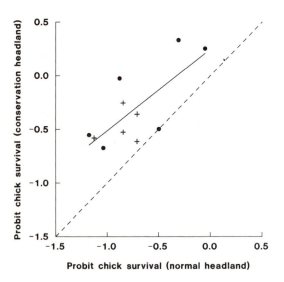

Fig. 18.3. Grey Partridge, Manydown, Hampshire (●) and East Anglia (+). Calculated survival rate (probits) of chicks at age 6 weeks in annual experiments with conservation headlands (selectively sprayed or unsprayed) (y axis) and control headlands (sprayed normally) (x axis).

cover (grass along the hedgerows) constitutes, at most, 2 per cent of their farmland habitat, the nesting density is 50 times the density of pairs expressed over the whole of the occupied habitat. Most egg and nest predators appear to locate nests by direct observation of the partridges (e.g. Carrion Crow *Corvus corone*) or by scent (e.g. Red Fox *Vulpes vulpes*). Although nests are carefully spaced out and well concealed, and although the hens are well camouflaged and furtive, partridge eggs and sitting hens represent, once found, a useful and substantial source of food for local predators.

The overall proportion of nests lost to predators was a positive linear function of nest density; the slope was not significantly different from zero, or was close to zero where gamekeepers controlled predator numbers (Potts 1986). Certain individuals of the most important egg predator, the Carrion Crow, are known to switch to, and then specialize on, partridges, thus providing a mechanism for the density-dependent response of at least one predator. It seems likely that the fox, a major predator of sitting hens, reacts in a similar way to prey density.

The experience of generations of gamekeepers indicated a considerable increase of partridge numbers with a decrease of predator numbers. Modelling showed clearly that this could be caused entirely by the removal of density dependence from the rate of nest predation (Potts 1980), but an experiment was needed to demonstrate the point conclusively.

Although planned to continue for a further 2 years, the Salisbury Plain experiment has already proved that predation alone can depress population density. The relationship between nest loss and predator removal was almost exactly as expected; the losses predicted by the model and the observed losses are compared in Fig. 18.4. Irrespective of area, predator removal resulted in lower losses, as predicted. We consider that this verifies the predation subroutines in the model given in Appendix 18.2. The fact that the observed points nearly all lie slightly below the 1 : 1 line suggests that a small amount of fine-tuning of the model may be necessary to reproduce the Salisbury Plain situation exactly. Similar variation was noted between individual farms in Sussex, whose combined data were used in the model.

Table 18.1 presents the outputs from runs of the model to show how different levels of nest predation influence the population dynamics of the population. When nest predation is low, the partridge population increases to a much higher level than when predation is high; at this higher level, it is limited by shooting or emigration because, if surplus numbers are not removed by shooting, the saturation of nesting habitat will force birds to emigrate. Obviously the trick in maximizing the bag is to raise numbers through predator control and to adjust the subsequent rate of shooting to a level that reduces density-dependent emigration in the following spring

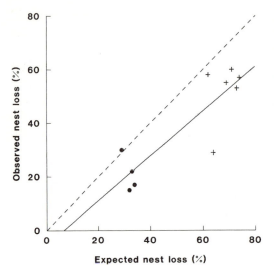

Fig. 18.4. Grey Partridge, Salisbury Plain. Relationship between observed nest losses (mostly predation) obtained by experimentation and expected losses predicted from the Sussex model of Appendix 1. Nest predators controlled (●), no predators controlled (+).

without at the same time reducing spring stocks. This is a juggling act easily optimized using a simple computer model.

The generality of these results is revealed by examining data from other studies of Grey Partridge populations, from 9 countries and averaging 13 years in length (Potts 1986). For the population at equilibrium, mean stock density was significantly related to the amount of adult mortality attribut-

Table 18.1 Reproduction gains and annual adult mortality rates in two simulated populations of Grey Partridges, (a) one with high and (b) one with low nest predation. The density-dependent relationships used in the simulation model were derived from the Sussex study, as were the chick survival rates. The model ran for 29 years; the outputs given below represent the average over the last 4 years of the run with optimal settings for herbicide use and nesting cover

	Population	
Model settings		
Predation control	No	Yes
Nest predation	high	low
Model output		
Equilibrium density	16 pairs km^{-2}	64 pairs km^{-2}
Percentage of stock recruited per year[1]	58%	76%
Percentage of adults lost per year[1]	55%	73%
Expectation of life of adults[2]	1.25 years	0.77 years

[1] Recruitment and losses differ by 3 per cent; although the population is at equilibrium it varies from year to year. The average over the last 4 years differs slightly from the true equilibrium.

[2] $-1/\ln(s)$ where s = annual adult survival rate, 'life' = life in the study area, i.e. excludes net emigration. The mean life-span of 14 captive Grey Partridges of wild stock was 2.9 years (based on the survival over 2 years, whereupon the survivors were released; S. D. Dowell, personal communication).

able to the forms of mortality found to be density-dependent in Sussex (k_{1+2}: predation on the nest and sitting hen, k_{4+5}: shooting and overwinter losses), by the following equation:

$$\ln(\text{equilibrium density}) = 2.80k_{1+2} + 1.92k_{4+5} + 2.49$$
$$(r = 0.663, \text{d.f.} = 29, P < 0.001)$$

Both forms of mortality contributed significantly to the regression ($P < 0.001$). The first two terms in the equilibrium equation were combined to provide an overall measure of density-dependent mortality in Fig. 18.5. As predicted from the Sussex data used in the model (illustrated in part by Table 18.1), the mean stock density in equilibrium populations is highest when the amount of density-dependent adult mortality is lowest, and vice versa.

Discussion

When the goal is to increase or maintain bird numbers, there is really only one test that matters: to use a popular aphorism—can words be turned into birds? In our case they were, but how crucial was the modelling process? Could we have turned words into birds without modelling?

Modelling is obviously not always essential. After all, it was not used by the gamekeepers of the 19th century who produced a sustainable yield at least ten times that achieved today. Nor was it available over 70 years ago, when Baranov (1918) gave a better description of the dynamics of compensation for harvesting mortality than is usually available today: 'a fishery, by thinning out a fish population, itself creates the production by

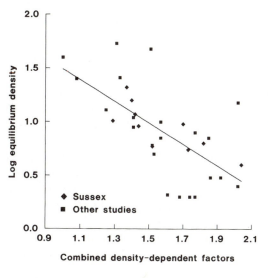

Fig. 18.5. Grey Partridge, stable populations. The equilibrium stock density (*y* axis) was negatively related to an overall measure of density dependent mortality obtained from a multiple regression equation relating stock density to k_{1+2}, and to k_{4+5} (see text).

which it is maintained'. This said, however, we believe that modelling is a very powerful way of investigating population dynamics, in particular for isolating, evaluating, and ranking the key parameters for experimentation. How else can one assess the long-term net effect of competing density-dependent factors or the relative benefits of management options without actually managing on a trial and error basis, with all the inefficiency and expense that this would entail?

The recent study by Robertson and Rosenberg (1988) illustrates the point well. They examined 67 scientific papers dealing with the harvesting by shooting of gamebirds over the years 1975 to 1985. Thirty-eight mentioned in varying detail the effects of shooting on total annual mortality or on the size and productivity of the breeding population. In only one study (Roseberry 1979) was it clearly understood that a reduction in the size of the breeding stock was a natural and inevitable consequence of shooting rather than of over-shooting, and that at equilibrium the effect of shooting on stocks would not be detectable without experimentally changing the level of shooting. If we substitute predation for shooting then we believe the same principles hold.

For some ecologists, compensatory mortality appears to have magic properties, which we believe would evaporate in the face of even a little modelling. The only way compensatory mortality can be invoked is by reducing the population level—other mortality rates can be reduced to compensate for shooting mortality *but population densities will be reduced, by however little, in the process*. In the partridge the critical point was not the extent to which predation and other mortalities were additive—at equilibrium they were not at all additive. The critical point was the extent to which the equilibrium levels were affected by the individual mortalities.

A simple example shows how this works. The example, a case of the Sussex model (Appendix 1), concerns the hypothetical partridge population with no shortage of nesting cover (9 linear km per km²), no use of pesticides, and minimum predation. With no shooting, such a population would show an overwinter survival rate, from the start of the period when shooting could have taken place to the time of stock determination in spring, of 36 per cent. Shooting at 41 per cent (the maximum sustainable yield) at the start of this period does not bring the overwinter survival rate down to the expected 21 per cent [i.e. $(100 - 41) \times 0.36$] because spring emigration losses are reduced, being density-dependent. The lower spring losses are, however, not sufficient to raise the overwinter survival rate of those birds not shot to 61 per cent, the value that would restore the overwinter survival rate including shooting to 36 per cent. Because of the imperfect compensation of shooting mortality by lowered overwinter losses, the stock declines. As the decline proceeds, the emigration rate of the non-shot birds also declines until the overwinter survival rate including shooting is restored to 36 per cent. At this new equilibrium level, the stock is almost 50 per cent lower than the non-shot stock.

Population dynamics have not often been modelled by ornithologists, even by those field-workers with most available data. This is partly because, we suspect, either the data were not collected with a synthesis in mind or because modelling was rejected. The main arguments in the latter case seem to be that modelling cannot prove anything, is too theoretical, oversimplifies the situation, is difficult mathematically, and involves tautologies. An underlying point may often be the concept of modelling as an alternative method. In our case the model was an integral and complementary part of an approach that involved long-term monitoring and experimentation, and was in no way an 'alternative'. Some students prefer the rigour of experiments; but how does one, without modelling, evaluate the repercussions of any experiment upon a population? In many ways, experiments without modelling are 'black boxes', with unquantified intermediate chains of cause and effect.

Today the environment is changing to such an extent and with such complexity that modelling may already have become an essential adjunct to research based on field studies. Monitoring studies will certainly be open to errors of interpretation and often there will be insufficient time for empirical (trial and error) approaches or long-term experiments. Modelling has one supreme advantage, in that only a fraction of the amount of time needed for other available avenues of exploration is required, and, moreover, there is nothing to lose even by complete failure. Modelling can readily explore complex interactions, by predicting the net outcome of changes in management and thus saving resources for those approaches most likely to succeed.

Summary

Populations of the Grey Partridge *Perdix perdix* were studied with special reference to the comparative roles of density-dependent and density-independent forms of mortality in determining equilibrium levels of stock and yield. Two sources of population data are used: a 21-year study on 29 km^2 of Sussex downland, which continues, and 39 studies reported in the literature.

Annual variation in population density is attributed to changes in chick mortality rates caused by fluctuation in the food supply (certain preferred groups of insects). The relationships were derived from field studies, ranging from extensive and long-term to very intensive studies. They were verified by large-scale experiments in which the insect food supply was improved by effective management of pesticide application.

Correlations suggesting that density-dependent nest predation was the main factor limiting population density were tested and verified in an experiment on Salisbury Plain. Mean equilibrium densities of Grey Partridges in spring ranged from 0.5 to 50 pairs km^{-2}. None of the variation in equilibrium densities of stable populations was attributable to differences in the mortality rates of chicks and none to overall mortality rates. Equilibrium densities were, however, shown to be determined largely by density-dependent predation and shooting.

It is demonstrated that research involving an integrated approach with long-term monitoring, simulation models, and experiments has been, for the Grey Partridge, extremely useful. The applicability of this approach may have some relevance for other species.

Acknowledgements

We are very grateful for the help of our colleagues, particularly Dr Stephen Tapper (Salisbury Plain Project), Dr Nick Sotherton (Cereals and Gamebirds Project), Stephen Moreby, Drs Mike Rands and Rhys Green for the data used in Fig. 18.2, and Lorraine Josling for typing the manuscript. We would also like to thank Dr Jean-Dominique Lebreton and the anonymous referees for their helpful comments.

References

Aebischer, N. J. and Potts, G. R. (1990). Sample size and area: implications based on long-term monitoring of partridges. In *Pesticide effects on Terrestrial Wildlife* (ed. L. Somerville and C. H. Walker), pp. 257–70. Taylor and Francis, London.

Baranov, F. I. (1918). On the question of the biological basis of fisheries. *Nauchnyi issledovatelskii iktiologicheskii Institute, Izvestiia*, **1**, 18–218.

Green, R. E. (1984). The feeding ecology and survival of partridge chicks (*Alectoris rufa* and *Perdix perdix*) on arable farmland in East Anglia, UK. *Journal of Applied Ecology*, **21**, 817–30.

Frank, H. (1970). Die Auswirkung von Raubwild-und Raubzeugminderung auf die Strecken von Hase, Fasan und Rebhuhn in einem Revier mit intensivster landwirtschaftlicher Nutzung. *Proceedings of the IXth International Congress of Game Biologists*, 1969, pp. 472–9. Moscow, U.S.S.R.

Moreby, S. J. (1988). A key to the identification of arthropold fragments in the faeces of gamebird chicks. *Ibis*, **130**, 519–26.

Pennycuick, L. (1969). A computer simulation of the Oxford great tit population. *Journal of Theoretical Biology*, **22**, 381–400.

Potts, G. R. (1980). The effects of modern agriculture, nest predation and game management on the population ecology of partridges *Perdix perdix* and *Alectoris rufa*. *Advances in Ecological Research*, **11**, 2–79.

Potts, G. R. (1986). *The partridge: pesticides, predation and conservation*. Collins, London.

Potts, G. R. and Vickerman, G. P. (1974), Studies on the cereal ecosystem. *Advances in Ecological Research*, **8**, 107–97.

Rands, M. R. W. (1985). Pesticide use on cereals and the survival of grey partridge chicks: a field experiment. *Journal of Applied Ecology*, **22**, 49–54.

Robertson, P. A. and Rosenberg, A. R. (1988). Harvesting Gamebirds. In *Ecology and management of gamebirds* (eds. P. J. Hudson and M. R. W. Rands), pp. 177–201. Blackwell Scientific Publications, Oxford.

Roseberry, J. L. (1979). Bobwhite population responses to exploitation: real and simulated. *Journal of Wildlife Management*, **43**, 285–305.

Marcström, V., Kenward, R. E., and Engren, E. (1988). The impact of predation on boreal tetraonids during vole cycles: an experimental study. *Journal of Animal Ecology*, **57**, 859–72.

Reitz, F., Scherrer, B., and Garrigues, R. (1988). La distribution de la taille des couvées de perdrix grise (*Perdix perdix* L.) et son utilisation pour l'estimation de parametres de réussité de la reproduction. *Gibier Faune Sauvage*, **5**, 411–26.

Tapper, S. C., Brockless, M., and Potts, G. R. (1988). The predation control experiment: the turning point, *The Game Conservancy Annual Review*, **19**, 105–11.

Vickerman, G. P. and O'Bryan, M. (1974), Partridges and insects. *The Game Conservancy Annual Review*, **9**, 35–43.

Appendix 18.1

Comparison of methods for estimating survival rates of Grey Partridge chicks

Given a random sample of partridge broods at age 6 weeks, the estimate of chick survival rate (*CSR*) proposed by Potts (1980) is calculated from the formula $CSR = 0.03665x^{1.293}$ where x is the geometric mean brood size; if $x > 10$ then $CSR = x/13.84$.

The method of Reitz *et al.* (1988) requires fitting a normal distribution, left-truncated at brood size 1, to the observed distribution of brood sizes. Their estimates of chick survival rate is obtained by calculating the mean of the fitted normal distribution truncated this time at brood size 0, and dividing this mean by the mean number of chicks hatched per successful nest.

We assessed both methods by applying them to autumn partridge counts from Damerham, Hampshire, in the years 1949–59. In these years, intensive studies by Terence Blank and John Ash resulted in an independent and accurate estimate of the true chick survival rates. The mean number of chicks hatched per successful nest at Damerham (1949–59) was 13.48. the results in Table 18.2 show that the method of Reitz *et al.* systematically overestimated chick survival rates, with a mean absolute difference between estimated and actual values of 14 per cent; the mean absolute difference was 4 per cent in the case of the Potts method. On average at Damerham, 73 per cent of barren pairs in the autumn count had hatched chicks but lost them; 36 per cent of pairs that successfully hatched

Table 18.2 Survival rates of Grey Partridge chicks to age 6 weeks at Damerham from 1949 to 1959: actual values and values estimated using the methods of Potts (1986) and Reitz et al. (1988)

Year	Chick survival rates		
	Actual	Potts (1986)	Reitz et al. (1988)
1949	0.77	0.71	0.81
1950	0.41	0.35	0.49
1951	0.30	0.35	0.50
1952	0.41	0.41	0.56
1953	0.21	0.24	0.38
1954	0.11	0.16	0.26
1955	0.26	0.28	0.41
1956	0.30	0.41	0.57
1957	0.38	0.36	0.49
1958	0.17	0.19	0.34
1959	0.45	0.44	0.59
Mean	0.34	0.35	0.49
s.e.	0.05	0.05	0.04

chicks subsequently lost them all. The corresponding, estimated, figures using the Potts method were 72 per cent and 33 per cent respectively, and using the method of Reitz et al., 7 per cent and 6 per cent. The latter method thus seriously distorted interpretation of the mortality processes affecting partridge reproduction.

Appendix 18.2

Computer simulation of the Sussex partridge population 1957–85. Adapted for Borland Turbo PASCAL from Microsoft BASIC (Potts 1986, p. 177).

```
program PartridgeModel (Input,Output);

const ChickMort: array [1957..1985] of real =
    (0.31,0.59,0.28,0.28,0.39,0.55,0.59,0.47,0.59,0.46,0.43,0.66,
    0.58,0.42,0.54,0.77,0.54,0.55,0.61.0.39,0.89,0.97,0.67,0.50,
    0.91,0.41,0.50,0.48,0.59); (* Chick mortality data from 1957 to
    1985 *)

var Year,Shoot:integer;
    Pairs,AugPop,Young,Females,OFemale,YFemale,LFem,Eggs,
    logEggs,ChMort,ChickSurv,k1,k2,k1Surv,k2Surv,k5,k5Surv,
    s2,s2Surv, Bag1,Bag2,Bag3,Bag,Hedge,ln10:real;
    Herb,Pred,Answer:char;
```

```
begin (* PartridgeModel *)
  ln10:=ln(10); writeln; writeln;
  (* Set up the run *)
  write('Number of partridge pairs per km2 in spring '); readln(Pairs);
  write('Kilometres of hedgerow per km2 ............. '); readln(Hedge);
  write('Were herbicides used on cereal crops (Y/N) .'); readln(Herb);
  write('Were predators controlled (Y/N) ............ '); readln(Pred);
  write('What shooting pressure (1, 2, or 3) ......... '); readln(Shoot);
  writeln; writeln; writeln('Year Spring August  Bag');
  for Year:=1957 to 1985 do
    begin
      Eggs:=Pairs*14/Hedge;
      logEggs:=ln(Eggs)/ln10;
      if (Pred='y') or (Pred='Y')
        then begin k1:=0.028+0.11*logEggs; k2:=0.59*k1 end
        else begin k1:=-0.217+0.578*logEggs;
             if k1<0.2 then k1:=0.2;
             k2:=0.31*k1;
             end;
      ChMort:=ChickMort [Year];
      if (Herb='n') or (Herb='N') then ChMort:=ChMort*0.53;
      k1Surv:=exp(-ln10*k1);
      k2Surv:=exp(-ln10*k2);
      ChickSurv:=exp(-ln10*ChMort);
      Young:=Pairs*14*k1Surv*ChickSurv;
      Females:=Pairs*k2Surv;
      AugPop:=Pairs+Females+Young;
      case Shoot of
        1: s2:=0;
        2: s2:=0.35/(1+exp(5.47-0.04*AugPop));
        3: s2:=0.35/(1+exp(5.47-0.023*(AugPop+150)));
      end;
      s2Surv:=exp(-ln10*s2);
      bag1:=1.43*(1-s2Surv)*Pairs;
      bag2:=0.91*(1-s2Surv)*Females;
      bag3:=0.87*(1-s2Surv)*Young;
      Bag:=Bag1+Bag2+Bag3;
      writeln(Year:4,Pairs:8:1,AugPop:8:1,Bag:7:1);
      OFemale:=Females-Bag2;
      YFemale:=(Young-Bag3)/2;
      Females:=0Female+YFemale;
      LFem:=ln((Females/Hedge)+1)/ln10;
      k5:=-0.07+0.39*LFem;
```

```
        if k5<0.11 then k5:=0.11;
        k5Surv:=exp(−ln10*k5);
        Pairs:=k5Surv*Females;
    end;
end.
```

19 Parasites, cuckoos, and avian population dynamics

A. P. DOBSON and R. M. MAY

Introduction

The last 10 years have seen significant increases in the understanding of the role that parasites and pathogens play in the regulation of host abundance. These advances are based, in part, on a theoretical framework developed to examine the population dynamics of parasite–host relationships (Anderson and May 1978, 1979, 1982; May and Anderson 1978, 1979). In this chapter we use these models and their multispecies extensions to examine the patterns of parasite infections found in birds. The models were developed initially to determine those attributes of the interaction between parasites and their hosts that are important in determining observed patterns of population dynamics and community structure. They are described and derived in detail elsewhere (Anderson and May 1978; May and Anderson 1978; Dobson 1985, 1988a, and in preparation). Rather than repeat these derivations, we give attention to the graphical illustration of different extensions of the models, while presenting only the more transparent results of any algebraic analysis. The final discussion outlines the implications of this work for the conservation of endangered species of birds.

Rates of parasitism in bird populations

Current estimates suggest that parasitism of one form or another may be the most common form of life-style in at least three of the five major phylogenetic kingdoms (May 1988; Toft 1986). Most bird populations support a diverse community of parasites (Table 19.1) and most individuals accumulate a steadily increasing burden of parasites throughout the course of their life (Fig. 19.1). As a high proportion of these parasites are ingested with the bird's food, host diet plays a significant role in determining the structure of the parasite community harboured by any species. Similarly, changes in the parasite community occur at different times of the year in migratory species and at different ages in species that show a pronounced change of diet between juveniles and adults.

Table 19.1 A comparison of the mean number of helminth species and mean numbers of worms per host in a number of duck, grebe, and shorebird species (after Stock and Holmes 1987*b*)

Host	No. of species	No. of worms
Grebes		
Eared Grebe	9.1	3641
Horned Grebe	6.7	1808
Red-necked Grebe	10.7	1808
Western Grebe	6.4	365
Ducks		
Blue-winged Teal	10.1	854
Lesser Scaup	14.3	22 231
Bufflehead	7.0	180
Canvasback	17.3	917
Gadwall	9.7	197
Mallard	10.2	266
Ring-necked	7.7	134
Ruddy	8.7	1680
Widgeon	5.2	85
White-winged Scoter	18.5	28 087
Shorebirds		
Bonaparte's Gull	5.5	51
Curlew	4.2	263
Godwit	3.4	509
Willet	8.6	654

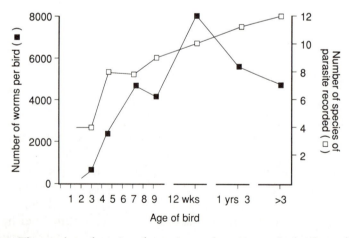

Fig. 19.1. The number of species of parasites and mean parasite burdens of brown pelicans *Pelecanus occidentalis* of different ages (data from Humphrey *et al*. 1978).

Although no studies have specifically addressed the relative importance of parasites among other potential sources of mortality, it is possible to assemble rough estimates of these figures from a variety of published studies (Table 19.2; see also Itamies *et al.* 1980). The data suggest that parasites may be a significant source of mortality. However, a more comprehensive understanding of the potential role of parasites in avian ecology is best obtained by examining the population dynamics of the interactions between birds and their different broad classes of parasites.

Table 19.2 Causes of attributed mortality in four large-scale post-mortem analyses of birds found dead. All figures percent of original sample

Reference	Jennings (1961)	Macdonald (1962a)	Macdonald (1962b)	Macdonald (1962c)
Environmental stress (extreme cold, starvation)	10.5	12.9	18.4	30.4
Physical injury	32.5	11.3	13.1	19.0
Organic disease (pneumonia)	15.5	*	*	*
Viruses	6.1	**	**	**
Bacterial and fungal infections	9.5	9.1	8.2	18.8
Parasitic diseases	11.6	24.7	8.7	11.0
Toxins and poisons	10.0	40.9	45.2	19.0

* The figures given by Macdonald (1962a, b, c) are underestimates, as only a few birds from a large kill were examined for organic disease.
** Viruses not checked for in these studies.

Macroparasites and microparasites

The enormous array of pathogens that infect humans and other animals may be conveniently divided on epidemiological grounds into *microparasites* and *macroparasites* (Anderson and May 1979; May and Anderson 1979). The former include most of the viruses and bacteria along with many protozoans and fungi and are characterized by their ability to reproduce directly within individual hosts, their small size and relatively short duration of infection, and the production of an immune response in infected and recovered individuals. Mathematical models examining the dynamics of these pathogens may, for most purposes, divide the host population into a few categories, such as susceptible, infected, and recovered classes. In contrast, the macroparasites (most parasitic helminths and arthropods) do not multiply directly within an infected individual, but instead produce infective stages which usually pass out of the host before transmission to another definitive host. Macroparasites tend to produce a limited immune response in infected hosts; they are relatively long-lived and usually visible to the naked eye. Mathematical models of the population dynamics of macroparasites have to consider the statistical distribution of parasites within the host population. Surveys of free-living populations suggest that macroparasites almost always have aggregated distributions, which can usually be described by the negative binomial distribution (Fig. 19.2).

Direct and indirect life cycles

A second division of parasite life histories distinguishes between those species with *monoxenic* life cycles, and those with *heteroxenic* life cycles. The former produce infective stages which can directly infect another definitive host individual; the latter utilize a number of intermediate hosts or vectors in their transmission between definitive hosts. The evolution of complex heteroxenic life cycles permits parasite species to colonize hosts from a wide range of ephemeral and permanent environments, while also permitting them to exploit host populations at lower poulation densities than would be possible with simple direct transmission (R. C. Anderson 1988; Dobson 1988b; Mackiewicz 1988; Shoop 1988). Parasites with heteroxenic life cycles are constrained to those areas where the distribution of all the hosts in the life cycle overlap. The heteroxenic life cycles of some parasite species often allow them sequentially to utilize hosts from both aquatic and terrestrial habitats.

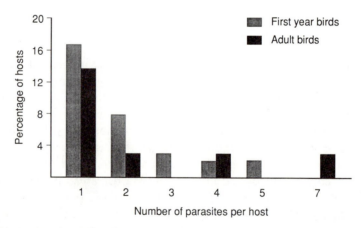

Fig. 19.2. The frequency distributions of the parasitic nematode *Porrocaecum ensicaudatum* in a sample of first-year/winter and adult Starlings found dead on the island of Skomer. Both distributions correspond to the negative binomial distribution (after James and Llewellyn 1967).

Cuckoos and brood parasitism

A third form of parasitism which is very important in birds is brood parasitism, occurring both within and between species (Payne 1977; Yom-Tov 1980). Coincidentally, the first scientific description of the cuckoo's behaviour was published exactly 200 years ago by Jenner (1788). In this type of parasitism, female cuckoos lay their eggs in the nest of another bird. This usually leads to partial or complete reduction of the brood reared by the parasitized bird, and is included in this review as there are important similarities between this and other types of parasitism (May and Robinson 1985).

Parasite life-history strategies

The complexities of the population dynamics of host–parasite associations may be reduced by deriving expressions which describe the most important epidemiological features of a parasite's life cycle (Anderson and May 1979; May and Anderson 1979; Dobson 1988; Dobson and Keymer 1985). Three parameters are important in describing the dynamics of a pathogen:

(1) the rate at which it will spread in a population;
(2) the threshold number of hosts required by the parasite to establish; and
(3) the mean levels of infection of the parasite in the host population.

Basic reproductive rate of a parasite, R_o

The models discussed below are primarily for macroparasites (the parasitic helminths and arthropods) living in hosts that are relatively long-lived, and considered to have dynamics that operate on a much longer time-scale than those of the parasites. It is thus assumed that host population size is relatively constant. By extending these models to examine communities containing more than one species of parasite, it should be possible to determine those features of parasite life cycles that may facilitate the co-existence of different species.

The basic reproductive rate, R_o, of a microparasite may be formally defined as the average number of new infections that a solitary infected individual produces in a population of susceptible hosts (Anderson and May 1979). In contrast, R_o for a macroparasite is defined as the average number of daughters that are established in a host population following the introduction of a solitary fertilized female worm. In both cases the resultant expression for R_o usually consists of a term for the rates of parasite transmission, divided by an expression for the rate of mortality of the parasite at each stage in the life cycle. Increases in host population size or rates of transmission tend to increase R_o, while increases in parasite virulence or other sources of parasite mortality tend to reduce the spread of the pathogen through the population. It is often the case that the same factors which increase virulence also enhance rates of transmission, so that the relation between R_o and virulence among different strains of a parasite species can be a tricky business (May and Anderson 1983).

Thresholds for establishment, H_T

The threshold for establishment of a parasite, H_T, is the minimum number of hosts required to just sustain the infection endemically within the host population. In general, an expression for H_T may be obtained from the equation $R_o = 1$, which applies to both micro- and macroparasites, with either simple or complex life cycles. The resultant expressions suggest that changes in the parameters that tend to increase R_o also tend to reduce H_T, and vice versa. Thus species that are more pathogenic require larger host populations to sustain them, while reductions in the mortality rate of transmission stages may allow parasites to maintain infections in populations previously too small to sustain them.

Mean prevalence and burden at equilibrium

It is also possible to derive expressions for the levels of prevalence (proportion of the hosts infected) and incidence (mean parasite burden) of parasites in the host populations. In general, parameters that tend to increase R_o, also tend to give increases in the proportion of hosts infected by a microparasite and increases in the mean levels of abundance of any particular macroparasite (Anderson and May 1979; May and Anderson 1979; Dobson 1988b). Most important, increases in the size of the host population usually lead to increases in the prevalence and incidence of the parasite population (Fig. 19.3).

The population dynamics of parasite–host associations

Microparasites

Simple models for host-microparasite associations show stable equilibria, with some constant proportion of hosts infected at any one time. Disturbances tend to induce oscillations which damp back to equilibrium (as in the more familiar Lotka-Volterra prey-predator models, the period of these oscillations is usually given roughly as the geometric mean of the characteristic time-scale for host and for parasite populations; Anderson

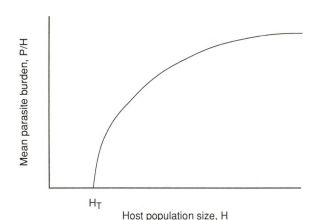

Fig. 19.3. The relationship between mean parasite burden and host population size for a basic Anderson and May macroparasite (after Dobson 1989).

Mean parasite burden, P/H

H_T

Host population size, H

and May 1982). A variety of realistic complications, such as stochastic fluctuations in birth rates (or other factors affecting host population size), seasonal changes in transmission rates, or the presence of free-living infective stages with long survival times, can 'pump' these intrinsic oscillatory tendencies, producing stable cycles or irregular, 'chaotic' fluctuations in the incidence and prevalence of infection.

The so-called 'interepidemic' cycles in the incidence of measles (2-year cycle), whooping cough (3-year cycle), and other childhood infections in developed countries, before the advent of vaccination, provide examples of such complexities in the dynamics of host–microparasite associations which are as yet not fully understood (Anderson *et al*. 1984; Schaffer and Kot 1985; Sugihara and May 1990). Other work on the relations between microparasites and invertebrates in general (and between forest insect pests and microsporidian protozoans or baculoviruses in particular) indicate that complicated dynamical phenomena—including long-term cycles—can and do arise (Anderson and May 1981). There have, however, been essentially no studies along these lines dealing with microparasitic infections of birds.

Macroparasites

The dynamical character (rough constancy, stable cycles, or irregular fluctuations) or the interaction between a macroparasite and its host is dependent upon the degree of aggregation of the parasites in the host population (Anderson and May 1978; Crofton 1971). The mechanisms causing this aggregation include behavioural and genetic differences in susceptibility of different host individuals to infection (Wakelin 1984; Wassom *et al*. 1974) and spatial and temporal heterogeneity in the distribution of parasite infective stages (Anderson and Gordon 1982). Although the relative intensity with which each process contributes to observed levels of aggregation varies between different host-parasite systems, increasing aggregation of the population (so that a smaller proportion of the host population harbours most of the parasites) tends always to increase the stability of the relationship. Essentially, this is achieved by parasite regulatory mechanisms, such as parasite-induced host mortality or density-dependent reductions in host fecundity and survival, which affect a higher proportion of the parasite population. Increased aggregation leads to reductions of the slope and asymptote of the line illustrated in Fig. 19.4, but has no effect on the position of H_T, the threshold number of hosts required to sustain an infection.

Brood parasites

For a cuckoo population or sub-population that parasitizes a single host species, the population density in year $t + 1$, C_{t+1}, is related to that in year t, C_t, by an expression derived by May and Robinson (1985):

$$C_{t+1} = sC_t + is'[1 - f(C_t)]H_t$$

Here s is the year-to-year survival probability for adult cuckoos ($s < 1$), i is the average number of female cuckoos fledged from each parasitized nest (if one cuckoo egg is laid and survives per nest, with a 50:50 sex-ratio, then $i = 1/2$) and s' is the probability of a cuckoo fledgling surviving to the next year. The population density of hosts in year t is H_t, and it is assumed that hosts produce only one clutch each year and that cuckoos attain sexual maturity at age 1 year; the above expression can easily be modified to represent more general assumptions about host egg-laying and

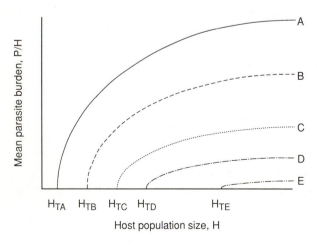

Fig. 19.4. The relationship between host population size and the burdens of five different parasite species (after Dobson 1985).

cuckoo life history (May and Robinson 1985). Finally, the function $f(C)$ represents the probability that any one nest will escape cuckoo parasitism, when the cuckoo population density is C (the so-called 'functional response').

The intrinsic reproductive rate of the cuckoo population, R_o, corresponds to the limit where cuckoo density is low: $R_o = C_{t+1}/C_t$ in the limit $C_t \rightarrow 0$. If cuckoos search for nests independently and randomly, then $f(C)$ is given by the zeroth term in a Poisson distribution, $f(C) = \exp(-\beta C)$, where β is the cuckoo 'searching efficiency' or 'area of discovery' (for further discussion see Hassell 1978). More generally, one can define the 'searching efficiency' by $1 - f(C) \rightarrow \beta C$ as $C \rightarrow 0$. It follows that R_o for cuckoos, parasitizing a host species of population density H, is given in terms of the parameters defined above as

$$R_o = s + is'\beta H$$

The threshold density of hosts, H_T, follows from the requirement that for cuckoos to persist $R_o > 1$:

$$H_T = (1 - s)/is'\beta$$

The dynamical behaviour of such associations—stable densities, or stable cycles, or chaotic or diverging fluctuations—is dependent on the details of the functional response, $f(C)$, of the cuckoos (and on the magnitude of the searching efficiency in relation to survival probabilities). Small changes in the parameters, and/or in the functional form of $f(C)$, can produce regular or erratic ('chaotic') oscillations in the density of both cuckoo and host populations (May and Robinson 1985).

Parasites and community structure

Multi-parasite systems

The model for macroparasites may be extended to include further species of parasite (Dobson 1985, 1989). Figure 19.4 illustrates this graphically for five hypothetical species of parasites. Two main points deserve attention. First, there appear to be a 'core' of species that are present in hosts at a wide range of population densities; at higher densities the community is supplemented by species that require large host populations to support them. Second, the mean burdens of all parasite species increase asymptotically as host population size increases. Thus, those species that establish at lower densities will tend to have higher burdens in large host populations than those species that require these larger populations to establish.

The models illustrated in Fig. 19.4 also provide a means of interpreting the patterns of relative abundance observed in surveys of the parasite species present in any host population. In many parasite communities, this mechanism may produce the approximately log-normal distribution of species abundance characteristic of many ecological communities (Fig. 19.5a, b). The frequency distribution of parasite species per host are essentially Poisson (Fig. 19.5c, d). As thresholds for establishment are, by definition, determined by transmission efficiency, those species that establish at low population densities will tend to be those whose abundance increases most rapidly as host population densities increase (Fig. 19.4). In contrast, those species with lower prevalence and incidence are likely to have lower transmission efficiencies, higher thresholds for establishment and slower increases of abundance with increasing host density. In 'species-rich' parasite communities, this will ultimately tend to give a distribution of abundances dominated by a few 'core' species and a range of less numerous, rare, or virulent 'satellite' species (Fig. 19.6).

These two patterns are of particular relevance to some of the recent discussions on parasite community structure. Present interest in Hanski's 'core and satellite' hypothesis of community structure in patchy environments (Hanski 1982) stems partly from its application to data for parasites of ducks (Bush and Holmes 1986a, b; Crompton and Harrison 1965) and grebes (Stock and Holmes 1987a, b). These latter authors define a 'core' species as one that is present in the majority of hosts examined (a more formal definition for such a species would be that its threshold for establishment is a host population of one individual!). Although this may initially appear a counter-intuitive definition, it is important to remember that many of the 'core' species in Bush and Holmes's (1986a, b) study were trematodes and cestodes highly adapted to ephemeral definitive hosts which only spend a proportion of their life-span in the habitat where transmission may occur; such species are likely to have thresholds for establishment that are effectively less than one individual (for discussion see Dobson 1988a; also reviews by Anderson 1988; Shoop 1988; Mackiewicz 1988).

Historical and stochastic events are also likely to have two major effects on the stucture of parasite communities. To a first approximation, 'historical' events are likely to be important in determining the species of parasites that are present in any habitat at any time, while the stochastic nature of the parasite transmission process is important in determining both the composition of the parasite community in any one host individual and the ability of these parasite species to coexist in a population of such hosts (Chesson and Case 1985; Ricklefs 1987). In the ideal, deterministic world of our model, the addition of species into parasite communities should occur as host population density increases and successive thresholds for establishment are crossed. However, most populations of

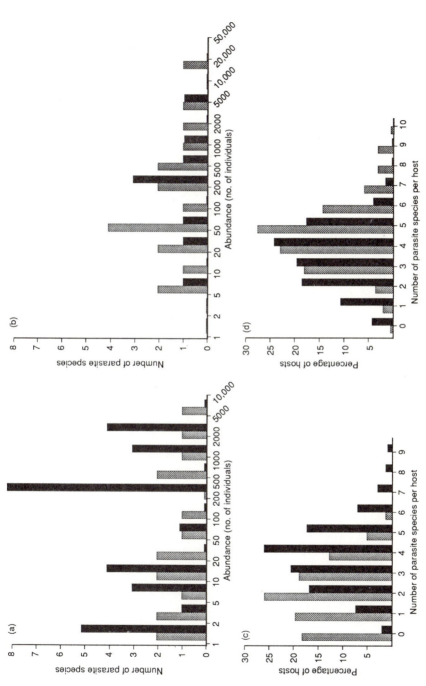

Fig. 19.5. The relative abundance of different parasite species in two host communities. (a) Herring Gulls (the solid bars are for adult birds, the shaded bars are for chicks) (data from Threlfall 1967). (b) Chickens in the Gambia (solid bars are for recently introduced experimental birds, shaded bars are for the resident native population) (data are from Hodasi 1973). In (c) and (d) the same data are replotted to illustrate the frequency distribution of numbers of species of parasite per host.

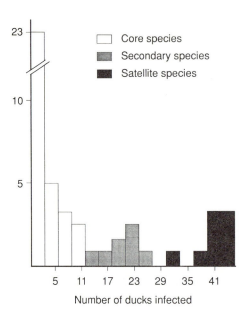

Fig. 19.6. The numbers of species of intestinal helminths (vertical axis) found in 1–3, 4–6, 7–9, ..., 43–45 members of a sample of Lesser Scaup Ducks (after Bush and Holmes 1986*a*).

hosts are fragmented into sub-populations of finite size, each of which exhibits pronounced variation around, but usually below, some carrying capacity.

The presence of any parasite species in a sub-population usually requires its introduction from a potential colonizing source, and these essentially stochastic events will lead to the absence of different parasite species from each isolated sub-population of the same host (Kennedy 1978*a*, *b*; Dobson and May 1986; Dobson 1988*b*). These absences are readily included into the model framework (Dobson 1989), where they may change the relative abundances of the species we expect to find actually present in the community.

A second level of stochasticity is encountered when we examine the number of parasite species harboured by each host. In the majority of studies for which data are available, this distribution is Poisson (Fig. 19.5c, d), which is the distribution we would expect if each individual host's probability of infection by any parasite species is essentially random. This point highlights an important paradox of parasite population dynamics. The stochastic nature of the infection process produces a random distribution of parasite species per host, essentially because any host has a relatively constant probability of being infected by each of the parasite species present in a finite pool of species. In contrast, the essentially random nature of the infection process for each individual parasite is compounded by variations in the susceptibility of each host individual to infection; this produces the aggregated distributions that are characteristic of most parasite species (Anderson *et al*. 1978; Anderson and Gordon 1982).

There are two further important consequences for the structure of parasite communities. First, although infection by any individual parasite infective stage may be random, many parasite species share common routes of entry into their hosts (e.g. many cestodes and acanthocephalans utilize the same species of arthropods as intermediate hosts). If individual hosts selectively feed on these intermediate host species, they will increase their probability of being simultaneously infected with two or more parasite species. Second, many (but not all) of the properties of parasite communities correspond to those required for 'lottery models' for co-existence of species living in a patchy environment (Chesson 1985; Chesson and Warner 1981; Warner and Chesson 1985). These lottery models suggest that stochastic variation, in the form of environmentally induced variations in rates of fecundity and mortality, may be sufficient to allow the coexistence of age-structured populations of species which would otherwise be unable to coexist in the absence of variation in these demographic rates.

Two-host/many parasite systems

The models can be further extended to examine some properties of communities that consist of a number of different parasite species and two different species of hosts (Dobson, in preparation). As expressions can be derived for R_o and H_T for each parasite species in each host, it is possible to produce a two-dimensional phase diagram based on the thresholds for establishment of each parasite in both host species (Fig. 19.7). Here the dynamics of the parasite-host interaction are likely to be determined by a number of factors:

(1) differences in the physiological and behavioural attributes of the host;
(2) different lengths of association in evolutionary time of the parasites with each host species; and
(3) different social organizations of the host species leading to different rates of transmission between conspecifics.

Thus, thresholds for establishment in these 'two-host' models are likely to vary, ranging from cases where the different host species are essentially substitutable, through to cases where the parasite is unable to develop in a host species, but may incidentally establish in this host owing to its presence in another species of host.

The composition of the parasite community in a simple two-host community may be examined graphically by joining the sequential

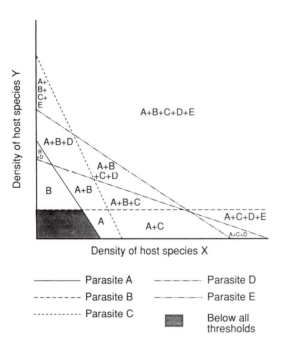

Fig. 19.7. The thresholds for establishment of five different hypothetical parasite species in a simple two-host community.

thresholds for establishment of each parasite species in each host species. This produces a phase diagram of combinations of host densities which can potentially support communities of different combinations of parasite species. An approximate estimate of the relative threshold for establishment of each parasite species is given by the number of hosts uninfected by that species. The most important point to emerge here is that parasite community structure is highly dependent upon the composition of the community of potential hosts. In particular, when the density of either host species is low, small changes in density can give rise to large changes in the structure of the parasite assemblage that these hosts can potentially support.

An example of how different susceptibilities to infection may lead to the local extinction of rare species appears to have occurred in Hawaii (Warner 1968; van Riper *et al*. 1986). Here the introduction of non-native birds has led to the establishment of avian malaria. This parasite is only mildly virulent to the populations of introduced avian species, with which the pathgoen has a relatively long evolutionary association. In contrast, the endemic Hawaiian land birds have only recently been exposed to avian malaria and exhibit high rates of mortality (van Riper *et al*. 1986). At low and medium elevations, malaria is endemic in the introduced species and readily transmitted to endemics by mosquitoes. This has led to a consider-

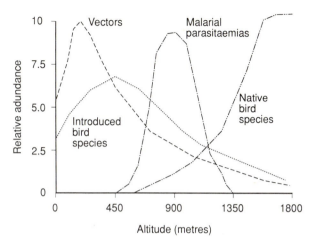

Fig. 19.8. Schematic representation of the relative densities of native and introduced bird species at different altitudes on Mauna Loa in the Hawaiian Islands. Also illustrated are the relative densities of the insect vector of avian malaria and the prevalence of malarial infection among avian host species. The scales of abundance are arbitrary but coarsely relative within each group, and prevalence is expressed on a scale where 10 represents 100% (after van Riper *et al.* 1986).

able reduction in range of many of the endemic bird species, which are now mainly restricted to mountain tops at altitudes too cold to support mosquitoes (Fig. 19.8).

Cuckoos and their host communities

Many brood parasites are generalists, utilizing many different host species. A notable example is the brown-headed cowbird *Molothrus ater*, in North America. Whereas it is difficult for a specialist brood parasite to drive its sole host species to extinction (because the parasite population itself declines as host densities fall), a generalist can extinguish one or more host species—locally or globally—so long as its population density is maintained by brood parasitism of other, more robust, hosts. May and Robinson (1985) discuss the demographic impact of such a generalized brood parasite upon any one host species, showing that a host can be driven to extinction if the probability of its nests experiencing brood parasitism exceeds a critical level, p_c, given by

$$p_c = (\gamma - 2\mu)/(\gamma - \gamma').$$

Here γ and γ' are the average numbers of offspring successfully reared to reproductive maturity (typically 1 year of age) from unparasitized and parasitized nests, respectively. The parameter μ represents the probability of an adult bird dying between one breeding season and the next (so that $1 - \mu$ is the survival probability, here treated as age-independent), and the sex-ratio at fledging is assumed to be 50:50. The parameters γ and γ' represent the combination of average clutch size, fledgling success, and probability of survival to reproductive age; as surveyed in detail by May and Robinson (1985), we tend to have less and less information about a particular species as we move along this chain (from clutch size to survival probability), especially for parasitized nests.

With the rapid clearing of forests for agriculture and cattle, over the past century in North America and at accelerating rates in the Neotropics today, cowbirds have expanded their ranges in significant ways. Nearly every issue of American bird journals contains at least one article on cowbirds, and there have been several reviews of the accumulating data (Rothstein 1975; Payne 1977). Most of the data are for the Brown-headed cowbird *M. ater* but there is increasing evidence that the Shiny Cowbird *M. bonariensis* is having a major impact on many neotropical species.

May and Robinson (1985) have summarized such data as are available, for brood parasitism by *M. ater* upon 11 species of North American birds (including Bell's Vireo and Kirtland's Warbler), and by *M. bonariensis* upon 3 neotropical species (see their Table 3). In no case were the data sufficient to make a definite judgement as to whether cowbird parasitism was driving a host to extinction or to some sustainable lower level. The data and theory are such that, if juvenile and adult mortality rates were known, it would generally be possible to predict the number of times females must renest in order to sustain the population. Conversely, if the number of nesting attempts per parasitized and per unparasitized female were known, it would be possible to predict the maximum mortality rates that would be consistent with maintaining the endangered population. Notwithstanding the cloudy nature of the current picture, best guesses at the parameters in the above equation do suggest that several species are indeed en route to extinction caused by generalist brood parasites (May and Robinson 1985). However it should be kept in mind that such parasitism is often the result of changes in the habitat of both host and parasite, and that the effects of brood parasitism may be difficult to distinguish from the effects of simple loss of suitable breeding habitat (Probst 1986).

Potential of using parasites to control introduced species

The absence of parasites from a population may also have significant consequences. Starlings and House Sparrows have established and spread across the North American continent in the hundred years since their initial introduction. Populations of both these species have depauperate

Table 19.3 The numbers of species of parasites recorded from surveys of Starlings and House Sparrows in Europe and the United States (after Dobson and May 1986)

Helminth group	Number of genera (species)			
	In Europe	In North America	Total Number	Number in common
A: Helminth parasites of European Starlings in Europe and America				
Trematoda	17 (26)	4 (4)	18 (28)	3 (2)
Cestoda	9 (12)	4 (5)	10 (14)	3 (3)
Nematoda	14 (26)	6 (10)	17 (30)	3 (6)
Acanthocephala	4 (6)	2 (3)	6 (9)	0 (0)
Total	44 (70)	16 (22)	51 (81)	9 (11)
B: Ectoparasites of House Sparrows in Europe and North America				
Acarina	8 (35)	5 (24)	10 (43)	3 (16)
Mallophaga	8 (18)	4 (9)	9 (22)	3 (5)
Siphonaptera	1 (7)	1 (4)	1 (9)	1 (2)
Total	17 (60)	10 (37)	20 (74)	7 (23)

parasite faunas when compared with populations in Europe (Table 19.3). No one has undertaken experiments to determine whether this relative absence of potentially regulating mechanisms has contributed to the success of these two species.

The relative absence of parasites from populations of rats, goats, and cats that have been introduced to oceanic islands may contribute to the success of these species as predators (Dobson 1988*b*). This success is usually to the detriment of the endemic bird fauna, which often consists of species with no intrinsic mechanisms of defence against predators. Introduced cats, rats, and goats are the major cause of extinction in endemic oceanic bird species (Boag 1983), and their control and removal may allow the reestablishment of locally extinct species and possible habitat regeneration. The relative absence of pathogenic parasite species from the introduced predators suggests the possibility of an economically inexpensive way of controlling the density of these species. A more detailed discussion of this problem is given in Dobson (1988*b*). Providing extreme care is taken to ensure that the introduced parasite will only infect the introduced pest species, the technique deserves consideration as a potential means of addressing this problem.

Discussion and implications for conservation

In the first section of this paper it was suggested that host population size is a major factor in determining both the size and diversity of the parasite community. As endangered or rare species often exist at low densities, specific pathogens may not present a significant threat to them. However, increasing the density of endangered species, by concentrating their populations into zoos or nature reserves, is likely to increase parasite transmission rates. This suggests that subdivision of the population may be a sensible policy to prevent epidemics which might infect all the members of an extant population.

Even when a population of an endangered species is subdivided, the most serious pathogenic threat to populations of endangered species are generalist parasites that are able to utilize a range of species as hosts. Introductions of pathogens from larger populations of common host species may drive the endangered species to extinction. Under these circumstances, management initiatives should be directed towards either reducing transmission rates between the common and rare species, or upon reducing levels of infection in the common species. These considerations will be particularly important when attempting to increase the size of wild populations by introducing captive-born individuals. As the latter may have been exposed to a range of pathogens when raised in captivity, it is essential that they be isolated for a suitable period of time (longer than the latent periods of the infections in question) before reintroduction into the wild, so that latent infections may be revealed.

The relative absence of parasites in the introduced pest species that cause high rates of extinction of endemic bird species on oceanic islands, suggests that the introduction of suitable parasites may be an effective way of reducing the abundance of these pest species. Although extreme caution must be used when determining the parasite species that might be introduced, a range of theoretical and empirical evidence suggests that parasites with low to intermediate virulence will produce the greatest decrease in population density of the pest species (Anderson 1979). As populations of introduced species on islands may exhibit reduced genetic variation due to founder effects, attempts to control them using parasites and pathogens have the additional benefit that they may provide important information on the interaction between parasite pathogenicity and levels of genetic variability in host populations. This information is unlikely to be obtained experimentally from endangered species, yet remains a potentially crucial aspect of their management.

The examples described in this chapter suggest that a full appraisal of the role of parasites in the management of endangered species requires both a better understanding of the evolution of speciality and 'host choice' in parasites, and of the population dynamics of multi-species parasite-host communities. However, preliminary models for the structure of parasite communities suggest that an understanding of parasite life-history strategies is crucial in determining our ability to dissect community structure.

Recent developments to our understanding of plant (Hubbell and Foster 1986; Tilman 1988) and intertidal communities (Roughgarden *et al*. 1988) similarly suggest that the life-history strategies of the component species play a very significant role in determining the community structure. The majority of avian extinctions of the next 50 years are likely to occur as tropical and temperate forests become further fragmented. Therefore it seems sensible to suggest that ornithologists should begin to utilize the vast knowledge now available on avian life histories as a tool to reconsider the structure of avian communities.

References

Anderson, R. C. (1988). Nematode transmission patterns. *Journal of Parasitology*, **74**, 30–45.

Anderson, R. M. (1979). Parasite pathogenicity and the depression of host population equilibria. *Nature*, **279**, 150–2.

Anderson, R. M. (1988). Reproductive strategies of trematodes. In *Reproductive biology of invertebrates*, Vol. 6 (eds. K. G. Adiyodi and R. G. Adiyodi). John Wiley, New York.

Anderson, R. M. and Gordon, D. M. (1982). Processes influencing the distribution of parasite numbers within host populations with special emphasis on parasite-induced host mortalities. *Parasitology*, **85**, 373–98.

Anderson, R. M. and May, R. M. (1978). Regulation and stability of host-parasite population interactions. I Regulatory processes. *Journal of Animal Ecology*, **47**, 219–47.

Anderson, R. M. and May, R. M. (1979). Population biology of infectious diseases. Part I. *Nature*, **280**, 361–7.

Anderson, R. M. and May, R. M. (1981). The population dynamics of microparasites and their invertebrate hosts. *Philosophical Transactions of the Royal Society [B]*, **291**, 451–524.

Anderson, R. M. and May, R. M. (1982). Directly transmitted infectious diseases: control by vaccination. *Science*, **215**, 1053–60.

Anderson, R. M., Whitfield, P. J., and Dobson, A. P. (1978). Experimental studies of infection dynamics: infection of the definitive host by the cercariae of *Transversotrema patialense*. *Parasitology*, **77**, 189–200.

Anderson, R. M., Grenfell, B. T., and May, R. M. (1984). Oscillatory fluctuations in the incidence of infectious disease and the impact of vaccination: time series analysis. *Journal of Hygiene*, Cambridge, **93**, 587–608.

Boag, P. T. (1983). More extinct island birds. *Nature*, **305**, 274–5.

Bush, A. O. and Holmes, J. C. (1986*a*). Intestinal helminths of lesser scaup ducks: patterns of association. *Canadian Journal of Zoology*, **64**, 132–41.

Bush, A. O. and Holmes, J. C. (1986*b*). Intestinal helminths of lesser scaup ducks: an interactive community. *Canadian Journal of Zoology*, **64**, 142–52.

Chesson, P. L. (1985). Coexistence of competitors in spatially and temporally varying environments: a look at the combined effects of different sorts of variability. *Theoretical Population Biology*, **28**, 263–87.

Chesson, P. L. and Case, T. J. (1985). Nonequilibrium community theories: Chance, variability, history and coexistence. In *Community ecology* (eds. J. Diamond and T. J. Case), pp. 229–39. Harper and Row, New York.

Chesson, P. L. and Warner, R. R. (1981). Environmental variability promotes co-existence in lottery competition systems. *American Naturalist*, **117**, 923–43.

Crofton, H. D. (1971). A model of host-parasite relationships. *Parasitology*, **63**, 343–64.

Crompton, D. W. T. and Harrison, J. G. (1965). Observations on *Polymorphus minutus* (Goeze, 1782) (Acanthocephala) from a wildfowl reserve in Kent. *Parasitology*, **55**, 345–55.

Dobson, A. P. (1985). The population dynamics of competition between parasites. *Parasitology*, **91**, 317–47.

Dobson, A. P. (1988*a*). The population biology of parasite-induced changes in host behavior. *Quarterly Review of Biology*, **63**, 139–65.

Dobson, A. P. (1988*b*). Restoring island ecosystems: the potential of parasites to control introduced mammals. *Conservation Biology*, **2**, 31–9.

Dobson, A. P. (1989). Models for multi-species parasite-host communities. In *Parasite communities* (eds. G. Esch, A. Bush, and C. R. Kennedy), pp. 261–88. Croom-Helm.

Dobson, A. P. and Keymer, A. E. (1985). Life history models. In *Acanthocephalan biology* (eds. D. W. T. Crompton and B. B. Nickol), pp. 347–84. Cambridge University Press, Cambridge.

Dobson, A. P. and May, R. M. (1986). Patterns of invasion by pathogens and parasites. In *Ecology of biological invasions of North America and Hawaii* (ed. H. A. Mooney and J. A. Drake), pp. 58–76. Springer-Verlag, New York.

Hanski, I. (1982). Dynamics of regional distribution: the core and satellite species hypothesis. *Oikos*, **38**, 210–21.

Hassell, M. P. (1978). *Arthropod predator–prey systems*. Princeton University Press, USA.

Hodasi, J. K. M. (1973). Comparative studies in the helminth fauna of native and introduced domestic fowls in Ghana. *Journal of Helminthology*, **43**, 35–52.

Hubbell, S. P. and Foster, R. B. (1986). Biology, chance, and history and the structure of tropical rain forest tree communities. In *Community ecology* (eds. J. Diamond and T. J. Case), pp. 314–30, Harper and Row, New York.

Humphrey, S. R., Courtney, C. H., and Forrester, D. J. (1978). Community ecology of helminth parasites of the brown pelican. *Wilson Bulletin*, **90**, 587–98.

Itamies, J., Valtonen, E. T., and Feagerholm, H. P. (1980). *Polymorphus minutus* infestation in eiders and its role as a possible cause of death. *Annales Zoologici Finnici*, **17**, 285–9.

James, B. L. and Llewellyn, L. C. (1967). A quantitative analysis of helminth infestation in some passerine birds found dead on the island of Skomer. *Journal of Helminthology*, **41**, 19–44.

Jenner, E. (1788). Observations on the natural history of the cuckoo. *Philosophical Transactions of the Royal Society of London*, , **78**, 219–35.

Jennings, A. R. (1961). An analysis of 1,000 deaths in wild birds. *Bird Study*, **8**, 25–31.

Kennedy, C. R. (1978*a*). The parasite fauna of resident char *Salvelinus alpinus* from Arctic islands, with special reference to Bear island. *Journal of Fish Biology*, **13**, 457–66.

Kennedy, C. J. (1978*b*). An analysis of the metazoan parasitocoenoses of brown trout *Salmo trutta* from British lakes. *Journal of Fish Biology*, **13**, 215–26.

Macdonald, J. W. (1962*a*). Mortality in wild birds with some observations on bird weights. *Bird Study*, **9**, 147–67.

Macdonald, J. W. (1962*b*). Mortality in wild birds. *Bird Study*, **10**, 91–108.

Macdonald, J. W. (1962c). Mortality in wild birds. *Bird Study*, **11**, 181–95.

Mackiewicz, J. S. (1988). Cestode transmission patterns. *Journal of Parasitology*, **74**, 60–71.

May, R. M. (1988). How many species are there on Earth? *Science*, **241**, 1441–9.

May, R. M. and Anderson, R. M. (1978). Regulation and stability of host-parasite population interactions. II. Destabilizing processes. *Journal of Animal Ecology*, **47**, 249–67.

May, R. M. and Anderson, R. M. (1979). Population biology of infectious diseases. Part II. *Nature*, **280**, 455–61.

May, R. M. and Anderson, R. M. (1983). Epidemiology and genetics in the co-evolution of parasites and hosts. *Proceedings of the Royal Society of London [B]*, **219**, 281–313.

May, R. M. and Robinson, S. K. (1985). Population dynamics of avian brood parasitism. *American Naturalist*, **126**, 475–94.

Moore, J. (1984). Parasites that change the behaviour of their host. *Scientific American*, **250**, 108–15.

Payne, R. B. (1977). The ecology of brood parasitism in birds. *Annual Review of Ecology and Systematics*, **8**, 1–28.

Probst, J. R. (1986). A review of factors limiting the Kirtland's warbler on its breeding grounds. *American Naturalist*, **116**, 87–100.

Ricklefs, R. E. (1987). Community diversity: Relative roles of local and regional processes. *Science*, **235**, 167–71.

Riper, III, C. V., Riper, S. G. V., Goff, M. L., and Laird, M. (1986). The epizootiology and ecological significance of malaria in Hawaiian land birds. *Ecological Monographs*, **56**, 327–44.

Rothstein, S. I. (1975). Evolutionary rates and host defences against avian brood parasitism. *American Naturalist*, **109**, 161–76.

Roughgarden, J., Gaines, S., and Possingham, H. (1988). Recruitment dynamics in complex life cycles. Science, **241**, 1460–6.

Schaffer, W. M. and Kot, M. (1985). Nearly one dimensional dynamics in an epidemic. *Journal of Theoretical Biology*, **112**, 403–27.

Shoop, W. L. (1988). Trematode transmission patterns. *Journal of Parasitology*, **74**, 46–59.

Stock, T. M. and Holmes, J. C. (1987a). *Dioecocestus asper* (Cestoda: Dioecocestidae): An interference competitor in an enteric helminth community. *Journal of Parasitology*, **73**, 1116–23.

Stock, T. M. and Holmes, J. C. (1987b). Host specificity and exchange of intestinal helminths among four species of grebes (Podicipedidae). *Canadian Journal of Zoology*, **65**, 669–76.

Sugihara, G. and May, R. M. (1990). Nonlinear forecasting as a way of distinguishing chaos from measurement error in time series. *Nature*, **344**, 734–41.

Threlfall, W. (1967). Studies on the helminth parasites of the herring gull, *Larus argentatus* (Pontopp.) in Northern Caernarvonshire and Anglesey. *Parasitology*, **57**, 431–53.

Tilman, D. (1988). *Plant strategies and the dynamics and structure of plant communities*. Princeton University Press, Princeton, NJ.

Toft, C. A. (1986). Communities of species with parasitic life-styles. In *Community ecology* (eds. J. Diamond and T. J. Case), pp. 445–64. Harper and Row, New York.

Wakelin, D. (1984). Evasion of the immune response: survival within low responder individuals of the host population. *Parasitology*, **88**, 639–57.

Warner, R. E. (1968). The role of introduced diseases in the extinction of the endemic Hawaiian avifauna. *Condor*, 70, 101–20.

Warner, R. R. and Chesson, P. L. (1985). Coexistence mediated by recruitment fluctuations: a field guide to the storage effect. *American Naturalist*, 125, 769–87.

Wassom, D. L., Dewitt, C. W., and Grundmann, A. W. (1974). Immunity to *Hymenolepis citelli* by *Peromyscus maniculatus*: genetic control and ecological implications. *Journal of Parasitology*, 59, 117–21.

Yom-Tov, Y. (1980). Intraspecific nest parasitism in birds. *Biological Review*, 55, 93–108.

20 Control of parasites in natural populations: nematode and virus infections of Red Grouse

PETER J. HUDSON and ANDREW P. DOBSON

Introduction

Many of the long-term studies on bird populations were stimulated by the pioneering publications of David Lack (1954, 1966, 1968). While Lack (1954) considered a range of density-dependent factors that could regulate populations and included a chapter on disease, he came to the conclusion that 'while further evidence is needed, it seems unlikely that disease is an important factor regulating numbers of most birds'. He came to this conclusion following the general view, held by most ecologists at that time, that parasites evolved to become benign symbionts that maintained high survival by reducing their deleterious effects on the host. Parasitologists, on the other hand, investigating parasite systems in man and domestic livestock viewed parasites as organisms that benefited at the expense of their host (Crofton 1971).

These apparently conflicting views were reconciled when the principles of epidemiology and ecology were combined and applied to wild animal populations. Empirical evidence suggested the relative effects of parasite increase with density, thereby decreasing growth rate of the host population by reducing the host's survival and fecundity (Anderson and May 1978; May and Anderson 1978). In evolutionary terms, the success of a parasite depends not on life expectancy alone but on successful transmission between hosts. Hence a successful parasite is one that produces transmission stages that become established and contribute to future generations, although in the process both host and parasite may die.

The synthesis produced by Anderson and May (*op. cit.*) has provided a sound framework for a series of studies on the population dynamics of disease agents in wild animals (e.g. Anderson 1982a; Rollinson and Anderson 1985). The essential model is of two linked differential equations which describe changes in the numbers of hosts and parasites. The attraction of these models over the simulation approach are threefold:

(1) They can often be solved analytically for equilibrium conditions;
(2) they can be developed and applied to a wide range of parasite-host systems whereas simulation models tend to be unique; and

(3) they provide an insight into stability and threshold properties and are thus useful in the development of control strategies.

Not surprisingly, the majority of studies relating to parasite control have concentrated on man and livestock. However, there is increasing awareness that parasites have a significant impact in problems relating to conservation (Dobson and May 1986; Dobson 1988; Scott 1988; Thorne and Williams 1988) and in the management of wild animals (Hudson and Dobson 1988). In this paper we describe two parasite systems of Red Grouse and the possible development of control techniques. We discuss the Red Grouse system because there is intensive and extensive information on changes in Grouse numbers and levels of parasite infection.

First we consider the intestinal nematode *Trichostrongylus tenuis* which causes rapid declines in the density of Red Grouse by reducing fecundity. Second we consider the tick-borne viral infection, Louping-ill which is principally a disease of sheep but has a marked effect on the survival of Red Grouse. The two systems are very different and raise different issues relating to management and conservation. In the control of *T. tenuis*, the main problem is simply to develop a technique for reducing the build-up of parasite numbers by applying an anthelmintic drug. The application of anthelmintics to domestic livestock is relatively simple but treating wild animals raises a number of technical problems. With the Louping-ill system the analysis indicates that the problem relates to livestock acting as a reservoir host and amplifying the disease, a problem not uncommon in other wildlife systems (e.g. Boshell 1969; Rogers 1988).

The impact of Trichostrongylus tenuis on Red Grouse

Life cycle of *T. tenuis*

T. tenuis inhabits the large caeca of Red Grouse and has a direct life cycle with no intermediate hosts. While the parasite is known to occur in a few alternative avian hosts, these are absent from the system under study. In essence, the life cycle of the nematode consists of two populations: (i) the mature worms living within the caeca of the Red Grouse and (ii) the population of free-living stages (Fig. 20.1).

Grouse do not appear to develop resistance to the parasite (Watson 1988) and mean worm burdens are high (Hudson 1986*a*). As with many macroparasites, *T. tenuis* has an aggregated distribution within the host population although the degree of aggregation is relatively low (Hudson 1986*a*). Production of eggs depends on the level of infection with a fall in the *per capita* egg production above burdens of 4500 worms. The parasite eggs leave the Grouse in the caecal droppings and their development is temperature dependent. Humidity is also important for hatching of eggs

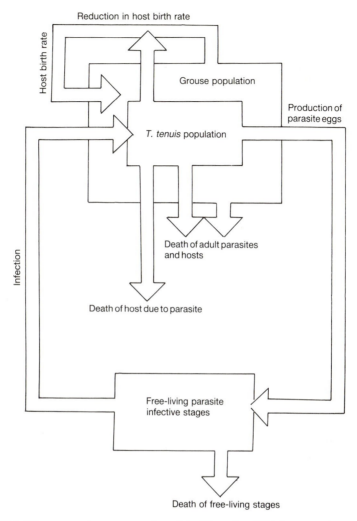

Fig. 20.1. Diagrammatic representation of the life cycle of *Trichostrongylus tenuis* in Red Grouse.

and is particularly important to the survival of free-living parasite stages. Should the caecal pat dehydrate the egg embryos fail to hatch and the free-living stages die (Hudson 1986*a*).

The third-stage larvae retain the cuticle of the second-stage larvae and migrate from the caecal pat to the growing tips of the heather. Humidity is also important to this stage and necessary for migration, although once at the heather tip the larvae may enter the in-rolled heather leaflet and obtain some protection from desiccation. Infection occurs when Grouse eat the growing tips of the heather plant. Active larvae can be found on the heather plant between June and September (Watson 1988), with a corresponding increase in worm burdens of the Grouse. By the end of the year ingested larvae do not develop into the adult stage but remain arrested, developing into adults in the following spring (Shaw 1988).

The impact of *T. tenuis* on Red Grouse

T. tenuis burrows into the caecal mucosa of the Grouse causing localized flattening, a disruption of the plicae, bleeding, and an increase in the size of the caeca (Watson *et al*. 1987). Although the function of caeca in birds is not clear, the parasites probably reduce brush-border enzyme activity and cause a decrease in absorptive area thus reducing digestive efficiency.

Experiments with wild Grouse have shown that reducing the size of the Grouse infection improves body condition, rate of weight gain in females prior to incubation, clutch size, hatching success, and chick survival (Table 20.1; Hudson 1986*a*, *b*). The parasite also reduces the survival of adult Grouse although this effect is small in comparison with its impact on female fecundity.

The reduction in condition of Red Grouse infected with *T. tenuis* may also reduce the competitive ability of the Grouse and increase their vulnerability to predation. Jenkins *et al*. (1963) found that Grouse without territories had higher worm burdens than individuals defending territories (discussed further by Hudson and Dobson 1990). More rigorous evidence is available to suggest that *T. tenuis* may make infected Grouse more vulnerable to predation. Hudson (1986*a*) has shown that pointing dogs selectively find the nests of hen Grouse with high worm burdens and suggests this is because parasites interfere with the bird's scent emission. Furthermore comparisons within and between areas found that Grouse killed by predators had higher worm burdens than Grouse that survived through the summer.

Table 20.1 The breeding production of female Red Grouse in relation to parasite burdens (after Hudson 1986*b*). High parasite burdens relate to natural levels of infection and low burdens are those that are reduced through oral chemotherapy

	1982		1983		1984	
Parasite burdens:	High	Low	High	Low	High	Low
Clutch size	7.9	8.3†	5.3	8.0**	7.6	8.8*
Hatching success	77%	97%†	38%	75%**	92%	96%†
Brood size						
(10 days)	3.6	5.5*	0.4	3.2**	4.9	7.4**

† Not significant, $P > 0.05$; * $P < 0.05$; ** $P < 0.01$.

Population model for *T. tenuis* in Red Grouse

An insight into the effects of *T. tenuis* on the dynamics of Red Grouse populations can be obtained through the development of a mathematical model. Essentially, the model is based on the framework developed by Anderson and May (Anderson and May 1978; May and Anderson 1978) but expanded into a three-equation model describing changes in the numbers of adult parasite, free-living parasitic stages, and Red Grouse (more details are provided by Hudson and Dobson 1989). The equations for the parasite and the Grouse are linked through the impact the parasites have on the survival and fecundity of the Grouse (Fig. 20.1).

The relative effects of the different processes on the size of each population can be found through examination of the properties of the model at equilibrium. This suggests that Grouse density varies inversely with parasite fecundity and rate of infection. Mean worm burden varies inversely with the parasite's ability to influence host survival and fecundity. The population of free-living stages varies directly with the pathogenicity of the parasite and the life expectancy of the free-living stages.

The dynamic properties of the model can be explored through stability analysis. In essence, the behaviour of the Grouse-*T. tenuis* system depends on four parameters:

(1) degree of parasite aggregation;
(2) life expectancy of the parasite's free-living and arrested stages;
(3) the parasite's effects on Grouse fecundity; and
(4) the parasite's effects on Grouse survival.

When the life expectancy of the free-living and arrested stages is long and the effects on Grouse fecundity are greater than on host survival, Grouse and parasite numbers oscillate in a cyclic manner. In the absence of any other constraint on host population growth, these fluctuations are unstable and sufficiently large to force one of the populations to extinction. In reality host density is likely to be limited at high density by other factors such as availability of nest sites or food. Inclusion of a term mimicking the effects of these factors through territorial behaviour (Davies 1978) produces stable limit cycles. The period of these cycles is influenced by the life expectancy of the free-living stages and the intrinsic growth rate of the Grouse population. Parameter estimates from field studies suggest the cycles will be between 3 and 10 years, similar to the 4.8 years recorded for Grouse populations by Potts *et al*. (1984).

In the model when the life expectancy of the free-living stages is low or infection rate is low, worm burdens fail to increase and the Grouse populations do not cycle (Fig. 20.2). Empirical data collected in the north of England by Hudson *et al*. (1985) showed that cyclic populations of

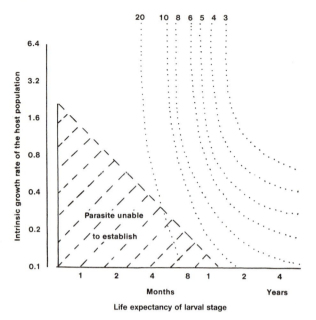

Fig. 20.2. The tendency of grouse populations to show population cycles with respect to life expectancy of larval stages and the intrinsic growth rate of the population.

Grouse were associated with significantly higher rainfall and parasite burdens than non-cyclic populations.

It is known that humidity influences the survival of free-living stages, and these environmental differences could therefore account for differences in worm burdens and patterns of population change.

The control of Trichostrongylus tenuis

Analysis of the model described in the previous section indicates the parameters that need to be altered to reduce the impact of *T. tenuis* on Red Grouse. Reducing the survival of free-living stages would make the greatest impact since this effectively reduces the rate of infection and prevents build-up of the infection.

Since the free-living stages are susceptible to desiccation, a reduction in humidity, perhaps through drainage, is an obvious suggestion for potential control and management of parasite populations. During the 1970s, 55 per cent of English Grouse moor owners (Hudson 1984) drained part of their estates to increase run-off and encourage heather growth, but this appears to have made no difference to standing water more than 2 m from the drain nor influenced humidity. At the interface of blanket bog and freely drained heather moorland it may be possible to convert blanket bog to the drier moorland, but this is limited in extent and is known to have a marked effect on the invertebrate fauna (Coulson 1988) and can be considered environmentally damaging.

Direct chemotherapy is a main component of disease control in man and domestic livestock and usually aims to reduce the adult worm population. The advantages of chemotherapy are immediate although reinfection can occur within a short time; the rates of reinfection can be high and hence direct chemotherapy is unlikely to be of lasting benefit. In the Grouse-*T. tenuis* system a large proportion of the population must be treated annually to make any lasting effect on the size of the parasite population (Fig. 20.3).

A technique for catching and treating Grouse orally has been developed and used successfully in both England and Scotland. On 4 occasions more than 1000 birds have been caught from an area of 28 km² within 5 months although this was still only 40 per cent of the population (cf. Fig. 20.3). While direct chemotherapy has been highly effective the technique is labour intensive and not of long-term benefit.

Most parasite control systems use an integrated control program combining chemotherapy and sanitation and husbandry to prevent reinfection. While sanitation is clearly impractical for wild animals, indirect chemotherapy is an approach that has been used in other systems. For instance, anthelmintics are incorporated in the feed of Pheasants to control *Syngamus trachea* and have been included in fermented apple

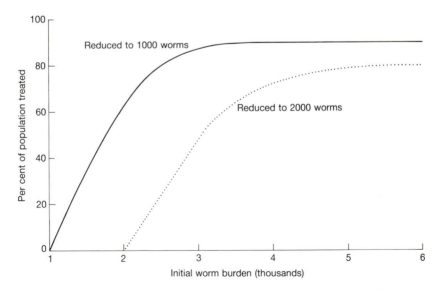

Fig. 20.3. The proportion of Grouse that need to be treated annually to reduce worm burdens to 1000 and 2000 worms respectively. (R_o taken as 5).

mash to control *Protostrongylus* sp. in Big Horned Sheep in America. Grouse do not take artificial feed although they do take grit to aid digestion. Strathclyde Chemicals have developed a technique for coating grit with a kernel fat and incorporating within this fat an anthelmintic (Fenbendazole, Panacur; Hoechst Animal Health). Wild Grouse will take up to 170 pieces of grit per day but normally take around 10. The concentration of the active drug can thus be set at a level sufficient to kill all parasites in hosts that regularly take medicated grit. Fenbendazole is known to be safe to wildlife and is not dangerous if taken in excess.

Initial results compared areas where medicated grit has been used with matched control areas with untreated grit, and demonstrated that the grit applications have almost halved worm burdens with a beneficial effect on production (Table 20.2). One problem with this kind of treatment, however, is that the parasites are exposed to a more constant but low level of anthelmintic than is possible with direct treatment and the parasites may evolve resistance to the drug. Resistance to fenbendazole is known to have occurred in other parasite systems.

Table 20.2 The effects of medicated grit on worm burdens of adult Grouse. Overall, worm burdens were significantly lower on treated areas ($P < 0.001$; from $x^2 = -\Sigma\ 2\log_e P$)

Region	Geometric mean worm burdens		Significance (P) 1-tailed t-test
	Untreated	Treated	
South Dales	2446	1004	0.010
South Dales	2454	1219	0.043
Borders	650	471	0.074
Borders	794	407	0.049
Perthshire	2188	1671	0.075
Inverness-shire	6907	3090	0.0002

In summary, the caecal nematode *T. tenuis* causes regular cycles in Grouse numbers by reducing host fecundity. Cyclic fluctuations in numbers occur when the life expectancy of larval stages of the parasite is long and this is associated with wetter ground where rainfall is high. The impact of the parasites on the Grouse population can be reduced through both direct and indirect chemotherapy.

The impact of Louping-ill on Red Grouse

Louping-ill (LI) is a viral infection of the central nervous system which can cause significant financial losses to the sheep and Grouse industry in Scotland and northern England. On moorland areas such as the North

Yorkshire Moors, LI is the second major cause of sheep mortality, after road accidents. The virus is transmitted between vertebrate hosts by the Sheep Tick, *Ixodes ricinus*, and while both the ticks and the virus have a patchy distribution they are widespread within the British Uplands (Fig. 20.4).

Although the ticks may feed on a wide range of hosts, only sheep and Grouse produce viraemias which exceed the threshold necessary for transmission (Reid 1978; Beasley *et al*. 1978). Both the larval and nymphal

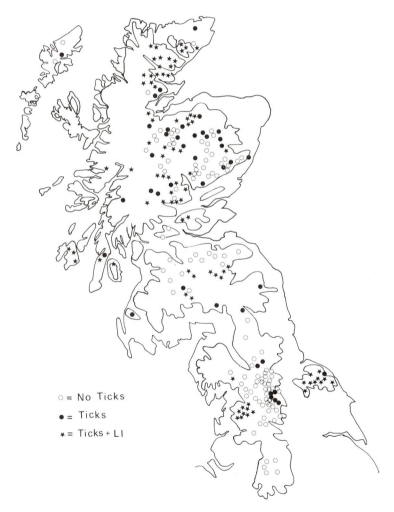

o = No Ticks

● = Ticks

✳ = Ticks + LI

Fig. 20.4. The distribution of ticks and ticks + Louping-ill on Grouse moors in England and Scotland. Data obtained from a questionnaire survey conducted in England, 1979 and in Scotland, 1983.

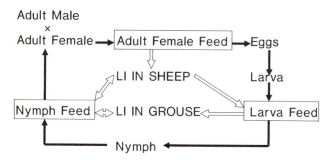

Fig. 20.5. Schematic representation of tick life cycle and the flow of Louping-ill between hosts; note that Grouse can only receive Louping-ill from nymphs while sheep receive it from both nymphs and adults.

stages of the tick feed on Grouse but the adult female ticks are rarely found. Since the virus is not transmitted vertically from the adult female tick via the egg to the larva and since larva only feed once, only a larva that becomes infected, moults into a nymph and subsequently feeds can infect a Grouse (Fig. 20.5) while both nymphs and adult female ticks can infect sheep.

Louping-ill and Grouse survival

Laboratory studies coupled with field experiments have found that Grouse chicks are highly susceptible to LI. In an experimental study of captive birds, Reid (1975) found 79 per cent ($n = 37$) of experimentally infected juvenile birds developed clinical signs of weakness and survived for only 6 days.

Comparative studies of chick survival in the north of England in relation to the presence of ticks and LI indicated that ticks alone were not detrimental to chick survival but in areas with both ticks and LI, chick survival was almost 50 per cent lower (Table 20.3). Ticks alone did not reduce growth rate of chicks (Hudson 1986*a*) although chicks infected

Table 20.3 The effects of ticks and Louping ill on the survival of Grouse chicks estimated from bag returns

	Moors with		
	No Ticks	Ticks	Ticks + LI
Number of ticks tagged	345	343	467
Percentage shot	15.1%	16.6%	7.9%

with LI did show reduced growth rates (Reid *et al*. 1978). The studies in Speyside, Scotland (see also Duncan *et al*. 1978) also found that infected chicks were more likely to die than uninfected chicks and overall chick survival was 89 per cent lower on an area with ticks and LI compared with an area with few ticks and consequently less LI.

In areas where ticks were abundant, 84 per cent of adult Grouse had antibodies to LI (Reid *et al*. 1978). The prevalence of antibodies increased with the age of chicks from 11 per cent in young chicks to an asymptote of about 30 per cent by week 4. Some of the chicks with antibodies were too young to have developed them themselves suggesting the antibody was maternal in origin.

The control of Louping-ill

Vaccination of sheep against Louping-ill

A commercially available vaccine against LI, developed by the Moredun Institute, has proved effective at reducing the rate of virus multiplication in sheep and cattle. Two doses of vaccine are required with an interval of 2 to 8 weeks between injections (Shaw and Reid 1981) to provoke an adequate antibody response. Adult ewes develop their own immunity which they pass to their lambs through the colostrum and generally farmers vaccinate only the susceptible sheep: hoggs (previous year's young) that have lost maternal antibody and adult ewes brought from other flocks.

In the epidemiology of the disease, vaccinating the sheep essentially reduces the abundance of the hosts that amplify LI and consequently alters the dynamics of the disease. Such changes may well alter the equilibrium conditions for the disease and could result in the elimination of the disease from the wild hosts.

The dynamics of Louping-ill and vaccination

An initial understanding of the dynamics of vector-transmitted micro-parasites can be attained by using an extension of the MacDonald/Ross malarial model, as described by Aron and May (1982). The model can also be extended (Rogers 1988) to examine features of the more complex two host-one vector systems such as Trypanosomiases.

The one host-one vector system is described by two linked differential equations that describe changes in the proportion of infected hosts and vectors. Aron and May (1982) provide an example for the malarial system (Fig. 20.6a) where the equilibrium conditions for the host and vector are determined separately and the equilibrium for the disease is identified at the intersection of the two equilibrium lines. This equilibrium is obviously

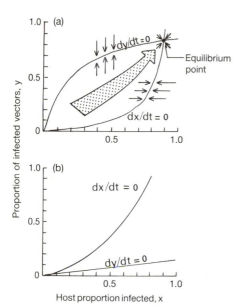

Fig. 20.6. Equilibrium conditions for host (x = proportion infected) and vectors (y = proportion infected) in a one host-one vector system; two phase planes are shown corresponding to values where x does not change along $dx/dt = 0$ and y does not change along $dy/dt = 0$ (after Aron and May 1982). Equilibrium conditions exist at the intersection of the lines and arrows indicate how dynamical trajectories will behave following perturbation. In (a) equilibrium conditions for the disease occur when the initial slope of the y isocline exceeds the initial slope of the x isocline; in (b) the initial slope of the y isocline is less than the x-isocline so the isoclines fail to cross and the disease cannot maintain itself.

produced only when the initial slope of the y-isocline is greater than that of the x-isocline. If the initial slope of the y-isocline is less than the x-isocline then the isoclines do not intersect so all trajectories converge at the origin and the disease is not maintained (Fig. 20.6b).

The parasite can only establish when the average number of new cases that arise from an infectious host (this is R_o, the basic reproductive rate of the disease) is greater than unity. For a vector-transmitted disease, R_o is dependent on the net transmission success from the host to the vector at the vector's biting rate and then back to the host at the same rate (i.e. biting rate squared), the ratio of vector to host numbers, the average survival time of the vectors, and the period the host is infectious (see Anderson 1982b for more details).

Obtaining an accurate estimate of R_o is not simple without careful verification of parameters and even then slight changes in certain parameters can greatly influence the estimate. However, in the case of the Grouse-LI-tick system because the biting rate of the ticks is low, the value of R_o is in the region of 10^{-2} to 10^{-4} and much lower than unity (Table 20.4). In the sheep system the value of R_o is more critical and is in the region of 1 depending on the relative proportion of bites that become infectious and the duration of the infection (Fig. 20.7). Empirical observations imply that R_o is greater than unity in the sheep system, equivalent to Fig. 20.6a while that in the Grouse system is equivalent to Fig. 20.6b.

Rogers (1988) extended the one host-one vector system to the two host-

Table 20.4 Variables and parameters for the Louping-ill system in Grouse and sheep (after Duncan *et al*. 1978; Hudson 1986*a*, unpublished data)

---

1. Variables and parameters

N_1	Density of Grouse km^{-2}	140
N_2	Density of sheep km^{-2}	200
V_e	Density of tick eggs km^{-2}	30×10^3
V_l	Density of tick larvae km^{-2}	1.14×10^3
V_n	Density of tick nymphs km^{-2}	0.22×10^3
V_a	Density of adult female ticks km^{-2}	0.04×10^3
m_1	Ratio of vectors: Grouse	10
m_2	Ratio of vectors: sheep	7
p_1	Proportion of tick bites on Grouse	0.10
p_2	Proportion of tick bites on sheep	0.90
d	Duration of tick feeding cycle	1 year
a_1	Tick biting rate on Grouse	P_1/day per year
a_2	Tick biting rate on sheep	P_2/day per year
b	Proportion of bites infectious	?0.1–0.8
$1/r_1$	Duration of infection in Grouse	?0.01–0.10 years
$1/r_2$	Duration of infection in sheep	?0.20–0.60 years
$1/u$	Life expectancy of ticks	1.23 years
c	Probability of tick infection	0.8
T	Incubation period in tick	0.02 years

2. Basic reproductive rate (R_o) in one host-one vector system. (After Anderson 1982*b*; Rogers 1988.)

$$R_o = \frac{a^2 mb}{ur}$$

For Grouse, $R_o = 10^{-2}$ to 10^{-4} For sheep R_o = approx. 1

3. Basic reproductive rate (R_o) in two host-one vector system. (After Rogers 1988.)

$$R_o = \frac{ce^{-uT}}{u} \left(\frac{a_1^2 b_1 m_1}{r_1} + \frac{a_2^2 b_2 m_2}{r_2} \right)$$

For Grouse + sheep, R_o = approx. 1

---

one vector system by considering first the fate of the infected vector and the rate at which it will bite individuals of both host species and secondly the rate at which infectious hosts infect the vectors. The sum of the rates at which vectors become infected from the two host species provides the value of R_o in the vector population and hence R_o of the total host population.

In the two host-one vector system the contribution from the Grouse is trivial and once again the value of R_o is dependent on accurate estimates of b_2 (proportion of infectious bites) and l/r_2 (duration of infection in sheep), but with the loss of virus from the sheep to the grouse the curve for $R_o = 1$ is moved to a slightly higher threshold (Fig. 20.7). In the phase plane diagrams of Aron and May (1982) this effectively elevates the equilibrium

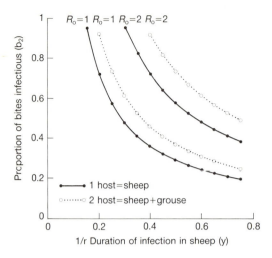

Fig. 20.7. The basic reproductive rate (R_o) of Louping-ill in the sheep-L1 system and the sheep-L1-grouse system in relation to the proportion of bites infectious and duration of infection in sheep. Grouse alone cannot sustain the disease (see text) but sheep can when the conditions allow $R_o \geqslant 1$; when both grouse and sheep are present some of the disease is lost from the system to the grouse and the threshold of $R_o = 1$ is increased.

prevalence line for the vector so producing a stable equilibrium for the disease (Fig. 20.8). Without sheep as a reservoir host the disease is unlikely to persist.

Vaccination effectively removes sheep as a reservoir host from the system and should result in the loss of LI from the Grouse population. Preliminary field studies show that this approach appears to have worked on moors sampled in England but not in Scotland (Table 20.5). According to the model, the reason for the disparity could be because of one or a combination of the following reasons:

1. *A higher proportion of the tick population bite Grouse in Scotland (a^2 greater)*. This could not be the sole reason since all the ticks would have to

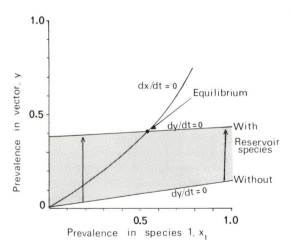

Fig. 20.8. Equilibrium conditions for a two host-one vector system after the introduction of a reservoir host (after Rogers 1988). The presence of the reservoir host increases the equilibrium prevalence line for the vector producing a stable equilibrium for the disease. Vaccination of the reservoir host should remove both the reservoir as an amplifier and the disease equilibrium.

Table 20.5 The effects of sheep vaccination on the prevalence of Louping-ill antibody in Grouse

	Sheep	
	Vaccinated	Not vaccinated
Percentage of Grouse with antibody:		
England	5%	35%
Scotland	60%	60%

feed on Grouse, which is presumably not so since a wider range of alternative hosts exist in Scotland.

2. *Scottish Grouse have developed a greater resistance to the disease than English Grouse and thus provide an alternative host reservoir that has the potential to remain infectious for a longer period (b large)*. It is possible that some Grouse populations have developed resistance to the disease and while there is anecdotal evidence to support this suggestion, the data is lacking. Furthermore this could not be the sole cause, since infections would need to last for several years and longer than the life expectancy of the Grouse.

3. *The ratio of vectors to Grouse is much higher in Scotland*. While Grouse populations are lower and tick populations are generally higher in Scotland this difference is insufficient to account for the difference.

4. *There is an alternative reservoir host not considered within the model*. While there is an association between the disease and the presence of Roe Deer in Scotland (Duncan, unpublished), tests conducted by Reid (1978) and Reid and Duncan (unpublished) at the Moredun Institute have eliminated Roe Deer, Red Deer, and Blue Hares as amplifiers of the disease.

5. *The vaccine is not eliminating the sheep as a reservoir host*. If the sheep population is still acting as a reservoir then the basic reproductive rate of the disease is greater than unity and the disease will have a stable equilibrium. The vaccination programme itself has been tested (Shaw and Reid 1981) although these workers found that double vaccination was necessary to provide complete protection, and a recent survey (Hudson unpublished) has shown that few sheep are vaccinated twice. A second source could be the lambs that amplify the disease after losing the protection of the maternal antibody.

In summary, the model indicates that Grouse cannot sustain the disease without a reservoir host. Vaccination should remove sheep as a reservoir host and result in the fall and final loss of the disease from the vector and

Grouse. This appears to have occurred in English populations but not in Scottish populations and we propose that in Scotland the disease may be amplified by sheep that are not vaccinated twice or through lambs after they lose the maternal antibody.

Implications for conservation

Parasites can reduce the fecundity and survival of their hosts in a way that can influence both population size and the pattern of population changes. The general approach taken in this paper has been to use analytical models to determine which of the life history parameters of parasite and host are important in determining the persistence and impact of a disease. The model is then tested experimentally and used to design control techniques which in turn provide the manager with further information about the system. Such models and the approach taken has interesting implications for conservation and management in general.

In the two systems presented the models, based on empirical data, have indicated effective control strategies which may not have been obvious without the application of the models. In the *T. tenuis* system the limitations of direct chemotherapy were quantified, and in the Louping-ill system the model implied that grouse alone cannot sustain the disease and by implication vaccination of the sheep, as reservoir host, should eliminate the virus from the Grouse. Naturally these models carry certain limitations, the major assumption being that food, predators, and other competitors show no regulatory effects. Furthermore the resolution of the host-parasite systems does not account for either financial costs or changes in community structure after the removal of a parasite (see Dobson and Hudson 1986). The development of any control strategy would need to balance ecological and economic costs of conservation against the benefits of either reducing prevalence or totally eradicating a parasite.

Some of the major problems to conservation of disease control relate to the presence of domestic animals acting as a reservoir host and amplifying the disease in wild, and in some instances endangered, animals. In India, Boshell (1969) described high mortality amongst monkeys (*Presbytis entellus* and *Macaca radiata*) after cattle were introduced into the monkeys' forest habitat; this increased abundance of the biting tick *Haemophysalis spingera* and consequently the rate of infection of the flavivirus causing Kysanur Forest Disease. As with the Grouse-LI system, control of ticks or vaccination in amplifying hosts could influence the dynamics within the wild host. Another example is provided by the arthropod-transmitted virus, Blue-tongue, which is present in domestic livestock limited the reintroduction of Arabian Oryx in parts of Oman (Jones 1982).

Distemper threatens what could be the last remaining population of

Black-Footed Ferrets (May 1986; Thorne and Williams 1988) and while managers make efforts to prevent transmission by vaccination there may be a case to look in more detail at the dynamics of this highly infectious disease. Since transmission occurs via direct contact, changes in group size and the behaviour of the reservoir hosts could help prevent transmission from dogs and other carnivores.

Man may also act as a reservoir host, for instance in Indonesia the presence of Tuberculosis, *Mycobacterium tuberculosis*, in Man has prevented reintroduction of Orang-Utans (*Pongo pygamaeus*) (Jones 1982). Agriculturalists often have different perspectives on disease and consider wild animals as the reservoir hosts which amplify the disease and cause problems to livestock. While this view is perhaps valid in the control of Bovine Tuberculosis (Anderson and Trewhella 1985) transmitted to Cattle from Badgers *Meles meles*, there are possibilities for developing alternative control methods which are less destructive to the wild host population than the gassing of Badgers.

In summary, parasites have a serious and important impact on the population dynamics of their host, and both ecologists and conservationists should consider parasitism as one of several regulatory factors affecting population dynamics. The development of analytical models has provided an insight into the dynamics of host-parasite relationships and also helped to identify areas for the development of control methods.

Summary

The development of analytical models describing the dynamics of host-parasite interactions have stimulated studies of infections of wild animal populations and indicated possible control techniques. It is apparent that parasites influence the size of the host population and the pattern of population changes and this is illustrated with reference to Red Grouse (*Lagopus lagopus scoticus*).

The caecal nematode *Trichostrongylus tenuis* causes a reduction in the breeding production of Grouse sufficient to account for the regular cyclic fluctuations in numbers. The control of *T. tenuis* has centred around chemotherapy; direct chemotherapy has proved successful but is of no long-term benefit while initial results from indirect chemotherapy, using grit coated with a fat and incorporating an anthelmintic, are encouraging.

Louping-ill is a viral infection of sheep and Grouse which can cause serious losses to both populations. Examination of the dynamics of the disease indicates that Grouse cannot sustain the disease in the absence of sheep which act as a reservoir host. The vaccination of sheep should remove Louping-ill from Grouse; this strategy has been successful in some areas but has failed as effective control measure in parts of Scotland. It is suggested that vaccination of sheep has failed to prevent sheep amplifying the virus to Grouse. The relevance of this approach to the conservation and management of wild animals is discussed.

Acknowledgements

We wish to thank the land-owners in England and Scotland for allowing us to collect samples and Dr Hugh Reid for conducting analysis of LI antibody at the Moredun Research Institute. It is a pleasure to thank David Newborn, John Renton, Graeme Dalby, and Flora Booth for field assistance. Kate Lessels greatly improved an earlier draft of the manuscript.

References

Anderson, R. M. (1982a). *Population dynamics of infectious diseases: theory and applications*. Chapman and Hall, London.

Anderson, R. M. (1982b). Epidemiology. In *Modern parasitology: A textbook of parasitology* (ed. F. E. G. Cox), pp. 204–51. Blackwell Scientific Publications, Oxford.

Anderson, R. M. and May, R. M. (1978). Regulation and stability of host-parasite population interactions: I Regulatory processes. *Journal of Animal Ecology*, **47**, 219–49.

Anderson, R. M. and Trewhella, W. (1986). Population dynamics of the badger (*Meles meles*) and the epidemiology of bovine tuberculosis (*Mycobacterium bovis*). *Transactions of the Royal Society of London [B]*, **310**, 327–81.

Aron, J. L. and May, R. M. (1982). The population dynamics of malaria. In: *Population dynamics of infectious diseases* (ed. R. M. Anderson), pp. 139–79. Chapman and Hall, London.

Beasley, S. J., Campbell, J. A., and Reid, H. W. (1978). Threshold problems for louping ill in *Ixodes ricinus*. *Proceedings of the International Conference on Tick-borne Diseases and their Vectors. Edinburgh*, pp. 497–500.

Boshell, M. J. (1969). Kysanur forest disease: ecological considerations. *American Journal of Tropical Medicine and Hygiene*, **18**, 65–80.

Coulson, J. C. (1988). The structure and importance of invertebrate communities on peatlands and moorlands, and effects of environmental and management changes. In *Ecological change in the uplands* (eds. M. B. Usher and D. B. A. Thompson), pp. 365–80. Blackwell Scientific Publications, Oxford.

Crofton, H. D. (1971). A quantitative approach to parasitism. *Parasitology*, **62**, 179–93.

Davies, N. B. (1978). Ecological questions about territorial behaviour. In *Behavioural ecology. An evolutionary approach* (eds. J. R. Krebs and N. B. Davies), pp. 317–50. Blackwell Scientific Publications, Oxford.

Dobson, A. P. (1988). Restoring island ecosystems: the potential of parasites to control introduced mammals. *Conservation Biology*, **2**, 31–9.

Dobson, A. P. and Hudson, P. J. (1986). Parasites, diseases and the structure of ecological communities. *Trends in Ecology and Evolution*, **1**, 11–15.

Dobson, A. P. and May, R. M. (1986). Patterns of invasions by pathogens and parasites. In *Ecology of biological invasions of North America and Hawaii* (eds. H. A. Mooney and J. A. Drake), pp. 58–76. Springer-Verlag, New York.

Duncan, J. S., Reid, H. W., Moss, R., Phillips, J. D. P., and Watson, A. (1979). Ticks, Louping-ill and Red Grouse on moors in Speyside, Scotland. *Journal of Wildlife Management*, **43**, 500–5.

Hudson, P. J. (1984). Some effects of sheep management on heather moorlands in the north of England. In *The impact of agriculture on wildlife and semi-natural habitats* (ed. D. Jenkins), pp. 143–9. I.T.E. Symposium No. **13**, Abbots Ripton.

Hudson, P. J. (1986*a*). *Red Grouse: The Biology and management of a wild gamebird*. The Game Conservancy Trust, Fordingbridge.

Hudson, P. J. (1986*b*). The effect of a parasitic nematode on the breeding production of Red Grouse. *Journal of Animal Ecology*, **55**, 85–92.

Hudson, P. J. and Dobson, A. P. (1988). The ecology and control of parasites in gamebird populations. In *Ecology and management of gamebirds* (eds. P. J. Hudson and M. R. W. Rands), pp. 98–133. Blackwell Scientific Publications, Oxford.

Hudson, P. J. and Dobson, A. P. (1989). Population biology of *Trichostrongylus tenuis*, a parasite of economic importance for red grouse management. *Parasitology Today*, **5**, 283–91.

Hudson, P. J. and Dobson, A. P. (1990). The direct and indirect impact of the caecal nematode *Trichostrongylus tenuis* on Red Grouse. In *Bird-parasite interactions: ecology, evolution and behaviour* (ed. J. Loye and M. Zuk), pp. 49–68. Oxford University Press, Oxford.

Hudson, P. J., Dobson, A. P., and Newborn, D. (1985). Cyclic and non-cyclic populations of Red Grouse: a role for parasitism? In *Ecology and genetics of host-parasite interactions* (eds. D. Rollinson and R. M. Anderson), pp. 79–86. Academic Press, London.

Jenkins, D., Watson, A., and Miller, G. R. (1963). Population studies on Red Grouse *Lagopus lagopus scoticus* (Lath.) in north-east Scotland. *Journal of Animal Ecology*, **32**, 317–76.

Jones, D. M. (1982). Conservation in relation to disease in Africa and Asia. *Symposium of the Zoological Society of London*, **50**, 271–85.

Lack, D. (1954). *The natural regulation of animal numbers*. Oxford University Press, Oxford.

Lack, D. (1966). *Population studies of birds*. Oxford University Press, Oxford.

Lack, D. (1968). *Ecological adaptations for breeding in birds*. Methuen, London.

May, R. M. (1986). The cautionary tale of the black-footed ferret. *Nature*, **320**, 13–14.

May, R. M. and Anderson, R. M. (1978). Regulation and stability of host-parasite interactions. II. Destabilising processes. *Journal of Animal Ecology*, **47**, 249–67.

Potts, G. R., Tapper, S. C., and Hudson, P. J. (1984). Population fluctuations in Red Grouse: analysis of bag records and a simulation model. *Journal of Animal Ecology*, **53**, 21–36.

Reid, H. W. (1975). Experimental infection of Red Grouse with Louping-ill virus (Flavivirus group). I. The viraemia and antibody response. *Journal of Comparative Pathology*, **85**, 223–9.

Reid, H. (1978). The epidemiology of Louping-ill. *Proceedings of International Conference on Tick-borne diseases and their vectors. Edinburgh*, pp. 501–6.

Reid, H. W., Duncan, J. S., Philips, J. D. P., Moss, R., and Watson, A. (1978). Studies on Louping-ill virus (Flavivirus group) in wild Red Grouse (*Lagopus lagopus scoticus*). *Journal of Hygiene*, **81**, 321–9.

Rollinson, D. and Anderson, R. M. (1985). *Ecology and genetics of host-parasite interactions*. Academic Press, London.

Rogers, D. J. (1988). A general model for African trypanosomiasis. *Parasitology*, **97**, 193–212.

Scott, M. E. (1988). The impact of infection and disease on animal populations: implications for conservation biology. *Conservation Biology*, **2**, 40–56.

Shaw, B. and Reid, H. W. (1981). Immune responses of sheep to Louping-ill virus vaccine. *Veterinary Record*, **109**, 529–31.

Shaw, J. (1988). Arrested development of *Trichostrongylus tenuis* as third stage larvae in Red Grouse. *Research in Veterinary Parasitology*, **45**, 256–8.

Thorne, E. T. and Williams, E. S. (1988). Disease and endangered species: the black-footed ferret as a recent example. *Conservation Biology*, **2**, 66–74.

Watson, H. (1988). The ecology and pathology of *Trichostrongylus tenuis* (Nematoda), a parasite of Red Grouse (*Lagopus lagopus scoticus*). Unpublished Ph.D. Thesis, University of Leeds.

Watson, H., Lee, D. L., and Hudson, P. J. (1987). The effect of *Trichostrongylus tenuis* on the caecal mucosa of young, old and anthelmintic-treated wild Red Grouse, *Lagopus lagopus scoticus*. *Parasitology*, **94**, 405–11.

21 Population and contamination studies in coastal birds: the Common Tern Sterna hirundo

PETER H. BECKER

Introduction

Since the industrial revolution mankind has been introducing substantial quantities of various foreign substances into the environment. This is done both intentionally, e.g. to eliminate undesired organisms by the use of biocides, and unintentionally or, rather, due to lack of proper forethought whereby various substances are emitted into the environment in the course of energy or chemical production: or through uncontrolled release of garbage or oil, perhaps from oil platforms or tankers. Nowadays no continent, sea or ecosystem is unaffected by pollution.

Nor are birds spared the effects of pollution. Birds have often played an important role as indicators of environmental problems. Thus plastics in the stomach of seabirds indicated the increased contamination of the sea with rubbish (e.g. Bourne 1976; Furness 1985). Similarly, mass mortality of marine birds through oiling drew attention to the marine oil pollution problem (e.g. Bourne 1976; Reineking and Vauk 1982; Dunnet 1982). Further, the massive mortality in several bird species, together with reproductive losses or population declines, led to public awareness of the chemical contamination problems. Furthermore, biocides can affect birds indirectly by the destruction of habitats or food supply, leading to reduction or extinction of bird populations (e.g. Morrison 1986).

Why is it that birds so often seem to provide the first warning of environmental problems? The answer to this question is that birds are conspicuous organisms within an ecosystem and relatively easy to observe; they are the object of considerable attention, so discernible changes in their biology seldom remain unnoticed. Furthermore, being top predators many birds are especially threatened by toxic chemicals and persistent compounds are accumulated to such high levels that they lower reproductive output or, even, cause death.

It was just after the second world war that environmental chemicals were observed to affect the populations of several bird species in many parts of the world. Persistent pesticides such as DDT and its metabolites, dieldrin, telodrin, endrin, or organic-mercury compounds, led to increased mortality or to reduced reproductive success and to declines in population size.

An immense number of papers and reviews report the effects of toxic chemicals on birds. These effects are known from field and laboratory studies to vary greatly from species to species (e.g. Ratcliffe 1970; Prestt and Ratcliffe 1972; Stickel 1973, 1975; Ohlendorf et al. 1978; Fimreite 1979; Koeman 1979; Newton 1979; Prinzinger and Prinzinger 1979; Peakall 1972, 1980; Risebrough 1986; Fleming et al. 1983; NERC 1983; Eisler 1986). In addition to direct lethal poisoning of adult birds by pollutants, many sublethal effects have been identified, in particular during reproduction. These include failed egg production, delayed breeding, lowered clutch size, reduced eggshell quality, egg cracking, lowered fertility, mortality of embryos or hatched young, and abnormalities or aberrant behaviour of adults or young, all of which result in reductions of reproductive output.

Several recent reviews deal with the observed correlations between toxic residues and changes in population size (e.g. Morrison 1986, Risebrough 1986; Evans and Nettleship 1985). In industrialized countries, the 1970s were generally characterized by a decline in the impact of a number of dangerous persistent pesticides, resulting in an improved environmental situation accompanied by a recovery in bird numbers. The situation in the Third World, however, where major pesticide problems are now being created, is vastly different (Risebrough 1986).

Relationships between environmental chemicals and population trends are well-documented in, for example, the Peregrine Falcon *Falco peregrinus* (Hickey and Anderson 1969), the Brown Pelican *Pelicanus occidentalis* (e.g. Blus 1982; Anderson and Gress 1983), the Double-crested Cormorant *Phalacrocorax auritus* (Gress et al. 1973; Anderson and Gress 1983; Pearce et al. 1979), the Northern Gannet *Sula bassana* (see Evans and Nettleship 1985), and—in Europe—in the Sparrowhawk *Accipiter nisus* (Ratcliffe 1970; Newton 1981, 1986; Wallin 1984), the Peregrine (Ratcliffe 1970, 1980), the White-tailed Eagle *Haliaeetus albicilla* (Helander 1977), and in several species breeding on the Wadden Sea coast (Koeman et al. 1967a; Koeman 1971; Swennen 1972; Becker and Erdelen 1987).

This chapter is concerned with examples of bird populations on the Wadden Sea. The first aim is to show the value of coastal breeding birds as indicators and monitors of the contamination of coastal waters; and secondly to highlight some obstacles to the long-term monitoring of avian contamination with environmental chemicals and of their effects. Such monitoring is urgently required for reasons of species protection. The problems referred to are thrown up in the analysis of (i) population trends, (ii) residue levels in birds, and (iii) their reproductive success.

Trends in bird populations on the Wadden Sea coast in relation to environmental chemicals

On the Dutch coast, from 1964 to 1968, an enormous mortality was observed in terns, Herring Gulls *Larus argentatus*, Eiders *Somateria mollissima*, and other bird species. Pesticides such as dieldrin, telodrin and endrin—cyclodienes of high toxicity (e.g. Ohlendorf *et al*. 1978; Hudson *et al*. 1984)—caused mortality of adults and young. Eggshell thinning probably induced by elevated levels of DDE (a DDT metabolite) also occurred but, at least in the Sandwich Tern *Sterna sandvicensis*, no obvious hatching failure was observed. Koeman *et al*. (1967*a*) and Koeman (1971) were able to show that the pesticides originated in effluents from a factory near Rotterdam.

The consequence of pollution by chemicals was a steep decline in Dutch seabird populations, especially of Sandwich Terns and Eiders (for reviews see Koeman 1971; Duinker and Koeman 1978; Rooth 1980*a*, *b*; Smit 1981). In Eiders only the females were affected, as a consequence of marked weight loss during incubation; mortality diminished with increasing distance of the breeding site from the source of the pollution (Swennen 1972). Likewise, in the area of the German Wadden Sea, tern eggs contained dieldrin and telodrin in concentrations comparable with those in the Dutch birds (Table 21.1; cf. Koeman *et al*. 1976*a*; Koeman 1971). In several eggs endrin and dieldrin were present at levels which, in other species, had proved detrimental to breeding success (endrin *c*. 0.3 ppm: e.g. Fleming *et al*. 1982; Blus 1982; Blus *et al*. 1983; dieldrin *c*. 0.5 ppm: e.g. Blus *et al*. 1974; Blus 1982; Lockie *et al*. 1969).

An analysis of the most important breeding sites along the German Wadden Sea coast in 1950–79 revealed that during the mid-1960s conspicuous decreases or low breeding populations were noted in most species of seabirds, waders, and ducks (Becker and Erdelen 1987; Fig. 21.1). These trends also seemed to be related to the pollution in the Netherlands, thus showing the wide-ranging effects of a local marine pollution incident.

Apparently terns are more threatened than other seabirds, as their bodies accumulate greater burdens of toxic pollutants due to their position in the food web, and also because they are more sensitive to pollutants (see below; Gilbertson *et al*. 1976; Pearce *et al*. 1979). The tern populations began to decline in the early 1950s (Koeman 1971; Rooth 1980*a*, *b*; Becker and Erdelen 1987; Fig. 20.1); in the Sandwich Tern higher mortality rates were recorded before the 1960s (Koeman 1971). These

Table 21.1 Organochlorines in eggs of terns (Common Tern *Sterna hirundo*, and Sandwich Tern *Sterna sandvicensis*, 1974 only) breeding on the Wadden Sea island of Minsener Oldeoog, 1965–86. Data (mg/kg fresh weight) are of geometric mean and range

Year	No. of eggs	Dieldrin	Endrin	p,p'-DDE	Reference
1965	14	0.70 (0.31–3.50)	0.24 (0.03–1.60)	0.54 (0.25–1.50)	Koeman et al. (1967a)
1966	6	0.29 (0.08–0.55)	0.08 (0.04–0.13)	0.44 (0.09–0.86)	Koeman et al. (1967a)
1968	8	0.58 (0.02–2.60)	0.07 (0.02–0.63)	0.32 (0.14–0.98)	Blaszyk (1972)
1974 S.T.	4	0.14 (0.09–0.27)	n.d.	0.54 (0.36–0.97)	Goethe unpublished
1981	10	0.017 (0.010–0.036)	n.d.	0.106 (0.053–0.206)	Becker et al. (1985a)
1986	10	0.017 (0.001–0.037)	n.d.	0.118 (0.051–0.191)	Becker et al. (1988)

S.T. = Sandwich Tern, n.d. = not detected (1981 and 1986 = < 0.001 mg/kg).

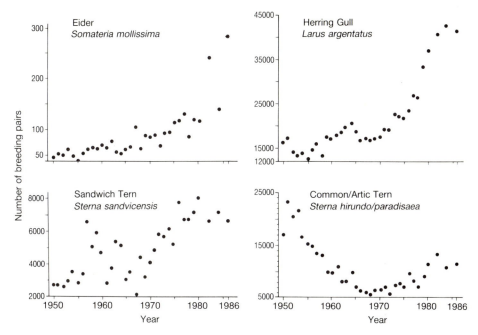

Fig. 21.1. Trends from 1950 to 1986 in numbers of pairs of seabird species breeding at 28 important sites on the German Wadden Sea coast. See Becker and Erdelen (1987) for location of sites, methods, data sources for 1950–80, and trends of other species until 1979. The population totals of 1982 and 1984 originate from Taux (1984, 1986), of 1986 from the Niedersächsisches Landesverwaltungsamt and the Landesamt für den Nationalpark Schleswig-Holsteinisches Wattenmeer. The trend for Common Tern corresponds to that of Common/Arctic tern.

earlier decreases seem to reflect the rapid expansion of the chemical industry, especially of pesticide production, after the Second World War. In addition, the continuing loss of breeding habitats due to natural succession and to direct human influences may have contributed to the decline in the Wadden Sea (Becker and Erdelen 1987).

In the Netherlands the pesticides industry reacted to the findings (published later by Koeman *et al.* 1967*a*), by stopping production of telodrin in the autumn of 1965 and building a new purification plant (Koeman 1971; Duinker and Koeman 1978). Thereafter the concentrations of these pesticides in the environment decreased gradually, mortality rates fell, and bird populations increased once more (Rooth 1980*a, b*); the same held true for the German Wadden Sea (Becker and Erdelen 1987; Table 21.1, Fig. 21.1). Since the early 1970s, the use and/or production of many of the persistent chemicals have been prohibited in Europe (e.g.

Clausing 1986) and filtration plants for the treatment of factory effluents have been installed or improved resulting, by the 1980s, in a lowering of the impact of pesticides on the environment (e.g. Coulson *et al*. 1972; Becker *et al*. 1980, 1985*a*) and in a recovery of the bird populations (raptors: e.g. Ellenberg 1981; Newton 1986; Risebrough 1986; coastal birds: Fig. 21.1; Becker and Erdelen 1987).

The continuing upward trends in the numbers of most species of seabirds and waders in the Wadden Sea area (Fig. 21.1) would seem to rule out permanent detrimental effects through present-day environmental contamination. Other factors now favour coastal breeding birds such as improvements in the food supply due to eutrophication (e.g. Beukema and Cadée 1986; Gerlach 1987) and increased protection through nature conservation measures (for further discussion see e.g. Vauk and Prüter 1987; Becker and Erdelen 1987; Dunnet *et al*. 1990). For terns, however, the scarcity of suitable breeding habitats seems to limit further increase in the breeding populations in the Wadden Sea.

These results illustrate how severe the impact of environmental chemicals have been on coastal bird populations. In the interests of conservation and to prevent similar crashes in bird numbers in the future, a great advantage would be the opportunity to recognize threats posed by environmental chemicals at an early stage. One possibility is to use the birds themselves as indicators of environmental pollution. However, how far the study of population changes is helpful in this respect is not known, especially those of the threatened tern species.

Population development is a function of natality, mortality, immigration, and emigration, each of which is influenced by a multitude of factors. Fluctuations in population size cannot be explained by any single factor without corroboration of simultaneous effects on each of the components that make up the structure of the population. The massive declines in raptor populations stimulated detailed enquiries into the causes and led eventually to the discovery of the pollution problems (see 'Introduction'). In other species adult mortality or irregularities in reproductive biology were the factors that prompted further investigations.

After reductions in natality of K-selected, long-lived species with a very late age of first breeding, the first alterations in population size may not be detectable for many years. Thus the effects of toxic pollutants may only be recognized *a posteriori* (Morrison 1986). The coincidental or correlative relationships that have been drawn between levels of toxic chemicals and population change can in most cases be regarded only as circumstantial evidence.

Population studies in migrating colonial species (e.g. terns)—characterized by high tendencies to move between breeding sites—are practically impossible because:

(1) there are problems in arriving at mortality rates from the analysis of ringing recoveries (e.g. Kakhani and Newton 1983; Anderson *et al.* 1985);

(2) mortality during wintering and migration is not sufficiently well-documented; and

(3) estimating mortality from the percentage of adults rebreeding in the following year is difficult for methodological reasons: extensive trapping and retrapping of ringed adults in several major colonies would cause unacceptable disturbance and lead to trap shyness (Nisbet 1978).

Methodological problems also hinder the recording of the annual size of the breeding population in colonial species (e.g. Nettleship 1976; Harris and Lloyd 1977; Becker and Nagel 1983; Wanless and Harris 1984; Hatch and Hatch 1988). In terns resettling creates large-scale population monitoring requirements (Becker and Erdelen 1987). The proportion of breeding to non-breeding adults is also difficult to establish and seems to depend, among other things, on the food supply (Furness 1987, 1988).

As a result, studies of population structure in themselves are not very helpful in detecting any on-going adverse effects of toxic chemicals without additional corroboration. Nevertheless, the recording of bird numbers can help to pinpoint any shifts in population parameters caused by environmental changes and research into the causes can be initiated in advance of population declines that are likely to be particularly destructive to the survival of the species.

The recent contamination of seabirds in the Wadden Sea

Intersite and interspecific differences

Despite the potential value of seabirds for providing advance warning of changes in the environment, very few North Sea studies are concerned with interspecific, intersite, or year-to-year differences in toxic contamination of seabirds. A large number of valuable but isolated results are not comparable with one another because of methodological inconsistencies (methods of chemical analysis, selection of compounds, species, and tissues, limited geographical coverage, sample size, etc.; see also NERC 1983). In 1981 we therefore performed the first intensive study covering the German coast to establish definitively an association between the contamination of eggs in sea- and shorebirds with the presence of organochlorines and mercury (Becker *et al.* 1985*a*, *b*). Particular care was taken to establish intersite and interspecific differences through a comparison of 7 species and 7 regions along the coast.

As expected from the trophic levels of their prey, terns had the highest

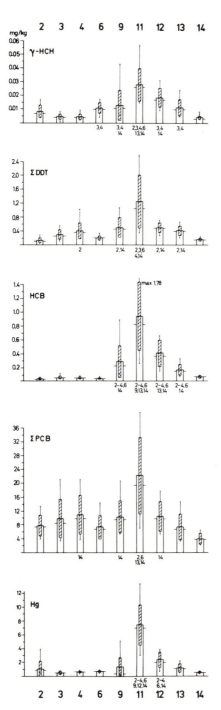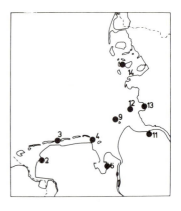

Fig. 21.2. Geographic variation in concentrations of organochlorines and mercury in eggs of Common Terns from the German North Sea coast in 1987. Residue levels are given in mg/kg wet weight as mean (column) $\pm$ SD (hatched column), geometric mean (horizontal line), and range (vertical line). Ten eggs were analysed per area (one egg/clutch). Area numbers below the columns indicate significant intersite differences (at least $P < 0.05$, Scheffé test with $\log_{10}$ of the residue levels; in each case the numbers are given only below the higher of the two contaminated areas compared).

concentrations of pollutants in their eggs (Fig. 21.3). Industrial chemicals, such as polychlorinated biphenyls (PCBs), hexachlorobenzene (HCB), or mercury were found in greatest quantities, as confirmed by other studies (e.g. Furness and Hutton 1980; Bergstrøm and Norheim 1986; Heidmann *et al.* 1987a; Focardi *et al.* 1988). The pesticide levels had been decreasing steadily since the mid-1960s and by 1981 were detected at considerably lower concentrations (Table 21.1). Pronounced regional differences clearly showed that the pollution of the birds originates to a great extent from the breeding area (see Fig. 21.2). Also migratory species, such as terns, are therefore likely to be valuable indicators of specific localized environmental pollution at the nesting colony (Custer *et al.* 1983; Nisbet and Reynolds 1984; Becker *et al.* 1985a; Struger and Weseloh 1985). Inshore feeding species such as terns and gulls, and shorebirds, such as the Oystercatcher *Haematopus ostralegus* take large amounts of food—essential for egg production—from the breeding grounds adjacent to their breeding sites (for courtship feeding in the Common Tern see Nisbet 1977). Thus the accumulation of pesticides in the eggs during their development reflects contamination of the local food web. In the Wadden Sea, the intersite variation of egg residue levels corresponds to that of marine invertebrates and fish (RSU 1980).

These results show that birds whether breeding, resting, or wintering in the inner German Bight are particularly threatened by toxic residues. It is precisely in this region that terns breed in significant numbers (Taux 1986), as do Kittiwakes *Rissa tridactyla* and Guillemots *Uria aalge* on the one and only cliff off the southern North Sea coast, on Helgoland.

Monitoring contaminants in seabirds

In view of the enormous problems with pollution in the North Sea it would seem sensible to use seabirds to monitor chemical contamination in the marine environment. However, few studies of seabirds in the North Sea allow a comparison of concentrations of toxic chemicals in birds between different years (NERC 1983; Moksnes and Norheim 1986; Becker *et al.* 1988; Becker 1989). These studies, like those of Olsson and Reutergårdh (1986) and Focardi *et al.* (1988), revealed great year-to-year variability showing that samples taken from a long run of years are required in order to detect trends in contamination.

As a result of these shortcomings, a 3-year monitoring project was set up in 1987 in cooperation with the Chemical Institute of the Tierärztliche Hochschule Hannover and funded by the Umweltbundesamt. The Common Tern and the Oystercatcher were selected as monitor species, being quite dissimilar with respect to their position in the food web and in their migration habits. The breeding biology of both species is well

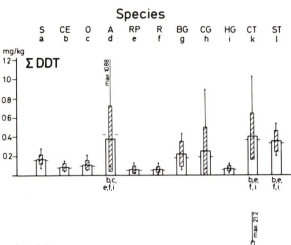

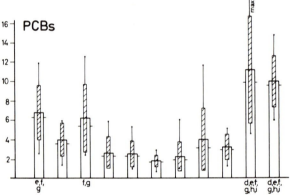

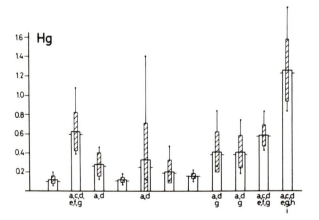

Fig. 21.3. Concentrations of Σ DDT, PCBs and mercury in eggs of 11 coastal bird species breeding on Jade Bay (areas 4–6, Fig. 21.2) in 1987. Residue levels are given in Fig. 21.2; 10 eggs were analysed per species. The numbers below the columns indicate significant differences (see Fig. 21.2) between the species: (a) Shelduck *Tadorna tadorna*, (b) Common Eider *Somateria mollissima*, (c) Oystercatcher *Haematopus ostralegus*, (d) Avocet *Recurvirostra avosetta*, (e) Ringed Plover *Charadrius hiaticula*, (f) Redshank *Tringa totanus*, (g) Black-headed Gull *Larus ridibundus*, (h) Common Gull *Larus canus*, (i) Herring Gull *Larus argentatus*, (k) Common Tern *Sterna hirundo*, (l) Sandwich Tern *Sterna sandvicensis*. In the case of the Avocet, the mean was calculated without the highest value.

documented, as is their feeding and population biology. Egg sampling was chosen for methodological and scientific reasons (e.g. Ohlendorf *et al*. 1978; NERC 1983; Mineau *et al*. 1984; Gilbertson *et al*. 1987; Becker *et al*. 1985*a*, 1989; Becker 1989). Nine to fourteen sample sites along the German coast were used such that the effects of the transport of pollutants into the North Sea by the three great rivers Elbe, Weser, and Ems could be recorded (Fig. 21.2). Ten eggs per species per area per year were analysed for mercury and organochlorine residues, including single congeners of the PCBs (for methods used see Becker *et al*. 1985*a*, *b*; Heidmann 1986; Heidmann and Büthe 1986).

The geographic variation of residue levels in Common Tern eggs sampled in 1987 is shown in Fig. 21.2. Eggs from the Elbe estuary and the inner German Bight were most contaminated owing to the input of the river Elbe. The high levels of PCBs and mercury in eggs from the Elbe are still giving cause for considerable alarm (see below). In all contaminants the input from the Elbe into the North Sea, its increasing dilution, and displacement owing to the currents flowing from west to north-east can be traced by the contamination levels of the terns' eggs (areas 11–14, Fig. 21.2). Thus the concentration in eggs from Norderoog (area 14) are at the same low level as in those from the East Frisian Islands. The geographic pattern of concentrations of polluting chemicals in the Oystercatcher is comparable to that of the Common Tern (see Becker 1989). However, additional sample sites at the Weser estuary provided information showing that the Weser, too, is contributing to the load of γ-HCH, ΣDDT, and PCBs in the German Bight. A comparison with the results of 1981 and 1986 revealed extreme variations in the residue levels among years and showed that 1987 was characterized by the highest levels of most contaminants (Becker 1989), emphasizing the necessity·of long-term studies to verify trends in contamination.

In addition to the large scale, 3-year monitoring programme, some other aspects of basic interest are also being studied. Thus the interspecific variation in contamination in 11 species was evaluated at the Jade (Fig. 21.3). For γ-HCH, HCB, and octachlorstyrol the Common and Sandwich Terns again are the most heavily-polluted species in accordance with their high position in the food chain. The ultimate aim of the project is to produce a basis for a programme monitoring environmental chemicals on the North European coast using suitable seabird species.

Problems of monitoring chemicals

One of the remaining problems for this method of monitoring chemicals is the restriction to a small number of selected chemicals, occasioned by the high cost of chemical analyses and the present state of toxicological knowledge (e.g. Koeman 1979). For example we still analyse for some 'old' pesticides, which were dangerous years ago, but are now obviously no longer of relevance to the environment in the case of the Wadden Sea (e.g. dieldrin or DDT). In consequence, the occurrence and accumulation in animals of new, potentially dangerous, chemicals threatens to remain undetected (e.g. organo-phosphate pesticides; White *et al*. 1983; Henny *et al*. 1985; polychlorinated napthalenes; Brinkman and Reymer 1976). If production or deployment of such toxic chemicals is known, a minimum number of samples certainly ought to be analysed as routine (NERC 1983; Fleming *et al*. 1983).

Another problem is the toxicological importance to birds of the present residue levels. Are the seabirds in fact adversely affected by the observed concentrations of toxic residues? In particular, is their mortality elevated or their reproductive success impaired? These questions cannot be answered by monitoring chemicals. According to information stemming mostly from species other than seabirds, from laboratory research, or from a small number of field studies, the concentrations of the insecticides DDT and its metabolites, dieldrin, endrin, and lindane found in eggs originating from the German coast in the 1980s were far below levels critically affecting breeding success (see e.g. Ohlendorf *et al*. 1978; Becker *et al*. 1985*a*).

Research into the effects of chemicals on the breeding biology of seabirds and waders in captivity is particularly difficult to carry out. However, there are few field studies to detect correlations between contamination and reproductive biology of such species in the North Sea area (e.g. Furness and Hutton 1979, 1980) and elsewhere. Harris and Osborn (1981) dosed North Sea Puffins *Fratercula arctica* with 30–35 mg PCB (Aroclor 1254) by implantation: survival and breeding performance were not impaired, although the PCB concentrations in fat, tissues, and eggs (9.6–81.3 mg/kg) were much higher for some years than in undosed conspecifics. Caspian Tern *Sterna caspia* reproduction was normal with PCB egg-levels in the range 24.7–27.8 ppm (Lake Michigan; Struger and Weseloh 1985). Hatching success seemed not to be reduced with 92 ppm PCBs in Herring Gull eggs (Jørgensen and Kraul 1974) but was reduced at 142 ppm (Gilman *et al*. 1977). These results are in contradiction to those of laboratory studies, although carried out on different species, which showed that concentrations of more than 5 ppm PCB in the egg (fresh weight) could depress breeding success (Bush *et al*. 1974; for further literature see Stickel 1975; Peakall 1972, 1980; Ohlendorf *et al*. 1978). In most eggs of sea- and shorebirds breeding on the Wadden Sea coast, PCB concentrations reached or exceeded this range (Common Tern, Elbe estuary 1987: 22.3 mg/kg, Fig. 21.2), so negative effects cannot be excluded. In all species studied, the mercury concentrations in most eggs exceeded 0.5 mg/kg, thus potentially endangering breeding success (see

Figs. 21.2, 21.3; in Becker *et al*. 1985*b*; cf. Ohlendorf *et al*. 1978; Fimreite 1979). However, it must be stressed that different species react differently to any particular contaminant, and extrapolations from one species to another must be regarded with considerable caution.

Unfortunately little is known about the interactive effects of different contaminants (Risebrough 1986), a problem that will probably remain with us, owing to the vast number of chemical combinations possible in the environment and in the birds. The limitations to contamination studies may be overcome to some extent by additional field-work on breeding biology in relation to concentrations of toxicants.

The breeding success of Common Terns in the Wadden Sea

The breeding success of birds is particularly sensitive to certain toxic residues (e.g. Robinson *et al*. 1967; Koeman *et al*. 1967*b*; Anderson and Hickey 1976; Newton 1979; Prizinger and Prinzinger 1979; Hartner 1981; Risebrough 1986). To study the effects of different environmental factors on breeding biology, a project on the Common Tern was initiated at the beginning of the 1980s. This species appeared ideal from several points of view:

(1) numbers of Common Tern declined during the 1950s and 1960s in the Wadden Sea area as a result of environmental chemicals;
(2) the diet is almost entirely fish, thus toxic residues are concentrated more highly than in some other species;
(3) Common Terns seem to react more strongly to toxic residues than gulls (e.g. DDE: Pearce *et al*. 1979);
(4) almost nothing was known about the reproductive success of Common Tern on the Wadden Sea coast although basic information was available from similar studies in other countries (e.g. Nisbet and Drury 1972; Morris *et al*. 1976; Erwin and Smith 1985; and
(5) over 9000 pairs breed along the German North Sea coast.

Five colony sites on the Jade Bay (Fig. 21.4) were selected for study and classified with respect to their geographic location—on Wadden Sea islands (Wangerooge, Minsener Oldeoog, Mellum) or on the mainland coast (Wilhelmshaven, salt-marsh of Augustgroden)—and in relation to differences in climate and food supply. The latter is enriched for coastal breeders by additional food from freshwater sources inland, which are not affected by the tides (Becker *et al*. 1987; Becker and Frank 1988). The colonies are exploited by different predators (Mellum: Herring Gulls; Wangerooge: Black-headed Gulls *Larus ridibundus*; Oldeoog; Short-eared Owl *Asio flammeus*; Wilhelmshaven harbour: Brown Rats *Rattus norvegicus*), suffer different amounts of flooding (salt-marshes of

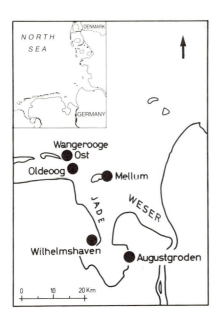

Fig. 21.4. The location of the Common Tern study colonies on Jade Bay (Lower Saxon Wadden Sea, Federal Republic of Germany).

Wangerooge and Augustgroden) and are of different sizes (Oldeoog 1500–3700 breeding pairs; other colonies <250 pairs). Data on breeding biology were gathered by checks either of sample areas (Oldeoog, Wangerooge) or of the whole colonies, generally every other day during the breeding season. To get information on fledging success, samples of clutches were fenced (Nisbet and Drury 1972; Erwin and Custer 1982; Becker and Finck 1986).

Table 21.2 shows the results of seven breeding seasons on Minsener Oldeoog as an illustration of the large annual fluctuations in reproduction, caused mainly by variations in chick mortality. Apart from 1987, when Brown Rats killed many of the chicks, losses were chiefly due to the effects of weather and food shortage (for details see Becker and Finck 1985, 1986). The hatching success was relatively constant apart from 1986 and 1987, the years of high predation.

Other than on Minsener Oldeoog and at Augustgroden, flooding had as much influence on chick production as did weather and food availability (Table 21.3), followed in importance by predation, chiefly on eggs (for details see Becker and Anlauf 1988*a*, *b*). Without the predation and flooding losses, the hatching success in Augustgroden from 1982 to 1986 would have been about 87–96 per cent. Comparable results were obtained from Wangerooge (cf. Großkopf 1989). Until 1983 Common Terns bred adjacent to a big Herring Gull colony on Mellum island. A very small number of fledged chicks was recorded during the breeding seasons of

Table 21.2 Breeding success of Common Terns on the Wadden Sea island of Minsener Oldeoog from 1981 to 1987. Late nests are excluded. 'Other causes' of chick losses were mainly weather and food shortage. For methods and details see Becker and Finck (1985, 1986)

Year	No. of clutches studied	Clutch size (n eggs)	Hatching success (n chicks)	Fledging success (n chicks >18 days)	Chicks fledged per clutch		Losses (% of eggs) Predation	Other causes
1981	94	2.9 (268)	90% (241)	18% (43)	0.46	eggs:	5	5
						chicks:	4	70
1982	74	2.6 (192)	86% (165)	43% (71)	0.96		3	11
							14	35
1983	153	2.5 (378)	91% (344)	67% (230)	1.50		2	7
							7	23
1984	92	2.8 (255)	91% (231)	3% (7)	0.08		4	5
							6	82
1985	92	2.9 (266)	90% (240)	27% (64)	0.70		3	7
							18	48
1986	68	2.6 (178)	80% (142)	77% (109)	1.60		7	13
							9	10
1987	60	2.6 (155)	77% (120)	2% (2)	0.03		14	9
							68	8

Table 21.3 Breeding success of three Common Tern colonies nesting on Jade Bay, averaged for the years 1980–7 (Oldeoog since 1981, Table 21.2). The data refer to sample nests or, in the case of Wilhelmshaven, to the total colony (approximate colony size: Oldeoog 1500–3700 pairs, Wilhelmshaven 75 pairs, Augustgroden 145 pairs). Late nests are excluded. 'Other causes' mainly refer to chick losses through weather and food shortage. (For further details see Becker and Finck 1985, 1986; Becker 1984b; Becker and Anlauf 1988a, b.)

| Colony | No. of clutches (eggs) | Clutch size | Hatching success (%) | Fledging success (%) | Losses (in % of eggs) | | | Chicks fledged per clutch |
					Flooding	Predation	Other causes	
Oldeoog	633 (1692)	2.7	88	35	0	20	50	0.8
Wilhelmshaven	446 (1247)	2.8	65	63	0	25	36	1.1
Augustgroden	458 (1180)	2.6	74	31	35	13	31	0.6

1979–83 (0.0–0.4 per pair), in consequence of heavy predation by Herring Gulls (42–96 per cent of chicks lost through predation; Becker 1984a, 1985, 1987).

The Common Terns breeding on a managed site in the harbour of Wilhelmshaven (Becker 1984b) were on average the most successful (Table 21.3). Despite a higher rate of predation (mainly by rats) than on Oldeoog and at Augustgroden, the fledging success was about 30 per cent better (Table 21.3). Discounting effects of predation, the hatching success in the breeding seasons 1985–7 would have been 89–90 per cent. The high fledging rates are associated with the additional availability of freshwater prey, resulting in a higher feeding rate in colonies on the mainland coast than on Minsener Oldeoog (Frank 1990).

The total number of breeding pairs being known, it was possible to calculate the average breeding success for the Common Tern population in Jade Bay between 1980 and 1987. The results give a hatching success of 86 per cent, a fledging success of 34 per cent and 0.79 chicks fledged/clutch. Egg losses (chick losses) amount to 1 per cent (2 per cent) of the eggs through flooding, 6 per cent (15 per cent) through predation, and 9 per cent (39 per cent) through other causes. In the chicks weather and food accounted almost entirely for the last mentioned 39 per cent and thus proved critical for the reproductive output of the population. The mean number of fledglings per clutch and year was more or less the same as that found in other studies (Nisbet 1973, 1978: 0.9–1.1; Greenhalgh 1974: 0.7–1.0; Courtney and Blokpoel 1983: 0.9; Erwin and Smith 1985: 0.6–0.8), but below the 1.1 fledgings per pair and year which seems to be the minimum necessary to secure population stability in North American

Common Terns (Nisbet 1978; DiCostanzo 1980). During the 1980s, the population of Common Terns in the German Wadden Sea seemed to be stable (cf. Fig. 21.1). In several years the colonies at Jade Bay fledged more than 1.1 chicks (Oldeoog; 2 years, Table 21.2; Augustgroden: 3 years, Wilhelmshaven: 5 years).

Several important points arise from this short overview with regard to detecting possible contamination effects. The high fluctuation in breeding output from year to year and between sites is striking. The ever-changing combinations of environmental and social factors influencing breeding success make it enormously difficult to discriminate between natural variability and the possible effects of pollutants. Consequently before the results can be used for extrapolation, a broad base of information is required on the variation in reproductive output and on the factors that affect it. This, however, presupposes long-term studies, with inherent problems of logistics, manpower costs, time, and funding—all reasons contributing to studies of seabirds nesting in different habitats not lasting for more than 1 or 2 years (Erwin and Smith 1985).

These points have not sufficiently been taken into consideration in some papers dealing with relationships between breeding success and environmental contamination. Often only one breeding site formed the basis for a study or the data were collected in the course of a few visits per breeding season (terns: e.g. Fimreite 1974; Blus and Prouty 1979; Blus *et al*. 1979; gulls: White *et al*. 1983). Such a basis may be questionable if one's aim is to detect small differences in breeding success associated with contaminant levels. Blus and coworkers recognize these restrictions and comment on their results with caution.

In particular a study of prefledging chick fate (mortality, growth rate) incurs high labour costs and is fraught with problems (e.g. Duffy 1979; Fetterolf 1983; Erwin and Custer 1982). Furthermore in Common Tern colonies the fledging success fluctuated more than the hatching rate (see above). According to Mineau *et al*. (1984) and Weseloh *et al*. (1988), net chick productivity was not adequate to indicate effects of the present-day pollutant levels in North American Herring Gulls. Hence, for monitoring purposes in larids, studies of the hatching success would seem to be more suitable: recording the necessary field data is easier, the variability lower, and analyses of embryo mortality, eggshell quality, and related contaminant levels can also be performed (see e.g. Blus 1984 for egg sampling technique). When possible, however, research on chick fate should be also carried out (see e.g. Gochfeld and Burger 1988). Fencing of tern broods is an adequate method which should be considered (see e.g. Erwin and Custer 1982; White *et al*. 1984; Custer and Mitchell 1987).

As revealed by many field studies, toxic chemicals can vastly reduce hatching, for example by impairing egg shell quality, causing embryonic mortality, or influencing the behaviour of breeding adults (e.g. Brown Pelican, Blus *et al*. 1974; Blus 1982; Anderson and Gress 1983; Double-crested Cormorant, Gress *et al*. 1973; Herring Gull 54–59 per cent, Keith 1966; 23 per cent, Teeple 1977; 41–62 per cent, Gilman *et al*. 1977; Common Tern 61 per cent, Fox 1976). In comparison with the reproductive success of Common Tern reported in other papers, hatching success in the Jade Bay colonies during the 1980s was good, an average of 86 per cent which, without predation, would have been about 92 per cent (cf. Nisbet and Drury 1982; Morris *et al*. 1976; Fox 1976; Burger and Lesser 1978). Hence, it can be concluded that hatching success was not depressed by environmental pollutants (Fig. 21.2, areas 4, 6) during the study.

Concluding remarks

The build-up of persistent contaminants in seabirds due to their high positions in the food chain, their sensitive reproductive success and population size, as well as the advantages of sampling from colonial animals (Heinz *et al*. 1979) make seabirds valuable monitors of the marine environment. Furthermore, the species themselves profit in terms of conservation from the warnings that such studies provide of imminent pollution problems. Hence, the maxim must be emphasized that long-term studies of the effect of toxic chemicals on seabirds should be encouraged (Ohlendorf *et al*. 1978; Fleming *et al.* 1983; Mineau *et al*. 1984; Evans and Nettleship 1985; Risebrough 1986; Clausing 1986; Morrison 1986; Furness 1987; Gilbertson *et al*. 1987; Becker 1989).

The Wadden Sea supports many seabird, wader, and waterfowl populations of international importance; these use the area for breeding, feeding, migrating, moulting, or as winter quarters (e.g. Knief 1987; Tasker and Pienkowski 1987; Prokosch 1988). For this reason and because of the vast pollution problems in the North Sea, a careful watch should be taken on contaminants in the Wadden Sea.

Terns as well as other migrating species depend for their reproductive performance on the resources in their breeding habitats on European coasts. In my opinion the states along the North Sea coast should be duty bound to study the effects of present environmental conditions on the breeding success of seabirds since this is so crucial for the maintenance of populations. Of the seabirds breeding around the Wadden Sea, the Common Tern seems to qualify as one of the most suitable monitors, as it also breeds along other North Sea coasts. We hope that our studies in monitoring toxic pollutants on the German coast will contribute to the establishment of a European monitoring programme with seabirds. Attention should be directed not only to contaminants of known toxicity, but also to other, new and potentially dangerous pollutants.

The monitoring of contaminants should be complemented by regular studies of possible effects on reproductive biology, especially at pollution 'hot spots' like estuaries. Investigations into hatching success, egg shell quality and, if possible, chick fate may indicate any severe problems arising from toxic substances. The data on hatching success and population trends of Common Terns presented in this paper indicate no negative effects of environmental chemicals at present in the Jade Bay. However, under unfavourable conditions, e.g. food shortage, contaminant concentrations can increase e.g. in blood, liver, or egg, and thus become toxic (e.g. Koeman *et al*. 1967*b*; Koeman 1971; Heidman *et al*. 1987*b*), and the breeding success in the most polluted areas of the Wadden Sea is not known.

Although long-term studies are often needed to provide the basic information necessary to detect the effects of toxic compounds, this is not always the view of public authorities responsible for funding or for providing permits for field work in protected areas (see e.g. Anderson *et al*. 1986). Investigations should cover feeding and breeding areas that coincide, for purposes of comparison, with known pollution 'hot spots', like the Elbe estuary or the inner German Bight, as well as sites characterized by low levels of contamination (NERC 1983; Evans and Nettleship 1985).

Furness and Monaghan (1987) argue that monitoring of breeding output provides information that is more valuable than can be obtained by monitoring population size. As pointed out in the second section on trends in seabird populations, population monitoring seems to provide little information that is useful as an early warning of the build-up of toxic pollutants. However, population trends should continue to be studied and recorded despite the obvious methodological difficulties. The examination of trends may be useful in detecting any changes in the quantity and quality of habitat (Morrison 1986) and changes in numbers could motivate detailed research into the causes of the population trends. In combination with other studies, counts can furnish additional knowledge about the impact of pollution on bird populations. Moreover, such hazards can be made clear to the public and presentation can be made of the positive effects on populations that can be achieved by lowering the levels of toxic pollutants, e.g. through restrictive legal measures.

Summary

In the 1960s populations of birds breeding on the Wadden Sea coast were adversely affected by toxic chemicals. In particular, the number of terns was reduced considerably (Fig. 21.1). Once the level of dangerous pesticides in the environment had decreased following purification measures and legal restrictions, bird populations increased again (Table 21.1, Fig. 21.1). This trend is still continuing so that at the present time toxic pollutants are probably of little importance to bird populations in most areas of the southern North Sea coast.

Monitoring of bird populations normally only reveals effects of contamination on the basis of data obtained over several years. Furthermore, many factors influence population size, of which the recording is difficult anyway. Hence population monitoring alone seems to be of little value as an early warning of any one specific factor, such as toxic contamination, and additional methods should be sought to indicate problems arising from chemicals in the coastal marine environment. Thus, on the German Wadden Sea coast (i) the contamination of coastal birds was investigated using egg analyses, and (ii) the breeding success of the Common Tern was examined together with its dependence on environmental factors.

Tern eggs contain the highest levels out of 11 seabird species studied of organochlorines and mercury (Fig. 21.3). However, geographical differences are pronounced, with the eggs of birds breeding in the Elbe estuary and the inner German Bight being particularly heavily contaminated with PCBs and mercury (Fig. 21.2). Long-term studies are necessary to reveal trends in the levels of contamination. One problem of monitoring contaminants is the enforced restriction (through high costs) to a small number of pollutants of known toxicological relevance.

At the Jade Bay the breeding success of Common Terns varies considerably between years and sites (Fig. 21.4). Climatic and food conditions during chick rearing account for most of the losses (Tables 21.2, 21.3), whereas breeding failure due to predation and flooding are of minor significance. The high hatching success at present suggests that contaminants are not having an important effect on hatching. The considerable variability in reproductive success and the factors that affect it make it difficult to detect any deleterious influence of environmental pollutants on breeding output in the field, in particular during the chick stage.

Even so, long-term studies providing background information on naturally occurring variation in reproductive success are urgently needed and should be encouraged. Such knowledge is an essential prerequisite for studies into the effects of toxic residues on reproductive success. It is indicated that studies monitoring contamination (egg sampling) and breeding success (hatching success) should be initiated to establish an early warning system against chemical pollution of coastal birds. Using an indicator species like the Common Tern with a wide distribution and on a high trophic level, long-term programmes must aim to cover sites near known pollution hot spots (estuaries) as well as less-polluted areas.

Acknowledgements

I thank all my collaborators in the Common Tern project for their help in gathering the data on breeding success. For chemical analyses and constant cooperation I am greatly indebted to A. Büthe, W. Heidmann, and H. A. Rüssel from the Chemical Institute of the Tierärztliche Hochschule Hannover. H. Heckenroth from the Niedersächsisches Landesverwaltungsamt Naturschutz, H. Brunckhorst from the Landesamt für den Nationalpark Schleswig-Holsteinisches Wattenmeer, and H. Thiessen from the Landesamt für Naturschutz und Landschaftspflege Schleswig-Holstein kindly provided recent breeding pair numbers. I am grateful also to C. Koepff and M. Wagener for assisting in the statistical analyses and in preparing the manuscript, which was improved by the helpful comments of I. Newton and another anonymous referee. K. Wilson checked the English. This work was supported by the Deutsche Forschungsgemeinschaft and the Umweltbundesamt.

References

Anderson, D. R., Burnham, K. P., and White, G. C. (1985). Problems in estimating age-specific survival rates from recovery data of birds ringed as young. *Journal of Animal Ecology*, **54**, 89–98.

Anderson, D. W. and Gress, F. (1983). Status of a northern pollution of California Brown Pelicans. *Condor*, **85**, 79–88.

Anderson, D. W. and Hickey, J. J. (1976). Dynamics of storage of organochlorine pollutants in Herring Gulls. *Environmental Pollution*, **10**, 183–200.

Anderson, D. W., Peterle, T. J., and Dickson, K. L. (1986). Contaminants: neglected and forgotten challenges. In *Transactions of the North American Wildlife and Natural Resources conference* (ed. R. E. McCabe), pp. 550–61. Washington DC.

Becker, P. H. (1984a). Wie richtet eine Flußseeschwalbenkolonie (*Sterna hirundo*) ihr Abwehrverhalten auf den Feinddruck durch Silbermöwen (*Larus argentatus*) ein? *Zeitschrift für Tierpsychologie*, **66**, 265–88.

Becker, P. H. (1984b). Umsiedlung einer Flußseeschwalbenkolonie in Wilhelmshaven. *Berichte der Deutschen Sektion des Internationalen Rates für Vogelschutz*, **24**, 111–19.

Becker, P. H. (1985). Common Tern breeding success and nesting ecology under predation pressure of Herring Gulls. *Acta 18th International Ornithological Congress*, **1982**, 1198–205.

Becker, P. H. (1987). Kann sich die Flußseeschwalbe auf Mellum vor Brutverlusten durch Silbermöwen schützen? In *Mellum—Portrait einer Insel* (ed. G. Gerdes, W. E. Krumbein and H. E. Reineck), pp. 281–92. Kramer, Frankfurt.

Becker, P. H. (1989). Seabirds as monitor organisms of contaminants along the German North Sea coast. *Helgoländer Meeresuntersuchungen*, **43**, 395–403.

Becker, P. H. and Anlauf, A. (1988a). Nistplatzwahl und Bruterfolg der Flußseeschwalbe (*Sterna hirundo*) im Deichvorland. I. Nestdichte. *Ökologie der Vögel*, **10**, 27–44.

Becker, P. H. and Anlauf, A. (1988*b*). Nistplatzwahl und Bruterfolg der Flußsee-schwalbe (*Sterna hirundo*) im Deichvorland. II. Hochwasser-Überflutung. *Ökologie der Vögel*, **10**, 45–58.

Becker, P. H. and Erdelen, M. (1987). Die Bestandsentwicklung von Brutvögeln der deutschen Nordseeküste, 1950–1979. *Journal für Ornithologie*, **128**, 1–32.

Becker, P. H., and Finck, P. (1985). Witterung und Ernährungssituation als entscheidende Faktoren des Bruterfolgs der Flußseeschwalbe (*Sterna hirundo*). *Journal für Ornithologie*, **126**, 393–404.

Becker, P. H. and Finck, P. (1986). Die Bedeutung von Nestdichte und Neststandor für den Bruterfolgs der Flußseeschwalbe (*Sterna hirundo*) in Kolonien einer Wattenmeerinsel. *Vogelwarte*, **33**, 192–207.

Becker, P. H. and Frank, D. (1988). Feeding strategies of Common Terns in the Wadden Sea. In *Seabird food and feeding ecology*. Proceedings of Third Inter-national conference of the Seabird Group (ed. M. L. Tasker), pp. 8–10.

Becker, P. H. and Nagel, R. (1983). Schätzung des Brutbestandes der Silbermöwe (*Larus argentatus*) auf Mellum, Langeoog und Memmert mit der Linientransekt-Methode. *Die Vogelwelt*, **104**, 25–38.

Becker, P. H., Conrad, B., and Sperveslage, H. (1980). Vergleich der Gehalte an chlorierten Kohlenwasserstoffen und PCB's in Silbermöwen (*Larus argentatus*)—Eiern von Mellum 1975 und 1979. *Vogelwarte*, **30**, 294–6.

Becker, P. H., Büthe, A., and Heidmann, W. (1985*a*). Schadstoffe in Gelegen von Brutvögeln der deutschen Nordseeküste. 1. Chlororganische Verbindungen. *Journal für Ornithologie*, **126**, 29–51.

Becker, P. H., Ternes, W., and Rüssel, H. A. (1985*b*). Schadstoffe in Gelegen von Brutvögeln der deutschen Nordseeküste. 2. Quecksilber. *Journal für Ornithologie*, **126**, 253–62.

Becker, P. H., Frank, D., and Walter, U. (1987). Geographische und jährliche Variation der Ernährung der Flußseeschwalbe (*Sterna hirundo*) an der Nordsee-küste. *Journal für Ornithologie*, **128**, 457–75.

Becker, P. H., Büthe, A., and Heidmann, W. (1988). Rückgänge von Schadstoff-gehalten in Küstenvögeln? *Journal für Ornithologie*, **129**, 104–6.

Becker, P. H., Conrad, B., and Sperveslage, H. (1989). Chlororganische Verbindungen und Schwermetalle in weiblichen Silbermöwen (*Larus argentatus*) und ihren Eiern mit bekannter Legefolge. *Vogelwarte*, **35**, 1–10.

Bergström, R. and Norheim, G. (1986). Persistente klorerte hydrokarboner 1 sjöfuglegg fra kysten av Telemark. *Fauna*, **39**, 53–7.

Beukema, J. J. and Cadée, G. C. (1986). Zoobenthos responses to eutrophication of the Dutch Wadden Sea. *Ophelia*, **26**, 55–64.

Blaszyk, P. (1972). Zur Frage der Gefährdung freilebender Vögel durch polychlorierte Biphenyle (PCB). *Berichte der Deutschen Sektion des Internationalen Rates für Vogelschutz*, **12**, 48–53.

Blus, L. J. (1982). Further inspection of the relation of organochlorine residues in Brown Pelican eggs to reproductive success. *Environmental Pollution*, **A28**, 15–33.

Blus, L. J. (1984). DDE in birds' eggs: comparison of two methods for estimating critical levels. *Wilson Bulletin*, **96**, 268–76.

Blus, L. J. and Prouty, R. M. (1979). Organochlorine pollutants and population Status of Least Terns in South Carolina. *Wilson Bulletin*, **91**, 62–71.

Blus, L. J. and Stafford, C. J. (1980). Breeding biology and relation of pollutants to Black Skimmers and Gull-billed Terns in South Carolina. *U.S. Fish & Wildlife Service Report*, **230**, 1–18.

Blus, L. J., Neely, B. S., Belisle, A. A., and Prouty, R. M. (1974). Organochlorine residues in Brown Pelican eggs: relation to reproductive success. *Environmental Pollution*, 7, 81–91.

Blus, L. J., Henny, C. J., Kaiser, T. E., and Grove, R. A. (1983). Effects on wildlife from use of Endrin in Washington State Orchards. *Transactions of the North American Wildlife and Natural Resource Conference*, 48, 159–74. Washington DC.

Blus, L. J., Prouty, R. M., and Neely, B. S. Jr. (1979). Relation of environmental factors to breeding status of Royal and Sandwich Terns in South Carolina, USA. *Biological Conservation*, 16, 301–20.

Bourne, W. R. P. (1976). Seabirds and pollution. In *Marine pollution* (ed. R. Johnston), Vol. 6. Academic Press, London.

Brinkman, U. A. T. and Reymer, H. G. M. (1976). Polychlorinated naphthalenes. *Journal of Chromatography*, 127, 203–43.

Burger, J. and Lesser, F. (1978). Selection of colony sites and nest sites by Common Terns (*Sterna hirundo*) in Ocean County, New Jersey. *Ibis*, 120, 433–49.

Bush, B., Tumasomis, C. F., and Baker, F. D. (1974). Toxicity and persistence of PCB homologs and isomers in the avian system. *Archives Environmental Contamination and Toxicology*, 2, 195–212.

Clausing, P. (1986). Chlororganische Insektizide in Europa—Kontaminationsgrad und Bestandsveränderungen bei Vögeln zehn Jahre nach dem DDT-Verbot. *Berichte aus der Vogelwarte Hiddensee*, 7, 47–53.

Coulson, J. C., Deans, I. R., Potts, G. R., Robinson, J., and Crabtree, A. N. (1972). Changes in organochlorine contamination of the marine environment of eastern Britain monitored by Shag eggs. *Nature*, 236, 454–6.

Courtney, P. A. and Blokpoel, H. (1983). Distribution and numbers of Common Terns on the Lower Great Lakes during 1900–1980: a review. *Colonial Waterbirds*, 6, 107–20.

Custer, T. W. and Mitchell, C. A. (1987). Organochlorine contaminants and reproductive success of Black Skimmers in South Texas, 1984. *Journal of Field Ornithology*, 58, 480–9.

Custer, T. W., Erwin, R. M., and Stafford, C. (1983). Organochlorine residues in Common Tern eggs from nine Atlantic Coast colonies, 1980. *Colonial Waterbirds*, 6, 197–204.

DiCostanzo, J. (1980). Population dynamics of a Common Tern colony. *Journal of Field Ornithology*, 51, 229–43.

Duffy, D. C. (1979). Human disturbance and breeding birds. *Auk*, 96, 815–16.

Duinker, J. C. and Koeman, J. H. (1978). Summary report on the distribution and effects of toxic pollutants (metals and chlorinated hydrocarbons) in the Wadden Sea. In *Pollution of the Wadden Sea area, Report*, 8. Wadden Sea Working Group, Leiden (eds. K. Essink and W. J. Wolff), pp. 45–54.

Dunnet, G. M. (1982). Oil pollution and seabird populations. *Philosophical Transactions of the Royal Society of London*, B316, 513–24.

Dunnet, G. M., Furness, R. W., Tasker, M. L., and Becker, P. H. (1990). Seabird ecology in the North Sea. *Netherlands Journal of Sea Research*, in press.

Eisler, R. (1986). *Polychlorinated biphenyl hazards to fish, wildlife, and invertebrates: a synoptic review*. Fish and Wildlife Service U.S. Dept. of the Interior, Biological Report, 85, 1–72.

Ellenberg, H. (ed.) (1981). *Greifvögel und Pestizide*. Ökologie der Vögel 3.

Erwin, R. M. and Custer, T. W. (1982). Estimating reproductive success in colonial waterbirds: an evaluation. *Colonial Waterbirds*, 6, 49–56.

Erwin, R. M. and Smith, S. C. (1985). Habitat comparisons and productivity in nesting Common Terns on the Mid-Atlantic Coast. *Colonial Waterbirds*, 8, 155–65.

Evans, P. G. H. and Nettleship, D. N. (1985). Conservation of the Atlantic Alcidae. In *The Atlantic Alcidae* (eds. D. N. Nettleship and T. R. Birkhead), pp. 427–88. Academic Press, London.

Fetterolf, P. M. (1983). Effects of investigator activity on Ring-billed Gull behaviour and reproductive performance. *Wilson Bulletin*, 95, 23–41.

Fimreite, N. (1974). Mercury contamination of aquatic birds in northwestern Ontario. *Journal of Wildlife Management*, 38, 120–31.

Fimreite, N. (1979). Accumulation and effects of mercury on birds. In *The bio-chemistry of mercury in the environment*, Chapter 22 (ed. J. O. Nriagu), pp. 601–27. Elsevier, Amsterdam.

Fleming, W. J., McLane, M. A. R., and Cromartie, E. (1982). Endrin decreases Screech Owl productivity. *Journal of Wildlife Management*, 46, 462–8.

Fleming, W. J., Clark, D. R. Jr., and Henny, C. J. (1983). Organochlorine pesticides and PCBs: a continuing problem for the 1980s. *Transactions of the North American Wildlife and Natural Resources Conference*, 48, 186–99.

Focardi, S., Fossi, C., Lambertini, M., Leonzio, C., and Massi, A. (1988). Long term monitoring of pollutants in eggs of Yellow-legged Herring Gull from Capraia islands (Tuscan archipelago). *Environmental Monitoring Assessment*, 10, 43–50.

Fox, G. A. (1976). Eggshell quality: its ecological and physiological significance in a DDE-contaminated Common Tern population. *Wilson Bulletin*, 88, 459–77.

Frank, D. (1990). Fütterrate und Nahrungszusammensetzung von Flußseeschwalben (*Sterna hirundo*) anhand automatischer Registrierung am Nest. *Proceedings International 100. DO-G* meeting, Bonn, 159–65.

Furness, R. W. (1985). Ingestion of plastic particles by seabirds at Gouch Island, South Atlantic Ocean. *Environmental Pollution*, 38, 261–72.

Furness, R. W. (1987). Seabirds as monitors of the marine environment. *International Council for Bird Preservation, Technical Publication*, 6, 217–30.

Furness, R. W. (1988). Changes in diet and breeding ecology of seabirds in Shetland 1971–1987. In *Seabird food and feeding ecology*, Proceedings of third International Conference of the Seabird Group (ed. M. L. Tasker), p. 20.

Furness, R. W. and Hutton, M. (1979). Pollutant levels in the Great Skua (*Catharacta skua*). *Environmental Pollution*, 19, 261–8.

Furness, R. W. and Hutton, M. (1980). Pollutants and impaired breeding of Great Skuas (*Catharacta skua*) in Britain. *Ibis*, 122, 88–94.

Furness, R. W. and Monaghan, P. (1987). *Seabird ecology*. Blackie, Glasgow and London.

Gerlach, S. A. (1987). Pflanzennährstoffe und die Nordsee—ein Überblick. *Seevögel*, 8, 49–62.

Gilbertson, M., Morris, R. D., and Hunter, R. A. (1976). Abnormal chicks and PCB residue levels in eggs of colonial birds on the Lower Great Lakes (1971–1973). *Auk*, 93, 434–42.

Gilbertson, M., Elliot, J. E., and Peakall, D. B. (1987). Seabirds as indicators of marine pollution. In *The value of birds* (eds. A. W. Diamond and F. L. Filion), pp. 231–48. International Council for Bird Preservation, Technical Publication, 6.

Gilman, A. P., Fox, G. A., Peakall, D. B., Temple, S., Carroll, T. R., and Haymes, G. T. (1977). Reproductive parameters and egg contaminant levels of Great Lakes Herring Gulls. *Journal of Wildlife Management*, 41, 458–68.

Gochfeld, M. and Burger, J. (1988). Effects of lead on growth and feeding behavior of young Common Terns (*Sterna hirundo*). *Archives of Environmental Contamination and Toxicology*, **17**, 513–17.

Greenhalgh, M. E. (1974). Population growth and breeding success in a salt marsh Common Tern colony. *Naturalist*, **931**, 43–51.

Gress, F., Risebrough, R. W., Andeson, D. W., Kiff, L. F., and Jehl, R. J. Jr. (1973). Reproductive failures of Double-crested Cormorants in southern California and Baja California. *Wilson Bulletin*, **85**, 197–208.

Großkopf, G. (1989). *Die Vogelwelt von Wangerooge*. Holzberg, Oldenbürg.

Harris, M. P. and Lloyd, C. S. (1977). Variations in counts of seabirds from photographs. *British Birds*, **70**, 200–5.

Harris, M. P. and Osborn, D. (1981). Effect of a polychlorinated biphenyl on the survival and breeding of puffins. *Journal of Applied Ecology*, **18**, 471–9.

Hartner, L. (1981). Wie schädigen die chlorierten Kohlenwasserstoffe die Vögel? *Ökologie der Vögel*, **3**, 33–8.

Hatch, S. A. and Hatch, M. A. (1988). Colony attendance and population monitoring of Black-legged Kittiwakes on the Semidi Islands, Alaska. *Condor*, **90**, 613–20.

Heidmann, W. A. (1986). Isomer specific determination of polychlorinated biphenyls in animal tissue by gas chromatography mass spectrometry. *Chromatographia*, **22**, 363–9.

Heidmann, W. A. and Büthe, A. (1986). Chlorpestizide und chlorhaltige Industriechemikalien. In *Rückstandsanalytik von Wirkstoffen in tierischen Produkten* (ed. H. A. Rüssel). Thieme, Stuttgart.

Heidmann, W. A., Beyerbach, M., Böckelmann, W., Büthe, A., Knüwer, H., Peterat, B., and Rüssel-Sinn, H. A. (1987*a*). Chlorierte Kohlenwasserstoffe und Schwermetalle in tot an der deutschen Norseeküste aufgefundenen Seevögeln. *Vogelwarte*, **34**, 126–33.

Heidmann, W. A., Büthe, A., Peterat, B., and Knüwer, H. (1987*b*). Zur Frage des Einflusses chemischer Rückstände auf das Sterben von Austernfischern (*Haematopus ostralegus*) an der niedersächsischen Küste im Winter 1986/87. *Vogelwarte*, **34**, 73–9.

Heinz, G. H., Hill, E. F., Stickel, W. H., and Stickel, L. F. (1979). Environmental contaminant studies by the Patuxent Wildlife Research Center. *In Avian and Mammalian Wildlife Toxicology, Special Technical Publication*, **693** (ed. E. E. Kenaga), pp. 9–35.

Helander, B. (1977). The White-tailed Eagle in Sweden. In *World conference on birds of prey* (ed. R. D. Chancellor). International Council for Bird Preservation, Basingstoke.

Henny, C. J., Blus, L. J., Kolbe, E. J., and Fitzner, R. E. (1985). Organophosphate cattle pour-on insecticide (Famphur) kills Black-billed Magpies and Red-tailed Hawks. *Journal of Wildlife Management*, **49**, 648–58.

Hickey, J. J. and Anderson, D. W. (1969). The Peregrine Falcon: life history and population. In *Peregrine Falcon populations: their biology and decline* (ed. J. J. Hickey), pp. 3–42. University of Wisconsin Press, Madison, Milwaukee and London.

Hill, E. F. and Camardese, M. B. (1986). Lethal dietary toxicities of environmental contaminants and pesticides to Corturnix. *U.S. Fish and Wildlife Service, Technical Report*, **2**, 147.

Hudson, R. H., Tucker, R. K. and Haegele, M. A. (1984). Handbook of toxicity of pesticides to wildlife. *U.S. Fish and Wildlife Publication no. 153*.

Jörgensen, O. H. and Kraul, I. (1974). Eggshell parameters, and residues of PCB and

DDE in eggs from Danish Herring Gulls (*Larus a. argentatus*). *Ornis Scandinavica*, 5, 173–9.

Keith, J. A. (1966). Reproduction of a population of Herring Gulls. *Journal of Applied Ecology*, 3, 57–70.

Keith, J. A. and Gruchy, I. M. (1972). Residue levels of chemical pollutants in North American birdlife. *Proceedings of the 15th International Ornithological Congress*, 437–54.

Knief, W. (1987). Die Bedeutung des Wattenmeeres für Vögel. *Seevögel*, 8, 23–8.

Koeman, J. H. (1971). Het voorkomen en de toxicologische betekenis van enkele chloorkoolwaterstoffen aan de Nederlandse kust in de periode 1965 tot 1970. Thesis, Utrecht.

Koeman, J. H. (1979). Chemicals in the environment and the effects on ecosystems. In *Advances in Pesticide Science: World Food Production—Environment—Pesticides*, Part 1 (ed. G. Geissbuhler), pp. 25–38. Pergamon, Oxford.

Koeman, J. H., Oskamp. A. A. G., Veen, J., Brouwer, E., Rooth, J., Zwart, P., Brock, E. v. d., and Genderen, H. van (1967*a*). Insecticides as a factor in the mortality of the Sandwich Terns (*Sterna sandvicensis*). *Mededelingen Rijksfactulteit Landbouwetenschappen Gent.*, 32, 841–54.

Koeman, J. H., Oudejans, R. C. H. M., and Huisman, E. A. (1967*b*). Danger of chlorinated hydrocarbon insecticides in birds' eggs. *Nature*, 215, 1094–6.

Lakhani, K. H. and Newton, I. (1983). Estimating age-specific bird survival rates from ringing recoveries—can it be done? *Journal of Animal Ecology*, 52, 83–91.

Lockie, J. D., Ratcliffe, D. A., and Balharry, D. A. (1969). Breeding success and organochlorine residues in Golden Eagles in West Scotland. *Journal of Applied Ecology*, 6, 381–9.

Mineau, P., Fox, G. A., Norstrom, R. J., Weseloh, D. V., Hallen, D. J., and Ellenton, J. A. (1984). Using the Herring Gull to monitor levels and effects of organochlorine contamination in the Canadian Great Lakes. In *Toxic contaminants in the Great Lakes* (ed. J. O. Nriagu and M. S . Simmons), pp. 426–52. Wiley, New York and London.

Moksnes, M. T. and Norheim, G. (1986). Levels of chlorinated hydrocarbons and composition of PCB in Herring Gulls (*Larus argentatus*) eggs collected in Norway in 1969 compared to 1979–81. *Environmental Pollution*, B11, 109–16.

Morrison, M. L. (1986). Bird populations as indicators of environmental change. In *Current Ornithology*, 3 (ed. R. F. Johnston), pp. 429–51. Plenum, New York and London.

Morris, R. D., Hunter, R. A., and McElman, J. F. (1976). Factors affecting the reproductive success of Common Tern (*Sterna hirundo*) colonies on the Lower Great Lakes during the summer of 1972. *Canadian Journal of Zoology*, 54, 1850–62.

NERC (1983). Contaminants in marine top predators. *National Environment Research Council, Report series C*, 23.

Nettleship, D. (1976). Census techniques for seabirds of arctic and eastern Canada. *Canadian Wildlife Service*, 25, 3–33.

Newton, I. (1979). *Population ecology of raptors*. Poyser, Berkhamsted.

Newton, I. (1981). Der Sperber und die Pestizide—ein Beitrag von den Britischen Inseln. *Ökologie der Vögel*, 3, 207–19.

Newton, I. (1986). *The Sparrowhawk*. Poyser, Carlton.

Nisbet, I. C. T. (1973). Terns in Massachusetts: present numbers and historical changes. *Bird Banding*, 44, 27–55.

Nisbet, I. C. T. (1977). Courtship-feeding and clutch size in Common Terns. In

Evolutionary Ecology. *Biology and Environment*, Vol. 2 (ed. B. Stonehouse), pp. 101–09. Macmillan, London.

Nisbet, I. C. T. (1978). Population models for Common Terns in Massachusetts. *Bird Banding*, **49**, 50–8.

Nisbet, I. C. T. and Drury, W. H. (1972). Measuring breeding success in Common and Roseate Terns. *Bird Banding*, **43**, 97–106.

Nisbet, I. C. T. and Reynolds, L. M. (1984). Organochlorine residues in Common Terns and associated estuarine organisms, Massachusetts, U.S.A., 1971–81. *Marine Environmental Research*, **11**, 33–66.

Ohlendorf, H. M., Risebrough, R. W., and Vermeer, K. (1978). Exposure of marine birds to environmental pollutants. *Wildlife Research*, *Report*, **9**. Washington DC.

Olsson, M. and Reutergårdh, L. (1986). DDT and PCB pollution trends in the Swedish aquatic environment. *Ambio*, **15**, 103–9.

Peakall, D. B. (1972). Polychlorinated biphenyls: occurrence and biological effects. *Residue Reviews*, **44**, 1–21.

Peakall, D. B. (1980). Pollutant levels and their effects on raptoral and fish-eating birds. *Proceedings of the 17th International Ornithological Congress*, **1978**, 935–41.

Pearce, P. A., Peakall, D. B., and Reynolds, L. M. (1979). Shell thinning and residues of organochlorines and mercury in seabird eggs, eastern Canada, 1970–76. *Pesticides Monitoring Journal*, **13**, 61–8.

Prestt, I. and Ratcliffe, D. A. (1972). Effects of organochlorine residues on European birdlife. *Proceedings of the 15th International Ornithological Congress*, pp. 486–513.

Prinzinger, G. and Prinzinger, R. (1979). Der Einfluß von Pestiziden auf die Brutphysiologie der Vögel. *Ökologie der Vögel*, **1**, 17–89.

Prokosch, P. (1988). Das Schleswig-Holsteinische Wattenmeer als Frühjahrs-Aufenthaltsgebiet arktischer Watvogelpopulationen am Beispiel von Kiebitzregenpfeifer (*Pluvialis squatarola*), Knutt (*Calidris canutus*) und Pfuhlschnepfe (*Limosa lapponica*). *Corax*, **12**, 273–448.

RSU (1980). *Umweltprobleme der Nordsee*. Rat von Sachverständigen für Umweltfragen, Sondergutachten 1980. Kohlhammer, Stuttgart.

Ratcliffe, D. A. (1970). Changes attributable to pesticides in egg breakage frequency and eggshell thickness in some British birds. *Journal of Applied Ecology*, 7, 67–115.

Ratcliffe, D. A. (1980). *The Peregrine Falcon*. Poyser, Carlton.

Reineking, B. and Vauk, G. (1982). *Seevögel—Opfer der Ölpest*. Niederelbe, Otterndorf.

Risebrough, R. W. (1986). Pesticides and bird populations. In *Current Ornithology*, 3 (ed. R. F. Johnston), pp. 397–427. Plenum, New York and London.

Robinson, J., Richardson, A., Crabtree, A. N., Coulson, J. C., and Potts, G. R. (1967). Organochlorine residues in marine organisms. *Nature*, **214**, 1307–11.

Rooth, J. (1980a). Common Tern (*Sterna hirundo*). In *Birds of the Wadden Sea*, Wadden Sea Working Group, Report 6 (eds. C. J. Smit and W. J. Wolff), pp. 258–65. Leiden.

Rooth, J. (1980b). Sandwich Tern (*Sterna sandvicensis* Latham). In *Bird of the Wadden Sea*, Wadden Sea Working Group, Report 6 (eds. C. J. Smit and W. J. Wolff), pp. 250–8. Leiden.

Smit, C. J. (1981). Distribution, ecology and zoogeography of breeding birds on the Wadden Sea Islands. In *Terrestrial and freshwater fauna of the Wadden Sea area*, Wadden Sea Working Group, Report 10 (eds. C. J. Smit, J. den Hollander, W. K. E. van Wingerden, and W. J. Wolff), pp. 169–231. Leiden.

Stickel, L. (1973). Pesticide residues in birds and mammals. In *Environmental pollution by pesticides* (ed. C. A. Edwards), pp. 254–312. Plenum, London.

Stickel, W. H. (1975). Some effects of pollutants in terrestrial ecosystems. In *Ecological toxicology research* (ed. A. D. McIntyre and C. F. Mills), pp. 25–74. Plenum, New York.

Struger, J. and Weseloh, D. V. (1985). Great Lakes Caspian Terns: egg contaminants and biological implications. *Colonial Waterbirds*, 8, 142–9.

Swennen, C. (1972). Chlorinated hydrocarbons attacked the Eider population in the Netherlands. *TNO-nieuws*, 27, 556–60.

Tasker, M. L. and Pienkowski, M. W. (1987). Vulnerable concentrations of birds in the North Sea. Nature Conservancy Council, Aberdeen.

Taux, K. (1984). Brutvogelbestände an der Deutschen Nordseeküste im Jahre 1982— Versuch einer Erfassung durch die Arbeitsgemeinschaft 'Seevogelschutz'. *Seevögel*, 5, 27–37.

Taux, K. (1986). Brutvogelbestände an der deutschen Nordseeküste im Jahre 1984— Zweite Erfassung durch die Arbeitsgemeinschaft 'Seevogelschutz'. *Seevögel*, 7, 21–31.

Teeple, S. M. (1977). Reproductive success of Herring Gulls nesting on Brothers Island, Lake Ontario, in 1973. *Canadian Field Naturalist*, 91, 148–57.

Vauk, G. and Prüter, J. (1987). *Möwen*. Niederelbe Verlag, Otterndorf.

Wallin, K. (1984). Decrease and recovery patterns of some raptors in relation to the introduction and ban of alkyl-mercury and DDT in Sweden. *Ambio*, 13, 263–5.

Wanless, S. and Harris, M. P. (1984). Effect of date on counts of nests of Herring and Lesser Black-backed Gulls. *Ornis Scandinavica*, 15, 89–94.

Weseloh, D. V., Elliott, J. E., and Olive, J. H. (1988). Herring Gull surveillance program. *Canadian Wildlife Service, Report*, D-2016 G.

White, D. H., Mitchell, C. A., and Prouty, R. M. (1983). Nesting biology of Laughing Gulls in relation to agricultural chemicals in South Texas, 1978–81. *Wilson Bulletin*, 95, 540–51.

White, D. H., Mitchell, C. A., and Swinford, D. M. (1984). Reproductive success of Black Skimmers in Texas relative to environmental pollutants. *Journal of Field Ornithology*, 55, 18–30.

Species Management

22 Control of bird pest populations

CHRISTOPHER J. FEARE

Introduction

Certain birds cause problems that necessitate preventive action. Birds are responsible for economic losses in agriculture (Wright *et al*. 1980), for problems associated with fouling in towns and cities (Feare 1985), for public and animal health hazards (Summers *et al*. 1984; Tosh *et al*. 1970), for dangers to aircraft (Blokpoel 1976), and for certain environmental problems where abundant predators threaten more 'desirable' species (Thomas 1972) or where exotics threaten indigenous species (Feare and Mungroo 1990; Douglas 1972; Zeleny 1969). Birds have been killed as attempts to resolve all of these problems.

The aim of any measures taken should be to prevent problems or to reduce them to acceptable levels and such measures should not necessarily include attempts to impose artificial ceilings on bird populations. However, in the past considerable effort has been devoted to killing as many pest birds as possible in the name of control, with little attention being paid to assessments of the extent to which this killing resolved the problems (Dolbeer 1988). Generally, killing birds is the most environmentally dangerous, unselective, and expensive option for preventing bird damage and is recommended only when other options have been tried and have failed (Dyer and Ward 1977; Feare 1986). Furthermore, increasing public concern for animal welfare, sometimes embodied in legislation, together with the realization that many attempts to reduce bird populations in the past have not resulted in damage reduction (Murton 1968), have led to the search for solutions that do not involve killing (Feare *et al*. 1988).

As a result, much can now be achieved using non-lethal damage prevention techniques, e.g. scaring (Inglis and Isaacson 1978, 1984), physical protection (Feare and Swannack 1978) and changes in husbandry (Dolbeer *et al*. 1984; Greig-Smith 1987; Twedt and Glahn 1982). Other options are also under investigation, e.g. chemical repellents (Greig-Smith *et al*. 1983; Rogers 1980) and chemosterilants (Cyr and Lacombe 1988). Nevertheless, birds are still killed in attempts to reduce damage, but it is appreciated that the mortality imposed by killing programmes generally fails to reduce bird pest populations (Ward 1979; White *et al*. 1985; Dolbeer 1988). There has been little critical examination of the effects that such killing has on those populations, but killing is now usually under-taken as an aid to scaring (ADAS 1987) or is aimed at reducing damage locally (Ward 1979; Elliott and Allan 1989). Here, I shall discuss the population processes important in considerations of control involving killing birds, present some case histories of attempts to limit populations, and discuss factors that contribute to success or failure.

The dynamics of bird pest populations

Fundamental to any carefully designed programme of population control is a thorough understanding of the population processes involved in both the undisturbed and controlled pest populations. The crucial parameters are fecundity, mortality rates of different age/sex classes, the timing and cause of density-dependent mortality and of density-independent mortal-ity where this imposes a ceiling on population size, movements of the age-and-sex-groups, and the scale of natural fluctuations. In populations whose numbers remain more or less constant, recruitment and immigra-tion are balanced by mortality and emigration, but most populations exhibit fluctuations that may involve changes in all of these parameters. Furthermore, many bird pest populations are undergoing marked increases or decreases, rather than remaining stable (Dolbeer and Stehn 1979; Feare 1989; Owen et al. 1986; Inglis, Thearle and Isaacson, see below for Woodpigeons). This is to be expected, since many bird pests of agriculture are dependent upon agricultural practices which themselves change according to political and economic pressures and with scientific and engineering advances.

In addition to overall variability in numbers, some bird pests vary geographically in their dynamics, with expansion in some regions, declines in others and apparent stability elsewhere (Dolbeer and Stehn 1979; Feare 1987). Any 'thorough understanding' of the population processes of an avian pest should therefore involve studies of those processes throughout the geographical range (or at least where damage occurs and, if the birds involved are migrants, where they originate) and either continuously or at regular intervals. In reality, we have comprehensive data for no bird pest and it is notable that pest birds do not figure among those species for which models of lifetime reproductive success are being constructed (Fitzpatrick et al. 1988; Newton 1988; Rowley and Russell 1988).

With this background, damage prevention strategies aimed at reducing pest populations must be based on assumptions. In the past, the most costly control, in terms of resources, has been based on the assumption that we could reduce overall numbers of certain bird pests; the Quelea Quelea quelea in Africa is the prime example here, but similar hopes have stimulated attempts to reduce the numbers of Red-winged Blackbirds

Agelaius phoeniceus and European Starlings *Sturnus vulgaris* in North America (White *et al.* 1985), European Starlings in summer in Belgium (Tahon 1980) and in winter in France (ACTA/INRA/SPV 1987), and Woodpigeons *Columba palumbus* in Britain (Murton 1965). Even the assumption that killing large numbers of birds locally could reduce damage in the vicinity has often not been borne out in practice (Murton *et al.* 1972; Feare *et al.* 1981).

To prevent waste of resources in the future, it is essential that we understand the reasons for the failure of so many control attempts to achieve their various objectives. Even though we do not have comprehensive data on the relevant aspects of the dynamics of any bird pest population, examination of some of those that have been studied reveals important aspects underlying failure.

Red-billed Quelea in Africa

In South Africa between 1956 and 1960, 400 million Quelea were estimated to have been killed by aerial spraying with avicides in order to prevent the birds eating prodigious quantities of cereal crops. By 1961 the population was thought to be under control. However, in 1962–3 spraying was extended because a much greater area of farmland had been damaged. In 1966–7 a record 112 million Quelea were estimated to have been killed, clearly indicating that the population was not being limited by this approach (Crook and Ward 1968). Overall, Ward (1979) estimated that up to 1000 million Quelea were killed annually throughout Africa; more recent estimates of the Quelea population of Africa (*c.* 1500 million; Elliott 1988) suggest that the early estimates of numbers killed were excessive, but nevertheless the mortality inflicted by man must have been enormous and costly, without achieving any reduction in population size. Ward (1965) concluded that Quelea populations were limited by food shortage, especially towards the end of dry seasons when seed stocks on the ground became depleted. Ward (1979) emphasized the practical difficulties of attempting to eliminate Quelea colonies and roosts. The species has a huge geographical range embracing semi-arid savannah south of the Sahara and many of the colonies and roosts are inaccessible, rendering discovery and spraying extremely difficult. Many Quelea probably never feed in agricultural land and control aimed at reducing the African Quelea population would thus kill many birds that did not contribute to damage; this represents a waste of resources and an unnecessary contamination of the environment with avicides. Ward (1979) also noted that after birds in roosts had been successfully destroyed by spraying, numbers often built up again in the roost within a few weeks, indicating a high turnover of birds in a given area after control.

Even in the absence of data on natural mortality rates and rates of population turnover, Crook and Ward (1968) and Ward (1979) concluded that the imposed mortality failed to approach the level of density-dependent natural mortality caused by competition for food. To achieve population limitation, a percentage kill much higher than the annual mortality rate is needed, otherwise, in a population limited in a density-dependent way, imposed mortality may simply substitute for a part of the natural mortality. Ward (1979) therefore proposed that the 'total population control strategy' should be replaced by an 'immediate crop-protection strategy', whereby only colonies *close to important* cereal-growing areas should be culled, and then only when the crops are at a vulnerable stage of development. In this strategy, the timing of control operations is governed by crop phenology, rather than by selecting times when the Quelea population is most vulnerable to lasting reductions in local populations. Clearly the closer, in time, that control is undertaken to the period when crops are vulnerable, the less time there will be for immigration to counter the effect of the cull.

Ward's proposals have been largely implemented by Regional Quelea Control teams. However, owing to the difficulties of studying such a widespread opportunist with movements that embrace itinerancy, nomadism, and migration, we are still no nearer to assessing annual mortality rates or turnover rates, or to being able to predict movements with accuracy. Elliott and Allan (1989) have highlighted the difficulties of defining 'close to' and 'important' in Ward's (1979) criteria for selecting roosts or breeding colonies for treatment. Hence, even though advances have been made in study techniques, for example radio-tracking and mass marking, unpredictable behaviour, such as failing to breed or breeding twice in one year, militate against the precise application of the strategy. It is also difficult to undertake experiments of the effectiveness of roost or colony destruction on damage levels, since it is impossible to find comparable test and untreated sites. Even the birds from a single colony or roost can change their foraging behaviour rapidly in response to the availability of natural foods and their phenology.

It is thus accepted that the destruction of Quelea concentrations is not a satisfactory solution to damage problems (Elliott 1988; Manikowski 1988), but in the absence of alternative practicable forms of damage prevention and of more precise knowledge of the factors that promote damage, aerial spraying with avicides is likely to remain the tactic to be employed in local damage alleviation for the foreseeable future. Since severe damage by Quelea is extremely localized and they do not threaten food production on a national scale (Elliott 1988), we must continue to rely on assumptions when predicting potential damage and accept that much control effort is of an insurance nature. With this strategy, the major requirements are to reduce costs, to reduce the quantities of chemical applied (Parker 1986; Manikowski 1988; Meinzingen *et al*. 1989) and to increase the resources available for research so that particularly the meaning of 'close to' and 'important' (Ward 1979) can be more clearly defined. Equally important is the search for non-lethal means of preventing damage (e.g. Elliott 1979).

The European Starling in Belgium

Starlings can inflict severe financial losses to soft fruit and in Europe cherries are especially vulnerable (Feare 1984). In Belgium, attempts were made to alleviate the problem by killing large numbers of Starlings in the cherry-growing region of Sint-Truiden by dynamiting the roosts. The expectation was that the Starling population would fall to acceptable levels and that losses incurred by the cherry growers would be minimized.

Each year, the Starling population of the cherry-growing area increases rapidly in the latter half of June (Stevens 1982). Ringing of birds in breeding colonies in Belgium has shown that some of the Starlings originate over distances at least 100 km from the cherry-growing area (Tahon 1980), and estimates of the Starling population of the cherry-growing area revealed that about 500 000 birds were present in summer. At first, birds assemble at night in a number of small roosts, but dynamiting cannot commence until these roosts coalesce into larger roosts. Consequently, dynamiting did not usually commence until early July (median date for 1972–8 was 9 July; Tahon 1980), by which time much damage had already been inflicted; thus the control measures protected only the later-ripening cultivars.

Roosts could only be dynamited if they formed in woods well away from human habitation; this constraint was at least partly responsible for limiting the number of birds killed annually to 28 000–280 000 (average 107 000) between 1972 and 1978 (Tahon 1980), on average representing about 20 per cent of the population. The percentage of juveniles in the birds killed in night roosts varied from 10 to 78, averaging 48 per cent (Tahon 1980), while daytime observations of birds in study areas within the cherry-growing area revealed that juveniles comprised 99.6 per cent of the population. Both techniques of estimating age composition are subject to bias. First, there is a spatial separation of age groups within roosts (Summers *et al*. 1986) such that dynamiting operations concentrated in the centre of roosts would produce a predominance of older birds. Second, adults and juveniles tend to segregate when feeding in summer (Feare 1984; Stevens 1985) so that the age groups may not be equally visible in field surveys. Generally, however, flocks of Starlings that feed in cherry orchards are heavily biased towards juveniles (Brown 1974; Feare 1984).

The number of birds that appeared each year did not fall (Tahon 1980), indicating that this summer population was not being reduced by the control operations. We can understand how this came about by making further assumptions about the age composition of the population in the area. First clutches of Starlings in north-western Europe average about 5 eggs, most of which are successful in terms of producing a fledgling. Second clutches are generally smaller and less successful and only a proportion, which varies annually, of breeding adults lays a second clutch (Feare 1984). At the end of the breeding season, therefore, pairs of adults may have produced on average about 7 flying young. Although adequate data are lacking, juvenile mortality is probably high in summer (Feare 1984) and if we assume that 30 per cent of the young birds die before the dynamiting operations began, this leaves 2 adults to every 5 young, or about 70 per cent young in the population. If these birds remained together during the summer, the 500 000 Starlings estimated to be in the cherry-growing area each summer (Tahon 1980) would comprise 150 000 adults and 350 000 juveniles. Of the 107 000 birds killed annually, 55 640 (52 per cent) would be adults and 51 360 would be juveniles. Based on these figures, we can calculate that the proportion of adults killed each year would be 37 per cent and the proportion of juveniles killed would be 15 per cent. However, after the breeding season, adults tend to be more sedentary than the young (Feare 1984) and the latter, as stated above, are attracted in large numbers to cherry orchards. Thus it seems likely that the proportion of juveniles in the 500 000 summer population would be higher than 70 per cent and the proportion of juveniles killed in the dynamiting operations would be correspondingly lower. Available estimates suggest that first-year mortality exceeds 60 per cent and that much of this occurs during the summer (Feare 1984), so that the mortality imposed in these control operations represents only a fraction of the natural mortality of juveniles. With the adults, on the other hand, the 37 per cent mortality inflicted by this form of control is around the level of the natural annual mortality (which in Britain is currently about 36 per cent; Feare 1984) and, as most of the adults killed are likely to be local birds, their numbers may well be reduced by the control.

The destruction of a roost did produce a reduction in numbers locally for a short period, but after a few days numbers increased again, indicating considerable bird mobility within the area (Stevens 1982), as observed also by Feare et al. (1981). The only long-term effect recorded was an increase in the proportion of juveniles in areas that were regularly dynamited (Tahon 1980; Stevens 1982); this would be expected if the small breeding population of the cherry-growing area was being reduced by the control operations, as suggested above. The control thus failed to reduce the overall population and reduced damage only to a limited extent, but may have reduced the small breeding population whose production made an insignificant contribution to the birds causing damage.

Feral Pigeons in towns

In urban areas, Feral Pigeons cause a number of problems associated with fouling, food contamination, and the potential of disseminating disease (Feare 1986). Increasingly, damage prevention is coming to rely on techniques for physically preventing pigeons from alighting on or entering buildings, since these provide more lasting solutions than have attempts to kill the offending birds. In addition to the failure to produce lasting population reductions, killing programmes have proved expensive because so many birds have to be removed. For example, Martin and Martin (1982) killed 582 pigeons at one site in 1.5 months, even though there were never more than 100 birds at that site. Similarly, Murton *et al.* (1972) killed over 9000 pigeons, over a 3-year period, but only managed to reduce the original population of 2600 birds by half.

These examples of attempts to control the Feral Pigeon indicate a high turnover of individuals through immigration and emigration at these sites, which was later verified by Lefebvre (1985) who monitored individually marked birds in flocks where no control was attempted. Turnover is probably enhanced during control operations, partly as a result of removing some of the birds, but also due to the techniques used. Feral Pigeons are generally killed by trapping, poisoning (Mix 1985), or by using stupefying baits (Thearle *et al.* 1971), all of which are techniques relying on the attraction of target birds to a bait. Murton *et al.* (1972) found that populations were limited by their food supply since, even though food was widely available, access was limited by frequent disturbances. Whenever pigeons did feed, competition at food patches was intense. Only about one-third of adults in the population bred, these being the dominant birds at food patches. These dominant individuals tended to feed at preferred sites where food was most reliably available and the laying of bait, which represented new food patches, attracted immigrants rather than diverting dominant birds from their existing food. Thus control operations of this kind cropped the more mobile non-breeding segment of the pest population.

Annual adult mortalities in different colonies are highly variable, ranging from 30 per cent to 85 per cent in undisturbed situations (Kautz 1985) and in the first year of life mortality is higher. Following harvest of 20–40 per cent of three colonies, Kautz (1985) found that egg and chick losses *declined* and, while there was no evidence of a reduction in the mortality of immatures, there was a weak suggestion of a decrease in adult mortality. Another compensatory reduction, this time in egg and chick mortality, was recorded by Preble and Heppner (1981) when they removed 20 per cent of eggs from a colony. Kautz's (1985) finding that different colonies experience widely differing mortalities and breeding success indicates that control programmes should be tailored to the colony in question. This is impracticable, however, as relevant data are not available for the colonies that are to be controlled. Furthermore, compensatory changes in production and mortality, immigration, and failure to kill the breeding segment of Feral Pigeon populations militate against the success of a control programme of this kind as a means of damage prevention.

Woodpigeons in England

The Woodpigeon is probably the most economically damaging bird pest in British agriculture and has been identified as that pest of oilseed rape crops which is most frequently cited as being responsible for driving farmers out of production (Lane 1984). In the 1960s, high populations, which were considered to cause serious damage to clover and sown cereals, stimulated studies by R. K. Murton and co-workers at Carlton, Cambridgeshire (Murton 1965; Murton et al. 1966). The monitoring of this population has been continued up to 1986 by A. J. Isaacson, and Murton's plus subsequent data are being analysed by I. R. Inglis, R. J. P. Thearle, and A. J. Isaacson (unpublished), who have kindly provided information relevant to this paper.

In the 1960s, the population of the Carlton study area reached a maximum in November, the total representing the number of adults plus the number of young produced in the year. The number of Woodpigeons that survived the winter was related to the amount of available clover (Murton 1968; Murton et al. 1966). Over-winter mortality was determined by a density-independent threshold level of clover availability in January and February and juveniles suffered particularly heavy mortality, most of them dropping below a critical 'starvation' weight of 450 g. In the 1960s, the cost of cartridges for Woodpigeon shooting attracted a government subsidy, the aim of this being to encourage shooting in order to reduce the bird population and thereby to reduce agricultural damage. Two kinds of shooting were employed: 'battue' shooting, where birds were shot as they returned to roost communally in woods at night, and 'decoy' shooting, where birds attracted to decoys placed in fields close to a hide were shot. Up to 60 per cent of the birds at Carlton were shot over the winter (Murton et al. 1974), but this did not influence the numbers present at the end of the winter, as the April population was still determined by clover availability. The population of the study area in spring was attained by mortality in excess of that imposed by shooting or, if clover stocks were high, by immigration that nullified the effects of shooting. The ineffectiveness of shooting was further demonstrated following its cessation for 3 years after the 1964–5 winter; this did not lead to an increase in the Woodpigeon population (Murton *et al*. 1974).

During Murton's studies, the amount of clover available in winter fell and, in response to this, the Woodpigeon population declined. Since the early 1970s, however, a dramatic change in agricultural practices, involving the introduction of oilseed rape as a break crop in cereal rotations, has had equally dramatic effects on the over-winter population processes of Woodpigeons. Although the population of the Carlton study area has increased over the period 1975–86, it is still lower than in the early 1960s. In the latter period, the breeding season has been advanced so that most young have fledged by September. In October, there was some density-dependent regulation related to the availability of waste grain in stubble fields and, in contrast to the 1961–70 period studied by Murton, there was no peak of numbers in November. Over-winter numbers were determined in a density-independent way by the availability of oilseed rape in November. Then, numbers remained constant from November to March and clover was not a determinant of numbers in any month, the birds having switched to a diet consisting almost entirely of oilseed rape leaves (Inglis found that the crops of shot birds contained 94 per cent of oilseed rape). Furthermore, the weights of juveniles in 1975–86 did not fall over winter and no birds were recorded below the starvation threshold of 450 g. Thus the growing of oilseed rape has removed the over-winter starvation and the population regulation that occurred because of this and, presumably in response to this, the Woodpigeon population has increased.

This current lack of regulation over-winter has implications with regard to damage prevention, for it means that any mortality imposed in a control programme may now be less likely to substitute for natural mortality. Thus it should now be possible to reduce Woodpigeon numbers in a given area artificially. However, it is not known whether this would lead to immigration, nor is the extent known to which the population would have to be reduced to lead to a significant diminution of damage. It is still, therefore, not possible to assess whether a reduction of the Woodpigeon population would be a cost effective way of alleviating damage to oilseed rape.

Failure and success of control

The foregoing examples have shown that attempts to reduce bird pest populations over large areas have failed in this objective. There are two basic reasons for this failure:

1. Most pest species have a wide geographical range and much of the population is inaccessible to control operations. Immigration from neighbouring areas where control is not practised readily counteracts the

temporary reductions caused by control. After most control attempts, rapid in-filling of depopulated areas occurs, especially where roosts have been the target of control. Such rapid turnover at roost sites may be a response to the removal of birds, but even in undisturbed situations it appears that population turnover in night roosts may be high (Morrison and Caccamise 1985).

2. Control attempts can be counteracted by compensatory changes in production and survival. A successful reduction in the size of a breeding colony of Herring Gulls *Larus argentatus* by killing nesting adults over a number of years resulted in a reduction of the age of first breeding and an increase in the breeding success of the remaining birds. In addition, these birds laid larger eggs and the breeding adults were larger than had been the breeders in the undisturbed colony. These changes were attributed to reduced intraspecific competition during the breeding season when the density of nesting birds had fallen (Coulson *et al*. 1982) and illustrate the complexity of population responses to artificial control.

The failure of strategies of total population reduction of pest species discussed above, together with an understanding of the reasons for failure, indicates that an overall population reduction strategy is impracticable. Furthermore, the rarely quantified but nevertheless large effort that is required to attain the kills achieved to date indicates that the application of even greater effort to achieve larger or more widespread mortality would not be cost-effective in terms of the damage alleviated. Also, uncontrolled killing would be environmentally undesirable due to the destruction techniques used and the unpredictable effects of removing so many birds. The reasons for failure outlined above do, however, provide pointers to situations where population limitation might reduce damage.

With small, isolated populations where a substantial reduction in numbers is achievable, lack of immigration would prevent the rapid re-population seen in the examples above. The House Crow *Corvus splendens*, which was introduced and is now well-established on Mauritius, is amenable to such control and annual culls could maintain the population at a low level (Feare and Mungroo, 1990). There are examples where populations of birds on islands have become pests and been subse-quently extirpated, e.g. the parakeet *Psittacula wardi* in the Seychelles (Penny 1974), although habitat change doubtless contributed to its demise. The maintenance of a reduced population of Herring Gulls on the Isle of May, Scotland, required annual culls as the reduced density of nest-ing birds proved attractive to settlers; these recruits consisted of many birds that had been raised in the colony but which would, in the absence of control, have dispersed to seek nest sites elsewhere (Duncan 1978; Coulson *et al*. 1982). This gull control was undertaken to prevent predation of tern colonies and to avoid degradation of the island's vegetation by the large numbers of nesting birds; here, it could be reasonably assumed that reducing the number of gulls would reduce the damage caused by them.

In North America, large numbers of blackbirds (Icteridae) and European Starlings have been killed in roosts with the specific aim of eliminating foci of infection of histoplasmosis, the causative agent of which proliferates in the faecal deposits beneath these roosts (Tosh *et al*. 1970). Elimination of roosts that become established near human habitation is considered preferable to dispersing the birds, which would simply establish new foci of infection. The strategy adopted here is thus akin to Ward's (1979) 'immediate crop protection strategy', rather than the unsuccessful attempts that were made to reduce the national populations of these birds in order to alleviate agricultural damage (White *et al*. 1985).

Most bird pests are short-lived with a high reproductive rate and high population turnover but some, like the Herring Gull mentioned above and several geese, have lower annual mortalities and reproductive rates and their populations are thus less able to respond to excessive mortality. Here, it should be possible to produce local or even more widespread and lasting reductions in numbers, as suggested by the recent history of the Dark-bellied Brent Goose *Branta bernicla bernicla* migrating through and wintering in Western Europe. An increase in the world population began in the early 1970s. This was attributed by Ogilvie and St Joseph (1976) to a run of years with above-average breeding success, combined with the cessation of hunting in Denmark. The geese migrate on a relatively narrow flyway and are thus vulnerable to shooting concentrated on their migration route, and in 1965 and 1966, 4.1 per cent and 8.4 per cent respectively, of the world population were shot in Denmark (Swift and Harrison 1979). The intensity of this shooting could probably be increased if the aim were to reduce the population to prevent agricultural damage. However, in this case the population reduction would have to take place a considerable distance from the area where most agricultural damage occurred, in south-east England, because here the birds are too widely dispersed to permit the required intensity of control. Clearly, a strategy such as this would require a high degree of international cooperation and complex funding, neither of which are very likely in a case such as this. The situation of the Dark-bellied Brent Goose is further complicated, however, by public concern and conservation interest which have resulted in protection under European legislation.

There are also dangers in applying a strategy of population reduction to a species with low reproductive rate, low annual mortality, and low population turnover. In such cases it might be relatively easy to impose culls sufficient to reduce population size. Moreover, natural mortality fluctuates and intensive control in a year when unusually heavy mortality occurred could effect a serious and long-lasting depletion of the population. This danger could be averted only by a continuous monitoring of relevant population parameters and a continuous adjustment of control effort to prevailing conditions, representing a highly sophisticated management of the population. However, it is unlikely that control could be regulated with such sensitivity and in any case, the amount of damage done by most bird pests would not warrant the high cost of the monitoring and associated control. Nevertheless, the dangers inherent with such vulnerable species constitute the most persuasive argument against allowing unregulated shooting of Dark bellied Geese, which hunting organizations wish to be returned to quarry lists (Swift and Harrison 1979).

Conclusion

When selecting a damage prevention strategy, the simplest, cheapest, and environmentally safest options should be considered first (Dyer and Ward 1977; Feare 1986). On account of its cost, potential environmental harm, and unpredictable outcome, killing should be regarded as a last resort. The examples given in this paper show that it is usually impracticable to produce sufficiently large and long-lasting reductions in populations to alleviate the damage problems identified. Killing can therefore only be expected to be a useful means of damage prevention in localized and somewhat special situations, so that the imposition of artificial limits on numbers is of very limited use in bird pest control.

It is thus essential that a damage prevention strategy involving lowering numbers has well-defined objectives relating to the damage to be prevented or reduced, and the feasibility of attaining these objectives may only be assessed through research. This research should address the extent to which numbers must be reduced, the practicality of achieving this reduction, any side effects of the techniques selected on the target population and on other habitat components, and the costs of the operations in relation to the benefits of reduced damage. Throughout the period of control, which may be prolonged, adequate monitoring should ensure that the objectives are being achieved; if they are not, the control operation should be adjusted or abandoned. At the same time, alternative means of damage avoidance should be sought in attempts to reduce costs and to identify more environmentally acceptable methods.

Unfortunately, this ideal approach is rarely followed because political pressures demand more instant solutions to damage problems and funding is often inadequate to allow the necessary research. For these reasons, it is inevitable that killing will continue to be advocated as a damage prevention tactic and will be practised, as in Quelea control, even in the absence of sound evidence that the agricultural damage is actually prevented or reduced.

Summary

Attempts are made to limit the damage done by bird pests by killing large numbers of individuals in the expectation that by so doing the population size will be reduced. This paper describes such attempts to reduce damage by Red-billed Quelea in Africa, European Starlings in Belgium, Feral Pigeons in towns, and Woodpigeons in England. These examples illustrate that the killing of large numbers of pest birds usually fails to reduce population size and offers only limited protection from damage. Failure results from rapid immigration into areas where birds have been killed and from compensatory changes in production, survival, and recruitment. At present, our knowledge of the population dynamics of most bird pests is inadequate to permit prediction of the effects of killing birds, usually at considerable expense, and we cannot therefore assess the cost-effectiveness of such control operations in reducing damage. Further research is needed to allow quantification of the parameters that militate against the success of attempts to alleviate damage by killing birds, and all such control operations should be closely monitored to assess costs, efficacy, and environmental side-effects.

Acknowledgements

I am grateful to Ian Inglis for allowing me to quote his unpublished data on the Woodpigeon and to Tony Hardy, Jim Nichols, and John Coulson for comments on the manuscript.

References

ACTA/INRA/SPV (1987). *Les Etourneaux Sansonnets en France*. Cyclostyled report of the ACTA/INRA/SPV working group 'Etourneaux'.

ADAS (1987). *Bird scaring*. Ministry of Agriculture, Fisheries and Food, Pamphlet **903**.

Blokpoel, H. (1976). *Bird hazards to aircraft*. Irwin and Company, Canada.

Brown, R. G. B. (1974). Bird damage to fruit crops in the Niagara Peninsula. *Canadian Wildlife Service Report*, **27**, 1–57.

Coulson, J. C., Duncan, N., and Thomas, C. (1982). Changes in the breeding biology of the Herring Gull (*Larus argentatus*) induced by reduction in the size and density of the colony. *Journal of Animal Ecology*, **51**, 737–56.

Crook, J. H. and Ward, P. (1968). The Quelea problem in Africa. In *The problem of birds as pests*, (eds R. K. Murton and E. N. Wright) Academic Press, London.

Cyr, A. and Lacombe, D. (1988). La stérilisation est elle une avenue fertile pour contrôler les populations d'oiseaux nuisibles? *Proceedings of the XIX International Ornithological Congress, Ottawa*, pp. 484–92.

Dolbeer, R. A. (1988). Current status and potential of lethal means of reducing bird damage in agriculture. *Proceedings of the XIX International Ornithological Congress, Ottawa*, pp. 474–83.

Dolbeer, R. A. and Stehn, R. A. (1979). *Population trends of Blackbirds and Starlings in North America 1966–1976*. United States Department of the Interior, Fish and Wildlife Service Special Scientific Report No. 214.

Dolbeer, R. A., Woronecki, P. P., and Stehn, R. A. (1984). Blackbird (*Agelaius phoeniceus*) damage to maize: crop phenology and hybrid resistance. *Protection Ecology*, 7, 43–63.

Douglas, G. W. (1972). Ecological problems caused by introduced animals and plants. *Victoria Research*, 14, 1–6.

Duncan, N. (1978). The effects of culling Herring Gulls (*Larus argentatus*) on recruitment and population dynamics. *Journal of Applied Ecology*, 15, 697–713.

Dyer, M. I. and Ward, P. (1977). Management of pest situations. In *Granivorous birds in ecosystems*, (ed. J. Pinowski, and S. C. Kendeigh). Cambridge University Press, Cambridge, pp. 267–300.

Elliott, C. C. H. (1979). The harvest time method as a means of avoiding Quelea damage to irrigated rice in Chad/Camaroun. *Journal of Applied Ecology*, 16, 23–35.

Elliott, C. C. H. (1988). Do bird pests contribute to famine in Africa? *Proceedings of the XIX International Ornithological Congress, Ottawa*, pp. 455–63.

Elliott, C. C. H. and Allan, R. G. (1989). Quelea control strategies in practice. In *The Quelea—Africa's bird pest*, (ed. R. C. Bruggers and C. C. H. Elliott). Oxford University Press, Oxford, pp. 317–26.

Feare, C. (1984). *The Starling*. Oxford University Press, Oxford.

Feare, C. J. (1985). Humane control of urban birds. In *The humane control of land pests and birds*, (ed. D. P. Britt) pp. 50–62. Universities Federation for Animal Welfare, London.

Feare, C. J. (1986). Pigeons: past, present and prerequisites for management. *Proceedings of the 7th British Pest Control Association Conference*.

Feare, C. J. (1987). Where have all the Starlings gone: or have they? *BTO News*, 149, 6.

Feare, C. J. (1989). The changing fortunes of an agricultural bird pest: the European Starling. *Agricultural Zoology Reviews*, 3, 317–42.

Feare, C. J. and Mungroo, Y. (1990). The status and management of the House Crow (*Corvus splendens*) in Mauritius. *Biological Conservation*, 51, 63–70.

Feare, C. J. and Swannack, K. P. (1978). Starling damage and its prevention at an open-fronted calf yard. *Animal Production*, 26, 259–65.

Feare, C. J., Isaacson, A. J., Sheppard, P. M., and Hogan, J. M. (1981). Attempts to prevent Starling damage at dairy farms. *Protection Ecology*, 3, 173–81.

Feare, C. J., Greig-Smith, P. W., and Inglis, I. R. (1988). Status and potential of non-lethal means of reducing bird damage in agriculture. *Proceedings of the XIX International Ornithological Congress, Ottawa*.

Fitzpatrick, J. W., Woolfenden, G. E., and McGowan, K. J. (1988). Sources of variance in lifetime fitness of Florida Scrub Jays. *Proceedings of the XIX International Ornithological Congress, Ottawa*, 1, 876–91.

Greig-Smith, P. W. (1987). Bud-feeding by Bullfinches: methods for spreading damage evenly within orchards. *Journal of Applied Ecology*, 24, 49–62.

Greig-Smith, P. W., Wilson, M. F., Blunden, C. A., and Wilson, G. M. (1983). Budeating by Bullfinches (*Pyrrhula pyrrhula*) in relation to the chemical constituents of two pear cultivars. *Annals of Applied Biology*, 103, 335–43.

Inglis, I. R. and Isaacson, A. J. (1978). The responses of Dark-bellied Brent Geese to models of geese in various postures. *Animal Behaviour*, 26, 953–58.

Inglis, I. R. and Isaacson, A. J. (1984). The responses of Woodpigeons (*Columba palumbus*) to pigeon decoys in various postures: a quest for a super-normal alarm stimulus. *Behavior*, 90, 224–40.

Kautz, J. E. (1985). Effects of harvest on Feral Pigeon survival, nest success and population size. Unpublished Ph.D. Thesis, Cornell University.

Lane, A. B. (1984). An enquiry into the response of growers to attacks by insect pests in oilseed rape (*Brassica napus* L.), a relatively new crop in the United Kingdom. *Protection Ecology*, 7, 73–8.

Lefebvre, L. (1985). Stability in flock composition of urban pigeons. *Auk*, 102, 886–8.

Manikowski, S. (1988). Aerial spraying of Quelea. *Tropical Pest Management*, 34, 133–40.

Martin, C. M. and Martin, L. R. (1982). Pigeon control: an integrated approach. *Proceedings of Vertebrate Pest Conference*, 10, 190–2.

Meinzingen, W. W., Bashir, El S. A., Parker, J. D., Heckel, J.-U., and Elliott, C. C. H. (1989). Lethal control of Quelea. In *The Quelea—Africa's bird pest*, (ed. R. L. Bruggers and C. C. H. Elliott) Oxford University Press, Oxford, pp. 293–316.

Mix, J. (1984). Public relations problems erupt when Sun tackles pigeon program. *Pest Control*, September 1984, 20–2.

Morrison, D. W. and Caccamise, D. F. (1985). Ephemoral roosts and stable patches? A radiotelemetry study of communally roosting Starlings. *Auk*, 102, 793–804.

Murton, R. K. (1965). *The Wood Pigeon*. London, Collins.

Murton, R. K. (1968). Some predator-prey relationships in bird damage and population control. In *The problems of birds as pests*, (ed. R. K. Murton and E. N. Wright) Academic Press, London, 157–69.

Murton, R. K., Isaacson, A. J., and Westwood, N. J. (1966). The relationships between wood-pigeons and their clover food supply and the mechanism of population control. *Journal of Applied Ecology*, 3, 55–96.

Murton, R. K., Thearle, R. J. P., and Thompson, J. (1972). Ecological studies of the Feral Pigeon (*Columba livia*) var. I. Population, breeding biology and methods of control. *Journal of Applied Ecology*, 9, 875–89.

Murton, R. K., Westwood, N. J., and Isaacson, A. J. (1974). A study of woodpigeon shooting: the exploitation of a natural animal population. *Journal of Applied Ecology*, 11, 61–81.

Newton, I. (1988). Individual performance in Sparrowhawks: the ecology of two sexes. *Proceedings of the XIX International Ornithological Congress, Ottawa*, 1, 125–54.

Ogilvie, M. A. and St. Joseph, A. K. M. (1976). Dark-bellied Brent Geese in Britain and Europe, 1955–1976. *British Birds*, 69, 422–39.

Owen, M., Atkinson-Willes, G. L., and Salmon, D. G. (1986). *Wildfowl in Great Britain*. Cambridge University Press, Cambridge.

Parker, J. D. (1986). A novel sprayer for the control of Quelea birds. *Tropical Pest Management*, 32, 243–5.

Penny, M. (1974). *The birds of Seychelles*. Collins, London.

Preble, D. E. and Heppner, F. H. (1981). Breeding success in an isolated population of Rock Doves. *Willson Bulletin*, 93, 357–62.

Rogers, J. G. (1980). Conditioned taste aversion: its role in bird damage control. In

Bird problems in agriculture, (ed. E. N. Wright, I. R. Inglis, and C. J. Feare) British Crop Protection Council, Croydon, pp. 173–7.

Rowley, I. and Russell, E. M. (1988). Lifetime reproductive success in *Malurus splendens*: A co-operative breeder. *Acta XIX Congressus Internationalis Ornithologici*, **1**, 866–75.

Stevens, J. (1982). The effect of TNT-actions on the starling (*Sturnus v. vulgaris* L) population in the Belgian cherry-growing area in the fruit-period. *Annales Société Royale Zoologique de Belgique*, **112**, 205–16.

Stevens, J. (1985). Foraging success of adult and juvenile starlings (*Sturnus vulgaris*): a tentative explanation for the preference of juveniles for cherries. *Ibis*, **127**, 341–7.

Summers, R. W., Prichard, G. C., and Brookes, H. B. L. (1984). The possible role of Starlings in the spread of TGE in pigs. *Proceedings of Bird Control Seminar*, **9**, 303–6.

Summers, R. W., Westlake, G. E., and Feare, C. J. (1986). Differences in the ages, sexes and physical condition of starlings *Sturnus vulgaris* at the centre and periphery of roosts. *Ibis*, **129**, 96–102.

Swift, J. A. and Harrison, J. G. (1979). Shooting of *Branta bernicla* in Europe. Proceedings of the First Technical Meeting on Western Palaearctic Migratory Bird Management, pp. 152–62.

Tahon, I. (1980). Attempts to control Starlings at roosts using explosives. In *Bird problems in agriculture*, (ed. E. N. Wright, I. R. Inglis, and C. J. Feare). British Crop Protection Council, Croydon, pp. 56–68.

Thearle, R. J. P., Murton, R. K., Senior, M. M., and Malam, D. S. (1971). Improved stupefying baits for the control of town pigeons. *International Pest Control*, March/April, 11–19.

Thomas, G. J. (1972). A review of gull damage and management methods on nature reserves. *Biological Conservation*, **4**, 117–27.

Tosh, F. E., Doto, I. L., Beecher, S. B., and Chin, D. T. Y. (1970). Relationship of Starling-Blackbird roosts and endemic histoplasmosis. *American Review of Respiratory Diseases*, **101**, 283–8.

Twedt, D. J. and Glahn, J. F. (1982). Reducing Starling depredations at livestock feeding operations through changes in management practices. *Proceedings of Vertebrate Pest Control Conference*, **10**, 159–63.

Ward, P. (1965). Feeding ecology of the Black-faced Dioch, *Quelea quelea*, in Nigeria. *Ibis*, **107**, 173–214.

Ward, P. (1979). Rational strategies for the control of Queleas and other migrant bird pests in Africa. *Philosophical Transactions of the Royal Society, London [B]*, **287**, 289–300.

White, S. B., Dolbeer, R. A., and Bookhout, T. A. (1985). Ecology, bioenergetics and agricultural impact of a winter-roosting population of Blackbirds and Starlings. *Wildlife Monographs*, **93**, 1–42. ·

Wright, E. N., Inglis, I. R., and Feare, C. J. (1980). *Bird problems in agriculture*. British Crop Protection Council, Croydon.

Zeleny, L. (1969). Starling versus cavity nesting birds. *Atlantic Naturalist*, **24**, 158–61.

23 The population dynamics of culling Herring Gulls and Lesser Black-backed Gulls

J. C. COULSON

Introduction

Since 1972, extensive culling of Herring Gulls *Larus argentatus* and Lesser Black-backed Gulls *L. fuscus* has taken place in Britain. The reasons for culling have varied. Some culling was justified on the grounds of nature conservation; to protect vegetation, to create space for terns *Sterna* to nest or to reduce predation on nests and young of other seabird species (Anon. 1972, 1981). Elsewhere the culling was introduced to reduce the risk of 'bird-strikes' on aircraft near airbases (Blokpoel 1976) or reduce the nuisance caused by gulls nesting in towns, e.g. Scarborough and South Shields (Local Environmental Officers, *personal communications*). Large culls have also been made to reduce the risk of pollution of water supplies and to protect the quality of grouse moors, e.g. Abbeystead, Lancashire (M.A.F.F. *personal communication*). As a result of these activities, nearly 100 000 Herring and Lesser Black-backed Gulls have been killed in Britain between 1972 and 1987.

It is implicit in all the documentation produced to justify culls that the reduction in numbers is solely within the colony being culled. Nowhere is there even a hint suggesting that culling could influence the numbers of the gull species beyond the limits of the colony. There is a fallacious belief in ornithology and conservation that a bird colony consists of a discrete population—a closed group of individuals whose reproduction is the only source of recruits. There is now much evidence that this is not the case, not only in seabirds but in birds generally. To express it differently, philopatry is only partial and a number of seabirds breed in colonies other than those in which they were born and reared. Clearly the degree of philopatry and the distribution of those individuals that do not exhibit philopatry is of major importance in evaluating or predicting the effects of a cull. Accordingly, this aspect is considered in detail in this paper.

Apart from the numbers of individuals killed, there are further potential effects of a cull. For example, does a cull affect the behaviour of the adults that survived the cull or of those individuals that are about to recruit to the breeding group? A cull could cause adults to emigrate or influence the place of nesting of birds about to breed for the first time, perhaps encouraging them to move elsewhere.

Another effect of a cull is to reduce the density of nesting birds. Thus a cull could be expected to produce a relaxation of density-dependent effects which may affect the age at first breeding, egg size, and breeding success.

The impact of culling on Herring and Lesser Black-backed Gull numbers has recently taken on a new importance because of the reported decline of the former, but not the latter, in Britain within the past 20 years. This dramatic change, after a continuous increase of about 13 per annum throughout this century, has raised the question of how much the decline of the Herring Gull has been influenced by culling. This paper presents data obtained from the cull of Herring and Lesser Black-backed Gulls on the Isle of May, Scotland and on the Farne Islands, N.E. England.

Philopatry and dispersal

Philopatry

Many Herring Gulls return and nest near the place of their birth. Tinbergen (1953) commented briefly on this, and many ringed Herring Gulls have been recovered as breeding adults in the colony of their birth. Having said this, it must also be stated that it is much more difficult to find individuals that have moved away from their place of birth; there is a much greater area to search for these individuals than that occupied by the natal colony. The degree of philopatry for Herring Gulls has been measured on the Farne Islands, Northumberland, and the Isle of May, Scotland (Fig. 23.1) and at the former for Lesser Black-backed Gulls.

Philopatry is marked in Herring Gulls. Of 1264 individuals ringed as chicks on the Isle of May and later culled there as adults, 65 per cent were killed in the same sub-colony in which they were ringed, although the expected value was only about 12.5 per cent (8 sub-colonies available). Philopatry was stronger in the males, with 77 per cent of those returning to the Isle of May breeding in the natal sub-colony compared with 56 per cent of females (Chabrzyk and Coulson 1976).

It has been possible to examine philopatry in both Herring and Lesser Black-backed Gulls on the Farne Islands, where the series of small islands readily separates breeding birds into discrete groups (Fig. 23.2). In both species, philopatry was high with 55 per cent of Herring Gulls and 53 per cent of Lesser Black-backed Gulls returning to breed on the specific island of birth (Table 23.1).

The Farne Islands can be divided into two main groups—the inner and outer groups of islands. Again, it is evident that there is a greater tendency

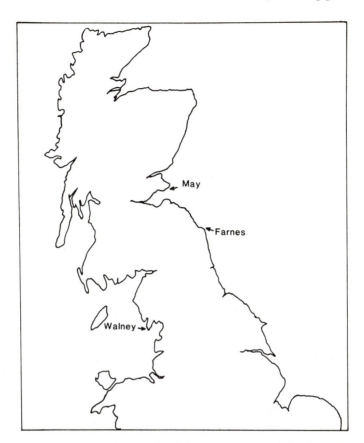

Fig. 23.1. Scotland and northern England showing positions of the Isle of May and of the Farne Islands. Walney is the nearest large colony of Lesser Black-backed Gulls to the Isle of May and the Farne Islands.

to return and breed on a neighbouring island rather than one in the group further away. Thus 92 per cent of Herring Gulls killed on the Farne Islands were culled on the same group of islands on where they were ringed as chicks. The comparable value for Lesser Black-backed Gulls was 82 per cent; the difference between the species is not significant.

Despite the highly significant tendency of Herring Gulls to nest close to their birth place, Chabrzyk and Coulson (1976) thought that only a minority of the surviving young did, in fact, breed in their natal colony. I have reanalysed the data using the additional recoveries that have been reported since 1974, and confirm this finding (Table 23.2). Probably less than 30 per cent of surviving young breed in the colony of their birth. Although we were aware of some Isle of May Herring Gulls nesting away

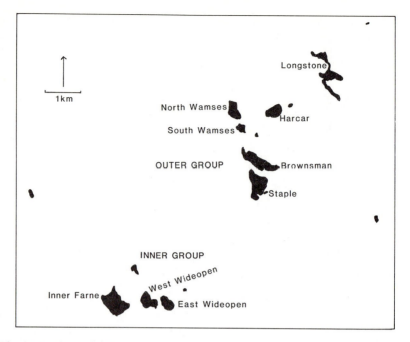

Fig. 23.2. The Farne Islands, showing the individual islands that form two groups—the inner and outer groups of islands. The shaded area is land at high tide.

from that island, the numbers involved were small. The information in Duncan and Monaghan (1977) increased the sample size, but the number of birds they reported were still nowhere near numerous enough to confirm the conclusion. Since the data in Table 23.2 depend upon estimates of the survival rates of immature and, to some extent, adult birds, it is necessary to consider whether the survival estimates used were too high. This is unlikely to be the case. Survival rates have to be high to produce the observed growth of the population during this century and the adult survival rate of 0.935 per year, rather than 0.90 or 0.87, gave the best results in the models of recruitment of young Herring Gulls into other colonies (see later).

Movement away from the natal area

Figure 23.3 shows the place of recovery of Herring Gulls ringed in the Firth of Forth (mainly the Isle of May) as chicks and recovered (other than in culls) more than 50 km away during the breeding season (May to mid-July) when 3 or more years old. It is assumed that these birds were recovered at or close to their chosen breeding area. Some individuals have been found up to 500 km away and most were found in or close to known colonies.

Table 23.1a Herring Gull philopatry on the Farne Islands, according to island of ringing and recovery

		Recovered							
		WW	EW	B	St	SW	NW	H	Total
Ringed	West Wideopens	2	1				1		4
	East Wideopens	2	14				2		18
	Brownsman			1	1	1	1		4
	Staple			2	2		1		5
	South Wamses					8	3	1	12
	North Wamses						9		9
	Harcar		2			3	6	1	12
	Longstone						1		1
	Total	4	17	3	3	13	23	2	65

	Inner	Outer
Ringed Inner	19	3
Outer	2	41

Table 23.1b Lesser Black-backed Gull philopatry on the Farne Islands, according to island of ringing and recovery

		Recovered								
		WW	EW	'W'	St	B	SW	NW	H	Total
Ringed	West Wideopens	89	17	—	2	1	8	3	0	120
	East Wideopens	24	50	—	1	1	6	11	1	94
	'Wideopens'	4	6	—	—	—	—	1	—	11
	Staple	nil								
	Brownsman	—	—	—	1	—	—	—	—	1
	South Wamses	5	4	—	1	—	28	17	—	55
	North Wamses	7	1	—	1	1	8	58	—	76
	Harcar	8	16	—	—	2	10	19	16	71
	Total	136	94	—	6	5	60	109	17	428

	Inner	Outer
Ringed Inner	190	35
Outer	41	162

There is a clear tendency for the birds to have recruited to breeding groups to the south of their natal area; this coincides with the observed winter dispersion of the Herring Gulls from the area (Parsons and Duncan 1978), suggesting that some of these birds may have remained and bred in the wintering area used whilst they were immature. This pattern of dispersal is similar to that reported by Duncan and Monaghan (1977) but is based on more data. Note, however, that only two recoveries were reported from the Farne Islands before culling started there.

Table 23.2 The number of young Herring Gulls carrying year-class rings which were estimated to be alive in 1973 compared with the numbers estimated to have returned to the Isle of May to breed

Year ringed	No. fledged*	Estimated no. survived to 1973†	Total no. culled or present on Isle of May	Percentage of survivors found on the Isle of May
1966	3943	2186	735	34
1967	4811	2852	872	31
1968	3896	2470	601	25
1969	360	244	56	23

* Number fledged from Parsons (1975).
† Based on life-table data.

Fig. 23.3. The distribution in the breeding season of Herring Gulls ringed in the Firth of Forth (mainly the Isle of May) and 3 or more years old. It is assumed that these birds are in their breeding areas. Birds recovered in culls are not shown.

Culling of gulls started on the Isle of May in 1972 and on the Farne Islands in 1976. It is of particular note that appreciable numbers of May-reared birds were culled whilst breeding on the Farne Islands, despite only 2 having been reported there as a result of natural mortality in previous years. Sine many May birds were killed in the first year of culling on the Farnes, it is evident that the birds had immigrated prior to, and not as a consequence of the cull. Ringing recoveries of gulls, other than those reported in culls, do not indicate the extent of the emigration and immigration.

In all, 117 Isle of May-reared chicks have now been culled on the Farne Islands. Table 23.3 presents these data in relation to the numbers ringed. Between 1962 and 1976, 4 per cent of the young gulls ringed on the Isle of May were subsequently culled on the Farne Islands. Allowing for mortality between the time of ringing and recovery, this probably represents 8 per cent of the production on the Isle of May. Since 1977, the proportion culled has increased appreciably reaching almost 14 per cent in the 1980–3 cohorts and corresponding to at least 20 per cent of the production of young. The value for 1980–3 is likely to increase further, since some of these birds are still alive and will be culled in future years.

Table 23.3 The proportion of Isle of May Herring Gulls moving to the Farne Islands as indicated by culling

Period of ringing	No. ringed on May	No. culled on Farnes	Recruits/1000 ringed
1962–67	10179	42	4.1
1968–72	6962	30	4.3
1973–76	1857	6	3.2
1977–79	535	5	9.3
1980–83	2468	34	13.8

Data includes 1987 cull.

The data in Table 23.4 also supports the existence of a change in the emigration rate, showing that there has been a large change in the ratio of Isle of May-reared chicks which were culled on the May and the Farnes. Little of this difference can be attributed to the culling intensity on the two islands. (Only 4 Herring Gulls ringed on the Farne Islands have been culled on the Isle of May but this low number is the result of only 250 young Herring Gulls being ringed on the Farnes from 1960 to date and not because of the lack of emigration.)

These data suggest that Herring Gulls reared on the Isle of May in the past 10 years are showing less philopatry than those produced previously but it is interesting to note that this change was not evident in the young

Table 23.4 The ratio of individual Herring Gulls from Isle of May cohorts culled on the Farne Islands and the Isle of May

Period of ringing	No. culled on May	No. culled on Farnes	Ratio May/Farnes
1962–67	1413	42	34 : 1
1968–72	747	30	25 : 1
1973–76	180	6	30 : 1
1977–79	36	5	7 : 1
1980–83	48	14	3 : 1

reared in the first 5 years of culling on the Isle of May but has developed more recently. One possible explanation is that the cull has resulted in a much higher proportion of the new breeding birds on the Isle of May being immigrants and that they, as immigrants, may be producing young which also lack the genetic requirements for philopatry.

Relatively few Lesser Black-backed Gulls have been ringed on the Isle of May. From those culled, about 2.7 per cent of ringed birds (53 out of 1949) had moved to the Farne Islands (*c*. 5 per cent of survivors) but there is no indication of the pronounced increase in emigration during the last 10 years found in Herring Gulls (Table 23.5). The movement of Lesser Black-backed Gulls reared on the Farnes to the Isle of May is about 0.9 per cent (26 out of 2936 ringed), which is one-third of the rate in the reverse direction. These values are lower but not significantly different from the equivalent rates for the Herring Gull.

Table 23.5 Proportion of Isle of May ringed Lesser Black-backed Gull chicks culled on Farnes

Period of ringing	No. ringed	No. culled	Percentage
1966–68	824	15	1.82
1969–76	292	9	3.08
1977–82	833	29	3.48

Note: recoveries up to and including 1987.

The results from the culls on the Isle of May and Farne Islands suggest that both Herring and Lesser Black-backed Gulls show marked philopatry, but that the majority of surviving young recruit away from the natal colony in decreasing numbers up to 500 km. Considering the many other colonies into which Isle of May-reared young Herring Gulls could and have colonized, rates of 70 per cent emigration and 30 per cent philopatry are not unreasonable.

The dynamics of the Isle of May cull

The cull on the Isle of May (Fig. 23.4) is the largest that has been carried out on a breeding colony (Duncan 1978). The culling has been in two phases; an intensive phase with large numbers culled and a lower intensity of culling to maintain the total numbers of both Herring and Lesser Black-backed Gulls at 3000 pairs. Figure 23.4 shows the numbers of Herring Gulls culled and the numbers breeding in each season prior to the annual cull. It is evident that the colony is being maintained at the new level by culling about 400 individuals each year.

Modelling the Isle of May cull

I have examined the effects of the Isle of May cull on Herring Gulls by taking the adult survival rate as 0.935 per annum (Chabrzyk and Coulson 1976; Coulson and Butterfield 1986), the initial population of 30 000

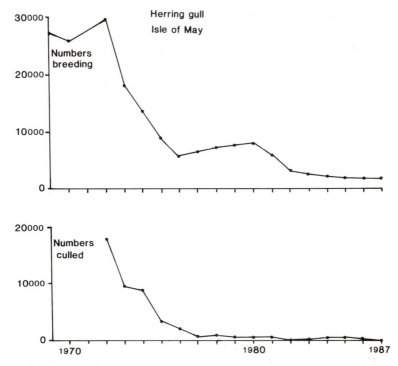

Fig. 23.4. The size of the Herring Gull colony on the Isle of May 1969–87, and the numbers of Herring Gulls culled from the start of the cull in 1972. The cull can be divided into two periods: the early, intensive culling and the later, less intensive culling used mainly to keep the colony size constant.

birds at the start of the 1972 breeding season, the year of the first cull, and counts of the numbers of breeding birds in each breeding season. The numbers of birds known to have been culled have been increased by 10 per cent to represent an unknown proportion of birds that died at sea and were not usually recovered (but some were washed ashore on adjacent coasts in years when easterly wind prevailed). The number of birds culled plus the natural mortality determines the numbers that survive to the next breeding season and these have been calculated. The number of recruits have been estimated from the difference between the survivors and the numbers nesting before each cull (Chabrzyk and Coulson 1976). Annual production for survival to breeding age (after applying estimates of the pre-breeding mortality) is about 0.195 per breeding adult. If the colony had been increasing at 13 per cent per annum (Chabrzyk and Coulson 1976) then in 1972 there would have been 5177 recruits, 0.195 of 26 549 (the 1971 numbers), to replace adult mortality and to increase the colony to 30 000 birds. Using this approach it has been possible to model the observed and the expected size of the colony as the cull progressed (Fig. 23.5a).

In the first 5 years of the cull, the observed decline was greater than expected. Therefore, either the input data in the model were inaccurate or the cull drove some gulls elsewhere. When the model is repeated using the later years of the cull (Fig. 23.5b), the agreement between observed and expected values is close, suggesting that the input survival and recruitment figures were realistic. Accordingly, it appeared that the cull drove 6000 breeders or potential breeders away. Colour-ringed breeding adults existed on the Isle of May and none of these were reported breeding elsewhere. It seems likely that the extensive culling activity drove away gulls about to recruit as breeding birds, thus making the reduction in breeding numbers greater than would have been expected from the size of the cull. This disturbance effect disappeared once the cull was reduced in intensity in the second half of the 1970s.

There are a number of implications of this model. First, the recruitment to the colony declined dramatically 4 years after the culling started, at which time the much reduced numbers of young, resulting from the first cull in 1972, were old enough to breed. However, since this recruitment is made up of only 30 per cent of birds born in the colony, the recruitment of immigrants into the colony must also have been affected. What is this effect? It may well be that culling deterred some of the normal immigration. It is likely that the numbers of immigrant recruits must be related in some way to colony size; small colonies attract fewer immigrants. Had immigration continued at the same numerical level as in 1972, then the colony could not have been maintained at about 3000 Herring Gulls without considerably higher culling levels than took place since 1975.

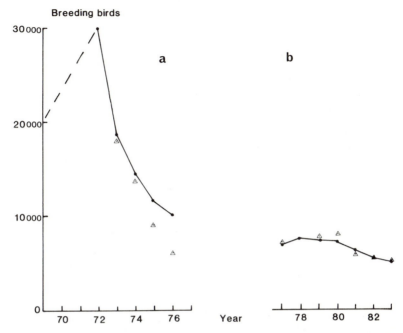

Fig. 23.5. (a) The observed and expected numbers of Herring Gulls on the Isle of May during the early years of the cull. The model is calculated from the numbers in 1972, before the start of culling. (b) The observed and expected numbers of Herring Gulls on the Isle of May. The model is calculated from the 1978 numbers.

Thus many of the recruits which, in the absence of the cull, might have been expected to have immigrated into the Isle of May had moved elsewhere. However, the size of the cull greatly reduced the number of Isle of May-reared Herring Gulls which emigrated even if the emigration rate remained unchanged.

These two effects, changes in emigration and immigration, tend to counterbalance each other, but overall the cull is acting as a sink, continuing to remove Herring Gulls from the whole area from which the immigrants originated. At the same time there is a reduction of emigration to the extensive area into which the large number of young formerly produced on the Isle of May contributed to the recruitment of adults. This matter can be better understood by modelling.

Model of inter-colony recruitment

1. Two discrete populations, A and B (Fig. 23.6).

This is the simplest of situations. There is not an exchange between colonies or populations A and B and hence each is dependent on its own

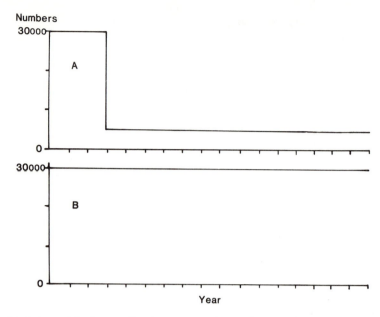

Fig. 23.6. A model of two gull colonies (A and B) with no exchange of birds, one of which (A) is culled.

production to replace adult mortality. If one is culled, this has no effect on the other.

2. Two colonies, C and D of the same size where there is an exchange between them of 50 per cent of their production of young (Fig. 23.7).

If the exchange is equal and other aspects are the same, e.g. survival rates, then if left undisturbed, each colony will remain stable or change at the same rate. However, if one is culled, reducing its size and recruitment follows the same pattern, then the size of both colonies will change, the first mainly because of the loss of adults and recruits, the second due to the loss of recruits only.

3. Two discrete colonies, E, and a group of colonies F–K (F–K total size twice that of E), all having 70 per cent emigration of young which move to the other colonies in direct proportion to their size. Colony E simulates the Isle of May situation (Fig. 23.8).

In this case, when colony E is culled to 10 per cent of its original size in one year and is kept at this level, there is a delayed but appreciable decrease in the total numbers breeding in E–K. For example, 13 years after the initial cull, reducing colony E to 10 per cent of its original size, the 'population' (colonies E to K) had declined to 60 per cent of the initial size, arising from the cull and also a 1.5 per cent decline each year in the size of colonies F–K after a time lag of 4 years.

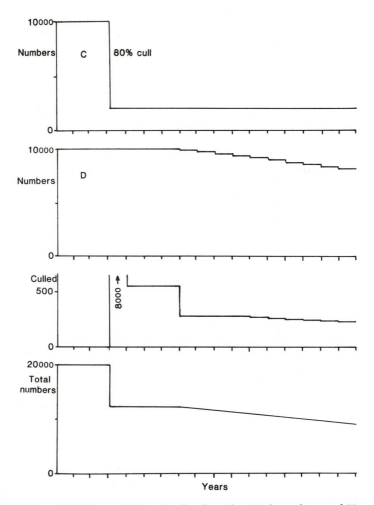

Fig. 23.7. A model of two colonies of gulls where there is the exchange of 50 per cent of the young production, of which one is culled.

It is possible to model many variations. For example, the effect of different areas being involved in immigration and emigration from the island, simulating a tendency to immigrate on to the Isle of May from more northern colonies but to emigrate mainly to the south. This will be the aim of future work.

The effects of reducing colony density

Following the intensive early culling on the Isle of May, the number of breeding gulls had been reduced by over 70 per cent. Despite this decline,

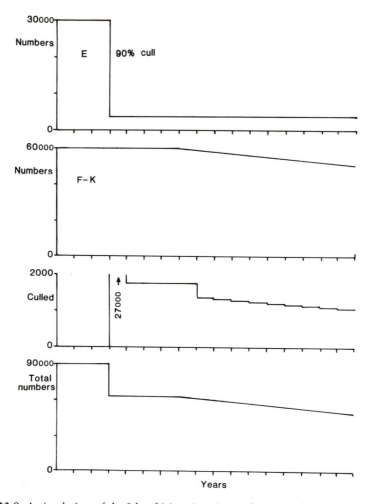

Fig. 23.8. A simulation of the Isle of May situation (colony E), where colonies F–K exist within the area of emigration from the May. All have 70 per cent emigration of the young production and these recruit into the other colonies in direct proportion to the colony size.

the area of the island occupied by nesting gulls only declined by about 10 per cent. Essentially the intensive culling reduced the nesting density to less than a quarter of the pre-cull density. In subsequent culling, the area occupied has been decreased to about 50 per cent of that previously used. This appreciably reduced density had the following effects on Herring Gulls (Coulson *et al.* 1982):

1. Age of first breeding has declined by over 1 year between 1966 to 1969 (mean of 6.0 years in 1966–70 cohorts and 4.8 in 1973–5 cohorts).

2. The average size of each egg laid increased by about 5 per cent. Spaans *et al.* (1987) has shown that egg size declines as his study colony increases in size and density.
3. It is probable that the production of young per pair would have increased because larger eggs are more successful (Parsons 1970) and also because predation and aggression directed towards the chicks is likely to have been reduced. Unfortunately data on breeding success on the Isle of May have not been collected, mainly because in many years egg removal and/or disturbance by culling took place.

These density-dependent responses by the gulls all tend to mitigate against the reduction aimed at in the cull.

Differences in the effects of culling on Herring and Lesser Black-backed Gulls

The policy of culling on the Isle of May was directed towards 'large gulls' and not specifically at one species. For example, baits were placed in nests irrespective of the species concerned. Although there was some segregation of Herring and Lesser Black-backed Gulls between rocky and grassy areas, all areas were culled, with the exception of one small islet, The Maidens, which contained about 100 pairs of Herring Gulls. Over a period of years, the cull showed no selection for areas favoured by one species or the other. At the start of the cull, there were about 11 Herring Gulls to every Lesser Black-backed Gull on the island. This ratio has progressively changed towards equality as the culling continued (Fig. 23.9).

It is evident that the population dynamics of the two species differ in at least one respect. At the moment, it is not possible to identify this difference, but several possibilities can be suggested. For example, the natural survival rate could be higher, or/and there is less colony philopatry in the Lesser Black-backed Gull than in the Herring Gull. It is important to examine these possibilities, as they probably contribute to the recent national decline in Herring Gulls, but not Lesser Black-backed Gulls, in Britain and the vast immigration of the latter species into the Netherlands in recent years.

Discussion

Culling of large gulls has been planned on the 'do it and adjust' principle. Few attempts have been made to apply population dynamics to assist with the planning of the programme. Those planning and advising on culling have tended to take a simplistic view and have assumed that a colony is a discrete population, whereas we now know that this is far from being the

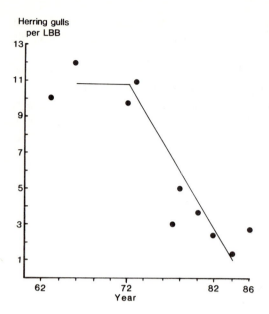

Fig. 23.9. Ratio of Lesser Black-backed Gulls (LBB) to Herring Gulls breeding on the Isle of May based on counts and numbers of young ringed in recent years when virtually all young were marked.

case. No consideration has been given to the effects of culls beyond the limits of the colony concerned. I hope it is now established that there is considerable inter-colony movement, not by the breeding adults, but by the birds about to breed for the first time. The extent to which philopatry operates is crucial to understanding and predicting the effects of culling.

Ecologically, there is much still to be learnt about recruitment in birds. How do those individuals that do not exhibit philopatry select their breeding area? To what extent do recruits respond to the size of a colony? It has been demonstrated in this paper that had the same numbers immigrated on to the Isle of May as occurred prior to the cull, many more birds would have had to be culled there in recent years to keep the numbers at 3000 pairs of large gulls. Thus there is strong evidence that immigration is dependent upon the size of the colony, suggesting that potential recruits 'share' themselves out between available colonies. We have little idea how this process operates. Chabrzyk and Coulson (1976) suggested that many young Herring Gulls select their colony in the breeding season before they first breed and that they may be influenced by the presence of unfledged young whose presence indicate a safe place to breed. There does not appear to be any information concerning the number of colonies visited by a prospecting gull, or at what point it becomes attached to a particular colony.

Little consideration has been given to the disturbance effects of culling. Large-scale culling is a highly disturbing operation. On the Isle of May, the large-scale culling in the first few years probably resulted in many young birds, which were about to breed, leaving the island without nesting. Whether these birds returned the next year or moved to other sites is not known, but the model of change in numbers suggests that there was increased emigration at this time, involving several thousands of birds. Since no established breeders were found to have emigrated, it seems likely that intensive culling has a bonus effect of reducing the colony size to a greater extent than that expected by the numbers culled alone. The problem arising from this effect is that some other colony received the additional emigrants. In the case of the Isle of May, the excess was received by colonies located up to 200 km away. Large increases in both Herring and Lesser Black-backed Gulls were noticed on the Farne Islands and Coquet Island, in Northumberland, and in towns in N.E. England and resulted in the introduction of culls in these areas in later years. Could the culling have contributed to the large emigrations of Lesser Black-backed Gulls to the Netherlands and the establishement of a large colony of both species in Suffolk? The timing of both phenomena in relation to the culling indicates this could be so, but few ringed birds have been identified to substantiate the hypothesis.

Culling gulls is not a pleasant task. If culling is going to be used in the future, then it should be linked with a better understanding of the population dynamics of the species. To do this, much more planning and preparation is necessary so that the cull will provide answers to ecological questions as well as reduce numbers. The fact that many ringed gulls existed on the Isle of May was the result of research studies directed towards other aims. In general, there is no preparation for culls by extensive ringing as a preliminary nor is there a co-ordinated monitoring programme to examine the effects other than to record the decrease in colony size. As a result, much information has been lost, which should be avoided if meaningful analysis of data is to be achieved in the future.

The fact that few, if any, persons considered that culling could influence gull numbers beyond the colony concerned is an indictment of our use of ecological knowledge. There is a much greater need to use ecological information in planning the management of animal populations.

Summary

1. The culling of Herring Gulls *Larus argentatus* and Lesser Black-backed Gulls *L. fuscus* in colonies has both direct and indirect effects on the numbers.

2. One of the most important factors influencing the direct effects of culling is the degree to which young birds return to breed in their natal colony. Whilst some individuals show marked philopatry, the majority

(70 per cent) of Herring Gulls breed in a colony other than that in which they hatched.

3. Dispersal of young Herring Gulls from their place of breeding extends up to 500 km, with appreciable numbers exceeding 100 km. Less detailed information exists for the Lesser Black-backed Gull.

4. Intensive culling seems to have the indirect effect of driving some potential recruits away from the colony. In the case of the Isle of May, this may have involved up to 6000 individuals. Adults that had already bred in the colony did not leave.

5. The extent of the philopatry resulted in the effects of the cull influencing the size of colonies within the range of normal dispersal. The cull acts as a sink, removing birds from a considerable distance. This could be an important factor in determining the recent decline in numbers of Herring Gulls in Britain.

6. Indirect effects of culling are induced through the reduction in the density of nesting birds, which relax density-dependent effects such as the age at first breeding, egg size, and breeding success.

7. Lesser Black-backed gulls have responded differently to the culling regime used on the Isle of May and population numbers have changed from 10 per cent to about 40 per cent of the total of the two species nesting there. The reason for this difference has not been identified.

8. A request is made for greater collaboration between bodies culling gulls and those making biological investigations on gull biology. Particular attention should be given to studying recruitment, particularly the selection by young birds of the colony in which they breed.

Acknowledgements

I wish to acknowledge the considerable assistance given throughout this work by the Nature Conservancy Council and in particular Ms Nancy Gordon, Peter Kinnear, and successive Regional Officers. I am indebted to Dr Jasper Parsons who undertook much of the gull study on the Isle of May prior to the cull and whose large-scale ringing activities have been so valuable in evaluating the effects of the cull. Subsequently, many people assisted in ringing and collecting data concerning the culled gulls and I owe particular thanks to Drs Jennifer Butterfield, N. Duncan, G. Chabrzyk, Pat Monaghan, C. S. Thomas, R. D. Wooller, and Sarah Wanless and to Mrs K. Evans, Ian Deans, and M. Tasker. We have greatly appreciated the facilities made available to us by the Isle of May Observatory whilst staying on the island and also to Mr B. Zonfrillo who has supplied details of the numbers of gulls ringed annually on the May.

The data on the culls on the Farne Islands were supplied by the National Trust and I am grateful to Mr P. Hawkey for the detailed information on the numbers of birds culled. The Natural History Society of Northumbria kindly made available all of the ringing recoveries from the Farne Islands and I must thank Mr D. Noble-Rollin and his assistants for seeking out many of the details. The British Trust for Ornithology supplied all of the ringing recoveries of Herring and Lesser Black-backed Gulls ringed or recovered in the breeding season in the Firth of Forth area.

Finally I wish to express my appreciation of support from NERC through research grants and studentships, which has made much of this study possible.

References

Anon. (1972). *Isle of May Gull control project*. Unpublished. Nature Conservancy policy document. Nature Conservancy Council, Edinburgh.

Anon. (1981). *Isle of May National Nature Reserve Third Management Plan 1981–1985*. Unpublished. Nature Conservancy Policy Document. Nature Conservancy Council, Edinburgh.

Blokpoel, H. (1976). *Bird hazards to aircraft*. Books Canada Limited, London.

Chabrzyk, G. and Coulson, J. C. (1976). Survival and recruitment in the Herring Gull *Larus argentatus*. *Journal of Animal Ecology*, **45**, 187–203.

Coulson, J. C. and Butterfield, J. (1986). Studies on a colony of colour-ringed Herring Gulls. I. Adult survival rates. *Bird Study*, **33**, 51–4.

Coulson, J. C., Duncan, N., and Thomas, C. (1982). Changes in the breeding history of the Herring Gull (*Larus argentatus*) induced by reduction in the size and density of the colony. *Journal of Animal Ecology*, **51**, 739–56.

Duncan, N. (1978). The effects of culling Herring Gulls (*Larus argentatus*) on recruitment and population dynamics. *Journal of Applied Ecology*, **15**, 697–713.

Duncan, N. and Monaghan, P. (1977). Infidelity to the natal colony by breeding Herring Gulls. *Ringing and Migration*, **1**, 166–72.

Parsons, J. (1970). Relationship between egg size and post-hatching chick mortality in the Herring Gull (*Larus argentatus*). *Nature*, **228**, 1221–2.

Parsons, J. and Duncan, N. (1978). Recoveries and dispersal of Herring Gulls from the Isle of May. *Journal of Animal Ecology*, **47**, 993–1005.

Spaans, A. L., de Witt, A. A. N. and Vlaardingen, M. A. (1987). The effects of increased population size of Herring Gulls on breeding success and other parameters. *Studies in Avian Biology*, **10**, 57–65.

Tinbergen, N. (1953). *The Herring Gull's world*. Collins, London.

24 Responses of North American duck populations to exploitation

JAMES D. NICHOLS

Introduction

The goals of managing any population of birds can usually be expressed in terms of a desired population size (number of birds) and, in the case of exploited species, a desired harvest (number taken via exploitation). Informed management of exploited species should thus be based on some knowledge of the effects of exploitation on population size. All changes in population size are determined by the four so-called fundamental demographic variables, survival, reproduction, immigration, and emigration, and any effects of exploitation (hunting) on population size must come about through changes in these variables. Hunting can certainly influence movement and distribution patterns of migratory birds. However, if we restrict our attention to sufficiently large geographic areas (e.g. North America), then we can effectively ignore immigration and emigration and focus on survival and reproduction. In this paper are reviewed studies on the effects of exploitation on North American duck populations.

Migratory bird hunting regulations in the United States

In the United States, regulations governing the hunting of migratory birds were legislated at state and local levels in the mid to late 1800s. The Migratory Bird Treaty Act of 1918 placed the responsibility and authority to manage migratory birds within the United States on the Federal government. Annual hunting regulations issued from 1918 until the 1930s were simple, liberal, and uniformly applied throughout the United States. The drought years of the 1930s brought restrictions, but regulations remained simple and uniform. During the 1930s and 1940s, reliable funding sources were established for migratory bird management, permitting the initiation of banding programmes and systematic waterfowl surveys designed to provide an information base for management decisions (Anderson and Henny 1972; Hawkins *et al.* 1984; U.S. Fish and Wildlife Service 1988). This increase in information, together with increased involvement and interest in migratory bird management by state agencies, led in 1947 to the division of the United States into four flyways for purposes of setting annual hunting regulations. Since that time hunting regulations have become increasingly complex and the process by which regulations are set has evolved to the current situation in which Flyway Councils representing member states have substantial input into the decision-making process (U.S. Fish and Wildlife Service 1988).

Annual migratory bird hunting regulations in the United States can be divided into two categories for the purpose of discussion: framework regulations and special regulations (see U.S. Fish and Wildlife Service 1988 for a detailed discussion of these regulations). Framework regulations are regarded as the core of annual regulations and include season length, daily bag limit, and the outside dates for the opening and closing of the season. Framework opening and closing dates for North American waterfowl are usually October 1 and January 20, respectively. The North American hunting season thus occurs just after the reproductive season during the fall migration and winter periods.

Special regulations are those that are developed in response to specific desires or management needs and which pertain to particular areas, species, or situations. Special regulations include zoning (partitioning a state into areas for which hunting regulations can be set independently), split seasons (division of the hunting season into two or three non-consecutive segments), special seasons (regulations permitting the additional take of a particular species beyond that permitted in the regular season; e.g. the September teal season designed to increase harvest of the blue-winged teal *Anas discors*, an early migrant exposed only to light hunting pressure during the regular waterfowl season), area closures (hunting of a particular species believed to need special protection is closed in concentration areas of the species, e.g. canvasback *Aythya valisineria*), and bonus birds (birds of designated lightly hunted species may be added to the regular daily bag limit).

Annual waterfowl hunting regulations in the United States have varied substantially over the last three decades. During most of the 1960s hunting regulations were adjusted each year in accordance with estimated spring and predicted fall population sizes. Beliefs about the effects of hunting on population size changed in the early 1970s, however, and annual regulations during this decade were generally more liberal and exhibited less year-to-year variation. Then, during the period 1979–84, bag limits and season lengths in the United States were held constant as part of the experimental Stabilized Regulations Programme (Brace *et al.* 1987). Nearly all analyses and inferences about the effects of hunting regulations on North American waterfowl populations are conditioned on this pattern of historical variation in the regulations.

Sources of North American waterfowl population data

Several data collection programmes (reviewed by Martin *et al*. 1979) have been implemented in North America for the purpose of providing information on waterfowl populations needed both for establishing annual hunting regulations and investigating long-term population dynamics. Population size of prairie-nesting duck species has been estimated in May of each year since 1955 via an aerial survey covering approximately 3 370 000 km² of breeding habitat in Canada and the United States. Transects are flown according to a systematic, stratified sampling design, and a double-sampling approach is used to estimate visibility (proportion of ducks present that is seen from the air) from intensive ground counts on a subsample (Pospahala *et al*. 1974; Martin *et al*. 1979). Estimates of the number of potential wetlands containing water are also obtained from this survey. Some of the transects flown in the May survey are flown again in the July Production Survey (Henny *et al*. 1972; Martin *et al*. 1979). This survey provides information on production of young, renesting activity and number of ponds. Finally, a winter survey is conducted every January and provides, at best, an index to waterfowl abundance on the wintering grounds (Martin *et al*. 1979).

Estimates of the number and composition of waterfowl legally harvested in the United States each year are obtained through the U.S. Harvest Survey (Martin and Carney 1977; Martin *et al*. 1979). This survey has two components. The Hunter Questionnaire Survey has been in operation since 1952 and involves mailing questionnaires to a sample of hunters after the hunting season each year. The completed questionnaires contain information on the number of ducks and geese harvested during the previous hunting season. The Parts Collection Survey involves the mailing of special envelopes to a sample of hunters who are asked to mail in wings of ducks (and tail feathers of geese) shot during the hunting season. These wings are identified to species, age, and sex by biologists. Data from these two components of the Harvest Survey are used to estimate the number of waterfowl harvested each year by species, age, and sex.

Large numbers of waterfowl of various species are individually banded each year throughout North America (Martin *et al*. 1979). Records of these bandings and of subsequent recoveries (for hunted species, most recoveries are from birds shot or found dead during the hunting season) are stored on computer files at the U.S. Fish and Wildlife Service Bird Banding Laboratory, Laurel, Maryland. For a given banding location (or group of similar locations combined to form a banding reference area), an effort is made to band birds of a particular species at the same time of year for each of a number (hopefully >5) of consecutive years. The number of birds banded each year and the subsequent numbers recovered in the first, second, etc., hunting seasons after banding constitute the data needed to estimate annual survival and hunting mortality rates.

Annual survival rate, S_i, is defined here as the probability that a bird alive at the mid-point of the banding period in year i will survive at least until the banding period in year $i + 1$. For most waterfowl species in North America, banding is conducted during either preseason (July–September) or winter (January–February) periods. Recovery rate, f_i, is defined as the probability that a banded bird alive in the banding period of year i is shot or found dead in the subsequent (year i) hunting season and its band reported to the Bird Banding Laboratory. Estimation proceeds by first focusing on a banded sample from one particular year and then by treating numbers of resulting recoveries in each hunting season after banding as multinomial random variables. Multinomial cell probabilities (probabilities of being recovered in particular hunting seasons) are modelled as functions of S_i and f_i. Different models are developed by incorporating different assumptions about the age- and time-specificity of the survival and recovery rate parameters. Estimation in accomplished using the method of maximum likelihood and model selection is based on the results of goodness-of-fit tests and likelihood ratio tests between models (Brownie *et al.* 1985; White 1983; Conroy and Williams 1984).

In addition to S_i and f_i, it is useful to define two parameters associated with hunting mortality. We define harvest rate, h_i, as the probability that a bird alive at the mid-point of the banding period in year i is shot and retrieved by a hunter in the subsequent (year i) hunting season. Harvest rate differs from recovery rate because not all birds shot and retrieved are reported by hunters to the Bird Banding Laboratory. We define band reporting rate, λ_i, as the probability that a banded bird shot and retrieved by a hunter in year i will be reported to the Bird Banding Laboratory. Reporting rate is sometimes estimated from reward band studies in which a monetary reward is offered for the return of specially marked bands (Henny and Burnham 1976; Conroy and Blandin 1984). Harvest rate is then estimated as:

$$\hat{h}_i = \hat{f}_i / \hat{\lambda}_i \tag{1}$$

Kill rate, K_i, can be defined as the probability that a bird alive at the time of banding in year i is killed by a hunter during the subsequent (year i) hunting season. Kill rate differs from harvest rate in that it includes birds that are shot by hunters but not retrieved. Unretrieved kill has been estimated from Hunter Performance Survey data obtained by state and federal biologists secretly observing waterfowl hunters (Martin and Carney 1977).

The above estimates and data are not equally available for all North American waterfowl species. For example, the May aerial breeding ground survey provides population size estimates for about 20 species of ducks. However, the survey area is not the principal breeding grounds for many species of ducks and for all geese, so except for special survey efforts, estimates of breeding population size are unavailable for these species. The Harvest Survey results in estimates of harvest for virtually all North American waterfowl species, but the precision, and hence the utility, of estimates varies greatly. Similarly at least some individuals of virtually all North American waterfowl species have been banded at one time or another. However, the large banded sample sizes needed for precise survival and recovery rate estimates have been obtained for only a handful of these species. The mallard *Anas platyrhynchos* is the most abundant duck species in North America and accounts for 30–45 per cent of the total duck harvest in the United States (Martin and Carney 1977). The various data collection programmes concentrate on mallards, with the result that better data are available for this species than for any other waterfowl species in North America. Synthetic summaries and analyses of mallard data from these various sources appear in the eight reports of the *Population ecology of the mallard* series (Anderson and Henny 1972; Pospahala *et al*. 1974; Anderson *et al*. 1974; Anderson 1975; Anderson and Burnham 1976; Martin and Carney 1977; Munro and Kimball 1982; Nichols and Hines 1987).

Regulations, harvest, and harvest rate

The translation of hunting regulations into an effect on the size of a managed population can be viewed as involving two steps. The initial step involves the manner in which regulations determine harvest rate. The second step then concerns the influence of harvest rate on subsequent population size. Although it is possible to treat these two steps as a 'black box' and look directly for effects of changes in hunting regulations on subsequent population size, most managers desire a mechanistic under-standing that requires some knowledge of both steps.

Total harvest and harvest rate can be estimated each year and both quantities have been used in analyses of the influence of hunting regula-tions (see review in U.S. Fish and Wildlife Service 1988). We can write total harvest, H_i, as the product of preseason population size, N_i, and harvest rate, h_i:

$$H_i = N_i h_i \qquad (2)$$

In most cases, when we implement changes in hunting regulations, we are trying to bring about changes in harvest rate. Certainly many factors other than hunting regulations are likely to influence harvest rate. These other factors include hunter effort and behaviour (Nieman *et al*. 1987), weather and other environmental conditions (Blohm *et al.* 1987), the physiological condition of individual birds (Hepp *et al.* 1986; Blohm *et al.* 1987; Reinecke and Shaiffer 1988), and bird abundance (Trost *et al.* 1987). To date there have been no comprehensive efforts to model harvest rate of any waterfowl species as a function of the major components of annual hunting regulations and these other variables of potential influence.

There have been two kinds of analyses directed at the relationship between hunting regulations and harvest rate. The first involves a comparison of estimated recovery rate (indices of harvest rate; Henny and Burnham 1976; Conroy and Blandin 1984) during years characterized by relatively liberal vs. relatively restrictive regulations. Martin *et al.* (1979) and Rogers *et al.* (1979) examined historical hunting regulations for mallards in North America over the approximate period 1960–75 and selected years of extreme liberal and restrictive bag limits and season lengths. Recovery rates for mallards of all age–sex classes banded pre-season throughout North America were significantly larger during the liberal than during the restrictive years. Krementz *et al.* (1988) conducted a similar analysis for American black ducks *Anas rubripes*, a species inhabiting eastern North America. The historical variation in hunting regulations was less marked for black ducks than for mallards, but nevertheless Krementz *et al.* (1988) found evidence of higher recovery rates in the more liberal years for three of the four age–sex classes.

The second type of analysis concerning the relationship between regulations and harvest rate contrasts estimated recovery rates immediately before and after a specific change in regulations. Delayed season openings and daily bag limit restrictions were imposed on mallard hunting in southern Manitoba during 1973–8. Recovery rates of mallards banded in Manitoba were found to be lower during this set of restrictive years than during years of more liberal regulations occurring immediately before and after implementation of the restrictions (Caswell *et al.* 1985). During 1973–9, the state of Minnesota restricted shooting hours by implementing a special afternoon closure in an effort to reduce harvest rates of mallards. Recovery rates of Minnesota-banded mallards were significantly lower during the years of afternoon closure than during years of traditional sunset closure but with otherwise similar regulations (Kirby *et al.* 1983). At the continental level, a recent analysis provided strong evidence of a decrease in mallard recovery rates in 1985–7, a period in which restrictive regulations followed 5 years (1979–84) of relatively liberal stabilized regulations (R. Trost, personal communication).

Two analyses of this type have also been used to evaluate the potential effects of special early seasons for wood duck *Aix sponsa* in southern states. Johnson *et al.* (1986*b*) compared recovery rates of wood ducks banded in the southern states of the Atlantic Flyway during the years preceding (1970–6) vs. following (1977–83) implementation of the special early season. Recovery rates of young male wood ducks were significantly lower during the years of the early season, but recovery rates for the other three age–sex classes were similar for the two periods. A similar analysis of wood ducks banded in Tennessee and Kentucky provided evidence of higher recovery rates during the years of the early season (Sauer *et al*. 1990).

Our knowledge of the relationship between hunting regulations and subsequent harvest rates is imprecise. We are currently unable to predict harvest rate for any species as a function of all the different components of hunting regulations. The analyses briefly reviewed above simply address directional changes in harvest rate, and even these studies are not without problems from a statistical design perspective. These analyses all suffer from potentially confounding covariation of regulations and population size; restrictive regulations were historically prescribed when predicted fall population size was low, and liberal regulations often corresponded to years of high population size. However, if anything, this covariation probably made the above analyses conservative. If population size influences harvest rate, then it is likely that a given set of hunting regulations will result in a higher harvest rate when population size is low (Trost *et al*. 1987). I believe that these analyses have provided evidence that we can implement changes in hunting regulations which produce changes in harvest rate of predictable direction. This is not at all surprising when continental regulations are considered. It is somewhat surprising, however, that local regulations such as the Manitoba restrictions and the afternoon closure in Minnesota also appear to have brought about desired changes in harvest rate.

Exploitation and population size

The analyses reviewed in the preceding section provide some evidence that we can use hunting regulations to bring about changes in harvest rate. It is very likely that this inference also extends to kill rate. Regarding the effects of changes in harvest rate and kill rate on population size, there is no question that when a bird is shot and killed, population size is reduced by one. The question of interest to population ecology and management involves the duration of this effect. For example, we are interested in whether this immediate reduction in population size in the fall of year i translates into a corresponding reduction in the spring or fall of year $i + 1$, or instead whether the population exhibits some compensatory response such that population size in year $i + 1$ is unaffected by the temporary reduction.

Two general approaches are available for studying the effects of variation in hunting regulations and kill rates on subsequent population size. The first approach is the more direct and involves comparing estimates of population size for years of differing hunting regulations or kill rates. The second approach focuses on the relationship between regulations and/or kill rate and the fundamental demographic variables responsible for population change, specifically survival rate and reproductive rate. Although this second approach is less direct, it provides at least the potential for developing a mechanistic understanding of population response to exploitation.

Regulations and/or kill rate vs. population size

The most rigorous application of this direct approach was completed recently for mallards by Reynolds and Sauer (submitted). They estimated the annual finite rate of increase of North American mallard populations as the ratio of successive population size estimates from May aerial survey data. They used multiple linear regression analysis to investigate the possible relationship between rate of increase and both reproductive rate (indexed by the harvest age ratio) and harvest rate (estimated as a weighted average based on banded samples throughout North America). The relationship to population rate of increase was positive for the reproductive rate index and negative for the estimated harvest rate and was significant in both cases.

Reeves (1966) and Bellrose (1976) reviewed historical data on hunting regulations and perceived population size of wood ducks. They concluded that 'excessive exploitation' in the 1800s was at least partially responsible for the very low numbers existing in the early 1900s and that subsequent season closures were largely responsible for a substantial recovery of the population.

Exploitation and reproductive rate

The influence of exploitation on number of young produced depends critically on the possible relationship between reproductive rate (the number of fledged young produced per breeding–age female alive in spring) and population size. For example, assume that harvest in the fall hunting season of year i reduces breeding population size in the spring of year $i + 1$ below that which would have existed in the absence of hunting. If reproductive rate is unaffected by density, then the proportional reduction in production of young is equal to the proportional reduction in breeding population size. However, if reproductive rate varies inversely with breeding density, then density-dependent changes in reproductive rate may effectively compensate for changes in breeding population size brought about by hunting. Density-dependent recruitment rates are thought to be the primary compensatory response of many vertebrates (e.g. fish, large mammals) to exploitation.

There is no study, of which I am aware, of a possible relationship between waterfowl reproductive rate and either hunting regulations or harvest rates. However, a few studies have been directed at possible density dependence of reproductive rates in North American waterfowl. There is evidence that several components of reproductive rate of wood ducks vary in a density-dependent manner. For example, Haramis and Thompson (1985) found that the proportion of eggs producing ducklings that left the nest box was much lower in years with large numbers of breeding females. High densities in breeding wood duck populations are thought to result in lower proportions of females breeding, higher rates of nest abandonment, and lower nest success (review in Nichols and Johnson 1990).

Results of investigations of possible density-dependent reproductive rates in prairie-nesting species are not nearly as conclusive as those for wood ducks. Pospahala *et al.* (1974) investigated possible density dependence in reproductive rates of mallards and other prairie-nesting species using long-term data from the May and July aerial surveys. Regression analyses provided evidence that numbers (and density) of duck broods were linearly related to numbers (and density) of breeding ducks, indicating density independence of production *rate* (e.g. expressed as number of ducklings or broods per adult female; Pospahala *et al.* 1974). They found little evidence of density dependence but suspected its occurrence, nevertheless. Kaminski and Gluesing (1987) investigated density-dependent reproduction in mallards using long-term data sets from both continental surveys and intensively studied local areas. Multiple regression analyses provided evidence that reproductive rate (as indexed by harvest age ratio) was negatively related to estimated breeding population size, and Kaminski and Gluesing (1987) concluded that density was an important determinant of mallard reproductive rate. However, recent reanalyses of the continental data used by Kaminski and Gluesing (1987) indicate that the inference of density dependence depends heavily on 2 years of very high population size, suggesting the possibility of a threshold response (U.S. Fish and Wildlife Service 1988).

It may well be that reproductive rate is essentially density independent below some threshold population size, and that density dependence occurs only when this threshold is exceeded. Such a threshold response would be entirely consistent with mechanisms hypothesized to produce density dependence in mallard reproductive rates, many of which specifically involve 'overcrowding' (Dzubin 1969).

Exploitation and survival rate

Historically, most investigations of the effects of exploitation on migratory bird populations in North America have focused on survival. Anderson and Burnham (1976) specified two extreme hypotheses about the effects of exploitation on survival, the completely additive mortality hypothesis and the completely compensatory mortality hypothesis. Although it is likely that neither of these hypotheses provides an exact portrayal of the real world, they have proven very useful as conceptual reference points from which predictions have been deduced and tested. These hypotheses are briefly described below and relevant tests and analyses reviewed.

The hypotheses

The additive and compesatory mortality hypotheses can be expressed in several equivalent ways (Anderson and Burnham 1976). Here, I will express them in terms of the relationship between kill rates and annual survival rates, two quantities that are annually estimated or indexed using data from operational banding programs. Figure 24.1 portrays this relationship under the completely additive mortality hypothesis. S_0 represents the annual survival rate in the absence of hunting mortality (i.e. when $K_i = 0$). As the risk of hunting mortality increases (as K_i increases), the annual survival rate decreases in an approximately linear manner (as K_i approaches 1, the linear model is a poor approximation; see Anderson and Burnham 1976).

Kill rate, as we have defined it thus far, does not represent the probability of dying as a result of hunting in the complete absence of nonhunting

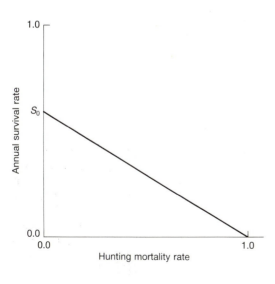

Fig. 24.1. Approximate relationship between annual survival rate and hunting mortality rate under the additive mortality hypothesis. S_0 represents the annual survival rate in the complete absence of hunting mortality (i.e. when hunting mortality rate = 0). Under this hypothesis, annual survival rate decreases in an approximately linear manner as hunting mortality rate increases.

mortality sources (i.e. K_i is not a net rate in a competing risk framework; see Chiang 1968). Instead, some nonhunting mortality likely occurs during the hunting season, and this has some influence on the magnitude of K_i. If we make the simplifying assumption that no nonhunting mortality occurs during the hunting season, then we can write the annual survival rate as:

$$S_i = S_0 (1 - K_i) = S_0 - S_0 K_i \qquad (3)$$

Equation (3) holds exactly in cases where K_i and $1 - S_0$ represent the probabilities of dying from one mortality source when no other mortality sources are operating. In the usual situation of incomplete temporal separation of mortality sources, our estimates of K_i are not obtained in the complete absence of nonhunting mortality. However, equation (3) provides a reasonable approximation of the additive mortality hypothesis even in this situation (Anderson and Burnham 1976; Nichols *et al*. 1984).

The completely compensatory mortality hypothesis is depicted in Figure 24.2. Under this hypothesis, annual survival rate (S_i) is independent of

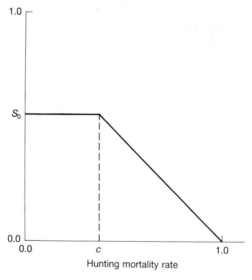

Fig. 24.2. Approximate relationship between annual survival rate and hunting mortality rate under the compensatory mortality hypothesis. S_0 represents the annual survival rate in the complete absence of hunting mortality (i.e. when hunting mortality rate = 0). Under this hypothesis, annual survival rate remains constant at S_0 for a certain range of values of hunting mortality rate (i.e. within this range, changes in hunting mortality rate bring about no corresponding changes in annual survival rate). However, after hunting mortality rate reaches some threshold level, c, further increases in hunting mortality rate do bring about decreases in annual survival rate.

variation in kill rate (K_i) whenever kill rate is less than some threshold value (c), i.e.

$$S_i = S_0 \text{ for } 0 < K_i < c \qquad (4)$$

At some point, further increases in K_i must result in a reduction in S_i. Note that $c \leqslant (1 - S_i)$, so that species with high survival rates in the absence of hunting have less potential to exhibit compensation than species with lower survival rates.

Consider again the situation of complete temporal separation of hunting and nonhunting mortality. In this case, $(1 - K_i)$ represents the probability of surviving the hunting season. Under the additive mortality hypothesis, S_0 represents the probability of surviving the period extending from the end of the hunting season to the beginning of the next year's season. However, under the compensatory mortality hypothesis for $K_i < c$, the probability of surviving the period after the hunting season is $S_0/(1 - K_i)$.

Predictions

Nichols *et al.* (1984) listed three predictions that should be useful in testing and distinguishing between these competing hypotheses:

1. The compensatory mortality hypothesis predicts that there is no relationship between annual survival rates and hunting mortality rates for $K_i < c$. The additive mortality hypothesis predicts a negative relationship between annual survival rates and hunting mortality rates for the entire possible range of hunting mortality rates $(0 < K_i < 1)$.

2. Under most reasonable scenarios, the compensatory mortality hypothesis predicts a negative relationship between hunting mortality rate (K_i) and nonhunting mortality rate after the hunting season. The additive mortality hypothesis predicts that these mortality rates are independent of each other.

3. The compensatory mortality hypothesis predicts a positive relationship between nonhunting mortality rate and population size or density at some time of the year. Under most reasonable scenarios, nonhunting mortality after the hunting season should be positively related to population size at the end of the hunting season, if the compensatory mortality hypothesis is true. The additive mortality hypothesis predicts no relationship between nonhunting mortality and population size.

Note that the two original competing hypotheses (e.g. as expressed in Figs 24.1 and 24.2) are phenomenological in nature and do not deal with mechanisms. This is reasonable because they were initially developed for the purpose of testing, and hence are expressed in terms of quantities that are readily estimated. However, as these hypotheses began to receive increased consideration, attention was directed to possible underlying mechanisms. The additive mortality hypothesis requires only that hunting and nonhunting mortality sources act as independent competing risks, a reasonable *a priori* assumption. The compensatory mortality hypothesis is not so easily explained. Most current mechanistic explanations of the compensatory mortality hypothesis involve a nonhunting mortality source that acts in a density-dependent manner (see prediction 3). In some cases the individuals of a population can exhibit heterogeneity in their probabilities of surviving both hunting and nonhunting mortality sources. If a positive correlation exists between the probabilities of surviving these sources, such that individuals with high probabilities of surviving hunting mortality also have high probabilities of surviving nonhunting mortality and vice versa, then mortality for the entire population can appear partially compensatory even when the additive mortality hypothesis holds true for individuals or groups of individuals characterized by similar survival rates (see Burnham and Nichols 1985; Johnson *et al.* 1986*a*).

Tests and evidence for prediction (1)

The large majority of tests designed to distinguish between these two hypotheses have been based on the relationship between annual survival rate (S_i) and either kill rate (K_i) or an index of kill rate (harvest rate, h_i, or recovery rate, f_i), and have thus addressed prediction (1) above. We can express both hypotheses in terms of a single equation:

$$S_i = S_0 (1 - bK_i) \tag{5}$$

Under the additive mortality hypothesis, we expect, approximately, that $b = 1$ (see equation 3). Under the compensatory mortality hypothesis, we expect $b = 0$ for $K_i < c$. Thus our primary interest is in testing the hypotheses that $b = 1$ and $b = 0$. In addition, we would like to estimate b whenever possible.

Anderson and Burnham (1976) reviewed studies conducted prior to 1975 which investigated the relationship between annual survival rates and kill rates of migratory birds. They noted two major problems with these early studies that led to unreliable inferences. First, the methods used to estimate the parameters on which the tests were based, were valid only under very restrictive and unlikely assumptions. Thus, parameter estimates were usually biased, often badly. The second problem involved the use of estimates with non-negligible sampling correlations to test for relationships of interest. For example, a common procedure involved the use of correlation analysis to test for an association between estimates of recovery rates (indices of kill rates) and annual survival rates. However, estimates of both parameters were based on the same banding data sets and exhibited a substantial negative sampling correlation. Analyses based on these estimators tended to show significant negative correlations which were interpreted as evidence for the additive mortality hypothesis. However, it is now believed that these results were largely attributable to the sampling correlations and probably did not reflect relationships between the true underlying parameters (Anderson and Burnham 1976).

Anderson and Burnham (1976) reanalysed banding, recovery, and other data for North American mallards using estimators from appropriate models (those of Brownie *et al*. 1985) and constructing tests that properly incorporated any sampling covariation. Their results favoured the compensatory mortality hypothesis. Because of the importance of this inference to waterfowl management, additional analyses have been conducted for mallards using new data and some additional test procedures. Interest in the generality of the Anderson and Burnham (1976) results has also led to similar investigations of several other species of North American ducks. Here, I provide only a brief review of methods used in recent tests of these hypotheses and of resulting conclusions. More detailed reviews are presented by Nichols *et al*. (1984) and U.S. Fish and Wildlife Service (1988).

Most of the published analyses addressing prediction (1) for North American waterfowl species have not been experimental in nature, but have relied on historical variation in hunting regulations or harvest rates over long time periods. With mallards, for example, we have good banding and recovery data for many different banding locations throughout North America extending from 1950 to the present. There has been substantial year-to-year variation in predicted fall population sizes and in management philosophy over this time period and this variation has resulted in corresponding variation in hunting regulations and in both harvest and kill rates. During this period, we estimate that North American hunters have killed from about 0.10 to 0.40 of the preseason mallard population in different years, with an annual average of about 0.20 to 0.25. Most analyses have addressed questions about the value of b in equation (5) by investigating temporal covariation between survival rate estimates and both kill rate indices and hunting regulations. Such analyses are carried out within each of a number of geographic areas of banding.

One of the statistical approaches used to analyse such long-term data sets involves fitting band recovery models representing competing hypotheses and testing between these models (e.g. using likelihood ratio tests). For example, models $M1$ (for adults) and $H1$ (for adults and young) described by Brownie *et al*. (1985) are general models permitting year-to-year variation in both survival and recovery rates. Models $M2$ and $H02$ are less general in that they permit year-to-year variation in recovery rates but assume that annual survival rate is constant over time (Brownie *et al*. 1985). Models $M2$ and $H02$ thus correspond to the compensatory mortality hypothesis for $K_i < c$. Likelihood ratio tests of $M1$ vs. $M2$ and $H1$ vs. $H02$ test the null hypothesis of constant survival and variable recovery rates against the general alternative of time-varying survival and recovery rates (Brownie *et al.* 1985). Failure to reject either M2 or H02 provides some evidence in favour of the compensatory mortality hypothesis, although the low power of these tests prevents strong inference. Rejection of M2 or H02 in favour of M1 or H1, however, indicates only that survival rates exhibit some temporal variation and does not provide evidence for the additive mortality hypothesis. Burnham and Anderson (1984) developed a model to correspond specifically to the additive mortality hypothesis (S_i was modelled as in equation 3). They then evaluated the likelihood function for both this additive model and the compensatory model, M2, for each of a number of data sets. The size of the likelihood was used to decide which model was more appropriate for a given data set, and the proportion of data sets for which each of the two models was selected provided information on which model was likely to be operating in nature (Burnham and Anderson 1984).

The obvious approach to testing hypotheses about b in equation (5) is to attempt its estimation. However, the sampling correlation between survival and recovery rate estimates prevents the use of simple linear regression techniques on vectors of survival and recovery rate estimates obtained from the same data sets. Anderson and Burnham (1976) developed a variance components approach which used estimated sampling variances and covariances of survival and kill rates and permitted estimation of b. Anderson *et al.* (1982) introduced an ultrastructural band recovery model in which survival was modelled as in equation (5), permitting direct estimation of b. Simpler approaches have involved attempts to obtain independent estimates of survival and kill rate for use with standard linear regression techniques. Anderson and Burnham (1976) used Harvest Survey and May Aerial Survey data to obtain an alternative estimate of harvest rate which exhibited no sampling covariance with survival rate estimates. Linear regression was then used to estimate and test hypotheses about b. Blandin (1982) used recovery rates estimated from preseason banding data and survival rates from winter banding data in conjunction with linear regression analysis. Nichols and Hines (1983) randomly partitioned band recovery data sets into two separate sets and used both to estimate survival and harvest rates. Survival rates from one data set were then used in regression analyses with harvest rates from the other data set to yield inferences about b.

Another approach to testing prediction (1) involves selecting years with extreme hunting regulations or harvest rates from historical data and testing the hypothesis of no difference between average survival rates for the two sets of years. Under the additive mortality hypothesis, we would expect higher survival rates in the years of restrictive regulations and low harvest rates, whereas under the compensatory mortality hypothesis, we would expect similar survival rates regardless of regulations or harvest rates. Anderson and Burnham (1976) first used such contrasts, selecting years of extreme restrictive and liberal hunting regulations. Nichols and Hines (1983) selected years of extreme harvest rates using estimates based on one data set from their random partitioning of data. They then contrasted survival rate estimates for the selected years obtained from the complementary data sets.

As indicated above, all of the described approaches to testing hypotheses about *b* have focused on possible temporal covariation between survival and kill rates within particular geographic areas. Trost (1987) selected an alternative approach using mean annual survival and harvest rate estimates for different geographical areas to focus on possible geographic variation (i.e. do birds in areas with higher harvest rates tend to exhibit lower survival rates?). Trost (1987) used the partitioned data set approach to obtain independent estimates of survival and harvest rates for use with standard linear regression across geographical locations.

All of the above approaches have involved attempts to utilize existing historical variation in harvest regulations and rates found in long-term data sets. In addition to these efforts, there have been a few studies at specific locations involving comparisons of survival estimates from periods preceding and following specific changes in hunting regulations. These studies can still not be considered experiments in the classical sense (e.g. they lack geographical replication and year effects are confounded with effects of regulations), but they may permit stronger inferences than analyses based solely on historical variation. These studies simply involve tests for differences between survival rates in the periods of differing regulations (Kirby *et al.* 1983; Caswell *et al.* 1985; Johnson *et al.* 1986*b*; Sauer *et al.* 1990; M. Anderson and J. Serie, personal communication).

Results of using these various approaches to testing prediction (1) for several different North American duck species are summarized in Table 24.1. This is simply an updated version of one that first appeared in Nichols *et al.* (1984) and was later updated in U.S. Fish and Wildlife Service (1988). The indicated support for the compensatory and additive mortality hypotheses and the label of 'inconclusive' merely reflect my own subjective opinion.

The analyses for mallards have produced some results that fall into each of the three categories of evidence (Table 24.1). However, the bulk of the evidence for mallards based on analyses of historical data sets has favoured the compensatory mortality hypothesis, and it is difficult to avoid the conclusion that at the continental level historical variation in annual survival rates of mallards has not been closely tied to variation in harvest rates and hunting regulations. Cautionary considerations regarding the different analyses of mallards have been discussed by Nichols *et al.* (1984)

Table 24.1 Summary of published results from studies addressing prediction (1) of the compensatory and additive mortality hypotheses about the relationship between kill rate, K_i, and annual survival rate, S_i.

| | | | Age–Sex class[a] | | | |
| | | | Adult | | Young | |
Species	Published reference		Male	Female	Male	Female
Mallard	Anderson and Burnham	(1976: 22–25)	C	C	C	C
		(1976: 25–30)	C	C	C	C
		(1976: 31–33)	C	C	—	—
	Rogers *et al.*	(1979: 119–122)	C	C	C	C
		(1979: 123–125)	C	C	—	—
	Nichols and Hines	(1983: 342–345)	C	C	C	?
		(1983: 345–346)	C	C	C	A
	Kirby *et al.*	(1983: 212)	?	?	?	?
	Burnham and Anderson	(1984: 108–110)	C	?	—	—
	Burnham *et al.*	(1984)	C	?	—	—
	Caswell *et al.*	(1985: 552)	A	A	C	C
		(1985: 553)	?	A	—	—
	Trost	(1987: 275–276)	A	C	—	—
		(1987: 276–277)	C	C	A	C
		(1987: 277–278)	C	C	C	C
		(1987: 279–281)	A	?	A	?
		(1987: 281)	?	?	—	—
Black duck	Blandin	(1982: 96–110)	?	?	?	—
		(1982: 110–111)	?	?	—	—
	Krementz *et al.*	(1988: 217–219)	A	?	A	—
		(1988: 219–222)	?	?	A	?
Wood duck	Johnson *et al.*	(1986b)	?	?	?	?
	Sauer *et al.*	(1990)	A	A	A	A
	Trost	(1990)	C	?	A	A
		(1990)	C	?	A	A
		(1990)	?	?	—	—
Canvasback	Nichols and Haramis	(1980: 169–171)	?	?	—	—
	M. Anderson and J. Serie	(personal communication)	—	A	—	?
Ring-necked duck	Conroy and Eberhardt	(1983: 133–136)	?	?	—	—

[a] 'C' indicates that the results generally supported the compensatory mortality hypothesis; 'A', results generally supported the additive mortality hypothesis; '?' results were inconclusive and provided no evidence that could be used to distinguish between the two hypotheses.

and include the reliance on an observational (correlational), rather than experimental, approach. It is interesting that one of the only studies on mallards contrasting survival rates in years preceding and following a specific change in hunting regulations found some evidence that the regulations change was associated with a change in survival rates (Caswell *et al.* 1985).

The analyses using species other than mallards are based on smaller data bases and more frequently yielded inconclusive results (Table 24.1). However, these analyses also yielded evidence in favour of the additive mortality hypothesis more frequently than did mallard analyses. Two of the wood duck analyses were quasi-experimental in that they contrasted survival and recovery rates occurring before and after the initiation of special early seasons in southern states. In the Atlantic Flyway, the special seasons did not result in a marked increase in recovery rates, so results of comparisons of survival rate were not very informative (Johnson *et al.* 1986*b*). However, special early seasons for wood duck in Kentucky and Tennessee did result in increased recovery rates and lower survival rates supporting the additive mortality hypothesis (Sauer *et al.* 1990).

The canvasback study of M. Anderson and J. Serie (personal communication) included one period of stable, restrictive hunting regulations (1974–80), a brief period of more liberal regulations (1984–6), and a recent period of restrictive regulations including season closure in the United States (1987–8). This study was based on return rates (proportion of marked birds seen on a breeding-ground study area in year i that is seen on or near the area in year $i + 1$) of female canvasbacks, and is the only study listed in Table 1 which did not utilize the band recovery models of Brownie *et al.* (1985) to estimate survival rate. Because of the high degree of philopatry of female canvasbacks, these return rates are thought to be very good indices of survival rate. Preliminary analyses indicated that return rates of adult females differed significantly among the three time periods and were lowest in the years of liberal regulations. Return rates of young birds exhibited the same pattern of variation, and differences among the three periods approached significance (M. Anderson and J. Serie personal communication).

Tests and evidence for predictions (2) and (3)

I am aware of no specific tests of prediction (2), but a few investigations have been directed at prediction (3). Proper tests of prediction (3) first require a guess about the period of the year during which nonhunting mortality acts in a density-dependent manner. We would then like the ability to estimate both nonhunting mortality during this interval and population size (or perhaps density relative to some critical resource) at the beginning of this interval. Some inferences would be possible if we could obtain such estimates for a series of years during which density either could be manipulated or perhaps exhibited substantial natural variation.

Estimates of annual survival rate are available from operational banding programs, whereas estimates of seasonal survival rate (corresponding to periods smaller than 1 year) require special efforts of data collection.

Annual survival rate estimates have thus been used to investigate questions about possible density dependence, despite the fact that they are not nearly as suitable for such investigations as seasonal survival estimates. Anderson (1975) and Nichols *et al.* (1982) investigated the possible relationship between annual survival rates of mallards and both spring mallard population size and mallards-per-pond ratios. The analyses of Nichols *et al.* (1982) provided some evidence of lower survival rates in years of high numbers of mallards per pond. Conroy and Eberhardt (1983) found some evidence of a negative relationship between annual survival rates (estimated from winter bandings) and spring population size of ring-necked ducks (*Aythya collaris*). These analyses do not argue strongly for the compensatory mortality hypothesis, but simply provide some evidence for the existence of a possible mechanism (density-dependent mortality) that might underlie this hypothesis.

Recent investigations have used radio-telemetry and multiple banding periods per year for the same population of birds to provide data for estimating seasonal survival rates. These studies were not coupled with efforts to manipulate density and did not cover periods experiencing substantial natural variation in density. The relevance of these studies to prediction (3) thus has been restricted to identification of periods of the year during which substantial nonhunting mortality (capable of compensating for variation in hunting mortality) occurs. Blohm *et al.* (1987) found that survival of female mallards was lowest during the spring–summer breeding season (also see Johnson and Sargeant 1977; Cowardin *et al.* 1985), whereas survival rates of male mallards were very high during this period. The main source of hen mortality during this period is mammalian predation (Johnson and Sargeant 1977; Sargeant *et al.* 1984) and it is possible that this predation operates in a density-dependent manner, although this has not been demonstrated to date. Fall–winter survival rates of males were lower and more variable than those of females, a finding consistent with the higher harvest rates of males (Blohm *et al.* 1987). Reinecke *et al.* (1987) estimated overwinter survival rates of female mallards in the Mississippi Alluvial Valley and found that hunting was the most important source of mortality. They found very low rates of mortality from nonhunting sources and this suggests that we may have to look for substantial density-dependent mortality at some other time of the year. It should be noted, however, that studies of winter mortality of mallards are currently underway in two other U.S. wintering locations and that results from one of these locations indicates that nonhunting mortality during winter may be of much greater importance in that area. Studies of winter mortality in other duck species (e.g. black ducks, canvasbacks) are either recently completed or now underway and may provide new insights.

These different studies directed at possible density-dependent mortality and at assessing the importance of mortality occurring at different times of the year have been interesting, but not definitive. We still lack evidence that nonhunting mortality at some time of the year acts in a strongly density-dependent manner. Because many hypotheses about density dependence involve resource limitation and because winter is a time of the year when resource limitation might be expected to occur, the winter was selected *a priori* by many of us as a likely period for the action of density-dependent mortality. However, recently completed studies indicate that nonhunting mortality rate during winter is not of sufficient magnitude to 'compensate' for substantial variation in hunting mortality, at least on some wintering grounds.

Environmental variation

There are several difficulties associated with testing the compensatory and additive mortality hypotheses and some of these involve environmental variation. Equations (3) and (4) clearly represent simplified views of the mortality process. Equation (3) asserts that all variation in annual survival rate results directly from variation in hunting mortality, and equation (4) depicts annual survival rate as a constant for $K_i < c$. Certainly, environmental variation must bring about year-to-year variation in S_0 in equations (3) and (4), increasing the difficulty in distinguishing additivity from compensation. Most mechanistic explanations of density-dependent non-hunting mortality involve more than simply numbers of birds (e.g. a birds/resources ratio or a birds/predators ratio), and environmental variation in these additional factors (e.g. resources, predator populations) would be expected to produce environmental variation in c (equation 4). This potential variation in c, coupled with our lack of knowledge of the location of this threshold, makes testing the compensatory mortality hypothesis very difficult.

Reflections on research strategy

As summarized above, until recently most studies of the mortality process in North American ducks have emphasized the relationship between hunting mortality rate and total annual survival rate (prediction 1) rather than between density and survival rate (prediction 3). It has been suggested (by a reviewer of this paper among others) that a preferable research approach would have been to bypass study of prediction 1 and to focus instead on possible density-dependent mortality relationships. I believe that three factors are important in explaining the historical development of the research approach that has evolved in North America. The first factor involves the expected results of studies of the effects of hunting mortality. For a variety of reasons (e.g. see Geis 1963), early researchers believed that increases in waterfowl hunting mortality produced decreases in annual survival rates. Results of early analyses (reviewed by Anderson and Burnham 1976) also pointed strongly toward the additive mortality hypothesis for ducks. Thus, there was thought to be at least a good possibility that there were no important density-dependent effects on waterfowl mortality. If analyses supported this expectation, then it would be unnecessary to give further consideration to density-dependent mortality.

The second factor concerns the directness of the management application of research results. From a management perspective, primary interest is in the relationship between either hunting regulations or hunting mortality rate, and total annual survival rate. It was thus natural to study this relationship directly, rather than to begin with a two-step approach studying the relationships between hunting mortality and population size (e.g. at the end of the hunting season) and then population size and nonhunting mortality rate.

The third factor concerns the kinds of historical data that were available for analysis, relative to data requirements for the two general types of analysis. Data available from the large-scale North American survey programmes permitted estimation of harvest rates, annual survival rates, and spring population size, and provided an index to winter population size. Initial analyses provided no evidence of a relationship between annual survival and spring population size (Anderson 1975), so it was natural to turn attention to the harvest rate vs. survival rate relationship, which was of direct management interest. Data were not available that would have permitted more detailed examination of questions about density-dependent mortality. For example, if mortality at some time of the year does act in a density-dependent manner, then this relationship might be expected to involve some resource in short supply. Reasonable tests for such a relationship would first require identification of the time of the year during which density-dependent nonhunting mortality was important and of the resource that was in short supply. The most informative tests would then require estimates of the ratio of population size to available resources at the appropriate time of the year, as well as corresponding estimates of seasonal (rather than annual) mortality. In addition, the scale at which such density-dependent effects can be perceived is not always clear. The local density that influences mortality may or may not be reflected in estimates of population size or density corresponding to larger geographical areas. It is also possible that density is best expressed in terms of a particular age–sex class or other 'type' of individual rather than simply as total individuals. In any case, the large-scale data bases available for some North American duck species are frequently inadequate for detailed investigation of density-dependent mortality relationships.

These three factors have contributed to the North American research emphasis on the relationship between the rates of hunting mortality and annual survival. When recent analyses provided evidence favouring the compensatory mortality hypothesis (at least in some species), research began to focus on density-dependent nonhunting mortality as a possible mechanism.

Discussion

Much of our information about population dynamics of North American duck species is based on studies of the mallard, and it is not clear how relevant such information is to other species. Patterson (1979) and Bailey (1981) emphasized the differences among the life history patterns of North American duck species and noted the relevance of these differences to management. Both Patterson (1979) and Bailey (1981) categorized ducks as relative '*r*-' and '*K*-strategists' (MacArthur and Wilson 1967; Pianka 1970). For example, mallards and blue-winged teal were hypothesized to exhibit relatively low annual survival rates and lack density-dependent mechanisms of population regulation (*r*-strategists), whereas canvasbacks and other diving ducks (genus *Aythya*) were hypothesized to have higher annual survival rates and exhibit density dependence (characteristics of *K*-strategists; see Patterson 1979; Bailey 1981).

Such a ranking, however, does not lead to unambiguous predictions about the relative ability of these species to respond to hunting mortality. Density-dependent reproductive rates represent one sort of 'compensatory response' to exploitation, and density-dependent nonhunting mortality rates have been described as a possible mechanism underlying the compensatory mortality hypothesis. Thus, we might be led to believe that diving ducks and other *K*-strategists might be most likely to exhibit compensatory responses to hunting, based on the hypothesized importance of density dependence to their population dynamics. However, it was noted above that the rate of nonhunting mortality in the absence of hunting is the maximum value that can be assumed by c, the threshold parameter in the compensatory mortality hypothesis. Thus, among species for which the compensatory mortality hypothesis is a reasonable approximation, those with the higher survival rates in the absence of hunting (*K*-strategists), probably should not be able to withstand as much hunting mortality as those with higher nonhunting mortality rates (*r*-strategists). So even if we were able confidently to rank duck species along an '*r*–*K* continuum', these rankings would not necessarily be useful in predicting population response to hunting.

Instead, inferences about the response to hunting of any particular species must rely on the specific evidence we may have of either density dependence in reproductive or survival rates, or additivity of hunting mortality rates. We noted weak evidence of density-dependent reproductive rates in mallards, at least for high population sizes, and strong evidence of density-dependent reproduction for wood ducks. We know very little about this potential compensatory response to hunting in any other North American duck species. With respect to responses in mortality rates, the majority of the evidence for mallards supports the historical operation of the compensatory mortality hypothesis, at least at the continental level. There are some indications of additivity (or exceeding the threshold) in specific areas and situations (Caswell *et al.* 1985). Regarding other species, however, analyses have provided at least some evidence supporting the additive mortality hypothesis for American black ducks, wood ducks, and canvasbacks.

The various operational data-collection programmes for prairie-nesting ducks in North America have produced an impressive data base permitting a good description of historical changes in population size for several important species. In addition to these descriptive uses, we have tried to use these data in the context of a 'natural experiment' to draw inferences about functional relationships important to population dynamics. Analyses of these historical data for mallards have led to the development of models for predicting reproductive rate as a function of pond numbers on the breeding grounds and of breeding population size (e.g. Martin *et al.* 1979). In this case, it was possible to investigate adequately the functional relationship of interest because the two 'independent' variables did indeed vary independently (breeding population size in year i was largely independent of the number of ponds present in year i). In this situation, the natural experiment is believed to have produced a reasonable inference, and a comparison of estimated mallard age-ratios with those predicted by our models indicates that the models have performed in a satisfactory manner (Martin *et al.* 1979).

The use of historical data to investigate factors, especially hunting regulations, influencing mallard survival rates has been more difficult. Part of the reason for the increased difficulty in investigating survival rates, as compared with reproductive rates, involves timing. Mallard reproduction occurs at a discrete time of the year and the relevant 'independent variables' correspond to that period. Mortality, however, occurs throughout the year from a variety of sources, and we can think of annual survival rate as the product of several seasonal survival rates, each influenced by a particular set of environmental and other factors. Whenever we focus on factors occurring at one time of the year, then factors in the other seasons become 'noise' making it difficult to distinguish a true 'signal'. However, another problem which makes the investigation of hunting effects especially difficult is the historical covariation of hunting regulations and mallard population size. In this case, we have purposely manipulated hunting regulations in concert with changes in population size, another factor suspected of influencing survival rate. Because of the historical dependence of hunting regulations and population size, our historical data cannot even be claimed to represent a good natural experiment. This realization has led to the plea for large-scale experimentation involving the manipulation of hunting regulations according to some *a priori* design (Anderson and Burnham 1976; Nichols *et al.* 1984; Anderson *et al.* 1987) and it is still my belief that we will never develop a good understanding of the effects of hunting, at the population level, without such experiments.

We must still set hunting regulations and attempt to manage North American waterfowl populations, even though our current knowledge of the effects of hunting is incomplete. I agree with U.S. Fish and Wildlife Service (1988: 96) that a reasonable approach to such management at the present 'would be a form of risk-aversive conservatism in which relatively restrictive regulations would be implemented for populations at low levels'. Thought it cannot be 'proven' that such an approach will yield the desired population increases, it nevertheless seems reasonable, given our current levels of understanding of the relevant processes.

Summary

Duck hunting regulations are set annually in the United States based on the idea that they may influence harvest rates and ultimately population size. Analyses of historical variation in North American hunting regulations have provided strong evidence that they do influence harvest rates. At least some North American duck populations are thought to exhibit density-dependent reproductive rates, and this provides one type of response to exploitation. The additive and compensatory mortality hypotheses express two competing views of how harvest mortality is translated into annual survival rates. Analyses of mallard data have tended to support the compensatory mortality hypothesis and have provided little evidence that historical variation in hunting regulations and harvest rates has been accompanied by corresponding variation in annual survival rates. Analyses for other species have provided more evidence that regulations and harvest have influenced survival rates and have thus favoured the additive mortality hypothesis. Large-scale harvest experiments are recommended as a means of increasing our understanding of the effects of exploitation on duck populations.

Acknowledgements

G. M. Haramis, G. H. Heinz, and two anonymous referees provided helpful reviews of the manuscript. I thank L. A. Hungerbuhler for help with the preparation of the manuscript.

References

Anderson, D. R. (1975). Population ecology of the mallard. V. Temporal and geographic estimates of survival, recovery and harvest rates. *U.S. Fish and Wildlife Service Resource Publication*, **125**, 1–110.

Anderson, D. R. and Burnham, K. P. (1976). Population ecology of the mallard. VI. The effect of exploitation on survival. *U.S. Fish and Wildlife Service Resources Publication*, **128**, 1–66.

Anderson, D. R. and Henny, C. J. (1972). Population ecology of the mallard. I. A review of previous studies and the distribution and migration from breeding areas. *U.S. Fish and Wildlife Service Resources Publication*, **105**, 1–166.

Anderson, D. R., Burnham, K. P., and White, G. C. (1982). The effect of exploitation on annual survival of mallard ducks: An ultrastructural model. *Proceedings International Biometrics Conference*, **11**, 33–9.

Anderson, D. R., Skaptason, P. A., Fahey, K. G., and Henny, C. J. (1974). Population ecology of the mallard. III. Bibliography of published research and management findings. *U.S. Fish and Wildlife Service Resources Publication*, **119**, 1–46.

Anderson, D. R., Burnham, K. P., Nichols, J. D., and Conroy, M. J. (1987). The need for experiments to understand population dynamics of American black ducks. *Wildlife Society Bulletin*, **15**, 282–4.

Bailey, R. O. (1981). A theoretical approach to problems in waterfowl management. *Transactions North American Wildlife and Natural Resources Conference*, **46**, 58–71.

Bellrose, F. C. (1976). The comeback of the wood duck. *Wildlife Society Bulletin*, **4**, 107–10.

Blandin, W. W. (1982). Population characteristics and simulation modelling of black ducks. Unpublished Ph.D. Thesis. Clark University, Worchester, Mass.

Blohm, R. J., Reynolds, R. E., Bladin, J. P., Nichols, J. D., Hines, J. E., Pollock, K. H., and Eberhardt, R. T. (1987). Mallard mortality rates on key breeding and wintering areas. *Transactions North American Wildlife and Natural Resources Conference*, **52**, 246–57.

Brace, R. K., Pospahala, R. S., and Jessen, R. L. (1987). Background and objectives on stabilized duck hunting regulations: Canadian and U.S. perspectives. *Transactions North American Wildlife and Natural Resources Conference*, **52**, 177–85.

Brownie, C., Anderson, D. R., Burnham, K. P., and Robson, D. S. (1985). Statistical inference from band recovery data: A handbook, 2nd ed. *U.S. Fish and Wildlife Service Resources Publication*, **156**, 1–305.

Burnham, K. P. and Anderson, D. R. (1984). Tests of compensatory vs. additive hypotheses of mortality in mallards. *Ecology*, **65**, 105–12.

Burnham, K. P. and Nichols, J. D. (1985). On condition bias and band-recovery data from large-scale waterfowl banding programs. *Wildlife Society Bulletin*, **13**, 345–9.

Burnham, K. P., White, G. C., and Anderson, D. R. (1984). Estimating the effect of hunting on annual survival rates of adult mallards. *Journal of Wildlife Management*, **48**, 350–61.

Caswell, F. D., Hochbaum, G. S., and Brace, R. K. (1985). The effect of restrictive regional hunting regulations on survival rates and local harvests of southern Manitoba mallards. *Transactions North American Wildlife and Natural Resources Conference*, **50**, 549–56.

Chiang, C. L. (1968). *Introduction to stochastic processes in biology.* Wiley, New York.

Conroy, M. J. and Blandin, W. W. (1984). Geographic and temporal differences in band reporting rates of American black ducks. *Journal of Wildlife Management*, **48**, 23–36.

Conroy, M. J. and Eberhardt, R. T. (1983). Variation in survival and recovery rates of ring-necked ducks. *Journal of Wildlife Management*, **47**, 127–37.

Conroy, M. J. and Williams, B. K. (1984). A general methodology for maximum likelihood inference from band recovery data. *Biometrics*, **40**, 739–48.

Cowardin, L. M., Gilmer, D. S., and Shaiffer, C. W. (1985). Mallard recruitment in the agricultural environment of North Dakota. *Wildlife Monographs*, **92**, 1–37.

Dzubin, A. (1969). Comments on carrying capacity of small ponds for ducks and possible effects of density on mallard production. Saskatoon Wetlands Seminar. *Canadian Wildlife Service Report Series*, **6**, 138–60.

Geis, A. D. (1963). Role of hunting regulations in migratory bird management. *Transactions North American Wildlife and Natural Resources Conference*, **28**, 164–72.

Haramis, G. M. and Thompson, D. Q. (1985). Density-production characteristics of box-nesting wood ducks in a northern greentree impoundment. *Journal of Wildlife Management*, **49**, 429–36.

Hawkins, A. S., Hanson, R. C., Nelson, H. K., and Reeves, H. M. (1984). Flyways. U.S. Govt. Printing Office, Washington, D.C.

Henny, C. J. and Burnham, K. P. (1976). A reward band study of mallards to estimate reporting rates. *Journal of Wildlife Management*, **40**, 1–14.

Henny, C. J., Anderson, D. R., and Pospahala, R. S. (1972). Aerial surveys of waterfowl production in North America, 1955–71. *U.S. Fish and Wildlife Service Special Scientific Report — Wildlife*, **160**, 1–48.

Hepp, G. R., Blohm, R. J., Reynolds, R. E., Hines, J. E., and Nichols, J. D. (1986). Physiological condition of autumn-banded mallards and its relationship to hunting vulnerability. *Journal of Wildlife Management*, **50**, 177–83.

Johnson, D. H. and Sargeant, A. B. (1977). Impact of red fox predation on the sex ratio of prairie mallards. *U.S. Fish and Wildlife Service Wildlife Research Report*, **6**, 1–56.

Johnson, D. H., Burnham, K. P., and Nichols, J. D. (1986a). The role of heterogeneity in animal population dynamics. *Proceedings of International Biometrics Conference*, **13**.

Johnson, F. A., Hines, J. E., Montalbano, F. III, and Nichols, J. D. (1986b). Effects of liberalized harvest regulations on wood ducks in the Atlantic Flyway. *Wildlife Society Bulletin*, **14**, 383–8.

Kaminski, R. M. and Gluesing, E. A. (1987). Density- and habitat-related recruitment in mallards. *Journal of Wildlife Management*, **51**, 141–8.

Kirby, R. E., Hines, J. E., and Nichols, J. D. (1983). Afternoon closure of hunting and recovery rates of mallards banded in Minnesota. *Journal of Wildlife Management*, **47**, 209–13.

Krementz, D. G., Conroy, M. J., Hines, J. E., and Percival, H. F. (1988). The effects of hunting on survival rates of American black ducks. *Journal of Wildlife Management*, **52**, 214–26.

MacArthur, R. W. and Wilson, E. O. (1967). *The theory of island biogeography*. Princeton University Press, Princeton.

Martin, E. M. and Carney, S. M. (1977). Population ecology of the mallard: IV. A review of duck hunting regulations, activity and success, with special reference to the mallard. *U.S. Fish and Wildlife Service Resource Publication*, **130**, 1–137.

Martin, F. W., Pospahala, R. S., and Nichols, J. D. (1979). Assessment and population management of migratory birds. *Statistical Ecology*, **S11**, 187–239.

Munro, R. E. and Kimball, C. F. (1982). Population ecology of the mallard VII. Distribution and derivation of the harvest. *U.S. Fish and Wildlife Service Resource Publication*, **147**, 1–127.

Nichols, J. D., Conroy, M. J., Anderson, D. R., and Burnham, K. P. (1984). Compensatory mortality in waterfowl populations: a review of the evidence and implications for research and management. *Transactions North American Wildlife and Natural Resources Conference*, **49**, 535–54.

Nichols, J. D. and Haramis, G. M. (1980). Inferences regarding survival and recovery rates of winter-banded canvasbacks. *Journal of Wildlife Management*, **44**, 164–73.

Nichols, J. D. and Hines, J. E. (1987). Population ecology of the mallard. VIII. Winter distribution patterns and survival rates of winter-banded mallards. *U.S. Fish and Wildlife Service Resource Publication*, **162**, 1–154.

Nichols, J. D. and Hines, J. E. (1983). The relationship between harvest and survival rates of mallards: A straightforward approach with partitioned data sets. *Journal of Wildlife Management*, **47**, 334–48.

Nichols, J. D. and Johnson, F. A. (1990). Population dynamics of wood ducks: a review. In *North American Wood Duck Symposium*, (ed. L. H. Fredrickson, G. V. Burger, S. D. Havera, D. A. Graber, R. E. Kirby and T. S. Taylor).(In press.)

Nichols, J. D., Pospahala, R. S., and Hines, J. E. (1982). Breeding-ground habitat conditions and the survival of mallards. *Journal of Wildlife Management*, **14**, 80–7.

Nieman, D. J., Hochbaum, G. S., Caswell, F. D., and Turner, B. C. (1987). Monitoring hunter performance in prairie Canada. *Transactions North American Wildlife and Natural Resources Conference*, **52**, 233–45.

Patterson, J. H. (1979). Can ducks be managed by regulation? Experiences in Canada. *Transactions North American Wildlife and Natural Resources Conference*, **44**, 130–9.

Pianka, E. R. (1970). On r- and K-selection. *American Naturalist*, **104**, 592–7.

Pospahala, R. S., Anderson, D. R., and Henny, C. J. (1974). Population ecology of the mallard. II. Breeding habitat conditions, size of the breeding populations, and production indices. *U.S. Fish and Wildlife Service Resource Publication*, **115**, 1–73.

Reeves, H. M. (1966). Influence of hunting regulations on wood duck population levels. In *Wood duck management and research: a symposium*, (ed. J. B. Trefethen), pp. 163–81. Wildlife Management Institute, Baltimore.

Reinecke, K. J. and Shaiffer, C. W. (1988). A field test for differences in condition among trapped and shot mallards. *Journal of Wildlife Management*, **52**, 227–32.

Reinecke, K. J., Shaiffer, C. W., and Delnicki, D. (1987). Winter survival of female mallards in the lower Mississippi Valley. *Transactions North American Wildlife and Natural Resources Conference*, **52**, 258–63.

Reynolds, R. E. and Sauer, J. R. (submitted). Changes in mallard breeding population in relation to variation in production and harvest rates.

Rogers, J. P., Nichols, J. D., Martin, F. W., Kimball, C. F., and Pospahala, R. S. (1979). An examination of harvest and survival rates of ducks in relation to hunting. *Transactions North American Wildlife and Natural Resources Conference*, **44**, 114–26.

Sargeant, A. B., Allen, S. H., and Eberhardt, R. T. (1984). Red fox predation on breeding ducks in midcontinent North America. *Wildlife Monographs*, **89**, 1–41.

Sauer, J. R., Lawrence, J. S., Carr, E. L., Cook, G. W., and Anderson, V. (1990). Experimental September duck hunting seasons in Kentucky and Tennessee. In *North American Wood Duck Symposium* (ed. L. H. Fredrickson, G. V. Burger, S. D. Havera, D. A. Graber, R. E. Kirby and T. S. Taylor). (In press.)

Trost, R. E. (1987). Mallard survival and harvest rates: a reexamination of relationships. *Transactions North American Wildlife and Natural Resources Conference*, **52**, 232–64.

Trost, R. E. (1990). An examination of the relationship between harvest and survival rates of North American wood ducks: 1966–85. In *North American Wood Duck Symposium* (ed. L. H. Fredrickson, G. V. Burger, S. D. Havera, D. A. Graber, R. E. Kirby and T. S. Taylor). (In press.)

Trost, R. E., Sharp, D. E., Kelly, S. T., and Caswell, F. D. (1987). Duck harvests and proximate factors influencing hunting activity and success during the period of stabilized regulations. *Transactions North American Wildlife and Natural Resources Conference*, **52**, 216–32.

U.S. Fish and Wildlife Service (1988). Supplemental environmental impact statement: issuance of annual regulations permitting the sport hunting of migratory birds. U.S. Department of the Interior, Washington, D.C.

White, G. C. (1983). Numerical estimation of survival rates from band-recovery and biotelemetry data. *Journal of Wildlife Management*, **47**, 716–28.

25 Science and craft in waterfowl management in North America

HUGH BOYD

Introduction

Waterfowl management is a species of political ecology, rather than of ecology itself. The distinction determines the kinds of information and argument that are relevant, and the kinds of verification or falsification that can be used to distinguish better information, judgements, and decisions from less good ones and how the crucial issue of forecasting should be approached.

In defining waterfowl management as 'what waterfowl managers do', there seem to emerge four principal tasks:

(1) proposing legislation, drafting regulations, and encouraging their enforcement (three very unequal contributions to the single aim of trying to stop people doing things thought by the regulators to be bad);
(2) site management—securing and managing wetlands for the greater good of waterfowl, or of wetlands themselves;
(3) monitoring variations in the abundance, productivity, survival, and distribution of waterfowl and in the activities of waterfowl hunters; and
(4) 'other related duties', such as planning, informing the public, co-operating or competing with other groups with interests in wetlands and waterfowl, supporting research and using its findings.

Some waterfowl management is practised on every continent. It is exceptionally prominent in North America because of the massive US refuge system, the international cohesion of regulatory activities under the umbrella of the Migratory Birds Convention of 1916, and the large-scale monitoring programmes that have been in operation in the USA and Canada for 22–34 years. The monitoring programmes were set up to achieve short-term aims, the annual adjustment of hunting regulations in the two countries, in response to perceived changes either in waterfowl population characteristics or in the activities of hunters or other disturbers of waterfowl. The runs of monitoring data are becoming long enough to permit their use for other longer-term and wider purposes, such as detecting and tracking the consequences of environmental changes, some due to human activities, and predicting what may happen, not just this year but for several years ahead.

In this paper I shall use some indices drawn from long-running monitoring programmes to raise questions that should be further addressed by researchers concerned about large-scale changes in the welfare of waterfowl and their uses by humans.

The management of long-distance migrants, such as most ducks, geese, and swans, which spend substantial parts of the year under several different jurisdictions, requires unusual attributes. Most people feel more comfortable in intensive studies of small areas. However, site-specific studies are largely irrelevant to the kind of macro-ecology needed to manage effectively at the national and international levels. Site management is, of course, important too, but is not our concern here.

Large-scale waterfowl management in North America

The aerial transect surveys of N.W. North America in May and July are subject both to the observational errors inherent in single-visit surveys conducted at high speed and to pressure to report the results very quickly, leading to errors in the recording, storage, and retrieval of data. Great efforts have been made to rectify the errors, including some substantial adjustments to earlier series in the light of later knowledge. Yet, at best, we are dealing with indices of abundance and breeding output that are of low reliability, not with precise measurements. Many of the estimates of hundreds of thousands of birds derive from sightings of very few individual birds.

For the regulatory purposes for which they were designed, the National harvest surveys (NHS) and the species composition surveys (SCS) in the USA and Canada have been successful, as have the spring and summer surveys of breeding numbers and breeding success. Because these surveys now provide runs of data for up to 34 years, obtained in standardized ways, they are potential sources of information about regional changes in environmental conditions, as well as waterfowl. The role of the surveys in regulatory decision-making relating to hunting seems likely to diminish in importance. Managers have abandoned their efforts at 'fine tuning' and there is a large and sustained decline in the number of people interested in hunting waterfowl, while many other forms of wetland-based recreation are becoming more popular.

In North America, as in Europe, quantitative information on use by waterfowl of specific sites is now used routinely in environmental assessment procedures. The challenge now is to extend the use of survey results into such politically fashionable but fuzzy enterprises as 'state of environment reporting' and of producing scenarios for the possible impacts of climatic warming.

Major shifts in the direction of scientific inquiries related to waterfowl will be required. Some preliminary efforts have been made (e.g. Boyd 1985*a*) and the pace of redirection is increasing, especially in connection with the prolonged drought that has reduced crops and ducks in the prairies and parklands. However, in the near future waterfowl management will be driven by the principles and goals of the North American Waterfowl Management Plan (hereafter, the Plan), adopted in principle by the governments of the USA and Canada in 1986. Because, unlike many other plans, this Plan is being taken seriously, I make it the starting point of my discussion, as a source of 'approved facts' and of guidance to managers and, implicitly, to scientists.

The general intent of the Plan is to perpetuate waterfowl populations and their supporting wetlands and other habitats by active management. The Plan uses as species population goals for the year 2000 numbers in the ranges observed in the 1970s, continuing an American tradition of looking backwards when setting targets. The Plan asserts that 'a fall flight of over 100 million birds (= ducks) during average years (and winter stocks of about 5.5 million geese) will meet the needs of hunters and other users'. Attention will remain focussed on the most popular and abundant game ducks. Three of the ten most common species are identified as decreasing—the Mallard, Pintail, and Blue-winged teal (*Anas platyrhynchos*, *A. acuta*, and *A. discors*)—the others showing no sustained trend over the period 1970–85, though nearly all have declined during the 1980s. Among 27 identified stocks of five species of geese only 2 are described as decreasing, with 9 'stable' and 16 increasing.

Such recognition that populations fluctuate and that restoring the fortunes of declining stocks is a particular concern of managers gives the impression that the Plan adopts a dynamic approach, which is true to only a limited extent. Little attention is paid to the implications of recent changes in the distribution of waterfowl and even less to the changes in the distribution and concerns of the interested public.

Perhaps understandably, in view of the political sensitivity of the subject, the allocation of the expected 'safe' kill of 20 million ducks and 2.3 million geese between the two countries is assumed to continue unchanged and no attention is paid to changing regional pressures. There can be few, if any, other international agreements for the management of heavily exploited renewable resources that take so casual a view of who should get what. Given that Canada produces more than 80 per cent of North American waterfowl and takes less than a quarter of them, it is understandable that Americans should take a complacent view of the existing situation. Given that more than 80 per cent of the waterfowl winter in the USA, it is also clear that the Canadian bargaining position is weak.

Recent changes in hunting effort and kill have been described elsewhere (Boyd 1983, 1985*a*, 1988) and will not be further pursued here. Further reductions in, and redistribution of, hunting demand will continue, for demographic and social reasons. These aspects of management are ignored in the Plan, though it calls for more research on the effects of hunting mortality on waterfowl populations.

Changes in the abundance of ducks and geese

The breeding surveys in May and July are low-level aerial transects flown over large parts of the north-west (Fig. 25.1). The procedures were described by Benning (1976) and the reliability of the results discussed by Martin *et al*. (1979). There are no comparable published maps or descriptions of the procedures used in the winter surveys. The latter are made in January using searches from the air, ground, and water by large numbers of observers, mostly organized by the state agencies: the numbers given here refer only to counts in the USA, those made in Canada having been fewer and much less consistent. Many counts have been made in Mexico, which is an important wintering area for several species, but, as in Canada, there are no annual series that can be used as an index.

Figure 25.2 suggests that, in comparison with 1955–80, the 1980s have been a bad time for ducks and a good time for geese. The return, called for in the Plan, of duck populations to their levels of the 1970s, will require a gain of at least 50 per cent on 1987 levels. The May surveys indicate that a comparable resurgence was achieved between 1965 and 1972, though the winter surveys did not reflect either that or earlier fluctuations in breeding numbers. Since 1980 the January counts have tracked the May surveys. The May and January surveys include many of the same individual birds but are not strictly comparable, as the May surveys do not cover large parts of the breeding range and the winter counts detect only a minority of the birds wintering in the USA. If the two estimates referred to the same stock, the numbers reported in January should be greater than those in the following May, as many ducks will have died in the intervening 4 months while hardly any will have been born.

January and May estimates of the numbers of some of the more abundant ducks are compared in Fig. 25.3. Though there have been general decreases since 1980, the fluctuations in 1955–80 varied widely in form. The disparity between the winter and spring estimates is relatively small for the Mallard, but larger for the Wigeon *Anas americana* and the pochards, at least partly because substantial numbers of them winter in Mexico.

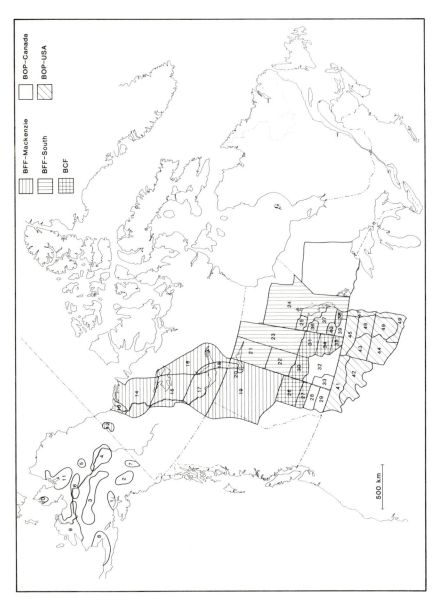

Fig. 25.1. The areas in Alaska, north-west Canada and the north-central USA surveyed for waterfowl in May and July. Shaded areas show waterfowl regions, after Lynch (1951) (see text for further details).

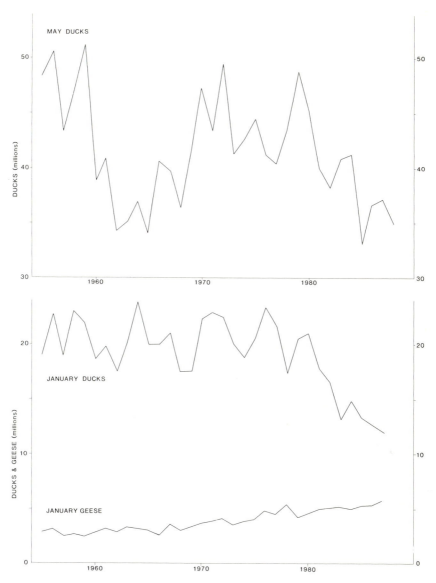

Fig. 25.2. Annual total numbers of ducks in the surveyed areas in May, and of ducks and geese found in the USA in January.

The numbers of ducks estimated to be in the surveyed areas in May are shown again in Fig. 25.4, grouped into the waterfowl regions identified by Lynch (1951, 1984) and named, from south to north, the Bald Open Prairie (BOP: here split into US and Canadian components), the Big Crow Factory (BCF), corresponding fairly closely to the Aspen Parkland Wetland

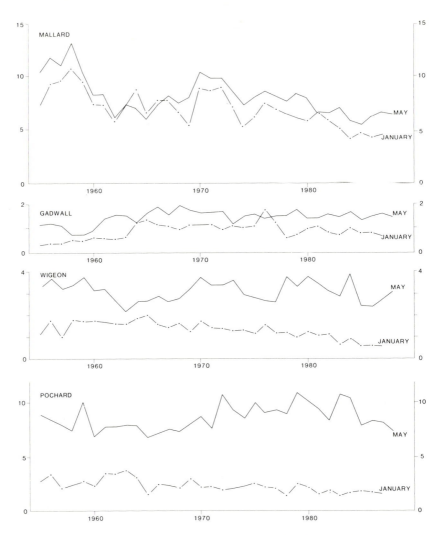

Fig. 25.3. Numbers of some species of ducks found in May and in January (in millions).

Region of the National Wetland Working Group (1986), and the Big Fish Factory (BFF), equivalent to the western parts of the Mid-Boreal, Continental High Boreal, and Sub-Arctic Wetland Regions. Here the BFF is split into Mackenzie and south-eastern sub-regions, because there is a long run of July breeding surveys in the latter but none for most of the Mackenzie. Lynch (*loc. cit.*) emphasized the likelihood of wide variations in duck use of the grasslands (BOP) and parklands (BCF), depending on the amount of surface water in the spring, much of it being impermanent. The lower part of Fig. 25.4 shows that duck numbers in the BOP and BCF

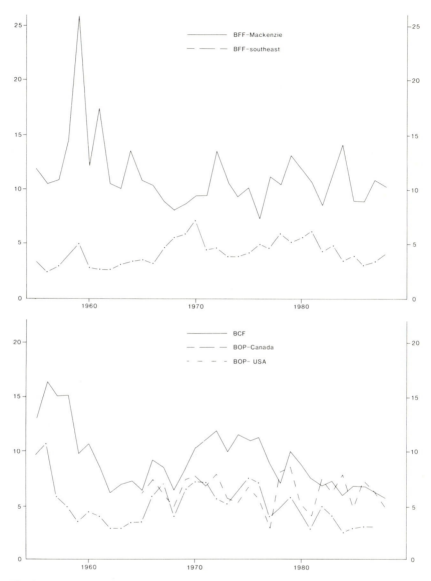

Fig. 25.4. Estimated number of ducks in May in waterfowl regions of north-western Canada and the north-central USA (in millions).

regions were exceptionally high in the 1950s, when water was unusually abundant. A period of drought in the early 1960s was followed by more plentiful water in the 1970s and a second prolonged drought lasting nearly continuously from the late 1970s until the present. The numbers of ducks in the BOP-USA region (Montana, North and South Dakota) have fluctuated widely since 1965, but without the long decline since the early 1970s seen in the Canadian part of the BOP and in the BCF. (The May survey system in the BOP-USA was incomplete before 1965.)

The most striking feature of the estimates of numbers of ducks in May in the BFF (upper part of Fig. 25.4) is the massive peak in the Mackenzie sub-region in 1959. This and the second peak in 1961 are believed to have resulted mainly from ducks moving north from the prairies and parklands as they dried up (Hansen and McKnight 1964; Smith 1970; Henny 1973). After those peaks in population size the numbers in the Mackenzie sub-region continued to fluctuate widely. Much of the variation may be due to the relatively low sampling effort in the north, where duck densities are lower than in the more intensively sampled BOP and BCF. Note that duck numbers in the BFF-south-east (the boreal regions of northern Saskat-chewan and Manitoba) were relatively low from 1955 to 1966, increased rapidly from 1966 to 1970, when numbers in the BOP and BCF were also increasing and have since fallen, more or less in parallel with the decline in the southern farmed areas.

The traditional wisdom of waterfowl biologists is that duck populations in the boreal regions are far less variable than those further south. The May surveys do not support this view. Nor do the July surveys, which are intended to detect broods and adult ducks that are still sexually active and liable to breed. In May most of the ducks seen can be identified to species level by the aerial observers. In July many of the broods cannot be identified from the air, even when the number of ducklings in a brood can be counted. Thus only estimates of the numbers of broods and of young seen (broods × mean brood-size) are obtained, without allocation to species.

The late-nesting index of single or paired adults unaccompanied by young (and not assembled into larger groups) is split between dabbling and diving ducks; but so much doubt exists about the numbers of ducklings likely to be reared by these birds that it seems best to exclude them from estimates of production. The observed regional mean brood-sizes have varied considerably from year to year. All four regional series show a downward trend over the entire period. The decline is less apparent in the BFF samples than in those from further south, though they may be due in part to the absence of BFF samples prior to 1969. The numbers of broods seen have declined, as has the mean brood-size. This is not due solely to the decreasing numbers of adults. Figure 25.5 shows that the ratio of young seen in July to adults in May has fallen in all three Canadian regions, though not in BOP-USA. This suggests that the ability to increase produc-tivity when the adult population has been reduced was lost or impaired in the prairie provinces during the 1970s, though proportional output appears to have risen again in more recent years.

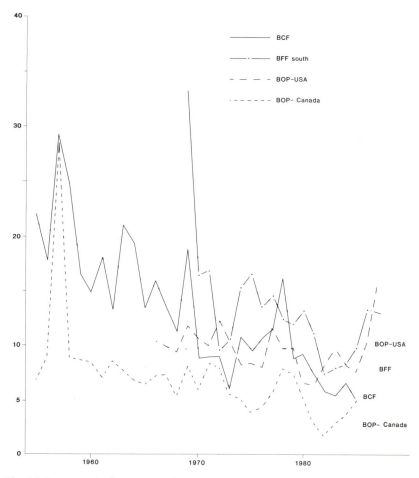

Fig. 25.5. Young ducks seen in July as percentage of adults seen in May in the four waterfowl regions.

Duck abundance and success and the amount of surface water

The general observation that ducks were more plentiful in the prairies in wet than in dry years led the USFWS to include 'pond counts' in their May and July aerial surveys. Boyd (1981), in a study of ducks in the Canadian prairies from 1955 to 1980, also used an index of soil moisture, based on weighted means of precipitation in the 21 months preceding 1 May, following a procedure developed by Williams and Robertson (1965) for the purpose of predicting wheat production. Rather higher correlations were found between duck numbers and the soil moisture index than between numbers of ducks and of ponds seen, although the latter were highly correlated.

These associations are also exhibited in BOP-USA (Fig. 25.6), where the numbers of young ducks in July have tracked the numbers of adults in May

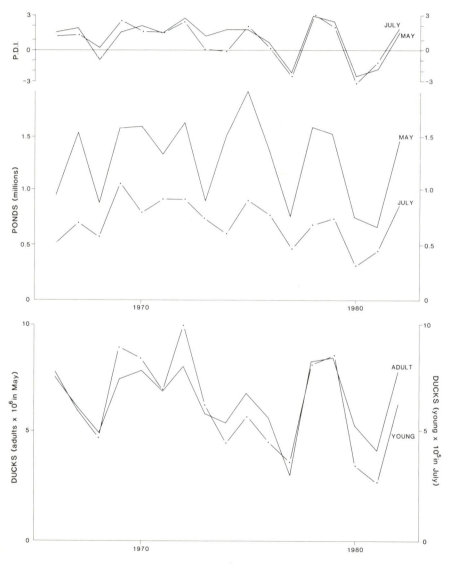

Fig. 25.6. Association between the numbers of adult ducks in May and young ducks in July, the numbers of ponds and the Palmer hydrological drought index in BOP-US (Montana, North and South Dakota).

very closely. The numbers of ducklings seen were only about one-tenth of the numbers of adults, not because this is a true measure of total production but because ducklings are much harder to detect from the air. Intensive studies from the ground suggest that the ratio of fledged young (in August) to adults is rather greater than 1.0.

In addition to duck numbers, Fig. 25.6 includes estimates of the numbers of ponds in May and in July and a 3-state May mean for the Palmer Hydrological Drought Index (PDI) calculated from tabular estimates for individual states prepared by the US National Climatic Data Center. The calculation of the tabled values was described by Karl and Knight (1985). As in the earlier study of ducks in the Canadian prairies, these data from BOP-USA show highly significant positive correlations between the numbers of adult ducks and of ponds in May ($r = 0.701$, $P < 0.01$) and the May Palmer Drought Index ($r = 0.854$, $P < 0.001$). The numbers of young in July are positively correlated with adult ducks in May ($r = 0.927$, $P < 0.001$), May ponds ($r = 0.614$, $P < 0.01$), July ponds ($r = 0.710$, $P < 0.001$), May PDI ($r = 0.815$, $P < 0.001$), and July PDI ($r = 0.844$, $P < 0.001$).

Regional shifts in midwinter distribution

From the points of view of hunters and refuge managers, the distribution of waterfowl during the hunting season is at least as important as their total numbers. No national surveys of waterfowl are carried out during the hunting season, so that in practice much of what we know about distribution in autumn and early winter has to be derived from harvest surveys of those that did not get away. The US winter surveys are intended to provide indices to the surviving stocks. What they tell us about distribution is, of course, affected by the severity of the season, and they underplay the role of the northern states, where there are important staging areas used by large numbers of geese and ducks in the fall and spring.

In examining changes in winter distribution over the last 30 years, I have grouped data supplied as state totals within each of the four flyways into northern, middle, and southern regions, a total of twelve in all. This is a cruder classification than that of the 18 regions used by Bellrose (1976) in studies of migration corridors and the chronology of migration. He was able to use weekly counts on national wildlife refuges, whereas in the present work access was available only to the state totals. The resulting loss of resolution does not wholly obscure some interesting findings. Differences in the behaviour of two dabbling ducks, the Mallard and the American Wigeon, are used as examples. Because of the erratic variations in counts in some regions it is convenient to use some averaging procedure, here 5-year means from 1955–9 to 1980–4 and a 3-year mean for 1985–7.

The numbers of Mallard found in winter decreased greatly between the 1950s and the 1980s. The main wintering strongholds of the Mallard lie in a diagonal from the Pacific north-west to the Mississippi delta states. In three of the four regions where Mallard are most abundant (Fig. 25.7), the proportions that the numbers represented of the national total have varied considerably, but without a sustained trend. The middle-tier states of the Mississippi Flyway held 26 per cent of the national sample in 1955–9 but only 13 per cent in 1985–7. There were slow proportional declines in the northern Central Flyway and, at a low absolute level, in the southern Atlantic Flyway states, while the small numbers in the north-east of the winter area, where they overlap with the Black Duck *Anas rubripes*, increased.

In winter the American Wigeon is largely concentrated in the Pacific Flyway and in the south of the Central and Mississippi flyways (Fig. 25.8). The general decline in the numbers of Wigeon found in January (which has not been matched by a decline in the May survey area, Fig. 25.3) is reflected in the regional totals, but not in a uniform way in the earlier years. The most obvious discrepancy was in the Mississippi delta states, where Wigeon increased greatly during the 1960s before declining steeply in the 1970s.

The general point to be made is that there may be great differences in the apparent well-being of stocks within different parts of the winter range of a

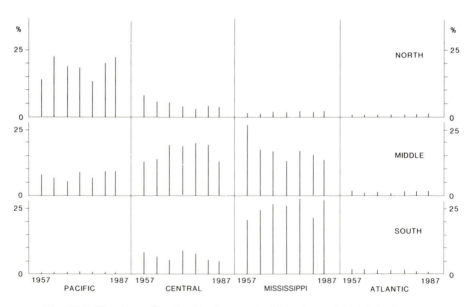

Fig. 25.7. Numbers of mallard in January in 12 regions of the USA; 5-years means expressed as a percentage of the national total.

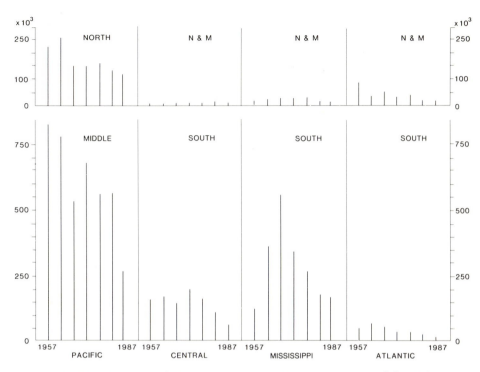

Fig. 25.8. Numbers of American Wigeon in January in regions of the USA: 5-year means expressed as a percentage of the national total.

single species. Amongst those ducks that are numerous and have extensive winter ranges it has proved difficult to identify sub-populations that could be managed as distinct biological entities. Managers are therefore compelled to use political boundaries, rather than natural ones.

Discussion

Johnson and Grier (1988) have discussed the determinants of the breeding distribution of ten species of ducks, using the May survey data, including the pond counts. They are less concerned than I have been with variations over time, but deal in greater detail with life history characteristics. They identify three patterns of settling: homing, opportunistic, and flexible. Homing is most pronounced amongst species that use more stable (and hence more predictable) wetlands, while opportunistic settling is more prevalent among species using less stable wetlands. These, and other points made by Johnson and Grier, are valuable reminders of the importance of interspecific differences in response to changes in resource availability, which are hidden when attention is restricted to 'total ducks'.

Turning from biology to policy advice, waterfowl managers, like politicians, must often say one thing while doing something quite different; their advisors must not do so. They must try to explore and present options in a disinterested yet imaginative way, expecting often to be ignored or overruled, for irrational reasons that may be sensible, however intellectually unpalatable.

To avoid ending on an immoralizing note, let me return to some of the questions posed by the breeding survey examples and ask you to consider how they might be followed up, whether with scientific or advisory aims in mind. First (Fig. 25.5), why should the mean brood-size have fallen in all the four regions where it has been estimated, in which breeding numbers have, until very recently, moved in different ways? Decline in brood-sizes has accompanied, but does not itself wholly account for, the declines in the ratios of ducklings to adults in the three Canadian regions. It is noteworthy that the ratio has not declined in the BOP-USA, which might indicate the possibility of a recovery in output in BOP-Canada and the BCF, given that Canadian agricultural practices tend to lag behind those in the USA by one or two decades.

The forecasts of climatic change, and events in recent years, suggest increased frequency and severity of drought in the BOP, and warming that may permit agriculture to move northward into some parts of the BFF, where soils permit. Are we able to predict how ducks may respond to such changes and what effects might they have on the implementation of the Plan, now seen to be largely in the hands of agricultural interests? The opportunities for further study and thought and for the framing of more searching questions continue greatly to exceed the scientific resources that can be mustered to address them. This means, as it has always done, that the tasks confronting waterfowl biologists working for governments are more, not less, intellectually demanding than those of unapplied ecology.

Summary

For more than 30 years the U.S. Fish and Wildlife Service and the Canadian Wildlife Service have carried out joint annual surveys of breeding numbers and success of waterfowl over a large part of North America, as well as assessing numbers in midwinter. They have also estimated the amount of waterfowl hunting and the reported kill in both countries. Duck numbers have been much lower in the 1980s than they had been earlier, while geese have been much more abundant and widespread than they were in the 1950s.

The decline in duck numbers has been particularly marked in the prairies and parklands of western Canada and the north-central USA and has been accompanied by a reduction in the production of young, which was greater than expected from the smaller numbers of breeding ducks. There have also been substantial changes in the winter distribution of ducks, although numbers in winter have varied less than those in summer. The difficulty of distinguishing 'signals' from 'noise', due to short-term variability and to the low precision and accuracy of large-scale monitoring, makes it difficult to establish the relative importance of the many factors affecting waterfowl numbers and distribution and to offer sound advice to managers. That difficulty is reflected in the North American Waterfowl Management Plan of 1986, which sets targets based on past performance, rather than on estimates of future needs and pressures.

References

Bellrose, F. C. (1976). *Ducks, geese and swans of North America*. Stacpole, Harrisburg, Pa.

Benning, D. S. (1976). Standard procedures for waterfowl populations and habitat surveys. *U.S. Fish and Wildlife Service Operating Manual*, Washington, D.C.

Boyd, H. (1981). Prairie dabbling ducks, 1941–1990. *Canadian Wildlife Service Progress Notes*, No. 119, 1–9.

Boyd, H. (1983). Intensive regulation of duck hunting in North America: its purpose and achievements. *Canadian Wildlife Service Occasional Paper*, No. 50, 1–22.

Boyd, H. (1985a). The large-scale impact of agriculture on ducks in the Prairie Provinces, 1956–1981. *Canadian Wildlife Service Progress Notes*, No. 149, 1–13.

Boyd, H. (1985b). The reported kill of ducks and geese in Canada and the USA, 1974–1982. *Canadian Wildlife Service Occasional Paper*, No. 55, 1–20.

Boyd, H. (1988). Recent changes in waterfowl hunting effort and kill in Canada and the USA. *Canadian Wildlife Service Progress Notes*, No. 175, 1–11.

Hansen, H. A. and McKnight, D. E. (1964). Emigration of drought-displaced ducks to the Arctic. *Transactions 29th North American Wildlife and Natural Resources Conference*, pp. 119–26.

Henny, C. J. (1973). Drought displaced movement of North American pintails into Siberia. *Journal of Wildlife Management*, 37, 23–9.

Johnson, D. H. and Grier, J. W. (1988). Determinants of breeding distributions of ducks. *Wildlife Monographs*, 100, 1–37.

Karl, T. R. and Knight R. W. (1985). Atlas of monthly Palmer hydrological drought indices (1931–1983) for the contiguous United States. National Climatic Data Center, *Historical Climatology Series*, pp. 3–7.

Lynch, J. J. (1951, 1984). Escape from mediocrity. US Fish and Wildlife Service Report, 1951, reprinted in *Wildfowl*, 35, 7–13.

Martin, F. W., Pospahala, R. S., and Nichols, J. D. (1979). Assessment and population management of North American migratory birds. *Proceedings 2nd International Ecology Congress, Parma, Italy*, August 1978, pp. 187–239.

Smith, R. I. (1970). Response of pintail breeding populations to drought. *Journal of Wildlife Management*, 34, 943–6.

Williams, G. D. V. and Robertson, G. W. (1965). Estimating most probable wheat production from precipitation data. *Canadian Journal of Plant Science*, 43, 34–47.

26 Florida Scrub Jay ecology and conservation

GLEN E. WOOLFENDEN and JOHN W. FITZPATRICK

Introduction

The Florida Scrub Jay *Aphelocoma c. coerulescens* has been the subject, for 20 years, of an intensive demographic and behavioural study at the Archbold Biological Station in Highlands County, Florida, USA. As a result, considerable information pertinent to conservation and management of this endangered species is available. As we explain below, we are convinced that long-term ecological studies are essential for successful management of any threatened or endangered organism.

In this paper we summarize aspects of life history, habitat, and demography that are pertinent to protection and management of a species that now exists only in small, isolated populations. We conclude with comments on the implications of our results to management of other similarly threatened species. Numerous publications provide more detailed summaries of the data presented here, especially Woolfenden and Fitzpatrick (1984, 1990) and references therein.

Conservation science, when properly applied to specific cases, identifies its goals in advance. We therefore propose some guidelines for delineating and managing isolated populations of Florida Scrub Jays. The goal of these guidelines is to maximize the probability of continued existence of this threatened jay, for at least the next 100 years, substantially throughout its present range.

Life history

Social organization

Florida Scrub Jays are permanently monogamous and permanently territorial. One to several nonbreeding jays live in the territories of about half the breeding pairs. Territorial groups vary in size from 2 (simple pair) to 8 adults, with an additional 1 to 5 juveniles. Large groups are rare; average group size is about 3.

Jays conduct foraging, predator surveillance and defence, territory defence, and most other daily activities in close proximity to one another within their group. While group members forage, one jay often sits on an exposed perch as a sentinel (McGowan and Woolfenden 1989). If an aerial predator is detected a distinctive warning call is given and all jays quickly dash for dense cover. Ground predators are mobbed with noisy scolds and even furtive pecks.

Florida Scrub Jays live in large, all-purpose territories throughout their lives except during dispersal, when prebreeders foray into neighbouring territories searching for breeding space. Territory defence is a group activity led by the breeders, especially the male, and sometimes the oldest male offspring. Territorial squabbles are most frequent in fall, and again in early sring when most territorial shifts take place.

Young jays remain in their natal territory for at least 1 year before attempting dispersal. Some, especially males, may remain nonbreeders for several years and, as 'helpers', assist the breeder in most activities. Although they do not build nests, incubate, or brood, helpers do feed nestlings and fledglings, mob predators, perform as sentinels, and defend the territory. Breeders assisted by helpers produce more independent young than pairs without helpers. Perhaps of equal importance, breeders show increased annual survival when helpers are present. Helpers also experience lower mortality while remaining within the group.

Reproduction

The nesting season extends from March through June, when a single brood of 1 to 5 fledglings is raised annually. Renesting after a nest fails is a regular event, especially early in the nesting season. The average number of attempts per pair per breeding season is 1.4, although about 25 per cent of pairs fail to produce fledglings during the nesting season. Nest failure almost always occurs through predation and increases rapidly through the season. We suspect that the sharp seasonal decline in expected nest success, plus increased risk of adult mortality during nesting, have helped select for a relatively brief nesting season.

Average production of young is about 2 fledglings per pair per year. Mortality of fledglings within the first few weeks of post-fledging is extremely high; on average only about 35 per cent of the fledglings survive to their first birthday, when they normally serve as helpers in the home territory.

Helpers

About 30–40 per cent of the jays alive at the onset of a nesting season are nonbreeders. Predominantly yearlings, these are adult-plumaged jays mostly living in their natal territory. By assisting related jays, helpers may further their own genetic representation in future generations, because helpers raise the average success of breeders and these extra young usually are genetic sibs or half-sibs of the helpers. Helpers also behave in ways to maximize their own opportunities to become breeders. The probability that colateral kin will be competitors for breeding vacancies is reduced through an intrafamilial dominance hierarchy in which males dominate females, breeders dominate helpers, and older helpers dominate younger ones.

Territory acquisition

Potential breeders greatly outnumber breeding vacancies and vigorous defence by all territory holders suppresses establishment of new territories by nonbreeding jays even when total population density is reduced. Helper males may remain nonbreeders for up to 5 years; helper females generally disperse after 1 or 2 years of helping. Both sexes follow the general strategy of living at home while monitoring the neighbourhood for breeding vacancies that arise through the death of an established breeder (Woolfenden and Fitzpatrick 1984). Established territories disappear after death of both breeders or when a widowed jay becomes paired in a neighbouring territory.

New territories are established through a process of so-called 'territorial budding' (Woolfenden and Fitzpatrick 1978). As family size grows, territory size expands at the expense of less successful neighbouring families. Frequently, following such a build-up, the dominant nonbreeding male takes over a segment of the enlarged family territory. Obtaining a mate from outside his family, he forms a pair on a new, 'budded' territory. Female helpers inherit breeding space only rarely. About 40 per cent of new breeders acquire breeding status through the budding process (Woolfenden and Fitzpatrick 1986).

Establishment of a new territory *de novo*, away from the home territory of both jays, is extremely rare (4 per cent of new breeders; Woolfenden and Fitzpatrick 1986). This fact provides direct evidence for the role of territorial behaviour as a proximate control of breeding density. As pointed out below, even when the density of potential breeders is low, existing territories encompass all available space. Weeks or months of daily territorial battles accompany any major boundary shift, especially attempted budding. The energy and risk involved in establishing completely new boundaries *de novo*, outside the safety of an adjacent natal territory and group, appear prohibitive.

Female nonbreeders also help at the nest. While helping, they monitor their neighbourhood for breeding vacancies or males that are budding a territory. As the territory in which they reside grows, the number of neighbouring territories that helpers can monitor without leaving home increases. Thus, the helping system benefits females even though they rarely inherit space directly.

New territories are rarely established in habitat previously unoccupied, because essentially no such land exists. This fact presents the most fundamental ecological constraint facing Florida Scrub Jay populations. *It also presents the fundamental conservation problem.* Extremely intense competition for breeding vacancies has given rise to a social system in which prebreeders queue for breeding space within existing territories. Permanent territoriality and group defence prohibit dramatic increases of breeding density from one year to the next. The process of working to become breeders involves a system of cooperation within the family group. The cooperative breeding system of the Florida Scrub Jay thus appears to be a natural evolutionary consequence of their restriction to a severely limited and sharply defined habitat.

Habitat requirements

The Florida Scrub Jay is totally restricted to a rare, relict habitat (Fig. 26.1) which depends upon periodic burning to persist. No conservation or management schemes will be successful without specific attention to this fact. Briefly stated, save the habitat in proper quantity and condition, and the jays are saved.

The essential habitat is oak-dominated scrub, a peculiar vegetation formation found only on extremely well-drained, sandy soils of old coastal dunes or paleodunes (Abrahamson *et al.* 1984; Laessle 1958, 1968). The essential plants in the habitat are four species of stunted, low-growing oaks: scrub oak *Quercus inopina*, scrub live oak *Q. geminata*, Chapman's oak *Q. chapmanii*, and myrtle oak *Q. myrtifolia*. In optimal habitat (i.e. where territory density and reproductive success are highest) most of the oak shrubs are between 1 and 3 m. tall. The oak thickets are interspersed with numerous patches of bare sand, where the jays bury thousands of acorns each fall for use throughout the year (DeGange *et al.* 1989). Trees and dense herbaceous vegetation are rare in naturally occurring habitats preferred by Florida Scrub Jays.

Fire

Fire is a frequent, natural event in scrub habitats throughout Florida. Lightning strikes are most frequent in the summer months and often ignite the scrub vegetation. For Florida Scrub Jays, a fire frequency averaging about once every 8 to 20 years is probably optimal (see below). Much greater frequency (e.g. every 2–3 years) consistently maintains the

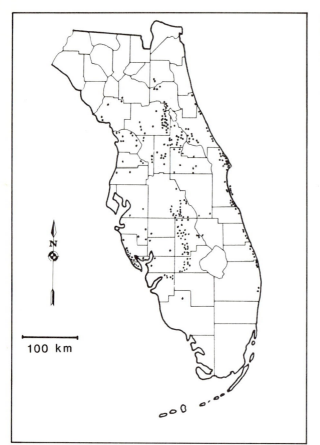

Fig. 26.1. Approximate locations of all remaining populations of Florida Scrub Jays as of early 1980s. Modified from Cox (1987).

100 km

principal oak species below acorn-bearing height and may slowly favour the spread of palmettos at the expense of oaks. Lower frequency produces tall, dense oak understories and pine forests. These habitats (referred to below as 'overgrown scrub') are shunned by the jays, and instead are occupied by a competitor-predator, the Blue Jay *Cyanocitta cristata*.

At Archbold Biological Station, fire suppression in a tract formerly supporting low, open oak scrub provided experimental documentation of the response of Florida Scrub Jays to advanced habitat succession. The original 400 ha property acquired in 1941 by Richard Archbold last burned extensively around 1927 (Myers and White 1987). Extensive portions of this tract were occupied by Florida Scrub Jays through the 1940s (Amadon 1944), 1950s, and 1960s (Woolfenden 1974). When our long-term study began in 1969, density of territories in a small portion of this tract where jays still remained was about 3.8 per 40 ha. By that time the habitat already had become tall and dense, with numerous sapling slash pines *Pinus elliottii*. During the 1970s we witnessed gradual decline of jays in this tract, and since 1980 the area has been colonized only sporadically by dispersing jays, and then abandoned. About 50 years post-fire was the time of final occupancy, but reproduction and survival had dipped below replacement levels even by 1969 (about 40 years post-fire; Fig. 26.2). We conclude that a fire period of 8 to 20 years is necessary for maintaining a healthy population of Florida Scrub Jays.

Population density and demography

Density

Within three habitat types, between 1971 and 1987, density varied across the study area from about 2 to about 6 territories per 40 ha of scrub (Fig. 26.3). Within the highest density areas in the core study tract mean density was about 5 territories per 40 ha throughout the 1970s. Density in this area was not altered by a major fire in 1977, but was reduced by a catastrophic epidemic in 1979. Density has not recovered since the epidemic,

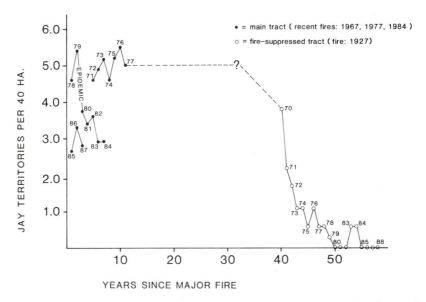

Fig. 26.2. Density of Florida Scrub Jay territories in several tracts of scrubby flatwoods at Archbold Biological Station, in relation to time since most recent major fire in the respective habitat patch. Precipitous decline after about 40 years post-fire reflects the point at which reproduction no longer kept pace with average mortality of adults and young in tall, overgrown scrub. Solid lines connect the data; the query signifies that the pattern of decline is unknown.

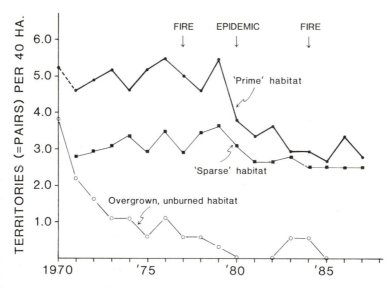

Fig. 26.3. Densities of Florida Scrub Jay territories in three habitats of scrubby flatwoods at Archbold Biological Station. Highest density occurs in periodically burned oak scrub atop well-drained sand ridges, while adjacent lower-lying habitat of sparser oak density and abundant palmetto-wiregrass areas supports larger territories, hence lower density. Overgrown scrubby flatwoods supports declining population 40 years after the most recent fire.

varying between 3 and 4 territories per 40 ha during the 1980s. The extensive 1984 fire also did not alter existing densities. Lower densities since 1979 coincide with generally elevated mortality among both juveniles and adults compared with the 1970s (juvenile mortality, 1970–8: 0.66; 1979–87: 0.71; adult mortality: 0.18 vs. 0.24, respectively). Higher mortality both reduced the populations of breeding-age helpers as potential recruits and opened more breeding vacancies within existing territories. Together these demographic factors reduced overall competition for space, presumably allowing territory holders to defend slightly larger territories and reducing breeder density. A positive but nonlinear relationship exists between density of potential recruits in autumn and territory density the following spring (Fig. 26.4). This suggests that territory density is somewhat elastic, but only when potential recruits are relatively sparse. A maximum apparently was reached during the 1970s, and the population has not recovered to that level since the 1979 epidemic.

Mortality was at or above the 18-year average for five consecutive years 1983–7, after fluctuating equally above and below the mean from 1970 through 1982 (Fig. 26.5). We have no explanation for this change. Large fires in 1977 and 1984 burned extensive areas within the study tract, but

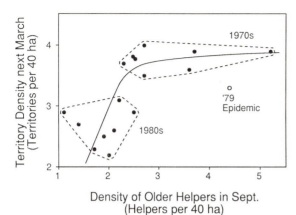

Fig. 26.4. Territory density at onset of breeding (March) plotted against density of older helpers the preceding September, 1972–88. Line is fitted by eye; dashed lines enclose density values for 1972–9 and 1981–8. Open circle shows values before and after the epidemic, which began in October 1979. Density has not recovered since.

territory boundaries did not shift in these areas and mortality was no higher within the burned habitat. Recovery from the 1979 epidemic may be involved somehow (see below).

Breeding density tends to be lower in scrub habitats at both ends of the habitat density spectrum. Jay density along the western edge of our study tract averaged between 2.5 and 3.5 territories per 40 ha during 1971–87 (Fig. 26.3). This area is characterized by scattered oak patches interspersed

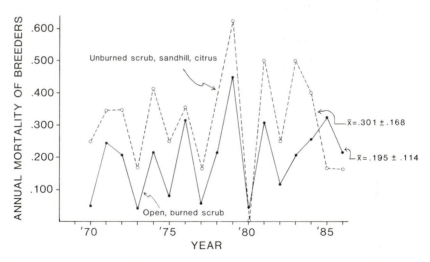

Fig. 26.5. Annual mortality of breeding Florida Scrub Jays at Archbold Biological Station, 1970–86. Mortality in open, periodically burned scrub in main study tract is lower than that of several other habitats, all unburned since about 1927. Dashed line includes small number of jays living at the edge of a citrus grove bordering tall, unburned ridge scrub for several years before going extinct.

with extensive areas of low palmetto and sparse oaks that are used less often by the jays. At the other extreme of oak density, in the area protected from burning since the 1930s, density of Florida Scrub Jays declined from about 3.8 territories per 40 ha to nearly zero and was occupied only sporadically since 1980 (Fig. 26.3). In general, Florida Scrub Jay densities even in excellent scrub habitat rarely exceed 5 to 6 territories per 40 ha and often are considerably lower.

Territory size

Most scrub is patchy and estimates of jay density in heterogeneous scrub do not provide exact estimates of territory size because the jays avoid many non-oak habitat patches. Our territory maps (1971 through to the present) reveal that average territory size, excluding major areas of unused habitat, is 9.0 ha (Woolfenden and Fitzpatrick 1984). Stable territories have ranged from 4 to 18 ha. For management purposes we consider 9.0 ha the average amount of scrub habitat required to support one breeding group of Florida Scrub Jays.

Considerable variation in territory size exists among groups, among habitats, and among years. Variables affecting territory size include habitat quality and group size (Woolfenden and Fitzpatrick 1978, 1984), as well as overall population density (as shown above). Of interest for management would be evidence of *minimum sustainable territory size*. To this end, we examined two samples of territories that were substantially smaller than average during the 1970s. Territories that persisted with a breeding pair for 3 or more consecutive years below the mean territory size ($\bar{x} = 5.4$ ha, SD = 1.6, N = 36) were significantly larger than those smaller-than-average territories that disappeared after only 1 or 2 years ($\bar{x} = 3.3$ ha, SD = 1.9, N = 13; $t = 3.9$, $P < .0003$). Many of the failures persisted for only a few months, as the breeders unsuccessfully attempted to expand their boundaries. This comparison suggests that minimum sustainable territory size on our study tract is probably between 4 and 5 ha of oak scrub.

Reproduction and mortality in several habitats

The balance between birth rates and death rates in a stable population depends upon a host of variables and can vary substantially from year to year. Any management scheme to protect Florida Scrub Jays must include efforts to reduce or avoid perturbations that reduce average reproductive success or increase average mortality beyond equilibrium values.

Table 26.1 summarizes reproduction and survival of Florida Scrub Jays at Archbold Biological Station from 1969 through 1986. Data are

Table 26.1 Mean survivorship and reproduction of Florida Scrub Jays in several habitats, 1969–86

	Optimal habitat		Suboptimal habitat					
	Periodically burned, open oak scrub		Unburned, overgrown scrubby flatwoods		Unburned Southern Ridge Sandhill (Slash Pine-Turkey Oak)		Mature Citrus bordering unburned scrub	
N (pair–years)	429		74		8		21	
Seasonal nest attempts	1.38	(593/429)	1.49	(110/74)	1.50	(12/8)	1.11	(20/18)
Fledglings/pair	1.97	(843/429)	1.58	(117/74)	1.38	(11/8)	2.00	(38/18)
Independent young/pair	1.17	(500/429)	0.80	(59/74)	1.13	(9/8)	1.56	(28/18)
Yearlings/pair	0.60	(259/429)	0.36	(27/74)	0.50	(4/8)	0.61	(11/18)
First-year survival	0.307	(259/843)	0.231	(27/117)	0.364	(4/11)	0.289	(11/38)
Breeder survival	0.789	(697/883)*	0.723	(107/148)	0.688	(11/16)	0.619	(26/42)
	Expected lifetime success/individual							
Breeding seasons	4.4		3.5		3.2		2.6	
Fledglings	4.3		2.8		2.2		2.6	
Independent young	2.6		1.4		1.8		2.0	
Yearlings	1.3		0.6		0.8		0.8	

*N = 883 breeder years for calculating breeder survival.

separated into four habitat categories pertinent to management concerns: periodically burned, open oak scrub (i.e. prime habitat); unburned and overgrown scrubby flatwoods; unburned southern ridge sandhill (Slash Pine/Turkey Oak formation); mature citrus groves bordering unburned scrub.

Long-term survival of a population of Florida Scrub Jays is impossible in all habitats we monitored except for periodically burned oak scrub (Fitzpatrick and Woolfenden 1986). In this habitat average annual reproductive success was 2.0 fledglings per pair; average survival of fledglings to age 1 year ('yearlings') was 0.31; average annual survival of breeding adults was 0.79. These values translate to an average lifetime production of about 1.3 jays of potential breeding age (yearlings) per individual. The surplus of 0.3 potential breeders per individual represents the source of the helper population—nonbreeders working their way into the breeding ranks.

In all other habitats annual reproductive success or survivorship, or both, were substantially lower than in periodically burned scrub. The worst of all habitats was overgrown, unburned scrubby flatwoods where reproductive success was 1.6 fledglings per pair, first year survival was 0.23, and annual breeder survival was only 0.72. Under these demographic conditions the average individual breeder does not replace itself, instead producing only 0.6 yearlings during its lifetime. This explains why the local populations inhabiting these unburned, late successional habitats became extirpated (Fig. 26.3). Ultimately, it also explains why Florida Scrub Jays are such habitat specialists (Fitzpatrick and Woolfenden 1986).

Annual variation

Natural environmental variability plays an important role in governing the probability that a given population can survive (MacArthur and Wilson 1967; Soule 1987). For Florida Scrub Jays, this aspect of population biology is especially crucial because of the patchy, 'island-like' distribution. Long-term protection of populations requires that they be large enough to withstand, or recover from, random fluctuations in birth rate or death rate.

Our long-term demographic data provide values for the variation surrounding mean death rates (Fig. 26.5) and birth rates (Figs 26.6, 26.7). Healthy, periodically burned scrub (Figs 26.5, 26.6) shows higher overall resilience than unburned scrub (Figs 26.5, 26.7). Three conclusions are paramount from these graphs:

(1) Reproduction and survival are consistently higher in the periodically burned scrub than in overgrown habitat. In only 4 of 18 years was average fledgling production slightly higher in the tall, unburned habitat. All were during relatively productive years in both habitats. Until 1985,

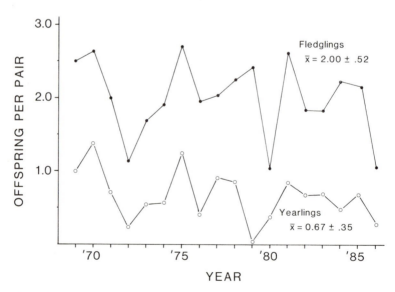

Fig. 26.6. Reproduction in open, periodically burned scrubby flatwoods on the main study tract at Archbold Biological Station, 1969–86 (cf. Fig. 26.7).

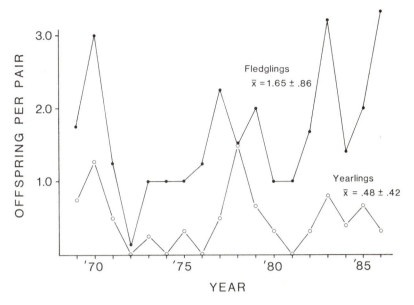

Fig. 26.7. Reproduction in tall, unburned scrubby flatwoods and tall ridge scrub with abundant scrub hickory at Archbold Biological Station, 1969–86 (cf.. Fig. 26.6).

when sample sizes in suboptimal habitat had become very small, only 1 year in 15 showed lower mortality in unburned habitat than in open, periodically burned habitat (Fig. 26.5). The single exceptional year was a year of unusually low mortality in both habitats (1980).

(2) Annual variation is substantial. Even in the optimal habitat, average reproductive success varied from 1.05 to 2.71 fledglings per pair (Fig. 26.6) and breeder mortality ranged from 0.04 to 0.45 (Fig. 26.5). Such wide swings in annual demographic parameters cause considerable local fluctuations in population size and structure. Assuming that these results are typical for Florida Scrub Jay populations throughout their restricted range, *the long-term implications of annual demographic variability must be built into any management or conservation plan*.

(3) Variation is especially pronounced in suboptimal habitats, suggesting yet another risk to the Florida Scrub Jay populations inhabiting them. Coefficient of variation in average fledgling production is lower in periodically burned scrub compared to unburned scrub (CV = 26 per cent vs. 52 per cent), and the same holds for production of yearlings (CV = 52 per cent vs. 88 per cent). Variation in annual mortality is similar between the two habitats (CV = 54 per cent vs. 55 per cent), but the total range is higher for the unburned habitat (including a year in which nearly two-thirds of the breeders died). Greater variation in demographic parameters

leads to greater annual fluctuations in population size. This leads to an important conclusion: *small, isolated populations of Florida Scrub Jays, already at risk because of low numbers, are even more likely to become extinct owing to normal demographic fluctuations if their habitat is not maintained in a low, open condition by periodic burning.*

Epidemics and other natural catastrophes

In addition to normal year-to-year fluctuations in birth and death rates, population numbers are affected by the frequency and severity of catastrophic mortality. All demographic models of extinction probabilities for small populations converge upon the conclusion that the relationship between population size and its probability of extinction depends very heavily upon the frequency and nature of catastrophes (e.g. Ewens *et al.* 1987; Shaffer 1987).

One catastrophic event has affected the Florida Scrub Jay population at Archbold Biological Station between 1969 and 1988. Between September 1979 and February 1980 an apparent epidemic swept through the study population, killing about half of all adult-plumaged jays and all but 1 of 93 juveniles of the 1979 breeding year (Fig. 26.8). Characteristics of this epidemic are described anecdotally in Woolfenden and Fitzpatrick (1984, pp. 351–5) and the demographic consequences are summarized here in Table 26.2. No epizootic agent was identified; we suspect that unprece-

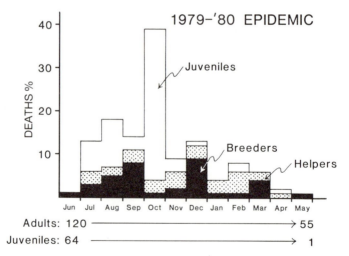

Fig. 26.8. Phenology of an apparent epidemic; nearly 70 per cent of the 184 Florida Scrub Jays alive in mid-June, 1979, were dead by the following May. Breeder and helper mortality approached 50 per cent, and juvenile mortality was essentially complete. These are two apparent peaks in breeder mortality.

Table 26.2 Effects of apparent epidemic in Florida Scrub Jay study tract, September 1979–February 1980

	Previous averages (1969–79)	1979–80
Annual survival:		
Breeders	0.81	0.55*
Helpers	0.80	0.49
Juveniles	0.34	0.01
Reproduction:		
(fledglings per pair)	2.04 ($\bar{x}$)	1.05 (1980)
	2.42 (1979)	
Breeding density		
(pairs per 40 ha)	5.0	3.8 (1980)

* Mortality of oldest breeders (age 9 or older, $N = 18$) significantly higher than that of youngest breeders (age 2 or 3, $N = 26$), with rates of 0.54 vs. 0.31, respectively ($x^2 = 6.32$, $P < .01$).

dented high water levels in late summer 1979 permitted a lethal arbovirus to incubate and spread, possibly borne by mosquitoes. In all, 128 out of 184 jays in the study tract died between June, 1979 and May, 1980. Fifty-three of these deaths (41 per cent) were noted during September and October (Fig. 26.8).

We have no data on how widespread the 1979 epidemic was in regions near to the study area, although we suspect it was region-wide. Moreover, we cannot yet estimate the frequency or timing of such epidemics in natural populations of Florida Scrub Jays. On our study site, epidemics are rare with respect to the average lifetime of a jay, as the study population has experienced only 1 in 20 years. The frequency of such catastrophic events presumably varies across the Florida peninsula. Even if we knew their local frequency, comparable data from several other Florida populations are required before the role of epidemics in determining minimum viable populations can be confidently modelled for this species.

Besides epidemic diseases, we know of no other *natural* catastrophes that affect Florida Scrub Jay populations. Fire, which rarely, if ever, burns all oak scrub over a large area, probably is never catastrophic in its long-term effects. Jays continue to use and defend burned oak scrub, and scrub regenerates extremely quickly after any fire (Abrahamson 1984). Because they do not migrate and live in a mild climate, Florida Scrub Jays are not subject to the large-scale, weather-related disasters that occasionally inflict heavy mortality on migratory bird populations or those restricted to cold, less predictable areas. Major tropical storms and hurricanes periodically batter portions of peninsular Florida, but the effects on oak scrub are minimal. No natural predator is known to be capable of systematically

devouring enough individual Scrub Jays in a local area to affect populations at catastrophic levels. In the absence of disease or habitat alteration
by humans, Florida Scrub Jay populations probably are as resistant to
natural catastrophe as any native bird population in Florida.

Dispersal characteristics

Florida Scrub Jays are extremely sedentary, a feature that has profound
influence on large-scale population structure, and upon habitat protection
measures. Jays remain in their natal territory for 1 to 5 years before
dispersing to become breeders, after which they remain on or near the
breeding territory until death. Only during a brief prebreeding period do
Florida Scrub Jays disperse over any appreciable distance. Figure 26.9
(from Woolfenden and Fitzpatrick 1986) shows documented dispersal
distances of 119 jays that became breeders in or near the study tract. As
usual for birds, females on average move farther from the natal ground
than the males. No male is known to have dispersed more than 1.5 km
from its birthplace. Average dispersal distance for females is 1.1 km, and
maximum confirmed dispersal distance is 7 km. Most Florida Scrub Jays

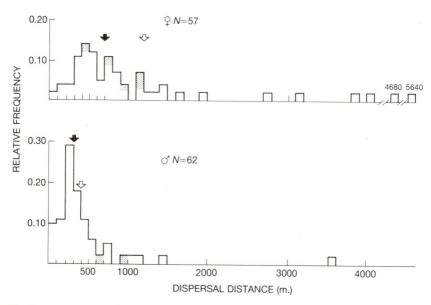

Fig. 26.9. Dispersal distances of male and female Florida Scrub Jays, measured
between nest last attended as a helper and first nest of first season as a breeder. Median
(solid arrows) and mean (open arrows) dispersal distances are indicated. Shaded
samples represent known minimum distances for known-age immigrants upon
entering the study tract. From Woolfenden and Fitzpatrick (1986).

become breeders within a few hundred metres of the patch of scrub in which they were hatched.

The sedentary nature of Florida Scrub Jays has important implications for conservation. Territorial inheritance and short-distance dispersal insure that protected patches of suitable habitat routinely pass among generations without requiring new colonization events. Successful lineages of jays probably retain 'possession' of good patches of habitat for many decades under natural circumstances. In addition, limited contact between local populations may reduce the probability that lethal diseases can spread across major portions of a regional population.

On the negative side, isolated patches of suitable scrub in which jay populations have drifted to extinction may go uncolonized indefinitely if located more than a few kilometres from the closest possible source population. Small populations are less likely to go extinct if located within the normal dispersal radius of other jay populations. For these reasons, clusters of small to intermediate-sized local populations may be the optimal configuration of Florida Scrub Jay meta-populations where the total area of preserved habitat is limited. Based on our dispersal data, we suggest that *isolated scrub habitat preserves within about 8 km of one another are sufficiently close to allow for occasional movement of Florida Scrub Jays among them*.

Survivorship of populations: simulation models

Our long-term data provide real-world values for the critical variables in demographic models of extinction probabilities as a function of population size. More important, our empirical values for annual variation in demographic parameters allow us to construct models that simulate quite precisely the natural fluctuations in population size (Goodman 1987). The only key variables we still lack are the frequency and relative severity of epidemics, because we have witnessed only one.

Appendix 26.1 summarizes procedures and assumptions of a preliminary Monte Carlo simulation model we developed to assess survival characteristics of different-sized Florida Scrub Jay populations under a variety of demographic conditions. The model population proceeds on a year-to-year basis from a specified initial size, which also defines the maximum allowable number of territories. Each year's fecundity and survival values are randomly chosen from a distribution matching that measured by us from 1969 through 1987 (Table 26.3). Epidemics strike at a specified average frequency and their characteristics match those we observed in 1979–80. Figure 26.10 illustrates one model population surviving at least 250 years and constrained to a maximum of 10 territories (= 20 breeders). The long-term role of epidemics is clearly visible.

Table 26.3 Demographic values and their annual variations used in computer simulation of Florida Scrub Jay populations

Mean annual fecundity:

Frequency	Mean annual fledglings/pair
.17	1.12
.06	1.73
.06	1.94
.32	2.04
.11	2.24
.11	2.45
.17	2.65

Mean annual survival:
Juveniles to age 1 year
    normal: .35
    epidemic: 0
Yearlings to age 2 years
    normal: .70
    epidemic: .50
Older helpers
    normal: .77
    epidemic: .50

Breeders:

Frequency	Annual survival
.16	.950
.11	.925
.06	.875
.05	.825
.33	.775
.06	.725
(.17)*	.68
(.05)*	.55 = epidemic

* Lowest values for mortality are varied according to specified frequency of epidemics.

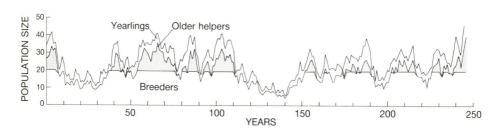

Fig. 26.10. Simulated fluctuations in a Florida Scrub Jay population beginning at carrying capacity of 10 territories (20 breeders), illustrating the behaviour of the model used to estimate persistence times (see text). Portions of this figure closely resemble the 18-year fluctuations documented in a real-world population at Archbold Biological Station.

Population sizes were varied from 1 to 100 territories, with a sample of 100 populations at each size allowed to persist until random extinction. Extinction times for each population size showed the expected, negative exponential distribution (Goodman 1987). Plotting and smoothing the resulting distributions on three axes (population or preserve size, survival probability, and time) depicts the fundamental management scenario for Florida Scrub Jay populations (Fig. 26.11).

These simulations, based on our long-term population data, indicate that for an isolated preserve to support a Florida Scrub Jay population with greater than a 90 per cent chance of persisting more than 100 years, the preserve should contain habitat sufficient to support 20 to 40 jay territories. Therefore, assuming that our observed range of territory sizes is representative, Florida Scrub Jay preserves should contain at least 200 to 400 ha of usable habitat. Networks of smaller preserves, in which individual populations are sufficiently isolated from one another to inhibit the spread of local diseases but close enough to permit recolonizations, would somewhat reduce the minimum sustainable size of each preserve.

Figure 26.12 schematically illustrates our conclusions of the viability of Florida Scrub Jay populations, assuming an epidemic frequency of once every 20 to 40 years. Such a scheme explicitly assumes that natural habitat characteristics, including frequency of fire, are allowed to persist and that

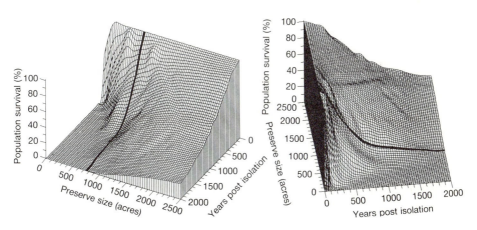

Fig. 26.11. Two views of the same three-dimensional graph showing relationships between preserve area, population survival, and time in simulated Florida Scrub Jay populations. Preserve area is expressed in ha and assumes a density of 4 Florida Scrub Jay territories per 40 ha. For a given population size, the graph plots proportion of such populations extant *N* years after isolation. Isobar for a 30-territory preserve is darkened for reference. Populations larger than 30 territories show 90 per cent or better probability of surviving at least 100 years. Graph assumes 0.05 annual probability of epidemic.

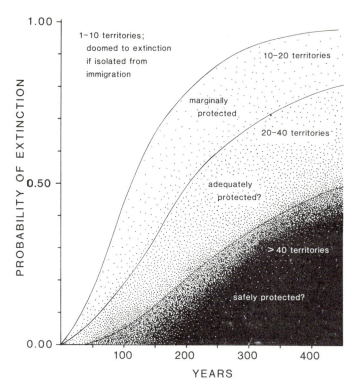

Fig. 26.12. Extinction probability as a function of time, for Florida Scrub Jay populations of different sizes. Degree of protection afforded by population size is indicated by shading.

man-induced mortality (from vehicles, house pets, poisons, wanton killing) is negligible. Excess mortality would have the effect of decreasing the expected persistence times for populations, and hence would demand increased preserve sizes to overcome. More serious long-term perturbations, especially fire suppression, would doom a population of Florida Scrub Jays to extinction regardless of its size.

Discussion

Several distinctive features of Florida Scrub Jay ecology explain why they are threatened with extinction. Similar features probably characterize many other threatened species around the world and so the management lessons learned from Florida Scrub Jays may have broader implications. Restriction to a severely limited and patchily distributed habitat places any but the most widely dispersing species at risk when the habitat is altered.

For Florida Scrub Jays, ecological specialization goes a step farther: only periodically- and recently-burned habitat patches can sustain a local population indefinitely. In presettlement times, such specialization was accommodated, and probably selected, by frequent natural fires. The result is a bird with low annual mortality that exhibits permanent territoriality, delayed breeding, group living, and limited dispersal all of which are behavioural adaptations to a rare habitat sharply defined by both soil characteristics and fire history.

Humans have caused two major changes in the ecological setting of Florida Scrub Jays. Destruction of habitat, of course, reduces or eliminates jay populations. Habitat patches become smaller and more widely spaced, accelerating local extinctions. Equally important, we have shown that fire suppression even where scrub is allowed to persist reduces the jays' ability to survive and reproduce. Fire-fighting in North America is a long and sacred tradition of their culture, yet numerous species of animals and plants depend entirely upon the early stages of post-fire succession. Not surprisingly, many of these species are declining rapidly. Their long-term protection must include fire management within suitable habitat preserves.

The potential for catastrophic epidemics has enormous implications in management design. In our unique example, 128 of 184 jays died in a 12-month period. Estimating persistence times of isolated populations requires values for annual reproduction and mortality, but calculation of 'minimum viable populations' depends even more strongly on the assumed frequency and nature of epidemics.

The long-term effects of epidemics on population viability are amplified in organisms, such as Florida Scrub Jays, in which the social system reduces the recovery rate. Our simulated populations routinely showed long delays in recovering to pre-epidemic levels. This was primarily a result of building delayed breeding into the 'behaviour' of 1- and 2-year-old helpers. In our real study population, a similar lag in recovery has taken place. Besides the effects of delayed breeding, we suspect that territorial behaviour itself is slowing the recovery. A reduced population of breeders continues to defend all the habitat, thus inhibiting the rate at which potential new recruits could establish themselves into the breeding ranks.

The profound effects of subtle changes in demography on overall breeding density, within an otherwise protected and stable population, provide an important lesson for management of isolated populations. Assumptions about density or average territory size are essential when calculating minimum sizes for habitat preserves. However, even within a highly stable population like Florida Scrub Jays, buffered against large fluctuations by territorial behaviour and delayed breeding, recruitment rates and breeding density are continuous variables that may change with subtle changes in average birth rates and death rates (Figs 26.4, 26.5). Long time-lags may routinely accompany population recoveries from catastrophic reductions in numbers.

Reduced dispersal, dependence upon periodic fires, susceptibility to epidemics, and slow post-epidemic recovery place isolated populations of Florida Scrub Jays at considerable risk of extinction, especially if fewer than 20 groups are involved. For this reason, the ideal solution for their long-term protection will be through clusters of habitat patches, in which a few very large, carefully preserved population centres are surrounded by smaller islands in which disease and fire are independent events. In such a system across Florida, the meta-population is preserved through occasional recolonization between centres, even as local populations occasionally flicker and die out.

Summary

Long-term demographic and life history data of Florida Scrub Jays provide ample evidence sufficient to recommend conservation and management procedures for this endangered species. The jay is endemic to a relict oak scrub habitat with an island-like distribution in central Florida. The habitat must burn every 8–20 years to remain optimal, and becomes unusable if unburned for 40–50 years. Pairs or family groups defend year-round territories, leaving no usable habitat undefended. Minimum territory size is about 5 ha, average size is 9–10 ha. Young jays live in the natal group for at least one full year before dispersing. Recruitment occurs by replacement following death of a breeder, or by territorial budding following several years of family growth. Most recruits disperse to within two territories of the natal ground. Maximum documented dispersal distance is 7 km (about 20 territory widths). Potential recruits (1-year-old jays) greatly outnumber breeding vacancies, creating intense competition for space. Territory size and breeder density are inversely correlated, and vary according to two highly variable demographic parameters: adult mortality (creates breeding vacancies) and juvenile mortality (reduces pool of potential recruits). One epidemic has occurred in 19 years (1979–80), killing all juveniles and about half the adults. Such catastrophic mortality profoundly influences subsequent age- and social-structure; more importantly, the probability of local extinctions in small populations is greatly increased. Considerable time-lags may ensue before the population returns to pre-epidemic densities. The time-lag is reinforced by permanent group territoriality, which inhibits rapid recruitment even when breeding density is reduced.

We have developed a simulation model to study the relationship between population size and extinction probability, using our empirically determined values of mean and variance for demographic variables and incorporating the effects of epidemics. For Florida Scrub Jay populations to have a better than 90 per cent probability of surviving more than 100 years, at least 30 contiguous territories must be protected. Protecting several local populations sited within about 8 km of one another would allow occasional colonization among them, and might protect portions of the meta-population from the spread of lethal epidemics.

Acknowledgements

We owe continuing thanks to the Archbold Biological Station for its extensive support of our research. In addition we thank Robert Curry, Debra Moskovits, Emily McGowan, Ronald Mumme, Arie van Noord-wijk, and Jan Woolfenden for important contributions to this study. Support of this work is gratefully acknowledged from the National Science Foundation (BSR-8705443), Florida Game and Fresh Water Fish Commission (GFC-70320), Conover Fund of the Field Museum of Natural History, and the University of South Florida.

References

Abrahamson, W. G. (1984). Post-fire recovery of the Florida Lake Wales Ridge vegetation. *American Journal of Botany*, 71, 9–21.

Abrahamson, W. G., Johnson, A. F., Layne, J. N., and Peroni, P. A. (1984). Vegetation of the Archbold Biological Station, Florida: an example of the Southern Lake Wales Ridge. *Florida Scientist*, 47, 209–50.

Amadon, D. (1944). A preliminary life history study of the Florida Jay, *Cyanocitta c. coerulescens*, *American Museum Novitates*, No. 1252.

Cox, J. (1987). Status and distribution of the Florida Scrub Jay. *Florida Ornithological Society Special Publication*, No. 3.

DeGange, A. R., Fitzpatrick, J. W., Layne, J. N., and Woolfenden, G. E. (1989). Acorn harvesting by Florida Scrub Jays. *Ecology*, 70, 348–56.

Ewens, W. J., Brockwell, P. J., Gani, J. M., and Resnick, S. I. (1987). Minimum viable population size in the presence of catastrophes. In *Viable populations for conservation*, (ed. M. E. Soule), pp. 59–68. Cambridge University Press, New York.

Fitzpatrick, J. W. and Woolfenden, G. E. (1986). Demographic routes to cooperative breeding in some New World jays. In *Evolution of animal behavior*, (eds M. H. Nitecki and J. A. Kitchell), pp. 137–60. Oxford University Press, New York.

Goodman, D. (1987). The demography of chance extinction. In *Viable populations for conservation*, (ed. M. E. Soule), pp. 11–34. Cambridge University Press, New York.

Laessle, A. M. (1958). The origin and successional relationships of sandhill vegetation and sand pine scrub. *Ecological Monographs*, 28, 361–87.

Laessle, A. M. (1968). Relationships of sand pine scrub to former shore lines. *Quarterly Journal of the Florida Academy of Sciences*, 30, 269–86.

MacArthur, R. H. and Wilson, E. O. (1967). *The theory of island biogeography*. Princeton University Press, Princeton, New Jersey.

McGowan, K. J. and Woolfenden, G. E. (1989). A sentinel system in the Florida Scrub Jay. *Animal Behaviour*, **37**, 1000–6.

Myers, R. L. and White, D. L. (1987). Landscape history and changes in sandhill vegetation in north-central and south-central Florida. *Bulletin of the Torrey Botanical Club*, **114**, 21–32.

Shaffer, M. (1987). Minimum viable populations: coping with uncertainty. In *Viable populations for conservation*, (ed. M. E. Soule), pp. 69–86. Cambridge University Press, New York.

Soule, M. E.(1987).*Viable populations for conservation*. Cambridge University Pess, New York.

Woolfenden, G. E. (1974). Nesting and survival in a population of Florida Scrub Jays. *Living Bird*, **12**, 25–49.

Woolfenden, G. E. and Fitzpatrick, J. W. (1978). The inheritance of territory in group-breeding birds. *BioScience*, **28**, 104–8.

Woolfenden, G. E. and Fitzpatrick, J. W. (1984). *The Florida Scrub Jay: demography of a cooperative-breeding bird*. Princeton University Press, Princeton, New Jersey.

Woolfenden, G. E. and Fitzpatrick, J. W. (1986). Sexual asymmetries in the life history of the Florida Scrub Jay. In *Ecological aspects of social evolution*, (eds D. I. Rubenstein and R. W. Wrangham), pp. 87–107. Princeton University Press, Princeton, New Jersey.

Woolfenden, G. E. and Fitzpatrick, J. W. (1990). The Florida Scrub Jay: a synopsis after 18 years of study. In *Cooperative breeding in birds: long-term studies of ecology and behavior*, (eds P. B. Stacey and W. D. Koenig), pp. 239–66. Cambridge University Press, New York.

Appendix 26.1

The following assumptions are incorporated into the simulation model, each of which represents a real biological feature of Florida Scrub Jay population:

1. Each territory contains one and only one pair of breeding jays.
2. The population is assumed to be at carrying capacity to start with and is not allowed to exceed the initial number of territories at any time during the 'life' of the population.
3. The initial population begins with one helper of potential breeding age in each territory.
4. A maximum average of three helpers per territory is permitted.
5. Yearlings are not allowed to breed.
6. The pool of older nonbreeders permits an equal chance for each individual to become a breeder (i.e. no age-specific differences in success probability after the first year).
7. All breeders have an equal probability of dying during a given year.
8. *Average* demographic values are set to a steady-state population (no intrinsic population growth or decline), but individual survival and reproduction is treated probabilistically.

9. Annual variation in mean birth rates and death rates reflects the measured annual values during the 18 years of our study.
10. Epidemics are incorporated as follows: when the highest observed value for breeder mortality is selected by chance (0.45 in our model), the program is flagged as an epidemic year, so that all juveniles are eliminated, death rate of older helpers becomes 0.50, and the following year's mean reproduction is set at its lowest observed value (1.1 fledglings per pair). Simulations for Fig. 26.12 assume a once-in-20-year average frequency of epidemics (0.05 probability each year).

27 Population dynamics and extinction in heterogeneous environments: the Northern Spotted Owl

RUSSELL LANDE

Introduction

Rapid destruction and fragmentation of natural environments caused by pressures from expanding human populations and economies, especially in developing tropical countries, is likely to cause the extinction of a large fraction of species recently existing on the earth (Meyers 1986; Simberloff 1986). To mitigate the loss of species diversity and the negative effects that this may have on future medical, industrial, and recreational development, and on regional and global climate change, it is necessary to educate the public and their political and economic leaders about the benefits of conserving biological diversity and restoring natural ecosystems. It is also increasingly important to understand the dynamics of small populations, and those in remaining fragmented habitats, in order for managers to reduce the risks of extinction.

Recent literature on conservation biology (Soulé and Wilcox 1980; Frankel and Soulé 1981; Schonewald-Cox *et al*. 1983; Soulé 1986) has emphasized the importance of population genetics, particularly inbreeding depression (reduced viability and fertility caused by matings between closely related individuals) and loss of genetic variation (for future adaptation to environmental changes) in small populations. This has led to the relative neglect of basic demographic factors which may pose a more serious threat to population persistence. Some conservation plans, such as for the Northern Spotted Owl and the Red-cockaded Woodpecker, (USDA Forest Service 1984; USDI Fish and Wildlife Service 1985) have been based primarily on principles of population genetics. It has even been suggested that managers should abandon populations not currently satisfying the genetic criteria, such as Harpy Eagles in Costa Rica (see Simberloff 1988).

Extinction is fundamentally a demographic event, to which genetic and environmental factors may contribute. For most species exposed to natural environmental fluctuations and artificial disturbance, it is likely that the population sizes required to promote a high probability of persistence for the forseeable future (e.g. 100 years) also are sufficient to avoid substantial inbreeding depression and loss of genetic variability, at least in quantitative (polygenic) characters that are highly mutable (Lande 1988a). Furthermore, gradual inbreeding or temporary population reductions do not necessarily cause substantial inbreeding depression (because natural selection can reduce the load of recessive lethal and sublethal mutations during inbreeding) and loss of genetic variability can be restored by mutation and immigration (Lande and Barrowclough 1987). Thus demographic factors are usually of more immediate importance than population genetics in determining the minimum viable population sizes for wild species. The main exceptions to this would occur for populations with moderate or high growth rates in a stable environment, for example in zoos. Population genetics should properly receive major emphasis in the propagation of captive animals, and in their reintroduction to natural environments, but even then demography should not be neglected.

In this paper are summarized methods for estimating the multiplication rate of an age- or stage-structured population, with particular reference to a typical avian life-history with constant adult fecundity and survivorship. I then review some of the major demographic factors that threaten the existence of wild populations and how they can be analysed. Finally, the application is described of these techniques to demographic and environmental data on the Northern Spotted Owl, and the conservation plan for this subspecies proposed by the U.S. Forest Service is evaluated.

Classical demography

Population multiplication rate

Leslie (1966) developed a model of a population with overlapping generations in which reproduction occurs at discrete intervals, such as a yearly breeding season. Assuming that the population is censused at the end of each breeding season, the following definitions apply. s_i is the probability of survival from age i to age $i + 1$, hence the probability of survival from fledgling (age 0) to age x is $l_x = s_0 s_1 \ldots s_{x-1}$. Fecundity, measured as the rate of production of female offspring per adult female of age x is b_x. The population multiplication rate, λ, is then the unique positive real solution of the Euler-Lotka equation,

$$\sum_{x=0}^{\infty} \lambda^{-x} l_x b_x = 1 \tag{1a}$$

If the vital rates do not fluctuate with time and there is a constant sex-ratio at fledgling, after a few generations the population asymptotically approaches a stable age distribution with population growth approximated by the geometric formula $N(t) = \lambda^t N(0)$ (Keyfitz 1977). A stable population therefore has $\lambda = 1.0$.

Bird species typically have adult survivorship and fecundity rates that are nearly independent of age (Deevey 1947). Although some evidence of senescence can be obtained from comparison of life-table analysis with records of maximum longevity of various species (Botkin and Miller 1974), it appears that natural mortality rates are sufficiently high for most birds that so few individuals survive to the age of senescence as to be demographically unimportant. We can therefore write the adult annual survivorship as s, and signifying the age of first reproduction as α yields $l_x = l_\alpha s^{x-\alpha}$. Similarly, the adult fecundity can be written as $b_x = b$ for $x \geq \alpha$, with zero fecundity for juveniles and sub-adults. The series in equation ($1a$) can then be summed to give

$$\lambda^\alpha (1 - s/\lambda) = l_\alpha b \qquad (1b)$$

Because the adult survivorship applies in the first year of reproduction, from ages $\alpha - 1$ to α, it should be noted that $l_\alpha = s_0 \ldots s_{\alpha-2} s$. If dispersal occurs during the first year, fledgling survivorship can be partitioned into the probability of surviving dispersal, s_d, and the probability of surviving to age one excluding the risk of dispersal, $s_0{}'$, so that $s_0 = s_0{}' s_d$. For example, a species that begins reproduction at age $\alpha = 3$ has $l_3 = s_0{}' s_d s_1 s$. Given values of the demographic parameters, equation ($1b$) can readily be solved to the desired degree of accuracy by trial and error using a hand calculator.

Generation time

The generation time, defined as the average age of mothers of fledglings in a population with a stable age distribution (Leslie 1966), is given by:

$$T = \sum_{x=0}^{\infty} x\lambda^{-x} l_x b_x \qquad (2a)$$

For the typical avial life history the generation time is:

$$T = \alpha + s/(\lambda - s) \qquad (2b)$$

Formulas ($1b$) and ($2b$) have been used to analyse population dynamics of Guillemots (Leslie 1966), California Condors (Mertz 1971), Everglade Kites (Nichols *et al*. 1980), and Northern Spotted Owls (Lande 1988*b*).

Sensitivity coefficients

Changes in the population multiplication rate with respect to small changes in the vital rates are called sensitivity coefficients (Goodman

1971). They indicate which of the vital rates have the largest influence on λ; such information is useful in suggesting which stages of the life cycle should be investigated ecologically to determine the most efficient management strategies for altering the population multiplication rate. The sensitivity coefficients are also important in calculating the standard error of λ and in determining the influence of environmental variation in the vital rates on the long-run multiplication rate of the population. Implicit differentiation of the Euler-Lotka equation (1b) yields the sensitivity coefficients for the typical avian life-history,

$$\partial\lambda/\partial\pi = \lambda/(\pi T), \partial\lambda/\partial s = \lambda(T - \alpha + 1)/(sT)$$
$$\partial\lambda/\partial\alpha = -\lambda[\ln(\lambda/s)]/T \qquad (3)$$

in which π represents any of the components of $l_a b$ (excluding s) such as juvenile or sub-adult survivorships or adult fecundity.

Standard error of λ

An approximate formula for the sampling variance of an estimated value of λ can be expressed in terms of the sensitivity coefficients (Kendall and Stuart 1977, Ch. 10.6).

$$\sigma_\lambda^2 = \Sigma_i \Sigma_j (\partial\lambda/\partial\pi_i)(\partial\lambda/\partial\pi_j)\sigma_{ij} \qquad (4)$$

where π_i here represents the ith of the life history parameters (the vital rates and the age at first reproduction). σ_{ii} is the sampling variance of an estimate of π_i, and σ_{ij} is the sampling covariance between estimates of π_i and π_j, which can be obtained from mark-recapture data using Jolly-Seber methods (Clobert and Lebreton, this volume). In the absence of extensive data on repeated recaptures, σ_{ii} and σ_{ij} can be approximated from the variances and covariances of the individual data divided by the appropriate sample sizes. Alternatively, an empirical sampling distribution for λ, and hence the bias and sampling variance of the estimated multiplication rate, can be derived from resampling schemes such as jackknifing or boot-strapping (Efron 1982; Meyer *et al.* 1986).

The magnitude of each term in this formula indicates the contribution of each of the demographic statistics to the sampling variance in the population multiplication rate. These can be used to design efficient programmes for future field work on the life history aimed at accurately estimating the population multiplication rate. The effort devoted to estimating each of the demographic statistics, π_i, should be inversely proportional to its net contribution to the sampling variance of λ, that is, to the second summation in equation (4).

Stochastic demography

Two kinds of random factors influencing population growth are usually referred to as demographic and environmental stochasticity (Shaffer 1981). Demographic stochasticity occurs because individuals of a given age or developmental stage have at any time certain probabilities of survival and reproduction; if these probabilities apply independently to each individual, the sampling variances of the vital rates are then inversely proportional to population size. Environmental stochasticity is caused by temporal changes in the vital rates that similarly affect all individuals of a given age or stage in the population; this produces sampling variances of the vital rates that are nearly independent of population size. Thus, because most populations undergo substantial fluctuations caused by changes in physical and biotic factors, environmental stochasticity is generally thought to be more important than demographic stochasticity in populations larger than about 100 individuals (Leigh 1981; Goodman 1987). This conclusion is supported by observations of birds on islands, which, except for populations initially less than 30 breeding pairs, become extinct much faster than predicted by demographic stochasticity alone (Leigh 1981).

In species with long-lived adults, a population may be maintained by occasional good years for reproduction and juvenile survival, interspersed between bad years. Fluctuations in the vital rates around their average values tend to decrease the long-run multiplication rate of a population, provided that there is no serial correlation (autocorrelation) in the fluctuations (Tuljapurkar 1982). Long-term weather records for the continental United States show low serial correlation in temperature and precipitation on a yearly time-scale (Namias 1978); for species with annual reproduction, this justifies in part the assumption of no autocorrelation in vital rates, although long-term fluctuations or cycles could arise from interaction with other species such as prey or parasites.

The long-run (geometric average) multiplication rate of a population subject to environmental fluctuations in vital rates with no serial correlation is approximately:

$$\Lambda = \lambda \, \exp\left\{ -\tfrac{1}{2} \lambda^{-2} \, \Sigma_i \Sigma_j \, (\partial\lambda/\partial\pi_i)(\partial\lambda/\partial\pi_j) \, V_{ij} \right\} \tag{5}$$

in which λ is the multiplication rate of the population in the average environment (calculated using the average demographic statistics in eqns. 1a or 1b) and V_{ij} is the covariance of fluctuations in the vital rates π_i and π_j (Tuljapurkar 1982; Lebreton and Clobert, this volume). This analytical formula is accurate for small or moderate fluctuations in the vital rates, with coefficients of variation (standard deviation divided by mean) up to 30 per cent (Lande and Orzack 1988). The formula should be accurate to within a few per cent even for large fluctuations in vital rates if their sensitivity coefficients are small, such that the bracketed term in (5) is much less than one in magnitude. For highly variable environments, it is necessary to estimate the long-run multiplication rate of a population by computer simulation.

There is insufficient data for most species to obtain accurate estimates of temporal fluctuations in the vital rates. The influence of fluctuating environments can then be evaluated only by choosing reasonable values for the variances and covariances of the vital rates, the V_{ii} and V_{ij}, based on the available information. For populations in which the multiplication rate is density independent, the probability of extinction up to a given time (e.g. 100 years) can be calculated analytically, or semianalytically with a minimum of computer simulation (Lande and Orzack 1988).

Habitat fragmentation and extinction thresholds

Edge effects and critical patch size

Destruction and alteration of natural areas often creates patches or islands of habitat, suitable for survival and reproduction of a species, surrounded by areas that are unsuitable or less suitable. This produces two types of edge effects. First, near an ecological boundary there is a deterioration of habitat quality, e.g. desiccation of forest patches near their edges, that renders the edges unsuitable for some species. Second, individuals may disperse across an ecological boundary into unsuitable regions where they will die or fail to reproduce.

Dispersal behaviour, especially individual dispersal distances and habitat preferences, along with classical demographic parameters determine the minimum size of a patch of suitable habitat that can sustain a population, known as the critical patch size. Kierstead and Slobodkin (1953) modelled random dispersal of individuals in a population on a single patch of suitable habitat surrounded by a region unsuitable for individual survival. For a two-dimensional region, the minimum diameter of a circular patch of suitable habitat necessary to sustain a population is

$$D > 3.4d/\sqrt{rT} \tag{6}$$

in which d is the square root of the mean squared individual dispersal distance, $r = \ln \lambda$ is the intrinsic rate of increase of the population within a large suitable region (away from edges), and T is the generation time defined above. When dispersal is restricted to a single dimension, such as along a river or a shoreline, the minimum length of suitable habitat in which a population can persist is about 35 per cent smaller than in the preceding formula. In either case, persistence of a population with a low intrinsic rate of increase per generation ($rT \ll 1$) requires that the patch of suitable habitat be much larger than the typical individual dispersal distance d.

The critical patch size model has been extended in several ways (Okubo 1980; Pease *et al.* 1989). Survival (and reproduction) of individuals outside of the patch, and nonrandom dispersal resulting from habitat preferences, act to decrease the critical patch size. Movement of the patch caused by climatic change, and Allee effects (declining survival and/or fecundity at low population densities, e.g. caused by difficulty in finding a mate), increase the critical patch size. An area with fixed boundaries containing suitable habitat may become unsuitable for a particular species because of long-term climatic change; this problem is especially serious for species that undergo long-distance seasonal migrations between widely separated patches of suitable habitat.

Local extinction and colonization

Levins (1969, 1970) introduced the concept of a meta-population composed of a set of local populations occupying disjunct patches of suitable habitat. Fragmentation of suitable habitat increases the risk of local extinction through reductions in local population size and immigration rate, and increasing edge effects. Species in patchy environments often persist on a regional basis through a dynamic balance between local extinction and colonization (Pickett and White 1985). Levins showed that a meta-population generally will not occupy all of the patches of suitable habitat available at any one time, but will persist on a regional basis in a continually shifting mosaic pattern of occupancy of the suitable patches; regional extinction may occur in the presence of suitable habitat if the rate of local extinction exceeds the rate of local colonization. The meta-population model, and the theory of island biogeography on which it was based, do not incorporate details of a species life history and dispersal behaviour that determine population dynamics within a patch of suitable habitat and migration rates between patches, which in turn determine the rates of local extinction and colonization.

These concepts can be extended to a territorial population by identifying the individual territory as the spatial unit. Local extinction then corresponds to the death of an individual inhabiting a suitable territory, and colonization corresponds to individual dispersal and successful settlement on a suitable unoccupied territory. Classical demographic theory for age- or stage-structured populations, and the dispersal behaviour of individuals, can thus be integrated with the theory of habitat occupancy in a patchy environment (Lande 1987). This facilitates investigation of the impact of continued habitat destruction and fragmentation, or habitat improvement, on the future occupancy of suitable habitat in a region.

For simplicity, a basic model was derived on the assumptions that patches of habitat, each the size of individual territories, are either suitable or unsuitable, and that suitable patches are randomly or evenly distributed in space (i.e. there is no large-scale clumping of suitable habitat over distances much greater than the individual dispersal distance, d). Juveniles disperse and are assumed capable of searching a finite number of potential territories before dying from predation, starvation, or other causes, if they do not find a suitable unoccupied territory. Increasing destruction and fragmentation of suitable habitat therefore increases juvenile mortality during dispersal in a density-dependent fashion.

The proportion of a large region composed of suitable territories is denoted as h, and the fraction of suitable territories that are occupied by adult females is signified as p. The information on life history and dispersal behaviour is contained in a composite parameter, k, termed the demographic potential of the population, which gives the equilibrium occupancy ($\hat{p}$) in a completely suitable region ($\hat{p} = k$ when $h = 1$). Increasing either the expected number of offspring an individual can produce, or the number of potential territories a dispersing juvenile can search, increases the demographic potential of the population and the equilibrium occupancy of suitable habitat. At demographic equilibrium the occupancy of suitable habitat is:

$$\hat{p} = \begin{cases} 1 - (1 - k)/h & \text{if } h > 1 - k \\ 0 & \text{if } h < 1 - k \end{cases} \tag{7}$$

This model reveals two general features of populations maintained through a balance between local extinction and colonization. First, as the proportion of suitable habitat in a region declines, the amount of the suitable habitat that is occupied also decreases; the equilibrium population density, which is proportional to the product of h and $\hat{p}$, therefore declines more rapidly than does the amount of suitable habitat during continued destruction and fragmentation of the habitat. Second, there is an extinction threshold or minimum proportion of suitable habitat in a region that is necessary to support a population. If the proportion of suitable habitat falls below $1 - k$ the population will become extinct. Extensions of this basic model demonstrate that a fluctuating environment, or edge effects due to the limited extent of the region containing suitable habitat, or an Allee effect caused by the difficulty of finding a mate, all increase the extinction threshold (Lande 1987).

Application to the Northern Spotted Owl

The Northern Spotted Owl is a monogamous territorial subspecies (*Strix occidentalis caurina*) that inhabits old-growth conifer forests in the Pacific

North-west region of the United States. Each pair utilizes for its home range about 1 to 3 square miles of forest more than about 200 years old below an elevation of roughly 4000 feet. They nest in old hollow trees and require an open understory, characteristic of old-growth forests, for efficient hunting of small mammals that comprise the bulk of their diet (Forsman *et al*. 1984; Gutiérrez 1985). Intensive logging in recent decades, especially on private land, has destroyed most of the old-growth forests on which these owls depend and the majority of that remaining is in areas managed by the U.S. Forest Service, with lesser amounts administered by the National Park Service and the Bureau of Land Management (Society of American Foresters 1984). The total population size of the Northern Spotted Owl has been recently estimated at about 2500 pairs (Dawson *et al*. 1987).

To comply with the National Forest Management Act of 1976, which requires that all native vertebrate species be maintained well-distributed throughout their range on federal land, the Forest Service formulated guidelines for the management of Spotted Owl habitat. The original plan was based on the supposition that protection from logging of suitable habitat for about 500 pairs, distributed throughout their range on national forests in western Washington and Oregon, would allow the population to maintain sufficient genetic variability for adaptation to future environmental changes and thus ensure its persistence (USDA Forest Service 1984). Later, demographic analyses of the population were conducted (USDA Forest Service 1986, 1988). The analyses performed by the Forest Service are seriously flawed, because of artifacts arising from arbitrary truncation of the life table at an age of 10 or 15 years; and their Final Supplemental Environmental Impact Statement does not analyse density-dependent habitat occupancy caused by continuing fragmentation of old-growth forest. For this reason, I summarize here only the demographic analysis of Lande (1988*b*).

Population multiplication rate

Northern Spotted Owls usually begin breeding at an age of $\alpha = 3$ years. Adults are long-lived but have low fecundity and juveniles experience high mortality, especially during dispersal in their first year. Young Spotted Owls are fledged in the summer and disperse long distances in the fall (dozens of kilometres). Based on recent demographic data, juvenile survivorship during dispersal is $s_d = 0.18$, the adult annual survival rate is $s = 0.942$, and adult females fledge on average $b = 0.24$ female offspring per year. The probability of juvenile survival to reproductive age is $l_3 = 0.0722$. From equations (1*b*) and (4), this information can be used to estimate the multiplication rate of the population in the average environment, and its standard error, as $\lambda = 0.96 \pm 0.03$. Although this is not significantly different from that for a stable population ($\lambda = 1.00$), it is in good agreement with the gradual declines of a few per cent per year in the population estimated from long-term surveys conducted since about 1970 in various areas by Forsman (USDA Forest Service 1986, Appendix C; unpublished data).

Sensitivity analysis (eqn. 3) indicates that the population multiplication rate is by far the most sensitive to changes in the adult survivorship, s, followed by juvenile survivorship during dispersal, s_d, and adult annual fecundity, b. Adult survivorship is already quite high and it may be difficult for management to increase it. The low survival of dispersing juveniles may result in part from recent habitat fragmentation, although this can be reversed in some areas by preserving the remaining old-growth as well as mature forest stands (100 to 200 years old) which will acquire the characteristics of old-growth within several decades.

The standard error of the population multiplication rate is attributable mainly to sampling variance in adult survivorship, because of its relatively large sensitivity coefficient, even though other vital rates had smaller sample sizes. This suggests that the most accurate estimate of current population multiplication rate can be obtained by concentrating the majority of future field work on measuring adult survivorship. Short-term demographic projections should of course be supplemented by field surveys to monitor long-term population trends.

The effect of environmental fluctuations in demographic statistics on the long-run multiplication rate of the population was investigated using formula (5). It was assumed that adult annual survivorship has a co-efficient of variation of 0.1, which indicates roughly the range of values obtained by different workers, and that the composite statistic $\pi = l_2b$ undergoes much larger fluctuations with a coefficient of variation of 5.0, representing observations of frequent poor years and occasional good years for reproduction and juvenile survival (e.g. Forsman *et al.* 1984). Under the pessimistic assumption that fluctuations in s and π are highly correlated, the estimated long-run multiplication rate of the population, $\Lambda = (0.98)\,\lambda$, is only slightly smaller than in a constant environment with the average demographic statistics. Because of the long adult life-expectancy, $1/(1 - s) \cong 17\frac{1}{4}$ years, substantial fluctuations in the vital rates have little influence on the long-run multiplication rate of the population, at least in the absence of serial correlation.

Habitat fragmentation and territory occupancy

Before the last century, roughly 60 per cent to 70 per cent of the Pacific North-west region of the United States consisted of conifer forest more than 200 years old, with the remaining areas of younger aged stands due to fire and other natural events (Franklin and Spies 1984). National forests in the Douglas-fir region of western Washington and Oregon are largely contiguous and currently contain a total of 7.6 million acres of commercial forest land, of which 2.9 million acres (38 per cent of the total) is more than 200 years old (Society of American Foresters 1984). Including forested land in wilderness areas and national parks in the region which are exempt from harvesting would not substantially alter this percentage of old growth; none of these areas contain unfragmented blocks of old growth much larger than the average individual dispersal distance of Northern Spotted Owls. Bureau of Land Management (BLM) holdings connect national forests on the Coast Range and the Cascade Mountains, but have a much lower proportion of old-growth and are interspersed with private land that is virtually devoid of old-growth; BLM and private lands are therefore likely to be population sinks for individuals dispersing from national forests, and are excluded from the analysis using the basic model (eqn. 7) which neglects edge effects.

Surveys of 46 Spotted Owl Management Areas (SOMAs) or prospective territories, designated by the Forest Service on national forests in western Washington and conducted during the years 1984–6, indicate that on average 22 per cent of SOMAs were confirmed to be occupied by a pair of spotted owls and a further 22 per cent were confirmed occupied by single owls (Allen *et al*. 1987), giving an annual average occupancy of 44 per cent (43 per cent of SOMAs were confirmed occupied by a pair in at least one of the years.) This figure probably overestimates the annual occupancy by females since it is likely that some of the single owls were males. SOMAs were confirmed occupied through daytime visual sighting after day or night response(s) to artificial calls.

The demographic potential of the population can be estimated from the basic model of territory occupancy (eqn. 7) using the formula $k = 1 - (1 - p)h$ with estimates of h and p, the current proportion of suitable habit in the region and the proportion of the suitable habitat occupied by females, and assuming the population is at a demographic equilibrium. If h is known exactly from complete habitat inventories, the standard error of an estimate of k is equal to h times the standard error of the occupancy. Using $h = 0.38$ and $p = 0.44$ yields the demographic potential $k = 0.79 \pm 0.02$. This gives an extinction threshold of $1 - k = 0.21 \pm 0.02$. This value is conservative (an underestimate) because of the assumptions that there is no environmental stochasticity, no edge effect from dispersal out of the region, and no difficulty in finding a mate (Lande 1987). The assumption of demographic equilibrium is also conservative, if the population is actually declining, since the current occupancy estimate would then be above the equilibrium value for present habitat conditions, and the demographic potential of the population would be an overestimate. The current suitability of SOMAs is postulated to be high and it is assumed that their suitability will not decline in the future due to increasing fragmentation of old forest within individual territories, or local extinction of prey species that are incapable of long-distance dispersal. The basic model of territory occupancy is optimistic, because violation of any of these assumptions would render population persistence more difficult. The calculated extinction threshold of 21 per cent is therefore likely to be a substantial underestimate.

Current Forest Service plans call for about 550 SOMAs distributed throughout the national forests in the region, in one-pair and three-pair units spaced respectively within 6 miles and 12 miles from the nearest SOMA. The 1986 draft plan specified that each SOMA contain a minimum of 1000 acres and a maximum of 2200 acres of (nearly) contiguous old-growth. (However, among six pairs of Northern Spotted Owls studied by radio-telemetry in north-western Oregon, 1008 acres was the minimum amount of old-growth included in a home range, and 2264 was the average.) The 1988 final plan calls for SOMA sizes to increase with latitude, as recommended by Dawson *et al*. (1987), from 1000 acres of suitable habitat in southern Oregon to 2200 acres in Washington (with 2700 acre SOMAs on the Olympic Peninsula) but redefines suitable habitat to include mature forest 100–200 years old. (However, the great majority of Northern Spotted Owls live and reproduce primarily in forests at least 200 years old.) Thus, if the Forest Service management plan is carried out, the proportion of old-growth on national forests will be between about $h = 16$ per cent in western Washington and about 7 per cent in south-western Oregon. Because these figures are below the conservative estimate of the extinction threshold for the population, it appears that the conservation plan developed for the Northern Spotted Owl is likely to cause its extinction. The inclusion of forested lands in wilderness areas and national parks would not alter this conclusion, which is also supported by computer simulations of Northern Spotted Owl dispersal and habitat occupancy conducted by Doak (1989).

Summary

Demographic data can be employed to estimate the multiplication rate of an age- or stage-structured population, λ, along with its standard error. Sensitivity coefficients (changes in λ in response to small changes in the vital rates) help to identify stages of the life cycle where management efforts will have the largest effect on population multiplication rate, and which of the vital rates should be the focus of future sampling efforts to obtain accurate estimates of λ. The influence of environmental fluctuations on the long-run multiplication rate of a population can be estimated from the pattern of temporal variability in the vital rates. Analytical models provide insights into the extinction dynamics of small populations in a fluctuating environment.

Dispersal behaviour and demographic characteristics determine the minimum size of a patch of suitable habitat that can support a self-sustaining population. Populations in fragmented habitats often persist on a regional basis through a balance between local extinction and colonization. Classical demography can be integrated with habitat occupancy and individual dispersal behaviour by identifying the unit of local extinction and colonization as the individual territory. Increasing fragmentation of suitable habitat leads to a lower equilibrium occupancy of the remaining habitat, and there is a minimum amount of suitable habitat randomly (or evenly) distributed throughout a large region which is necessary for persistence of a population.

Models of these processes are applied to data on the Northern Spotted Owl (*Strix occidentalis caurina*) inhabiting old-growth forests in the Pacific North-west region of the United States. Demographic analysis indicates that the conservation plan for this subspecies, originally established by the U.S. Forest Service on principles of population genetic alone, is likely to exterminate the population. This illustrates the claim that for wild populations, demographic factors are usually of more immediate importance than population genetics in conservation programmes.

Addendum: In May 1990, the Interagency Scientific Committee to Address the Conservation of the Northern Spotted Owl analyzed more recent and extensive demographic data and demonstrated values of population multiplication rate significantly less than unity in most regions. They recommended preservation (and restoration) of substantially more suitable habitat in much larger blocks than in the previous Forest Service plan. In July 1990, the US Fish and Wildlife Service listed the Northern Spotted Owl as 'threatened' under the Endangered Species Act, which provides legal protection for threatened and endangered taxa. As of August 1990, the US Congress was considering what management plans various federal agencies would be required to carry out with regard to the subspecies.

References

Allen, H. L., Dixon, K., Knutson, K. L., and Potter, A. E. (1987). Progress report No. 3. Cooperative administrative study to monitor spotted owl management areas in national forests in Washington. Washington Dept. of Game, Wildlife Management Division, Olympia, Washington.

Botkin, D. B. and Miller, R. S. (1974). Mortality rates and survival of birds. *American Naturalist*, **108**, 181–92.

Dawson, W. R., Ligon, J. D., Murphy, J. R., Myers, J. P., Simberloff, D., and Verner, J. (1987). Report of the scientific advisory panel on the spotted owl. *Condor*, **89**, 205–29.

Deevey, E. S. (1947). Life tables for natural populations of animals. *Quarterly Review of Biology*, **22**, 283–314.

Doak, D. (1989). Spotted owls and old growth logging in the Pacific Northwest. *Conservation Biology*, **3**, 389–96.

Efron, B. (1982). *The jackknife, the bootstrap and other resampling plans*. Society for Industrial and Applied Mathematics, Philadelphia, Pennsylvania.

Forsman, E. D., Meslow, E. C., and Wight, H. M. (1984). Distribution and biology of the spotted owl in Oregon, *Wildlife Monographs*, **87**, 1–64.

Frankel, O. H. and Soulé, M. E. (eds) (1981). *Conservation and evolution*. Cambridge University Press.

Franklin, J. F. and Spies, T. A. (1984). Characteristics of old-growth Douglas-fir forests. In *Proceedings 1983 Convention of the Society of American Foresters*, pp. 328–34. Society of American Foresters, Bethesda, Maryland.

Goodman, D. (1987). The demography of chance extinction. In *Viable populations for conservation*, (ed. M. Soulé), pp. 11–34. Cambridge University Press.

Goodman, L. A. (1971). On the sensitivity of the intrinsic growth rate to changes in the age-specific birth and death rates. *Theoretical Population Biology*, **2**, 339–54.

Gutiérrez, R. J. (1985). An overview of recent research on the spotted owl. In *Ecology and management of the spotted owl in the Pacific Northwest* (General Technical Report PNW-185), pp. 39–49. USDA Forest Service, Portland, Oregon, Pacific Northwest Forest and Range Experiment Station.

Kendall, M. and Stuart, A. (1977). *The advanced theory of statistics*, vol. 1. *Distribution theory*. MacMillan, New York.

Keyfitz, N. (1977). *Introduction to the mathematics of population, with revisions*. Addison-Wesley, Reading, Massachusetts.

Kierstead, H. and Slobodkin, L. B. (1953). The sizes of water masses containing plankton blooms. *Journal of Marine Research*, **12**, 141–7.

Lande, R. (1987). Extinction thresholds in demographic models of territorial populations. *American Naturalist*, **130**, 624–35.

Lande, R. (1988*a*). Genetics and demography in biological conservation. *Science*, **241**, 1455–60.

Lande, R. (1988*b*). Demographic models of the northern spotted owl (*Strix occidentalis caurina*). *Oecologia*, **75**, 601–7.

Lande, R. and Barrowclough, G. F. (1987). Effective population size, genetic variation, and their use in population management. In *Viable populations for conservation*, (ed. M. Soulé), pp. 87–123. Cambridge University Press.

Lande, R. and Orzack, S. H. (1988). Extinction dynamics of age-structured populations in a fluctuating environment. *Proceedings of the National Academy of Sciences USA*, **85**, 7418–21.

Leigh, E. G., Jr. (1981). The average lifetime of a population in a varying environment. *Journal of Theoretical Biology*, **90**, 213–39.

Leslie, P. H. (1966). The intrinsic rate of increase and the overlap of successive generations in a population of guillemots (*Uria aalge* Pont.). *Journal of Animal Ecology*, **25**, 291–301.

Levins, R. (1969). Some demographic and genetic consequences of environmental heterogeneity for biological control. *Bulletin of the Entomological Society of America*, **15**, 237–40.

Levins, R. (1970). Extinction. In *Some mathematical questions in biology*, Lectures on

Mathematics in the Life Sciences, (ed. M. L. Gerstenhaber), pp. 75–107. American Mathematical Society, Providence, Rhode Island.

Mertz, D. B. (1971). The mathematical demography of the California condor population. *American Naturalist*, 105, 437–53.

Meyer, J. S., Ingersoll, C. G., McDonald, L. L., and Boyce, M. S. (1986). Estimating uncertainty in population growth rates: jackknife vs. bootstrap techniques. *Ecology*, 67, 1156–66.

Meyers, N. (1986). Tropical deforestation and a mega-extinction spasm. In *Conservation biology, the science of scarcity and diversity*, (ed. M. Soulé), pp. 394–409. Sinauer, Sunderland, Massachusetts.

Namias, J. (1978). Persistence of U.S. seasonal temperatures up to one year. *Monthly Weather Review*, 106, 1557–67.

Nichols, J. D., Hensler, G. L., and Sykes, P. W., Jr. (1980). Demography of the everglade kite: implications for population management. *Ecological Modelling*, 9, 215–32.

Okubo, A. (1980). *Diffusion and ecological problems: mathematical models*. Springer-Verlag, New York.

Pease, C. M., Lande, R., and Bull, J. J. (1989). A model of population growth, dispersal and evolution in a changing environment. *Ecology*, 70, 1657–64..

Pickett, S. T. A. and White, P. S. (eds) (1985). *The ecology of natural disturbance and patch dynamics*. Academic Press, New York.

Schonewald-Cox, C. M., Chambers, S. M., MacBryde, B., and Thomas, L. (eds) (1983). *Genetics and conservation: a reference managing wild animal and plant populations*. Benjamin-Cummings, London.

Shaffer, M. L. (1981). Minimum population sizes for species conservation. *BioScience*, 31, 131–4.

Simberloff, D. (1986). Are we on the verge of a mass extinction in tropical rain forests? In *Dynamics of extinction*, (ed. D. K. Elliot), pp. 165–80. Wiley, New York.

Simberloff, D. (1988). The contribution of population and community biology to conservation science. *Annual Reviews of Ecology and Systematics*, 19, 473–511.

Society of American Foresters (1984). Scheduling the harvest of old growth. Society of American Foresters, Bethesda, Maryland.

Soulé, M. E. (ed.) (1986). *Conservation biology, the science of scarcity and diversity*. Sinauer, Sunderland, Massachusetts.

Soulé, M. E. and Wilcox, B. A. (eds) (1980). *Conservation biology, an evolutionary-ecological perspective*. Sinauer, Sunderland, Massachusetts.

Tuljapurkar, S. D. (1982). Population dynamics in variable environments. III. Evolutionary dynamics of r-selection. *Theoretical Population Biology*, 21, 141–65.

USDA Forest Service (1984). Final regional guide and final environmental impact statement for the Pacific Northwest region. Standard and guideline 8-1, 8-2, Appendices B to F. USDA Forest Service, Portland, Oregon.

USDA Forest Service (1986). Draft supplement to the environmental impact statement for an amendment to the Pacific Northwest regional guide, vols 1 and 2. USDA Forest Service, Portland, Oregon.

USDA Forest Service (1988). Final supplement to the environmental impact statement for an amendment to the Pacific Northwest regional guide, vols 1 and 2. USDA Forest Service, Portland, Oregon.

USDI Fish and Wildlife Service (1985). Red-cockaded woodpecker recovery plan. USDI Fish and Wildlife Service, Atlanta, Georgia.

28 Conserving threatened birds: an overview of the species and the threats

M. R. W. RANDS

Introduction

This paper presents a short summary of bird species currently believed to be threatened globally. I describe how threatened bird species are currently identified and present an overview of major threats to such species. The following account is intended as an introduction to the scale of the problems facing bird conservationists and to draw to the attention of population biologists our knowledge (or lack of it) of the ecology of these species.

Overview of globally threatened bird species

Numbers of threatened species

There are about 9000 species of birds worldwide, of which 1029 are currently believed to be threatened, i.e. are at risk of global extinction (Collar and Andrew 1988). Since 1958 there have been four reasonably comprehensive attempts to assess the number of globally threatened species (Table 28.1). These estimates suggest a steady increase in the proportion of the world's avifauna that is currently coming under threat: 1 per cent in 1958 rising to 3 per cent in 1979 with a more rapid increase to 11 per cent between 1979 and 1988.

There are a number of possible explanations for this apparent increase in the proportion of species threatened. It might be explained in part by the adoption of a broader definition of the term 'threatened' by IUCN (1980) and by the description of new species with highly restricted ranges. It also reflects an increase in the amount of data available and the coverage of this data in the review process. Most importantly however, the increased threat appears to be a consequence of the speeding up of environmental degradation (Collar and Andrew 1988). IUCN's redefinition accounted for 18 of the 172 threatened species described for Africa by Collar and Stuart (1985) and the discovery of new species added a further 2 (since

Table 28.1 Numbers of threatened bird species identified by four surveys between 1958 and 1988

Year	No. of threatened species	Source
1958	95	Greenway (1967)
1971	220	Vincent (1971)
1979	290	King (1978/9)
1988	1029	Collar and Andrew (1988)

1985 another 2 species have been discovered). Assuming that similar proportions apply elsewhere, these two explanations together probably therefore account for about 11.5 per cent of the 1029 threatened species and thus still only a small proportion of the total increase between 1979 and 1988.

Taxonomic distribution of threatened species

A breakdown of threatened species by family reveals that the danger of extinction is not just confined to certain highly specialized or adapted groups of birds. Appendix 28.1 lists those families of bird containing at least one threatened species and shows the number and proportion of threatened species per family. Table 28.2 summarizes this information. Over half the world's bird families contain at least one threatened species and in 24 families over 30 per cent of the species are threatened (see Table 28.3). Of these 'highly threatened families' the majority are non-passerines, the reason being that perhaps such species tend to be larger and live at lower densities than passerines and are thus more vulnerable to habitat destruction and other possible threats. Being more visible, non-passerines are also easier to study and so their status, i.e. whether threatened to a critical extent, may be assessed more accurately.

Table 28.2 Numbers of bird families including threatened species

% threatened species per family	No. of families
1–10	43
11–20	30
21–30	16
31–40	10
41–50	3
over 51	10

Table 28.3 Birds families where 30 per cent or more of the species are considered to be globally threatened

Families		No. of species in world	No. of threatened species	%
Apterygidae	Kiwis	3	1	33
Podicipedidae	Grebes	20	6	30
Fregatidae	Frigatebirds	5	2	40
Balaenicipitidae	Shoebill	1	1	100
Phoenicopteridae	Flamingoes	6	2	33
Anhimidae	Screamers	3	1	33
Megapodiidae	Megapodes	12	8	67
Cracidae	Curassows, Guans	44	16	36
Meleagridae	Turkeys	2	1	50
Mesitornithidae	Mesites	3	3	100
Pedionomidae	Plains-wanderer	1	1	100
Gruidae	Cranes	15	7	47
Heliornithidae	Sun-grebes	3	1	33
Rhynochetiae	Kagu	1	1	100
Tytonidae	Barn Owls	11	6	55
Nyctibiidae	Potoos	5	3	60
Coraciidae	Rollers	11	4	36
Phytotomidae	Plantcutters	3	1	33
Pittidae	Pittas	26	8	31
Xenicidae	New Zealand Wrens	3	1	33
Atrichornithidae	Scrub-birds	2	2	100
Rhabdornithidae	Philippine Creepers	2	1	50
Drepanididae	Hawaiian Honeycreepers	15	14	93
Callaeidae	Wattlebirds	2	2	100

Geographical distribution of threatened species

The number of threatened species per region is illustrated in Fig. 28.1. Although there are now threatened species throughout the world, concentrations are greater in the tropics than elsewhere. Of the 1029 threatened species, 884 occur in the developing nations of South and Central America, the Caribbean, Africa, Asia, and the Pacific, while 66 are only found in North America, Europe, the Middle East, the Soviet Far East, and Australia (79 being shared between the two sets of countries). Thus the burden of threatened species conservation lies with developing nations where resources are usually less available for conservation measures to be applied.

Islands versus continents

The proportion of threatened species found on islands rather than continents is shown for the surveys of King (1978–9) of Collar and

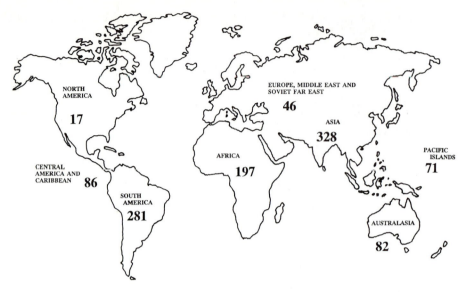

Fig. 28.1. World distribution of threatened birds.

Andrew (1988) in Table 28.4. Although nearly half the world's threatened birds are still found on islands, there is an indication that the proportion on continents is increasing, presumably as the rate of habitat destruction on continents increases.

Habitats of threatened species

Table 28.5 shows the habitats used by threatened species and reveals a number of interesting differences between the situation in 1979 and 1988. In both surveys, three major habitats were found to contain over 80 per cent of threatened species, with tropical forests far surpassing the other two major habitat types for content of threatened species. However, between 1979 and 1988 the relative importance of tropical forest has diminished, while both wetlands and grassland/savanna habitats have come to hold an increasing proportion of the world's threatened species;

Table 28.4 Proportions of threatened bird species on islands versus continents (after King 1978–9; Collar and Andrew 1988)

	Island	Continental
1978/9	53%	47%
1988	46%	54%

Table 28.5 Numbers of threatened bird species inhabiting different habitats

	A King (1978–9)	B Collar and Andrew (1988)	B/A
Tropical Forest	186	442	2.38
Wetland	38	216	5.69
Grasslands/Savannah	20	195	9.80
Other/No data	46	175	3.80
Total	290	1029	3.55

the increase from 7 per cent (20 out of 290) in 1979 to 19 per cent (195 out of 1029) in 1988 for grassland/savanna dwelling species is particularly striking. This does not imply a reduction in the importance of tropical forest as a habitat for threatened species (in 1988 it was critical for 256 more species than in 1979), but rather that further habitats have come under similar threat over the last decade.

Using the data collated by Collar and Stuart (1985) for Africa and related islands, it is possible to gain some insights into the biology of threatened species. For example, the food eaten by 131 of the 172 threatened species found in Africa (Table 28.6) indicates a predominance of invertebrate feeders. This may reflect simply the relative abundance of insectivores amongst the whole avifauna or may be a genuine indication that birds dependent on invertebrates are more likely to be threatened.

Overview of the threats

Assessing which species are threatened

For those concerned with conserving species at risk of global extinction one of the first tasks is to identify which species are threatened. Ideally such an assessment requires quantitative information of current population size, trends, vulnerability, and therefore of the likelihood of extinction. Such data are not very often available for even the well-studied temperate species and are far from complete for species which, because they are scarce, are difficult to find and study.

To focus attention on species believed to be threatened, the concept of 'Red Data Books' was developed (e.g. see Scott *et al*. 1987 for a historical review). These compilations draw together published and unpublished information on distribution, population size, ecology, threats, and conservation measures taken. On the basis of this synthesis further conservation measures can then be proposed.

Table 28.6 Diet of threatened bird species in Africa (data from Collar and Stuart 1985)

Foods	No. of threatened species	%
Invertebrate	60	35
No data	41	24
Fruit/Seed/Shoots	30	17
Omnivores	19	11
Fish	8	5
Birds/Mammals/Reptiles	5	3
Nectar	6	3
Molluscs	2	1
Carrion	1	1

Each species is assigned to a category of threat depending on the extent of the risk to which it is thought to be exposed. The categories used in the latest example of a bird red data book (Collar and Stuart 1985) are Extinct, Endangered, Vulnerable, Rare, Indeterminate, Insufficiently Known, and Out of Danger (Table 28.7). The placement of a species in one of the two most threatened categories requires a knowledge, or at least an assessment, of the causal factors leading to a species demise. Here it is assumed that numbers are declining (or that the range is contracting) and that this will lead to extinction unless the causal factors cease to operate. Species categorized as Rare (i.e. with small or highly restricted world populations) are not believed to be declining to the extent that they should be considered Endangered or Vulnerable. The remaining categories, Indeterminate and Insufficiently Known, do not call for data on the causes

Table 28.7 Definitions of categories of threat

Category	Definition
Extinct	Species not definitely located in the wild during the past 50 years
Endangered	Species in danger of extinction and whose survival is unlikely if the *causal* factors continue operating
Vulnerable	Species believed likely to move into the Endangered category in the near future if the *causal* factors continue operating
Rare	Species with *small world populations* that are not yet Endangered or Vulnerable but are at risk
Indeterminate	Species known to be Endangered, Vulnerable, or Rare but *not enough information* exists to say which of the three categories is appropriate
Insufficiently known	Species that are suspected but not definitely known to belong to any of the above categories, because of *lack of information*

of species' decline but reflect, perhaps more realistically, the lack of information. A review of any red data book species account (other than those for some large mammals and a few conspicuous birds) reveals that the judgement of how threatened a species is must be a subjective one based on the information available; there are simply no quantitative data on which to base the judgement.

Table 28.8 shows for threatened species in Africa the number in each category in 1979 (King 1978–9) and 1985 (Collar and Stuart 1985). In the seven intervening years the Endangered category increased by 9 species and 49 more were included in the Rare category. This reflects both an increase in the quality of data and research time, as well as increasing environmental degradation. For the 9 endangered species, it is the latter that has brought them closer to extinction.

Table 28.8 Number of threatened species in each category of threat for Africa in 1978 and 1985

Category	King (1979)		Collar and Stuart (1985)	
	No. of species	%	No. of species	%
0	0	2	1	
Endangered	19	29	28	16
Vulnerable	13	20	15	9
Rare	29	45	78	45.5
Indeterminate	4	6	31	18
Insufficiently known	—	—	18	10.5

Identifying the threats

The reasons why a species is threatened with extinction are many and varied. In some cases the range may be being reduced, by for example habitat destruction, in others population density may be declining due to increased mortality through over-exploitation.

The primary threats identified by Collar and Stuart 1985, for the threatened species of Africa and related islands are summarized in Table 28.9 (see also Stuart and Collar 1989). Habitat destruction, especially of forests, is clearly the largest single threat to species suvival in the region; introduced predators and competitors, direct human utilization (hunting, collection, etc.), and persecution are also major threats. Predation by exotic species is particularly prevalent on islands (Johnson and Stattersfield 1990). Where 'restricted range' is cited as a threat it is known that this has not come about directly as a result of any one of the foregoing threats

Table 28.9 Major threats to threatened bird species in Africa (data from Collar and Stuart 1985)

	Number of species	%
Habitat destruction		
Forest	70	41
Wetland	6	3
Grassland	3	2
Introduced predators/competitors	26	15
Over-exploitation	15	9
Persecution/Human disturbance	7	4
Bird trade/Zoo collections	3	2
Exotic fish	2	1
Hybridization	1	1
Natural (drought)	1	1
Restricted range	22	13
No data	37	22

listed above in Table 28.9. Together with the 37 species listed under 'no data', this category serves to highlight the level of ignorance surrounding threatened species conservation.

Conservation action required

In order to prevent the extinction of a species it is obviously necessary to remove the threat or threats. In some cases simply stopping the cause of threat may no longer be enough, and some remedial action may also be required. The actions possible can be divided into two types: interventionist and non-interventionist. The latter do not in themselves require any direct interaction with the birds or their habitat; examples include legislation/enforcement, habitat preservation through the development of a protected area network, and conservation education (attempting to change public attitudes and action towards wildlife).

Interventionist conservation actions involve either restoration of the habitat or restoration of the bird populations themselves. They may be short-term solutions that simply serve to prevent imminent extinction; for example the provision of supplementary food or perhaps in the most extreme case captive breeding for reintroduction. Long-term solutions require the removal of the ultimate threat and include such actions as habitat restoration (e.g. reafforestation) or total eradication of introduced predators.

Conclusions

In the brief overview given above the diversity of bird species currently under threat is described. Based on this review it is suggested that those persons wishing to conserve biological diversity by preventing extinctions must concentrate their efforts in a number of ways. First they should seek to conserve areas where the distributions of threatened species overlap the most. For example Collar and Stuart (1988) showed that 93 per cent of the threatened Afrotropical avifauna could be saved by protecting just five large forest blocks which together make up less than 10 per cent of the continent.

Second it is suggested they should concentrate on those families where a high proportion of the species are threatened (see Table 28.5). There may be families containing just one or two species like the Shoebill or the Turkeys or possibly of higher priority still, larger families like the Megapodes, Cranes, Pittas, or Hawaiian Honeycreepers where several species are threatened.

Given the priorities, population biologists can play a critical role in conserving threatened species. In order accurately to assess whether or not a species is in danger of extinction and, if so, to determine how threatened it is, requires data on distribution and numbers, and changes in both over time. There is a need to develop a set of simple repeatable survey techniques that can be applied to as many candidate species as possible (see Green and Hirons, this volume). Even a single reliable figure for population size would be beneficial in determining those species that may be in most need of further attention. While a great many ornithologists are willing to carry out such surveys, there is a real need to ensure that data are obtained in comparable ways.

Where species are judged to be threatened, conservation action is most likely to be successful if the threat can be identified. While the preservation of large tracts of forest or wetlands may ensure a species' survival, this is often no longer feasible given the human pressures and needs for the same natural resources. To save birds under these circumstances there ecological requirements must be determined so that the essential elements of the habitat can be retained (or artificially provided) while human exploitation is allowed to continue. An alternative, and more productive, role for the population biologist, may be to monitor experimentally the impact of conservation actions that are undertaken to save species. If population biologists had monitored a fraction of the conservation actions that have been attempted, we might now be in a position to predict the outcome of similar actions for other threatened species. Such rigour in conservation can provide insights into biological theory as well as aiding conservation practice.

References

Clements, J. (1981). *Birds of the World: a checklist*. Facts on File, New York.

Collar, N. J. and Andrew, P. (1988). *Birds to watch: the ICBP world checklist of threatened species*. Cambridge: ICBP.

Collar, N. J. and Stuart, S. N. (1985). *Threatened birds of Africa and related islands*. IUCN & ICBP, Cambridge.

Collar, N. J. and Stuart, S. N. (1988). *Key forests for threatened birds in Africa*. International Council for Bird Preservation. Monograph No. 3, ICBP, Cambridge.

Greenway, J. C. (1967). *Extinct and vanishing birds of the world*, 2nd revised edn. Dover publications, New York.

Johnson, T. H. and Stattersfield, A. J. (1990). A global review of island endemic birds. *Ibis*, **132**, 167–80.

King, W. B. (1978–9). *Red data book, 2. Aves*. 2nd edition. IUCN, Morges, Switzerland.

Morony, J. J., Bock, W. J., and Farrand, J. (1975). *Reference list of birds of the world*. American Museum of Natural History (Department of Ornithology), New York.

Scott, P., Burton, J. A., and Fitter, R. (1987). Red Data Books: the historical background. In: *The road to extinction*. IUCN/UNEP Gland, Switzerland.

Stuart, S. N. and Collar, N. J. (1989). *Birds at risk in Africa and related islands: the causes of their rarity and decline*. Proceedings VI Pan African Ornithological Congress, pp. 1–25.

Vincent, J. (1966–71). *Red data book. 2 Aves*. IUCN, Morges, Switzerland.

Appendix 28.1 The number of species, and number of threatened species per bird family, and the percentage of each bird family threatened. (Taxonomy from Morony *et al.* (1975), incorporating updating by Clements (1981))

Families		No. of species in world	No. of threatened species	%
Apterygidae	Kiwis	3	1	33
Tinamidae	Tinamous	46	8	17
Spheniscidae	Penguins	18	3	17
Podicipedidae	Grebes	20	6	30
Diomedeidae	Albatrosses	13	2	15
Procellariidae	Petrels, Shearwaters	66	19	29
Hydrobatidae	Storm-petrels	20	4	20
Pelecanoididae	Diving-petrels	4	1	25
Pelecanidae	Pelicans	7	2	29
Sulidae	Gannets, Boobies	9	1	11
Phalacrocoracidae	Cormorants	31	3	10
Fregatidae	Frigatebirds	5	2	40
Ardeidae	Herons, Egrets, Bitterns	64	7	11
Balaenicipitidae	Shoebill	1	1	100
Ciconiidae	Storks	17	5	29

Families		No. of species in world	No. of threatened species	%
Threskiornithidae	Ibises, Spoonbills	33	7	21
Phoenicopteridae	Flamingoes	6	2	33
Anhimidae	Screamers	3	1	33
Anatidae	Ducks, Geese, Swans	149	20	13
Cathartidae	New World Vultures	7	1	14
Accipitridae	Hawks, Eagles	217	37	17
Falconidae	Falcons, Caracaras	62	6	10
Megapodiidae	Megapodes	12	8	67
Cracidae	Curassows, Guans	44	16	36
Meleagridae	Turkeys	2	1	50
Phasianidae	Pheasants	183	42	23
Numididae	Guineafowl	7	1	14
Mesitornithidae	Mesites	3	3	100
Turnicidae	Button-quails	15	4	27
Pedionomidae	Plains-wanderer	1	1	100
Gruidae	Cranes	15	7	47
Rallidae	Rails, Coots	123	29	24
Heliornithidae	Sun-grebes	3	1	33
Rhynochetiae	Kagu	1	1	100
Otididae	Bustards	24	6	25
Haematopodidae	Oystercatchers	8	2	25
Recurvirostridae	Avocets, Stilts	13	1	8
Glareolidae	Coursers, Pratincoles	15	1	7
Charadriidae	Plovers	65	7	11
Scolopacidae	Sandpipers, Snipe	85	12	14
Laridae	Gulls, Terns	90	9	10
Alcidae	Auks	22	1	5
Columbidae	Doves, Pigeons	298	50	17
Loriidae	Lories	51	10	20
Cacatuidae	Cockatoos	18	5	28
Psittacidae	Parrots	256	56	22
Musophagidae	Turacos	23	2	9
Cuculidae	Cuckoos, coucals	129	10	8
Tytonidae	Barn Owls	11	6	55
Strigidae	Owls	129	15	12
Podargidae	Frogmouths	13	1	8
Nyctibiidae	Potoos	5	3	60
Caprimulgidae	Nightjars	76	8	11
Apodidae	Swifts	78	8	10
Trochilidae	Hummingbirds	316	30	9
Trogonidae	Trogons	37	3	8
Alcedinidae	Kingfishers	90	8	9
Momotidae	Motmots	9	1	11
Coraciidae	Rollers	11	4	36
Bucerotidae	Hornbills	46	7	15

Appendix 28.1 *cont.*

Families		No. of species in world	No. of threatened species	%
Galbulidae	Jacamars	15	1	7
Capitonidae	Barbets	79	4	5
Indicatoridae	Honeyguides	16	1	6
Ramphastidae	Toucans	41	1	2
Picidae	Woodpeckers	208	7	3
Eurylaimidae	Broadbills	14	1	7
Dendrocolaptidae	Woodhewers	48	1	2
Furnariidae	Ovenbirds	212	19	9
Formicariidae	Antbirds	221	31	14
Conopophagidae	Gnateaters	10	1	10
Rhinocryptidae	Tapaculos	27	2	7
Cotingidae	Cotingas	79	14	18
Pipridae	Manakins	54	2	4
Tyrannidae	Tyrant-flycatchers	366	16	4
Phytotomidae	Plantcutters	3	1	33
Pittidae	Pittas	26	8	31
Xenicidae	New Zealand Wrens	3	1	33
Philepittidae	Asitys	4	1	25
Atrichornithidae	Scrub-birds	2	2	100
Alaudidae	Larks	78	7	9
Hirundinidae	Swallows, Martins	81	4	5
Motacillidae	Wagtails, Pipits	54	4	7
Campephagidae	Cuckoo-shrikes	68	7	10
Pycnonotidae	Bulbuls	124	9	7
Laniidae	Shrikes	75	6	8
Vangidae	Vanga	12	2	17
Cinclidae	Dippers	5	1	20
Troglodytidae	Wrens	66	6	9
Mimidae	Mockingbirds, Thrashers	31	2	6
Muscicapidae	Thrushes, Warblers, etc.	1106	148	13
Paridae	Tits, Chickadees	48	2	4
Sittidae	Nuthatches	23	6	26
Rhabdornithidae	Philippine Creepers	2	1	50
Dicaeidae	Flowerpeckers	56	2	4
Nectariniidae	Sunbirds	115	8	7
Zosteropidae	White-eyes	82	19	23
Meliphagidae	Honeyeaters	169	12	7
Emberizidae	Buntings, Tanagers	323	49	15
Parulidae	New World Warblers	123	14	11
Drepanididae	Hawaiian Honeycreepers	15	14	93
Vireonidae	Vireos	40	1	2
Icteridae	New World Blackbirds	94	11	12
Fringillidae	Finches	124	8	6
Estrildidae	Waxbills	127	6	5
Ploceidae	Weavers, Sparrows	143	16	11
Sturnidae	Starlings	106	7	7
Oriolidae	Orioles	28	2	7
Dicruridae	Drongos	20	2	10

Families		No. of species in world	No. of threatened species	%
Callaeidae	Wattlebirds	2	2	100
Ptilonorynchidae	Bower-birds	17	2	12
Paradisaeidae	Birds of Paradise	42	5	12
Corvidae	Crows, Jays	109	12	11
			1029	

29 The relevance of population studies to the conservation of threatened birds

R. E. GREEN and G. J. M. HIRONS

Introduction

The existence of many bird species is now threatened as human activities increasingly damage or destroy their habitats or directly affect their survival or reproductive success. The challenge faced by bird conservationists is to ensure the persistence of as many species as possible by devising ways in which they can coexist with human activity, often within a small remnant of their former range. What role should studies of avian population processes play in this task?

The aim of most population studies of birds has been to improve understanding of the dynamics of animal populations in general (Lack 1966; den Boer and Gradwell 1971; Klomp and Woldendorp 1980). Particular attention has been paid to the question of whether bird populations are regulated and, if so, by what mechanisms. Different authors have stressed the importance of intraspecific competition for food, the effects of predators and diseases, and territorial behaviour. Species were selected for study because of their abundance in places accessible to the investigators and because of certain traits that made them convenient to work on, such as a willingness to breed in nestboxes. The conceptual framework and methodology derived from these classical population studies provide the scientific basis for the management of pest and quarry species and are increasingly being applied to the management of scarce or declining birds.

In this paper we review current knowledge of demographic parameters of threatened birds and examine the contribution that studies of population processes could make to their conservation. We identify two main areas of research: assessment of the risk of extinction, as a basis for assigning priorities for scarce conservation resources; and the identification of the causes of changes in demographic parameters and the consequences of such changes for population size and trend. We see the latter as a prerequisite for the development of appropriate techniques for the recovery of populations of threatened species.

Assessment of endangerment

It is obvious that many of the world's 9000 bird species are in no immediate danger of extinction; indeed the population size or geographical range of some are increasing. Bird conservationists need to direct their energy and limited resources into measures that will protect those species at greatest risk. A prerequisite for an effective programme of conservation action is research to identify bird species with small, declining, or vulnerable world populations. Such an assessment usually takes the form of a red data book (e.g. Collar and Stuart 1985) in which information on distribution, abundance, and immediate threats is compared across species.

The estimates of population size included in red data book entries are derived from a wide variety of survey methods (see Appendix 29.1) and are of variable quality. Furthermore, there is no standard procedure for combining factors such as small population size, rapid rate of decline, and vulnerability of habitat into an overall index of extinction risk.

Ideally, the identification of priority species for conservation would be based on a comparison of formal estimates of the probability of extinction within a specified time period across a large number of species. However, although explicit estimates of extinction risk have been made for a few species of endangered mammals (Shaffer 1983; Harris *et al*. 1988), we know of no reliable estimates of extinction risk for birds. The theory of population viability analysis is in the early stages of development (Soule 1987) and so it will probably be some time before definite assessments of the risk of extinction can be made for large numbers of bird species, even supposing that the parameters needed for these calculations can be estimated accurately.

For which threatened bird species have population sizes and trends been measured?

The basis of our review is the list of 1029 bird species considered by Collar and Andrew (1988) to be threatened with extinction. They were unable to devise a comprehensive set of objective criteria for the inclusion of a species in this checklist, but population size and trend, restricted distribution, and vulnerability to human activities all contributed to the assessment. Most of the species which were included, and for which a population estimate is available, have world populations of less than 1000 individuals (Fig. 29.1, Appendix 29.1).

We searched the literature on the 1029 species for estimates of population size, breeding success, survival rates, age at first breeding, and pre-breeding dispersal. In a few cases we have drawn on unpublished documents in the library of the International Council for Bird Preservation.

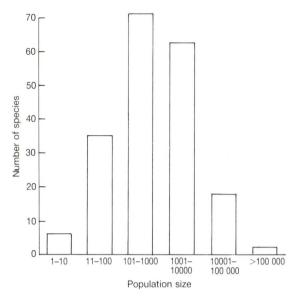

Fig. 29.1. Frequency distribution of estimated population size of threatened bird species. Data from Appendix 29.1. Estimates of numbers of pairs, breeding units, or singing males have been multiplied by two.

We have no doubt that some relevant material has been overlooked, particularly that in journals of restricted circulation or in languages other than English, but it is likely that the great majority of reliable estimates will have come to our attention. Parameter estimates were classified according to the fieldwork or analytical methods used. A checklist of species for which we found relevant information and details of our criteria are given in the Appendices.

We found estimates of population size for only 202 (20 per cent) of the 1029 threatened bird species, in spite of the fact that we were prepared to accept estimates in which only the breeding population was counted or for anything greater than 50 per cent of the species range. The proportion of species for which information was available varied greatly between zoogeographical regions (Fig. 29.2). We considered a good population estimate to be available where there was a total count or where allowance could be made for birds missed during the census (e.g. by mark-recapture analysis). Estimates of total population based on densities in sample plots were included if they met this criterion and if the area to which the extrapolation was applied was well-defined and uniform. According to this definition, there was at least one good estimate of population size available for 10 per cent of threatened species (51 per cent of those for which there was any estimate). Estimates that relied on arbitrary assumptions about census efficiency or on extrapolations from data relevant to less than half of the world distribution were available for a further 7 per cent of species. Population estimates with no indication as to how they

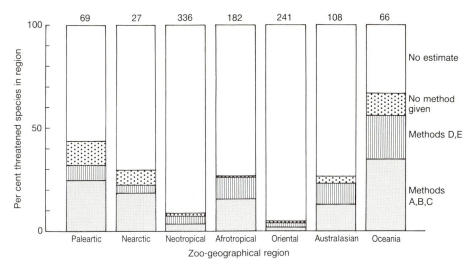

Fig. 29.2. Proportions of threatened bird species for which population estimates are available according to the zoogeographical region in which the main part of the population occurs. For details of the classification of the method used to arrive at the population estimates see text and Appendix 28.1. The number of threatened species in each region is indicated by the figure at the top of each column.

were obtained were the best that we could find for 3 per cent of threatened species.

For 51 per cent of the species for which estimates were available, the most recent was made more than 5 years ago (i.e. prior to 1984) (Fig. 29.3). No doubt there are more recent data for some species presently being prepared for publication. However, the fact that the most recent estimate was more than 10 years out of date for 16 per cent of the species for which there was any estimate, suggests that populations of many threatened birds whose status has been assessed at some time are not being monitored. This impression is reinforced by the small proportion of species for which more than one comparable estimate was available; 7 per cent of the 1029 species or 37 per cent of those for which there is any population estimate (Fig. 29.4). Hence even the crudest assessments of rate of population change are impossible for most threatened species. Only for a small number are long runs of census data available (Fig. 29.5).

Assessment of priorities for further research and conservation action

Conservationists face a number of difficult decisions as a result of the time taken to devise and implement a conservation programme once the need

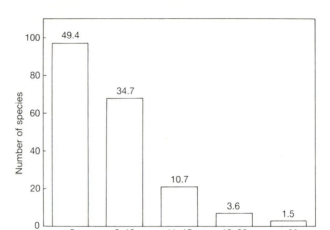

Fig. 29.3. Frequency distribution of the date of the most recent published population estimate for threatened bird species. Compiled in 1988.

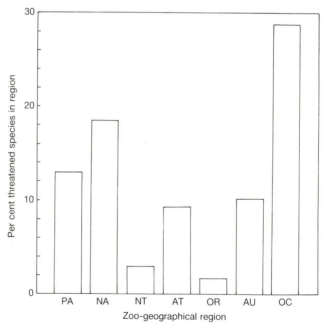

Fig. 29.4. Proportions of the total number of threatened bird species within a zoogeographical region for which more than one comparable population estimate is available. Estimates were required to be separated by at least 2 years. For classification of zoogeographical regions see Fig. 29.2 and Appendix 29.1.

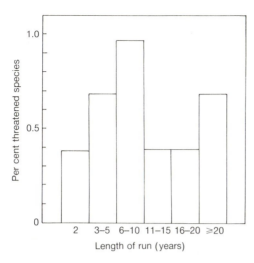

Fig. 29.5. Percentages of the total number of threatened bird species for which runs of comparable census data are available for consecutive years. Runs of at least 2 years are available for 30 species.

has been recognized for one. For example, suppose that a bird species has been found to have a world population of 1000 individuals. Although the population and range were formerly much larger, counts indicate that numbers have been approximately stable for 10 years. Should research begin immediately to devise methods for increasing the population? Would it be better to monitor population size regularly and only begin detailed research if a further decline takes place? If so, how often should the counts be carried out? Suppose that another species has a current world population of 10 000 individuals, but that there were twice as many 10 years ago. Does this species merit a higher or lower priority for conservation measures than the first? As far as we know, there are as yet no standard methods available for answering these questions. However, it is clear that decisions should allow for both the time that it would take for the species to decline to an irretrievably low level, once a decline was under way, and for the time required to devise and implement a conservation programme.

Among threatened bird species in which population declines have been monitored, rates of decline have varied considerably. We calculated annual population multiplication rates (λ) for all threatened and recently extinct species for which we could find a series of counts (Fig. 29.6). It was necessary to standardize the interval between counts used in this analysis. However, if we had required comparable counts at yearly intervals, data would only have been available for a handful of species. Therefore we calculated an average annual rate of change from any pair of counts separated by 5–10 years (mean 6.7 years). We were able to make 100 estimates of this type for 45 species (1–12 estimates per species). We have

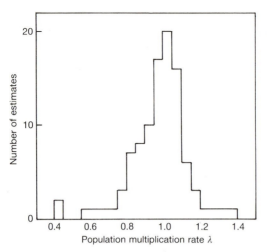

Fig. 29.6. Frequency distribution of 100 estimates of annual average population multiplication rate for 45 threatened bird species calculated from pairs of counts at 5- to 10-year intervals.

assumed that population size was measured without error. Errors of estimation would tend to increase the scatter of the estimates of rate of change, but should not lead to spurious trends.

A high proportion of monitored population changes occurred rapidly: 24 per cent of rates of change were declines at rates of loss in excess of 10 per cent per year, which is equivalent to a reduction by 40 per cent in the approximately 5-year periods chosen for study (Fig. 29.6). The most rapid documented decline ($\lambda = 0.13$) was that of the Guam flycatcher *Myiagra freycineti*. The world population was estimated at 450 individuals in 1981. The species was known to be extant in 1983, but was extinct by 1984 (Engbring and Pratt 1985). This population change was measured over too short a period to be eligible for inclusion in Fig. 29.6.

There was no tendency for rates of population change to be more rapid or variable for small rather than large birds (Fig. 29.7). We expected a correlation with body size because mortality rate and maximum rate of increase tend to be higher for small rather than large birds. There was more than one estimate of population multiplication rate available for 19 populations of 18 threatened bird species. In this data there was strong evidence for differences among populations in the extent of variation of multiplication rate between successive 5-to-10-year periods (Bartlett's test for homogeneity of variances; $\chi^2_{18} = 40.53$, $P = 0.002$). However, there was no significant correlation between within-population variance in λ and body weight ($r_{17} = -0.255$, $t_{17} = 1.09$, $P > 0.30$, both variables log transformed). Hence, rapid population changes have been recorded in birds from many taxa and of a wide range of body size. Even populations of large species, which might have been expected to have high survival rates and low rates of population change, can decline quickly.

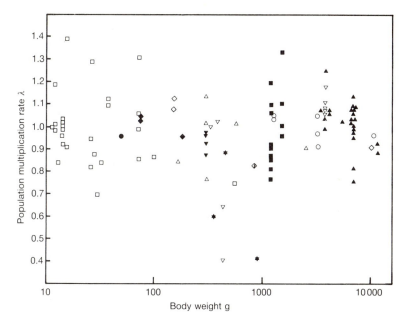

Fig. 29.7. Average annual population multiplication rate (see Fig. 29.6) of threatened bird species in relation to mean body weight. ★ Podicepidiformes, ▽ Procellariiformes, ○ Pelicaniformes, ■ Ciconiiformes, ◇ Falconiformes, ◇ Galliformes, ▲ Gruiformes, ◆ Charadriiformes, ▼ Columbiformes, △ Psittaciformes, ● Strigiformes, □ Passeriformes.

The time taken for conservationists to devise and implement a conservation programme varies enormously. In Fig. 29.8 we present some case histories of declining species in which the process of identifying conservation problems and taking countermeasures has been documented and in which the population eventually responded by showing a definite increase. We have excluded from consideration several species for which research and conservation action have been in progress for more than 20 years but with clear evidence of recovery still lacking.

Although this is by no means a random sample of recovery programmes for endangered birds, it is clear that a period of about 10 years, and often much longer, can be expected to elapse between the start of research and subsequent recovery. Long delays frequently involve an initial misdiagnosis of the cause of the species' plight, followed by ineffective conservation action. This highlights the importance of adequate monitoring and review of the efficacy of any conservation measures taken. There was no obvious relationship between the quantity of manpower and resources devoted to the task and the time taken to accomplish it.

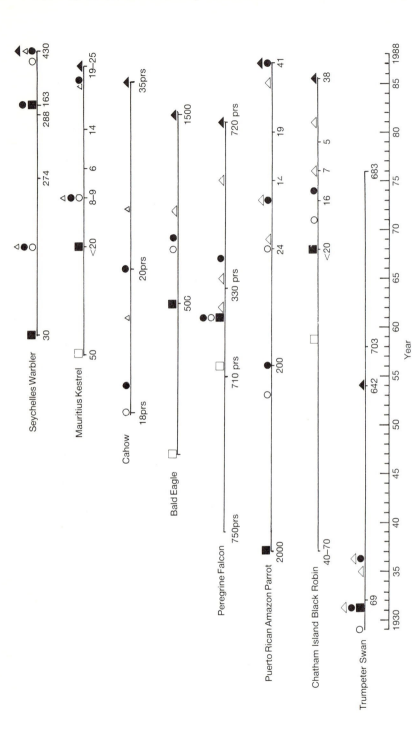

Fig. 29.8. Case histories of some declining bird species illustrating the time taken for conservationists to devise and implement successful conservation programmes. The timing of major events are indicated by symbols showing: the probable start of the population decline, □ ; decline noticed, ■ ; start of research to identify the cause of the decline, ○ ; diagnosis of the cause of the species' plight, ● ; implementation of remedial conservation action, △ ; recovery of the population, defined as population showing clear evidence of an increase, ▲ . The numbers below the lines for each species are estimates of the size of the wild population at the time shown. For further details, see Appendix 29.1.

These findings have implications for population monitoring programmes that are intended to signal the need for more detailed investigation of the cause of a decline once it has been detected. The combination of a long delay involved in devising and implementing appropriate conservation action with a rapid rate of decline might well result in extinction before the remedial action could have effect.

The investment of a small amount of research effort into a short-term study of the ecological requirements and breeding success of a species with a small but stable population would frequently reduce the time taken to identify the causes of any subsequent decline. It would also permit the monitoring programme to include environmental factors to which the species was likely to be sensitive and allow the use of comparative methods of diagnosis of the causes of decline (see below).

The case of the California Condor *Gymnogyps californicus* illustrates the need to obtain information on the basic ecology of a species at an early stage of research to identify possible conservation measures. Monitoring showed that the population was small but stable in the 1940s (Koford 1966), but was known to be declining by 1963 (Miller *et al.* 1965). However, detailed research on the ecological requirements and hazards faced by wild condors did not begin until October 1982 (Ogden 1985). By then the research was hampered by the small number of birds (about 20) available for study and by public fears about possible adverse effects of the research methods themselves. Valuable information was obtained, but was too late to avert the extinction of the California Condor in the wild in 1985. A similar research project in the 1960s, when it would have been technically feasible and raised fewer objections from environmentalists, might well have saved the condor.

Another consequence of the long delays involved in devising and implementing conservation measures is that predictions of the long-term security of a small population of birds calculated by current methods of population viability analysis are likely to be over-optimistic. We illustrate this by adapting an example of a birth-and-death model developed by Goodman (1987) with demographic and stochastic variation in population growth rate.

Goodman calculated the distribution of times to extinction for a model population with a small maximum size, assuming that the average growth rate was positive when the population was below the ceiling. We altered this model by supposing that a 20-year period during which the growth rate was negative occurred during the simulation. This was intended to represent a period of continuous adverse circumstances for a rare species terminated by successful diagnosis and remedial action. It can be seen that this modification had a drastic effect on the distribution of persistence times (Fig. 29.9). Hence models of minimum viable population size should

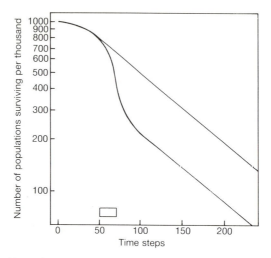

Fig. 29.9. The effect of a period of adverse conditions on the persistence of a small population. A birth-and-death model devised by Goodman (1987) was used to generate two survivorship curves representing the proportion of model populations remaining extant after various lengths of time. The upper curve shows the persistence of populations with an initial size and upper limit of 20 individuals. Births do not occur when this ceiling is reached. Otherwise the population growth rate is r = 0.05. The variance of r has an environmental component $V_e = 0.5$ and an individual component $V_1 = 0.25$. Otherwise this model is identical to that shown in Figs. 2.5 and 2.6 of Goodman (1987). The lower curve shows the effect on persistence of altering this model by imposing a 20-year period beginning in year 50 (shown by bar) during which r = −0.05.

explore the effects of long periods of adverse conditions as well as those of the severe but transient impact of events such as storms and epizootics (Ewens *et al.* 1987).

The satisfactory fit of some retrospective extinction data to models that do not incorporate the effects of long periods of adverse conditions (Belovsky 1987) should not argue against taking them into account in future conservation planning. Human activities resulting in changes such as pollution, habitat fragmentation, and exposure to new predators and diseases seem particularly likely to bring about long-term effects in the future. The realism of the models would be increased by the admission of events of this type, but the unpredictability of the incidence, duration, and severity of such events would be likely to decrease the confidence that could be put on any estimate of a secure minimum population size. If this proves to be the case the wisdom of using such estimates to guide conservation planning and reserve design would be doubtful.

Identification of management techniques for population recovery

The perception that a bird species is threatened often involves a preliminary, usually subjective, assessment of the most likely cause of its endangerment. However, initial ideas about the reasons for the rarity or decline of a species tend to be biased towards the obvious or simple and frequently turn out to be ill-founded when examined in detail. When the cause of a decline is correctly diagnosed it may seem that conservation measures should be directed towards those factors considered to have caused the species' predicament. However, in practice it may not be feasible to remove an introduced predator or restore a damaged habitat, at least in the short-term, and the manipulation of some factor unrelated to the original decline may be a better option. These difficulties argue for a more objective approach to the identification of appropriate techniques for the recovery of populations of threatened species.

The diagnosis of causes of population decline

Simulation modelling

For the purposes of this paper we follow Maynard Smith (1974) in defining a simulation model as a mathematical description of a population in its particular environment which must at least include estimates of age-specific birth and death rates and knowledge of how these rates vary with population density and other features of the environment which are likely to alter in future. The construction of a simulation model has been the result of several long-term studies of bird populations. The general approach is to collect life-table data within a well-defined study area and then to determine the relationships of survival and fecundity rates to intrinsic variables such as the age structure and density of the population and extrinsic variables such as weather, abundance of predators, and the availability of food. The data available are rarely sufficient to unambiguously define the precise form of the underlying relationships between variables, so assumptions are usually made which are mathematically convenient and biologically plausible. A much-used method is k-factor analysis (Varley and Gradwell 1960) in which the logarithm of a mortality rate is modelled as being linearly related to population density or external variables. This and other modelling and analytical techniques are dealt with elsewhere in this volume by Lebreton and Clobert.

If models could be constructed which would accurately describe the population dynamics of threatened bird species they would be of great value to conservationists. It would be possible to quantify the effects of past changes in productivity or survival on population size and also to predict the stages of the life cycle at which conservation action could be most effectively directed.

To illustrate this point we consider as an example the Tawny Owl *Strix aluco*, a strictly territorial, woodland predator of small mammals studied near Oxford, UK over the period 1947–59 (Southern 1970). Southern observed that the numbers of young owls fledged per pair varied widely between years and that this variation was related to fluctuations in the abundance of small woodland rodents. However, the proportion of young birds that was recruited to the territorial population, in their first autumn, changed from year to year in such a way as to compensate for the fluctuations in production. Variation in the number of recruits to the territorial population was almost entirely determined by variation in the number of territory-holding adults that had died during the previous year and not by the fluctuations in the availability of potential recruits. As a result the population density of adult owls was very stable over the entire period of the study.

Had the Tawny Owl been a threatened species, conservationists who observed the great variability in breeding success and complete failure to breed in some years might have concluded that measures to increase the number of young fledged, such as captive breeding or supplemental feeding during the breeding season, would be appropriate methods to increase the population. However, Southern's analysis would have shown that, provided productivity remained above a threshold, any further increase in the number of fledged young would have no effect on the size of the adult population. It would also have suggested that the determinants of the ceiling population level or carrying capacity should be investigated. In fact further research showed that stable owl population densities varied five-fold between types of woodland and that this was related to the average availability of prey (Hirons 1985). Therefore a conservation plan for the Tawny Owl would have included the management of existing woodland reserves to increase the abundance and availability of small rodents or the creation of new reserves with optimal characteristics for the birds.

If population models could be expanded to incorporate information on the costs of conservation action, an optimal conservation strategy could then be designed for optimum utilization of the funds available. However, the response of demographic parameters to conservation measures is rarely quantified in cost-benefit terms. Figure 29.10 shows the effect of employing wardens on the productivity of the threatened British population of the Stone Curlew *Burhinus oedicnemus*. The birds breed on semi-natural grassland and also on arable farmland where nests and young are frequently destroyed by farming operations. The wardens cooperate with

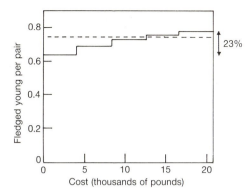

Fig. 29.10. Predicted mean productivity (fledged young per pair per season) for the entire population of Stone Curlews *Burhinus oedicnemus* in Britain (160 pairs) in relation to expenditure (at 1988 prices) on nest and chick protection by the employment of wardens. The maximum feasible expenditure is estimated to increase the number of young fledged by 23 per cent. The dashed line shows an estimate of the productivity required to maintain a stable population based on survival rate estimates for colour-ringed birds (R. E. Green, unpublished data), assuming that survival rates do not vary in relation to productivity.

farmers to protect nests and are thus able to increase Stone Curlew productivity on farmland to a level comparable with that for semi-natural grassland (Green 1988). These measures increase productivity sufficiently to maintain the population, but only at a substantial cost. Such an analysis allows the cost-effectiveness of wardening to be compared with that of alternative or complementary measures, such as the creation of semi-natural grassland reserves.

Unfortunately the use of simulation modelling as a diagnostic tool for the management of threatened species has serious drawbacks. Intensive fieldwork is required to establish fecundity and survival rates and this needs to take place over a long period in order to accumulate a sufficiently large sample (at least 10 years even for a bird with a short generation time). Estimates of any of the demographic parameters that would be required by the simplest population models have been published for only 58 threatened species and are available for only a tiny minority of threatened parrots and songbirds (Table 29.1, Appendix 29.2). For only 15 species are estimates of age at first breeding, overall productivity, and survival all available. For many of these parameters, and particularly for survival rates, there is only a single estimate of the average value with no indication of variation over time or between study areas.

Research funds and trained personnel are certainly not sufficient at present to carry out long-term population studies on more than a handful

Table 29.1 Numbers of threatened bird species for which estimates of demographic parameters have been published

Order	Demographic parameter									
	1	2	3	4	5	6	7	8	9	10
Sphenisciformes (3)	1	2	2	2	—	1	2	2	2	2
Procellariiformes (25)	—	1	3	3	—	—	1	3	—	1
Pelecaniformes (8)	1	2	3	3	1	—	1	4	2	1
Ciconiiformes (22)	1	2	1	1	1	1	—	2	—	—
Anseriiformes (21)	1	—	1	1	1	—	—	—	—	—
Falconiformes (44)	4	5	5	5	—	2	2	8	2	5
Galliformes (68)	—	—	1	—	—	—	—	—	—	—
Gruiiformes (52)	2	3	8	4	2	—	2	7	2	3
Charadriiformes (33)	2	2	4	4	2	2	—	6	2	2
Columbiformes (50)	1	—	1	1	—	—	—	1	—	—
Psittaciformes (71)	3	1	2	2	1	1	—	3	—	1
Strigiformes (21)	—	—	—	—	—	—	—	1	1	1
Coraciiformes (34)	1	—	—	—	—	1	—	1	1	—
Passeriformes (497)	5	4	10	9	1	3	3	4	3	5
Total species	22	22	41	35	9	11	11	42	15	21
% Threatened species	2.1	2.1	4.0	3.4	0.9	1.1	1.1	4.1	1.5	2.0

Note: The figures in brackets are the numbers of threatened species in each Order. Demographic parameters: 1, proportion of population that attempts to breed; 2, age at first breeding (mean or minimum in the wild); 3, hatching success (any method); 4, proportion of hatched young that fledge (any method); 5, Mayfield estimate of the proportion of eggs that give rise to fledged young; 6, frequency of re-nesting after breeding failure; 7, time to re-nesting after successful breeding; 8, overall estimate of productivity (young fledged per pair per season); 9, mean prebreeding dispersal distance; 10, annual survival rate (any method). See Appendix 29.3 for further details.

of the world's threatened bird species. Even if the necessary resources were to become available, the studies would take so long that many rare or declining species would be likely to vanish while under investigation. A possible alternative would be to carry out long-term population studies leading to simulation models of common species related to or otherwise representative of threatened birds. For example, the population model of the tawny owl, discussed above, might have been used to guide conservation measures for the threatened Spotted Owl *Strix occidentalis*. At present there are too few comparable simulation models of bird populations for it to be clear whether results from one species can be extrapolated to other ecologically similar species. It seems likely that species- or environment-specific factors would usually necessitate additional research, but population models for similar species might at least be used to suggest fruitful areas for short-term research on threatened birds.

Short-cut diagnostic methods

Temple (1985) suggested that the immediate causes of a population decline might be readily identified by measuring productivity and comparing it with the value expected for the species concerned. If productivity seems sufficiently high to balance the expected level of adult mortality then the cause of the decline is identified, by elimination, as low survival. Survival rates themselves are considered to be too difficult and time-consuming to be measured directly.

We have several criticisms of this approach. The first is that the method for obtaining an expected value for the productivity and survival rates of a species is not specified. We can only assume that parameter estimates for similar species with stable or increasing populations are to be used. While parameters of life history can be predicted to some extent from body weight and taxonomy, the errors of estimation are likely to be large for any individual species (Saethe 1988). A more fundamental criticism is that a bird species does not have fixed vital rates. There can be great differences in productivity and other demographic parameters between stable populations living in different environments. For example, Blondel (1985) found that Blue Tits *Parus caeruleus* and Coal Tits *P. ater* reared three times as many young per season in southern France as they did on the island of Corsica. However, tit populations in both areas are approximately stable.

Another problem is that many published studies of avian demography are carried out on species living on the mainland in the temperate zone, whereas many threatened birds live on islands and/or in the tropics. It would, therefore, often be difficult to decide whether to compare a threatened species with a congener from a different region or a distantly related species from a similar environment.

An example of the use of the approach advocated by Temple can be found in the diagnosis of the cause of the population decline of Kirtland's Warbler *Dendroica kirtlandii* between 1961 and 1971. Studies of nest success in the period 1931–71 indicated that productivity was low compared with other species of North American warblers, mainly because of brood parasitism by the Brown-headed Cowbird *Molothrus ater* (Mayfield 1975; Walkinshaw 1972, 1983). A cowbird control programme was begun in 1972 and about 3000 cowbirds per year were removed from Kirtland's warbler breeding areas. Rates of brood parasitism soon dropped to a low level and the average number of fledglings produced per nest tripled (Kelly and DeCapita 1982); however the number of singing males has not increased (Fig. 29.11). Although the cowbird control programme may have arrested a continuing decline to extinction (Mayfield 1975), it has clearly failed to enable the population to recover from the 60 per cent decline of the 1960s.

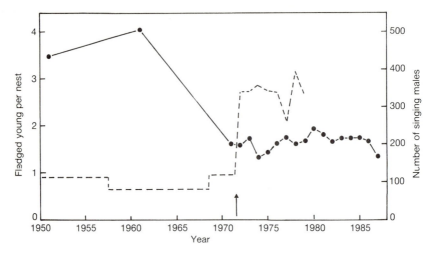

Fig. 29.11. Recent trends in nesting success (dashed line) and the world population of singing male Kirtland's warblers *Dendroica kirtlandii* (circles). The arrow indicates the beginning of a programme to remove brood-parasitic cowbirds *Molothrus ater* from the warblers' breeding grounds. Data from Ryel (1981) and Weinrich (1988).

Sensitivity analysis

Exploration of the properties of Leslie matrix models of animal populations with age-specific but time- and density-invariant fecundity and survival rates have revealed that the multiplication rate of a model population is most sensitive to variation in adult survival rate for long-lived species and most sensitive to variation in fecundity for short-lived species (Young 1968; Goodman 1971; Lebreton and Clobert, this volume). Sensitivity analysis would appear to offer a useful rule-of-thumb for conservationists in cases where average rates of survival and fecundity are known, even if the estimates are only approximate. For example, for a species with a high adult survival rate, conservation measures that increased adult survival would be expected to be more effective than measures that increased fecundity by the same amount.

There are important limitations on the usefulness of such an assessment. The first is that the sensitivity coefficients only establish the effect of a fixed absolute change in a parameter on the population multiplication rate. In devising a conservation plan it would be necessary to know the extent to which different demographic parameters could be influenced by management and the comparative costs of doing so. For example, survival rates of immature birds are typically substantially lower than those of adults. Suppose that the mean annual survival rate of adult birds in the population was 90 per cent, that 50 per cent of young females survived from fledgling to breeding age (5 years), and that the mean adult fecundity was 0.2 females per adult female per year. A sensitivity analysis by the method of Goodman (1971) shows that population multiplication rate is much more sensitive to a small change in adult survival rate than to a change of the same absolute magnitude in adult fecundity. The latter would in turn have a greater effect than the same change in the proportion of young surviving to breeding age (sensitivity coefficients 0.7937, 0.3571, and 0.1429 respectively).

In practice it might be expected that achievable modifications of demographic parameters might scale in proportion to the initial mortality and fecundity rates. Hence it might be possible to reduce a high mortality rate by a larger absolute amount than a low mortality rate. In the hypothetical example cited above, halving the adult mortality rate, halving the proportion of young that died before breeding age, and increasing adult fecundity to 50 per cent above its original value would all have a similar effect on the population multiplication rate.

Further complications arise if sensitivity analysis based on the assumption of constant values for demographic parameters is applied to populations which are in fact regulated by density-dependent processes. Elsewhere in this volume Johnson, Green, and Hirons examine the properties of a simple model of the Greater Flamingo *Phoenicopterus ruber* population of the western Mediterranean region. Sensitivity analysis using average values of adult and immature survival rates and fecundity indicates that population multiplication rate is most sensitive to variation in adult survival rate, as would be expected for a long-lived bird. However, the effect of changes in parameter values on equilibrium population size was examined for a model in which the number of adults that can breed is limited by the availability of suitable sites, as we believe to be the case. Survival rates were assumed to be density independent. It was then found that altering the availability of breeding sites has a much larger influence on equilibrium population size than would be expected from sensitivity analysis of a model without density dependence.

Hence, inferences drawn from the results of sensitivity analyses can be misleading and will not always indicate the best management options. They are best regarded as a guide for the allocation of sampling effort if the aim of measuring demographic parameters is to estimate the multiplication rate (Lande 1988).

Vacant habitat

If surveys of the distribution of a threatened bird species persistently show that a substantial proportion of the apparently suitable habitat is not occupied, it might be supposed that habitat availability is not limiting population size. Measures to create new areas of suitable habitat might then be predicted to have no effect and further research might focus on other factors. This argument might misdirect the research effort. It is likely to be difficult to identify accurately those attributes of a habitat that make it suitable, not just for survival, but also for successful breeding. The spatial pattern of different habitat components can be important in this. For example, sufficient foraging habitat may need to lie within a certain distance of a safe nesting or roosting place. If the habitat is fragmented the patches may be too small or widely dispersed to sustain a pair of birds. As a result, 'vacant' habitat may be unsuitable for the birds in ways that would be difficult to detect by superficial investigation.

This approach also assumes that birds are readily able to locate and occupy any suitable area from which they are not excluded by conspecifics. However, in territorial species, young birds dispersing in search of vacant habitat experience high mortality (Southern 1970; Hirons 1985). This may result from the difficulty of finding food and evading predators in unfamiliar surroundings made even more hostile by the presence of territorial conspecifics. This problem will be magnified if suitable habitat is fragmented, forcing dispersing birds to traverse unsuitable areas in their search for a vacant territory. Therefore it is likely that the area over which a dispersing bird is able to search before succumbing is limited. Under circumstances in which suitable unoccupied territories are scarce and dispersed it is also likely that some birds will fail to attract or locate mates.

Lande (1987, 1988, this volume) has developed models of a territorial species in which the capacity to search for vacant territories and mates is restricted. These models demonstrate that a substantial proportion of habitable territories can be vacant even in circumstances where an increase in the amount of habitat would lead to an increase in population and that birds can remain unmated because of the difficulty of finding one another. These problems increase as the proportion of the species' range covered by suitable habitat decreases. Thus in the Spotted Owl, where less than half of the apparently suitable habitat may be unoccupied in any one year (Allen *et al*. 1987), Lande's model predicts that increasing the proportion of the species' range covered by suitable habitat would increase both the owl population and the proportion of habitat occupied (Lande 1988).

The location of suitable habitat patches and mates is likely to be particularly difficult for long-distance migrants in which the young often do not return accurately to their natal area (Mead 1979; Buckland and Baillie 1987). Where the species' breeding range consists of large tracts of continuous suitable habitat, this will be of little consequence. However, for scarce species inhabiting a heavily fragmented habitat that forms only a tiny fraction of the geographical range, lack of natal philopatry may lead to difficulties in finding suitable habitat patches and/or mates.

Kirtland's Warbler, a long-distance migrant breeding in temporarily suitable patches of regenerating pine forest in Michigan, USA, typically occupies less than half the apparently suitable habitat in any one year (estimated at 38 per cent in 1980; Ryel 1981), and 15 per cent of singing males do not obtain mates (Probst 1986). Although availability of winter habitat in the Bahamas may limit the population, it has been suggested that difficulty in locating habitat patches and mates may also contribute to the rarity of the species (Mayfield 1983). This seems a reasonable suggestion since apparently occupiable habitat covers only 1–2 per cent of the total geographical range of the species, and isolated males not infrequently establish territories over 500 km outside this area (Ryel 1981).

The value of intraspecific comparisons

Another approach to investigating the cause of a population decline is to compare estimates of demographic parameters with similar estimates made at another time or place when a population of the same species was either stable or increasing. Differences in survival or fecundity associated with population declines are taken to indicate the mechanism of population change and correlations of rate of population change or vital rates with external variables are taken to indicate the causal factors.

An example of the use of this method can be found in the elucidation of the effects on raptor populations of contamination of their food chain by organochlorine pesticides. Poor breeding performance, in particular an unusually high rate of breakage of eggs, was noticed during the decline of raptors in Europe and North America in the 1950s and early 1960s (Newton 1979). Measurements of the shell thickness of raptor eggs in collections revealed that there had been a rapid and substantial reduction in shell thickness coinciding with the onset of the population declines and the beginning of generalized use of DDT. It was also observed that the decline in abundance of the Peregrine *Falco peregrinus* was greatest in parts of Britain where this pesticide was most frequently used in agriculture (Ratcliffe 1980). The inference that effects of pesticide residues on breeding success were the cause of the population decline was supported by toxicological studies of captive falcons.

A serious difficulty with the comparative approach is that it relies on inferences from correlation and association. As with any method involving correlation, misleading associations may occur by chance. If the biology of the species under study is well known it may be possible to identify and discount these spurious correlations. However, this is often not the case with threatened birds, for which there may be so many external variables which could conceivably have caused a decline that a mistaken diagnosis is almost inevitable. In such instances confidence in the correctness of the diagnosis would be reinforced if several lines of evidence pointed to the same causal mechanism for the decline.

These problems are well-illustrated by a long-term investigation of the causes of the decline of the Grey Partridge *Perdix perdix* in Britain. The population fell sharply in the early 1960s and has failed to recover. The use of chemicals for weed and pest control in agriculture was growing steadily during the most rapid phase of the decline and therefore seemed a likely cause. In this species variation in the proportion of chicks that survive from hatching to their first autumn is an important factor determining population changes from year to year (Jenkins 1961; Blank *et al.* 1967). It was known that insects associated with the weeds of cereal crops were the main food of partridge chicks (Southwood and Cross 1969). Hence the co-incidence of the decline in breeding populations and the increasing use of herbicides seemed to point to an indirect effect of these chemicals on partridge productivity via the food supply of the young. However, chick survival did not vary between the periods immediately before, during, and after the most rapid phase of the decline (Fig. 29.12; Potts 1980) a finding that is difficult to reconcile with this explanation.

Various chemicals came into widespread use in agriculture at about the same time as herbicides, including cyclodiene insecticides which caused large-scale direct mortality of partridges (Ash 1965). There have also been many changes in cropping and other agricultural practices which would be likely to affect demographic parameters other than productivity (O'Connor and Shrubb 1986). Had the decline in partridge population density been accompanied by a reduction in productivity then the association of both changes with the increased use of herbicides would have been strong evidence for a causal link. As this is not the case, it seems unlikely that much of the population decline was caused by effects of herbicides on productivity (Potts 1980). None the less, experimentally restricting herbicide use at the edges of cereal fields, where partridge broods spend much time foraging, has been shown to substantially increase both chick survival and the density of breeding pairs (Rands 1985, 1986).

In spite of the problems that we have identified in the application of comparative methods to the diagnosis of causes of population decline, we nevertheless suggest that this approach offers the most effective compro-mise between rigorous long-term analysis of population dynamics, which is excessively expensive in time and money, and short-cut methods which are unlikely to provide the correct diagnosis. The case of the Lord Howe Island Woodhen *Tricholimnas sylvestris* is an encouraging example of what can be achieved by the use of the comparative method (Fullager 1985; Miller and Mullette 1985). The woodhen is a flightless rail found

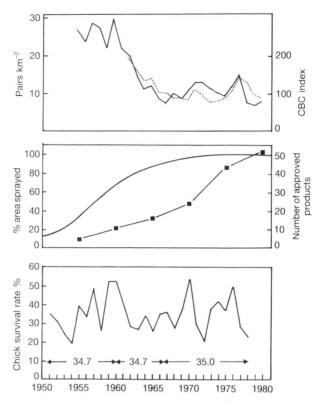

Fig. 29.12. Changes in the density of breeding populations of Grey Partridge *Perdix perdix* on farmland in Britain (upper diagram: full line, density from Game Conservancy National Game Census redrawn from Potts 1983; dotted line, British Trust for Ornithology Common Birds Census Index, for references see O'Connor and Shrubb 1986). A decline occurred coinciding with the increased use of herbicides in cereal crops—the main habitat during brood rearing (Green 1984). The middle diagram shows the estimated proportion of the cereal area treated with herbicides (line; from Woodford 1964) and the number of herbicides approved for use in cereals (squares; O'Connor and Shrubb 1986, Table 9.1). Indirect effects of herbicides on the food supply and hence the survival of partridge chicks are suggested. However, the estimated proportion of young that survived from hatching until the first autumn (lower diagram) showed no relationship with the prevalence of herbicide use (data from Potts 1980, Table 8). Mean chick survival rates are shown for the period of most rapid decline in relation to those immediately before and after the decline.

only on a small oceanic island between Australia and New Zealand. It was abundant all over the island in the 18th century, but was considered to be declining in numbers by 1853. By the time intensive research was begun in 1978 the total population was only 23 birds. The investigators mapped the distribution of woodhens, measured survival rates of individually marked birds, and were also able to measure productivity by making counts of surviving juveniles at the end of each breeding season. Studies of feeding behaviour showed that soil invertebrates were the main prey and surveys showed that food was plentiful in many areas from which woodhens were absent.

Introduced cats, rats, goats, and Masked Owls *Tyto novaehollandiae* were present on the island and feral pigs, which were introduced in the late 18th century, ranged over most of the low-lying areas. Significant numbers of woodhens were believed to remain only in moss forest near the summit of the island's highest mountain where they survived poorly and produced few young. However, in the late 1970s woodhens were discovered within a small area of lowland forest from which pigs were excluded by a steep ridge and which had not been surveyed before. Within this enclave, were three pairs of surviving woodhens that showed evidence of good breeding success.

Within 2 years of the onset of research the evidence was considered sufficient to indicate that predation by pigs was reducing both survival and breeding success by excluding the birds from the most favourable lowland habitat. A programme to eliminate pigs was started in 1979 and completed in 1981. Of the surviving eight pairs of woodhens, three were taken into captivity from the areas where survival and productivity had been poor. A successful captive breeding programme resulted in 85 birds being released into the wild in the period 1981–3. The first introductions were made into another area of lowland vegetation from which pigs were naturally excluded by cliffs. By 1984–5 it was estimated that there were 200 woodhens in the wild, distributed throughout their historical range. It was also known that captive-bred birds were breeding successfully in the wild. The key to this successful recovery was the demonstration, by research, that the woodhens thrived in the absence of pigs in spite of the continued presence of some of the other introduced predators, such as rats, which would have been more difficult to eliminate.

Discussion

The status of the majority of the world's threatened bird species is so inadequately known that no reliable assessment can be made of the extent of their endangerment. Many of these birds live in remote places that are rarely visited by ornithologists. However, even knowledge of accessible species depends to a surprising extent on anecdotes and guesswork rather than on survey methods whose accuracy can be assessed. The literature on bird census methodology is extremely voluminous (see Ralph and Scott 1981 and references therein), but this body of experience has had little influence on the design of surveys of the abundance of many threatened species. Additionally, the methods used to arrive at published estimates may often not be adequately documented.

For those species whose population size and distribution is known, it would be desirable to allocate scarce conservation resources on the basis of some rigorous estimate of the risk of extinction. Population viability analysis is designed to provide a framework for such assessments and to estimate the minimum level to which a population could be safely allowed to decline without incurring an unacceptably high risk of becoming extinct (Soule 1987). However, these methods are still being developed and as yet do not incorporate the impact of the long periods of adverse conditions commonly encountered in case histories of avian declines and extinctions.

Even when theories of species extinction are more fully developed their predictions should be used with caution. Estimates of extinction risk and minimum viable population size might eventually be used to allocate scarce resources for the purposes of conservation, but they can also be of immediate value, even when incorrect, to those persons whose economic interests conflict with proposals to maintain large tracts of undisturbed habitat for a threatened bird species. Conservation biologists need to be aware of the possible misuse of their findings as justification for habitat destruction.

We have already discussed the relative merits of various approaches to the diagnosis of causes of population decline and the identification of effective conservation measures. On balance, methods that exploit intraspecific comparison of population density, trends and demographic parameters are likely to be the most useful in practice. For all methods it is clear that the measurement of demographic parameters alone will seldom lead to proposals for successful conservation action. Information on environmental factors affecting the birds is also required and can only be properly interpreted if the ecological requirements of the species have been studied. Such studies also reduce the chance of misdiagnosis caused by spurious correlation when the comparative method is employed. In several cases where long-term, expensive efforts have been made to recover a declining population, a detailed but short-term study of the basic natural history of the species would have greatly strengthened the scientific basis of the conservation programme and saved a considerable amount of time.

It will often be necessary to implement conservation measures for a rapidly declining species when the available evidence that they are likely to be effective is still weak. In such cases adequate monitoring of the effects of the measures is particularly important.

Depressingly few success stories emerge from a review of the literature on the conservation of threatened birds. Even in these cases, success has often owed more to common sense and good fortune than to a rigorous scientific analysis of the species' problems. Several species that have been rescued are endemic to small islands and have been helped by the control of introduced predators or transfer to predator-free islands (Merton 1978; Flack 1978; Diamond 1984).

An increasing proportion of threatened species formerly inhabited large tracts of continuous mainland habitat, but are now restricted to small fragments of their historical range because of habitat destruction (Rands, this volume). Birds confined to habitat islands can be adversely influenced by new predators and competitors that colonize the modified surrounding areas (Wilcove *et al.* 1985). These effects are superficially similar to effects of introduced animals on endemic birds of oceanic islands and this parallel, indicates that similar conservation methods might be successful. However, mainland species in fragmented habitat are likely to be more difficult to manage than island birds because permanent eradication of predators and competitors from mainland habitat islands is unlikely to be feasible. Furthermore, the narrower ecological niche and greater habitat specificity of mainland species compared with similar island birds would be expected to make them more susceptible to the effects of new competitors and herbivores or other influences on their habitat.

A given amount of habitat degradation might have a greater direct impact on a mainland than an island species. Sensitivity to subtle habitat differences might also reduce the chance of successful transfers to predator-free areas outside the previous range. In mainland species that evolved in continuous habitats, individuals frequently disperse from natal areas before settling to breed. When confined to habitat fragments this may lead to mortality during dispersal, attempts to breed in unsuitable habitat, and failure to locate mates. These problems would be expected to occur more often in mainland than island species and to be particularly severe for long-distance migrants.

We conclude that the conservation of the increasing numbers of mainland species that are becoming threatened with extinction will require research of an unprecedented scale and effectiveness in order to diagnose adverse effects arising from habitat loss, fragmentation, and degradation and to devise effective countermeasures. Success in meeting this challenge is the yardstick by which the value of the new science of conservation biology should be judged.

References

Allen, H. L., Dixon, K., Knutson, K. L., and Potter, A. E. (1987). Progress Report No. 3. Cooperative administrative study to monitor Spotted Owl management areas in national forests in Washington. (unpublished) Washington Department of Game, Wildlife Management Division, Olympia, Washington.

Ash, J. S. (1965). Toxic chemicals and wildlife in Britain. *Proceedings VIth International Congress of Game Biology*, pp. 379–88.

Banko, W. E. (1960). *The Trumpeter Swan*. North American Fauna, No. 63. Fish and Wildlife Service, Washington DC.

Bartonek, J. C., Blandin, W. W., Gamble, K. E., and Miller, H. W. (1982). Number of Swans wintering in the United States. In *Proceedings Second International Swan Symposium*, (ed. G. V. T. Mathews and M. Smart), pp. 19–25. IWRB, Slimbridge.

Belovsky, G. E. (1987). Extinction models and mammalian persistence. In *Viable populations for conservation*, (ed. M. E. Soule), pp. 35–58. Cambridge University Press.

Blank, T. H., Southwood, T. R. E., and Cross, D. J. (1967). The ecology of the Partridge I. Outline of population processes with particular reference to chick mortality and nest density. *Journal of Animal Ecology*, **36**, 549–56.

Blondel, J. (1985). Breeding strategies of the Blue Tit and Coal Tit (*Parus*) in mainland and island Mediterranean habitats: a comparison. *Journal of Animal Ecology*, **54**, 531–56.

den Boer, P. J. and Gradwell, G. R. (ed.) (1971). *Dynamics of populations*. Pudoc, Wageningen.

Buckland, S. T. and Baillie, S. R. (1987). Estimating bird survival rates from organised mist-netting programmes. *Acta Ornithologica*, **23**, 89–100.

Collar, N. J. and Andrew, P. (1988). *Birds to watch: a checklist of the world's threatened birds*. ICBP, Cambridge.

Collar, N. J. and Stuart, S. N. (1985). *Threatened birds of Africa and related islands*. ICBP & IUCN, Cambridge.

Diamond, J. M. (1984). Back from the brink of extinction. *Nature*, **309**, 308.

Engbring, J. and Pratt, H. D. (1985). Endangered birds in Micronesia: their history, status and future prospects. *Bird Conservation*, **2**, 71–105.

Ewens, W. J., Brockwell, P. J., Gani, J. M., and Resnick, S. I. (1987). Minimum viable population size in the presence of catastrophes. In *Viable populations for conservation*, (ed. M. E. Soule), pp. 59–68. Cambridge University Press.

Flack, J. A. D. (1978). Interisland transfers of New Zealand Black Robins. In *Endangered birds: Management techniques for preserving endangered species*, (ed. S. A. Temple), pp. 365–72. University of Wisconsin Press, Madison, Wisconsin.

Fullagar, P. J. (1985). The Woodhens of Lord Howe Island. *Aviculture Magazine*, **91**, 15–30.

Goodman, D. (1987). The demography of chance extinction. In *Viable populations for conservation*, (ed. M. E. Soule), pp. 11–34. Cambridge University Press.

Goodman, L. A. (1971). On the sensitivity of the intrinsic growth rate to changes in the age-specific birth and death rates. *Theoretical Population Biology*, **2**, 339–54.

Green, N. (1985). The Bald Eagle. In *Audubon Wildlife Report 1985*, (eds A. S. Eno and R. L. Di Silvestro), pp. 509–31. National Audubon Society, New York.

Green, R. E. (1984). The feeding ecology and survival of Partridge chicks (*Alectoris rufa* and *Perdix perdix*) on arable farmland in East Anglia. *Journal of Applied Ecology*, **21**, 817–30.

Green, R. E. (1988). Stone-Curlew conservation. *RSPB Conservation Review*, **2**, 30–3.

Green, R. E. and Hirons, G. J. M. (1985). The population dynamics of Camargue Flamingos. (unpublished) Station Biologique de la Tour du Valat, Le Sambuc, 13200 Arles.

Harris, R., Clark, T. W., and Shaffer, M. (1988). Estimating extinction probabilities for the black-footed ferret. In *Reproductive biology of black-footed ferrets and small population biology as they relate to conservation*, (ed. U. S. Seal), Yale University Press.

Hirons, G. J. M. (1985). The effects of territorial behaviour on the stability and dispersion of Tawny Owl (*Strix aluco*) populations. *Journal of Zoology, London* Series B, **1**, 21–48.

Jenkins, D. (1961). Population control in protected partridges (*Perdix perdix*). *Journal of Animal Ecology*, **30**, 235–58.

Jones, C. G. (1987). The larger land birds of Mauritius. In *Studies of Mascarene Island Birds*, (ed. A. W. Diamond), pp. 208–300. Cambridge University Press.

Kelly, S. T. and DeCapita, M. E. (1982). Cowbird control and its effect on Kirtland's Warbler reproductive success. *Wilson Bulletin*, **94**, 363–5.

Klomp, H. and Woldendorp, J. W. (ed.) (1980). The integrated study of bird populations. *Ardea*, **68**, 1–255.

Koford, C. B. (1966). *The California Condor*. Dover, New York.

Lack, D. (1966). *Population studies of birds*. Oxford University Press.

Lande, R. (1987). Extinction thresholds in demographic models of territorial populations. *American Naturalist*, **130**, 624–35.

Lande, R. (1988). Demographic models of the Spotted Owl (*Strix occidentalis caurina*). *Oecologia*, **75**, 601–7.

Mayfield, H. F. (1975). The numbers of Kirtland's Warbler. *Jack-pine Warbler*, **53**, 39–47.

Mayfield, H. F. (1983). Kirtland's Warbler, victim of its own rarity? *Auk*, **100**, 974–6.

Maynard Smith, J. (1974). *Models in ecology*. Cambridge University Press.

MacPherson, S. L. and Tilt, W. C. (1987). *Amazona vittata*. A status report for 1987. *Parrotletter* **1**, 8–9.

Mead, C. J. (1979). Colony fidelity and interchange in the sand martin. *Bird Study*, **26**, 99–106.

Merton, D. V. (1978). Controlling introduced predators and competitors on islands. In *Endangered birds: Management techniques for preserving endangered species*, (ed. S. A. Temple), pp. 121–30. University of Wisconsin Press, Madison, Wisconsin.

Miller, A. H., McMillan, I. I., and McMillan, E. (1965). *The current status and welfare of the California condor*. National Audubon Society Research Report **6**.

Miller, B. and Mullette, K. J. (1985). Rehabilitation of an endangered Australian bird: The Lord Howe Island Woodhen *Tricholimnas sylvestris* (Sclater). *Biology and Conservation*, **34**, 55–95.

Mountfort, G. (1988). *Rare birds of the world*. Collins, London.

Newton, I. (1979). *Population ecology of raptors*. Poyser, Calton.

O'Connor, R. J. and Shrubb, M. (1986). *Farming and Birds*. Cambridge University Press.

Ogden, J. C. (1985). The California Condor. *Audubon Wildlife Report*, 1985, pp. 388–99. National Audubon Society, New York.

Potts, G. R. (1980). The effects of modern agriculture, nest predation and game management on the population ecology of partridges (*Perdix perdix* and *Alectoris rufa*). *Advances in Ecological Research*, **11**, 1–82.

Potts, G. R. (1983). The Grey Partridge situation. *Game Conservancy Annual Review*, **14**, 24–30.

Probst, J. R. (1986). A review of factors limiting the Kirtland's Warbler on its breeding grounds. *American Midlands Naturalist*, **116**, 87–100.

Ralph, C. J. and Scott, J. M. (1981). Estimating numbers of terrestrial birds. *Studies in Avian Biology*, **6**.

Rands, M. R. W. (1985). Pesticide use on cereals and the survival of Grey Partridge chicks: a field experiment. *Journal of Applied Ecology*, **22**, 49–54.

Rands, M. R. W. (1986). Unsprayed headlands: the answer for gamebirds? *Game Conservancy Annual Review*, **17**, 56–60.

Rands, M. (1989). Saving the Seychelles Brush Warbler. *Oryx*, **23**, 3–4.

Ratcliffe, D. A. (1980). *The Peregrine Falcon*. Poyser, Calton.

Ratcliffe, D. A. (1984). The Peregrine breeding population of the United Kingdom in 1981. *Bird Study*, **31**, 1–18.

Ryel, L. A. (1981). Population change in the Kirtland's Warbler. *Jack-pine Warbler*, **59**, 76–91.

Saethe, B.-E. (1988). Pattern of covariation between life-history traits of European birds. *Nature*, **331**, 616–17.

Shaffer, M. L. (1983). Determining minimum viable population sizes for the grizzly bear. *International Conference Bear Research Management*, **5**, 133–9.

Snyder, N. F. R., Wiley, J. W. and Kepler, C. B. (1987). *The parrots of Luquillo: natural history and consrvation of the Puerto Rican Parrot*. Western Foundation of Vertebrate Zoology, Los Angeles.

Soule, M. E. (1987). Where do we go from here? In *Viable populations for conservation*, (ed. M. E. Soule), pp. 175–84. Cambridge University Press.

Southern, H. N. (1970). The natural control of a population of Tawny Owls (*Strix aluco*). *Journal of Zoology, London*, **162**, 197–285.

Southwood, T. R. E. and Cross, D. J. (1969). The ecology of the Partridge. III. Breeding success and the abundance of insects in natural habitats. *Journal of Animal Ecology*, **38**, 497–509.

Temple, S. A. (1985). The problem of avian extinctions. *Current Ornithology*, **3**, 453–85.

Varley, G. C. and Gradwell, G. R. (1960). Key factors in population studies. *Journal of Animal Ecology*, **37**, 25–41.

Walkinshaw, L. H. (1972). Kirtland's Warbler—endangered. *American Birds*, **26**, 3–9.

Walkinshaw, L. H. (1983). *Kirtland's Warbler, the natural history of an endangered species*. Cranbrook Institute of Science, Bloomfield Hills, Michigan.

Weinrich, J. A. (1988). Status of the Kirtlands Warbler 1987. *Jack-Pine Warbler*, **66**, 155–8.

Wilcove, D. S., McLellan, C. H., and Dobson, A. P. (1985). Habitat fragmentation in the temperate zone. In *Conservation biology*, (ed. M. E. Soule), pp. 237–56. Sinauer, Sunderland.

Wingate, D. B. (1978). Excluding competitors from Bermuda Petrel nesting burrows. In *Endangered birds: Management techniques for preserving threatened species*, (ed. S. A. Temple). University of Wisconsin, Madison, Wisconsin.

Woodford, E. K. (1964). Weed control in arable crops. *Proceedings 7th British Weed Control Conference*, **3**, 944–62.

Wurster, C. F. Jr. and Wingate, D. B. (1968). DDT residues and declining reproduction in the Bermuda Petrel. *Science*, **159**, 979–81.

Young, H. (1968). A consideration of insecticide effects on hypothetical avian populations. *Ecology*, **49**, 991–3.

Appendix 29.1

Case histories of some declining bird species (see Fig. 29.8)

Trumpeter Swan *Cygnus cygnus buccinator* in Montana, Idaho, and Wyoming. Decline in early 1900s attributed to hunting, habitat destruction, and recreation. National Parks Service began research in 1929, organized autumn counts in the three states from 1931 and promoted the welfare of swans by various methods at the most important sites. Red Rocks Lake Refuge established in 1935; artificial feeding within the Refuge initiated in 1936–7 to encourage swans to remain in an unfrozen area safe from hunting (Banko 1960; Bartonek *et al.* 1982).

Chatham Island Black Robin *Petroica traversi*. Originally occurred on four islands in the Chatham group, but had disappeared from two by 1871 and from a third (Mangere) following the introduction of cats in the late 19th century. In 1937, 20–35 pairs remained on Little Mangere and a similar number was present in the 1950s. Following habitat changes the population declined to less than 20 individuals by 1968. Research started in 1972 diagnosed the cause of the decline as poor survival of adults during the breeding season, low reproductive output, and low survival of immatures after autumn due to poor habitat. In 1973 new habitat was planted on Mangere, by now cleared of cats. By 1976 only 7 birds remained on Little Mangere, and these were all transferred to Mangere. Since 1981 eggs have been successfully cross-fostered to Chatham Island Tomtits *Petroica macrocephala* on Rangatura Island. The total population on the two islands in 1985/6 was 38 (Flack 1978; Mountfort 1988).

Puerto Rican Amazon Parrot *Amazona vittata*. A rapid decline from about 100 000 birds in the late 19th century to about 2000 in the late 1930s coincided with the removal of primary forest. Initial research (1953–6) estimated 200 birds and added nest robbing by Man, shooting, and disturbance from recreation as reasons for decline. Research restarted in 1968. A continuing decline was attributed (1973) to lack of nest sites, low nesting success caused by competition with Pearly-eyed Thrashers *Margarops fuscatus* for nest holes, and parasitism of nestlings by warble flies. Captive propagation to compensate for poor breeding began in 1969 and the first releases were made in 1985. Nest boxes were provided to divert thrashers and interventions were made against warble flies from 1973. Recovery Plan drafted in 1982. The slow recovery rate after implementation of conservation measures was attributed (1987) to high mortality of juveniles and non-breeders and failure to breed by naive birds (Snyder *et al.* 1987; MacPherson and Tilt 1987).

Peregrine falcon *Falco peregrinus* in Britain. By 1955 the population had recovered from a decline caused by official control measures to protect military carrier pigeons during 1939–45. Research showed that the population was 68 per cent of pre-war levels in 1961, declining to 44 per cent by 1963–4. The decline was attributed to unusually heavy mortality of adults and reduced reproductive success from about 1956 onwards caused by persistent residues of organochlorine pesticides. The first report of the effect of pesticides was published in 1961, but the effect on eggshell thickness was not demonstrated until 1967. Numbers gradually recovered to pre-war levels following progressive restrictions on the use of pesticides introduced in 1962, 1965, and 1975 (Ratcliffe 1980, 1984).

Bald Eagle *Haliaetus leucocephalus* in USA. The number of nesting birds was first counted in 1962–3. Retrospective analysis suggested a sudden, catastrophic decline in breeding success beginning around 1947. Research (started 1968) attributed this to DDT residues. The use of DDT in the USA was banned in 1972 (Green 1985).

Cahow *Pterodroma cahow*. A rapid decline in the 17th century was certainly caused by overharvesting by Man. The species was believed extinct until rediscovered in 1951. Low breeding success was attributed to competition for nest sites with White-tailed Tropicbirds *Phaethon lepturus* (1954). Baffles were fitted to nest holes to exclude tropicbirds (1961). The population remained low (20 pairs in 1966) and reproductive success declined, apparently because of DDT contamination. The use of DDT in the USA was banned in 1972. By 1985 breeding success had trebled (35 pairs fledged 21 young) (Collar and Andrew 1988; Wurster and Wingate 1968; Wingate 1978).

Mauritius Kestrel *Falco punctatus*. Initial decline was caused by a 99 per cent reduction in the area of native forest; there were thought to be less than 10 pairs in 1968. A sharp decrease in numbers in the late 1950s was probably caused by organochlorine pesticide contamination. Field research and a captive breeding programme began in 1973. The main threats were diagnosed as continuing forest destruction, persecution by humans, and nest predation by monkeys. Slow failure to respond to remedial action was attributed to habitat degradation caused by the spread of exotic vegetation which reduces the availability of prey. Supplemental feeding and egg manipulation experiments to increase productivity began in 1984/5. (Collar and Stuart 1985; Jones 1987).

Seychelles Warbler *Acrocephalus sechellensis*. Confined to Cousin Island (27 ha); believed to be in no danger in 1953, but only 30 birds were counted in 1959. Research showed that territories in scrub were smaller than those in other habitats, indicating that this was the most suitable habitat. Management was undertaken from 1968 to remove coconut palms and encourage the spread of native scrub (*Pisonia*). Numbers increased until 1975, then levelled off and fell sharply in 1982. After the decline further research diagnosed the cause as suboptimality of mature *Pisonia* woodland as warbler habitat since territory sizes in it were larger than in scrub. Later research showed that breeding success was highest in mature *Pisonia* woodland, so these two criteria conflict as to the best habitat. The current population is considered to be the largest that the island can support, so some birds were successfully transferred to another island in 1988 where they have bred successfully (Collar and Stuart 1985; Rands 1989).

Appendix 29.2 Population estimates of threatened birds

	Zoogeo. region	Estimate	Popn. compnt.	Range	Year	Method	Prev. est.	Remarks refs.
Apterygiformes								
Little Spotted Kiwi *Apteryx owenii*	6	500–600	A	A	(1981)	N		
Sphenisciformes								
Yellow-eyed Penguin *Megadyptes antipodes*	6	1200–1800	BP	A	1981	C2		
Jackass Penguin *Spheniscus demersus*	4	134000	BA	A	1978	B2	A	
Peruvian Penguin *S. humboldti*	3	12100–15000	BA	A	1981; 83	B2	B	
Podicipediformes								
Atitlan Grebe *Podilymbus gigas*	3	210	A	A	1973	B2		extinct? 1980
Columbian Grebe *P. andinus*	3	300	A	A	1968	B2		extinct? 1977
Junin Grebe *P. taczanowskii*	3	c.100	A	A	1987	B2	A	
Hooded Grebe *P. gallardoi*	3	5000+	A	A	1984	D		
Procellariiformes								
Amsterdam Albatross *Diomedia amsterdamensis*	4b	30–50	A	A	1983	(B2)*	A	
Short-tailed Albatross *D. albatrus*	1	146	BA	A	1986	B2	A	
Cahow *Pterodroma cahow*	2	35	BP	A	1985	B2	A	
Magenta Petrel *P. magentae*	6	50–100	A	A	1985/86	B1		
Gon-gon *P. feae*	4b	several 100	BP	A	1960s; 81	E		
Freira *P. madeira*	4b	20	BP	A	1987	D	B	
Dark-rumped Petrel *P. phaeopygia*	7	c.3500	BP	B	1981; 85	D	B	
Cook's Petrel *P. cooki*	6	10000–50000	BP	A	(1982)	D		
Chatham Island Petrel *P. axillaris*	6	<500	A	A	(1981)	D		
Pycroft's Petrel *P. pycrofti*	6	<1000	BP	A	(1982)	D		
Black Petrel *Procellaria parkinsoni*	6	550–1100	BP	A	1978; 82	D	B	
Westland Black Petrel *P. westlandica*	6	1000–5000	BP	A	1982	C2	A	1
Newell's Shearwater *Puffinus newelli*	7	4000–6000	BP	A	(1982)	E		
Pelecaniformes								
Spot-billed Pelican *Pelecanus philippensis*	5	1300	BP	B	1982; (84)	B2		
Dalmatian Pelican *P. crispus*	1	514–1368	BP	B	(1987)	B2	B	
Abbott's Booby *Sula abbotti*	5	<2000	BP	A	1983–86	B2	A	
New Zealand King Cormorant *Phalacrocorax carunculatus*	6	<7000	BP	A	(1982)	D		

	Species	Population			Date			Notes
1	Pygmy Cormorant *Halietor pygmeus*	18250–21000	BP	B	1967–87	B2		
3	Galapagos Flightless Cormorant *Nannopterum harrisi*	800–1000	A	A	1986	B2	A	
4b	Ascencion Frigatebird *Fregata aquila*	8000–10000	BA	A	1957–59	B2		2
5	Christmas Frigatebird *F. andrewsi*	1620	BP	A	1984	B2	A	3
Ciconiiformes								
1	Chinese Egret *Egretta eulophotes*	270	BP	A	(1988)	B2		
5	Milky Stork *Mycteria cinerea*	6000+	A	B	1984	D		
1	Oriental White Stork *Ciconia boyciana*	c.3000	A	A	1985	B2		
5	Lesser Adjutant *Leptoptilos javanicus*	1100–1450	BA	B	1982–86	B2		
1	Northern Bald Ibis *Geronticus eremita*	c.400	A	A	1982; 88	A	A	
4	Southern Bald Ibis *G. calvus*	4600+	BA	A	(1985)	D	B	
1	Crested Ibis *Nipponia nippon*	40	A	A	(1987)	B2	A	
3	Andean Flamingo *Phoenicoparrus andinus*	150000	A	A	(1973)	E		
3	Puna Flamingo *P. jamesi*	50000	A	A	(1973)	D		
Anseriformes								
7	Hawaiian Goose *Branta sandvicensis*	390	A	A	1980	C4	A	
1	Red-breasted Goose *B. ruficollis*	27500	A	A	(1984)	N		
6	Freckled Duck *Stictonetta naevosa*	<19000	A	A	1983	D		
6	New Zealand Brown Teal *Anas aucklandica*	1500	A	B	1970s; 82/83	N; B1		
7	Hawaiian Duck *Anas wyvilliana*	3000	A	A	(1967)	E		
7	Laysan Duck *A. laysanensis*	400	A	A	1984	B1	A	
1	White-headed Duck *Oxyura leucocephala*	10920	A	B	1980s	B2	A	
Falconiformes								
2	California Condor *Gymnogyps californianus*	16	A	A	1984	A	A	extinct in wild 1985
1	Red Kite *Milvus milvus*	5500–15000	BP	A	(1986)	N	<B	
4a	Madagascar Fish Eagle *Haliaeetus vociferiodes*	40	BP	A	1982–86	B2		
1	White-tailed Eagle *H. albicilla*	2850–2925	BP	B	(1983; 86)	N		excludes China
1	Steller's Sea Eagle *H. pelagicus*	6000–7000	A	A	1985/86	C2		
4	Cape Vulture *Gyps coprotheres*	3700	BP	A	1983	B2		
3	Galapagos Hawk *Buteo galapagoensis*	130	BD	A	1981	B2	A	
7	Hawaiian Hawk *B. solitarius*	1400–2500	A	A	(1984)	N		
1	Spanish Imperial Eagle *Aquila adalberti*	105–106	BP	A	1981–86	B2		
1	Lesser Kestrel *Falco naumanni*	60000	BP	B	(1986)	N		
4b	Mauritius Kestrel *F. punctatus*	19–25	A	A	1985/86	B2	A	west Palaearctic only
Galliformes								
7	Niuafo'ou Megapode *Megapodius pritchardii*	200–400	A	A	1980s	N		

Appendix 29.2 *cont.*

	Zoogeo. region	Estimate	Popn. compnt.	Range	Year	Method	Prev. est.	Remarks refs.
Maleo *Macrocephalon maleo*	5	3000	BA	B	1970s	N		
Djibouti Francolin *Francolinus ochropectus*	4	1500	A	A	1984/85	N	A	
Gruiformes								
Black-necked Crane *Grus nigricollis*	1	1400–1500	A	A	(1988)	C2		
Hooded Crane *G. monacha*	1	6100–6500	A	A	1983–87	B2	B	
Red-crowned Crane *G. japonensis*	1	c.1450	A	A	(1988)	C2	B	
Whooping Crane *G. americana*	2	154	A	A	1987	A	A	
White-naped Crane *G. vipio*	1	3000–3500	A	A	(1988)	C2	A	
Siberian Crane *G. leucogeranus*	1	1857	A	A	1983; 87	A		
Wattled Crane *Bugeranus carunculatus*	4	6000–7500	A	A	1978–83	D		
Okinawa Rail *Rallus okinawae*	5	>1500	A	A	1983	E		
Guam Rail *R. oustoni*	7	<100	A	A	1983	N	A	
Inaccessible Rail *Atlantisa rogersi*	4b	10000	A	A	1982/83	C2		
Lord Howe I. Woodhen *Tricholimnas sylvestris*	6	200	A	A	1984	C2	A	
Gough Moorhen *Gallinula comeri*	4b	2170–3260	P	A	1973; 83	C2		
Takahe *Notornis mantelli*	6	180	A	A	(1988)	B1/C2	A	
Kagu *Rhynochetos jubatus*	6	500–1000	A	A	(1986)	N		
Great Bustard *Otis tarda*	1	16654–24689	A	A	(1985; 87)	E/N	B	
Great Indian Bustard *Ardeotis nigriceps*	5	1000–1500	A	A	(1987)	N		
Charadriiformes								
Chatham Island Oystercatcher *Haematopus chathamensis*	6	80	A	A	(1988)	N	A	
Piping Plover *Charadrius melodus*	2	3535–4147	BA	A	1977–84	B2		
St. Helena Plover *C. sanctaehelenae*	4b	200–300	A	A	1984	C2		
Hooded Plover *C. rubricollis*	6	>1800	A	A	early 80s	C2		
N.Z. Shore Plover *Thinornis novaeseelandiae*	6	140	A	A	1982	B1/B2	A	
Black Stilt *Himantopus novaeseelandiae*	6	c.50	A	A	(1987)	B2	A	
Slender-billed Curlew *Numenius tenuirostris*	1	<1000	A	A	(1988)	N		
Spotted Greenshank *Tringa guttifer*	1	<1000	A	A	(1988)	E		
Spoon-billed Sandpiper *Eurynorhynchus pygmeus*	1	2000–2800	BP	A	(1983)	E		

Species							
White-eyed Gull *Larus leucophthalmus*	1	5000–6500	BP	A	1985	E	
Audouin's Gull *L. audouinii*	1	5500–6000	BP	A	(1986)	N	
Relict Gull *L. relictus*	1	2000	BP	B	(1984)	N	
Kerguelen Tern *Sterna virgata*	4b	198+	BP	B	(1982)	B2	
Black-fronted Tern *S. albostriata*	6	1000–5000	BP	A	(1982)	E	
Damara Tern *S. balaenorum*	4	1000–2000	BP	A	(1982)	C2	
Japanese Murrelet *Synthliboramphus wumizusume*	1	1650	BA	A	1975–82	E	
Columbiformes							
Madeira Laurel Pigeon *Columba trocaz*	4b	>1000	A	A	1985	C4	
Pink Pigeon *Nesoenas mayeri*	4b	5–6	BP	A	1988	B2	A
Grenada Dove *Leptotila wellsi*	3	105	A	A	1987	D	
Marquesas Ground-dove *Gallicolumba rubescens*	7	225	A	B	1987	D	
Tooth-billed Pigeon *Didunculus strigirostris*	7	4800–7200	A	A	1978–81	C2	
Rapa Fruit-dove *Ptilinopus huttoni*	7	30	A	A	1984	N	A
Society I. Imperial-pigeon *Ducula aurorae*	7	100–1000	A	A	1984	D	A
Marquesas Imperial-pigeon *D. galeata*	7	200–400	A	A	1975	D	A
Psittaciformes							
Scarlet-breasted Lorikeet *Vini kuhlii*	7	1200	A	B	1979	E	
Blue Lorikeet *V. peruviana*	7	580–680	P	B	(1988)	D	
Ultramarine Lorikeet *V. ultramarina*	7	335–405	P	A	1975; 87	D	B
Antipodes Parakeet *Cyanoramphus unicolor*	6	2000–3000	A	A	1978	D	
Orange-bellied Parrot *Neophema chrysogaster*	6	132	A	A	1987	B2	A
Mauritus Parakeet *Psittacula egues*	4b	8	A	A	1987	B2	A
Hyacinth Macaw *Anodorhynchos hyacinthinus*	3	2500–5000	A	A	mid 80s	E	
Indigo Macaw *A. leari*	3	60–200	A	A	1983	D	
Blue-throated Macaw *Ara glaucogularis*	3	500–1000	A	A	1981/82	N	
Red-fronted Macaw *A. rubrogenys*	3	<5000	A	A	1981/82	N	
Maroon-fronted Parrot *Rhyncopsitta terrisi*	2	c.2000	A	A	1975–77	C4	
Rufous-fronted Parakeet *Bolborhynchus ferrugineifrons*	3	1000–2000	A	A	1986	A	A
Puerto Rican Amazon *Amazona vittata*	3	41	A	A	1987	A	
Yellow-shouldered Amazon *A. barbadensis*	3	500–550	A	B	(1988)	N	
St. Lucia Amazon *A. versicolor*	3	250	A	A	1986	D	A
Red-necked Amazon *A. arausiaca*	3	200–250	A	A	1988	C2	A
St. Vincent Amazon *A. guildingii*	3	420	A	A	1982	D	A
Imperial Parrot *A. imperialis*	3	60	A	A	1988	C2	A
Kakapo *Strigops habroptilus*	6	50	A	A	1987	B1/A	A

Appendix 29.2 *cont.*

	Zoogeo. region	Estimate	Popn. compnt.	Range	Year	Method	Prev. est.	Remarks refs.
Strigiformes								
Sokoke Scops Owl *Otus ireneae*	4	1000	P	A	1983	C2	A	
Grand Comoro Scops Owl *O. pauliani*	4b	several 10s	P	A	1985	D		
Seychelles Scops Owl *O. insularis*	4b	80	P	A	1975–76	C2		
Usambara Eagle Owl *Bubo vosseleri*	4	200–1000	A	A	(1985)	E		
Blakiston's Fish Owl *Ketupa blakistoni*	1	325–425	P	B	1984; (84)	B2; E		excludes China
Spotted Owl *Strix occidentalis*	2	4000–6000	A	B	mid 80s	D		
Caprimulgiformes								
Puerto Rican Whippoorwill *Caprimulgus noctitherus*	3	400–500	P	A	1971	C2		
Apodiformes								
Seychelles Swiftlet *Collocalia elaphra*	4b	<1000	A	A	(1984)	D		
Tahiti swiftlet *Aedromas leucophaeus*	7	200–500	A	A	(1988)	E		
Juan Fernandez Firecrown *Sephanoides fernandensis*	3	200–500	A	A	1987	D		
Coraciiformes								
Tuamoto Kingfisher *Halcyon gambieri*	7	400–600	A	A	1974	N		
Marquesas Kingfisher *H. godeffroyi*	7	350–550	P	A	1974	N		
Narcondam Hornbill *Aceros narcondami*	5	400	A	A	1972	E	A	
Red-cockaded Woodpecker *Picoides borealis*	2	3000–4400	A	A	(1984)	N	B	
Okinawa Woodpecker *Sapheopipo noguchii*	5	20–60	P	A	1973	E		
Passeriformes								
Masafuera Rayadito *Aphrastura masafuerae*	3	500–1000	A	A	1986	D		
White-browed Tit-spinetail *Leptasthenura xenothorax*	3	25	P	B	1987	N		
Apurimac Spinetail *Synallaxis courseni*	3	250–300	P	B	1983	D		
Gurney's Pitta *Pitta gurneyi*	5	>16	P	A	1987	N		
Rufous Scrubbird *Atrichornis rufescens*	6	172	P	A	1980	B3		
Noisy Scrubbird *A. clamosus*	6	156	C	A	1983	B3	A	
Raso Lark *Alauda razae*	4b	>150	A	A	1985	D	A	
Sokoke Pipit *Anthus sokokensis*	4	2000	P	B	1983	D	B	
Mauritius Cuckoo-shrike *Coracina typica*	4b	210–220	P	A	1974/75	C3		
Reunion Cuckoo-shrike *C. newtoni*	4b	100–150	P	A	1974	C3		

Species								
Mauritius Black Bulbul *Hypsipetes olivaceus*	4b	500	A	A	1974–75	C2		
White-breasted Thrasher *Ramphocinclus brachyurus*	3	60	P	B	1987	N	B	
Usambara Ground Robin *Dryocichloides montanus*	4	28000	A	A	(1981)	D		
Thyolo Alethe *Alethe choloensis*	4	1500	P	B	1983–84	D		
Seychelles Magpie-robin *Copsychus sechellarum*	4b	23	A	A	1987	A	A	
Kamao *Myadestes myadestinus*	7	24	A	A	1981	C4	A	
Olomao *M. lanaiensis*	7	19	A	A	1976–83	C4		
Puaiohi *M. palmeri*	7	20	A	A	1981	C4	A	
Fuerteventura Stonechat *Saxicola dacotiae*	4b	650–850	P	A	1985	C2		
Taita Thrush *Turdus helleri*	4	90–190	A	B	1985	D		
Aquatic Warbler *Acrocephalus paludicola*	1	1330–1540	P	B	(1988)	N		
Nihoa Reed Warbler *A. familiaris*	7	198	A	A	1973	C4	A	
Rodrigues Warbler *A. rodericanus*	4b	30–45	A	A	1983	C2	A	
Seychelles Warbler *A. sechellensis*	4b	430	A	A	1987	B2	A	
Grand Comoro Flycatcher *Humblotia flavirostris*	4b	several 1000	P	A	1985	D		
Chatham I. Black Robin *Petroica traversi*	6	38	A	A	1987/88	A	A	
Seychelles Paradise-flycatcher *Tersiphone corvina*	4b	72–75	A	A	1988	B2	A	
Rarotonga Monarch *Pomarea dimidiata*	7	24	A	A	1987	A	A	
Tahiti Monarch *P. nigra*	7	2	A	A	1985	A	A	extinct?
Marquesas Monarch *P. mendozae*	7	400–550	P	A	1975	E		
Iphis Monarch *P. iphis*	7	several 100	P	A	1975	E		
Fatu Iva Monarch *P. whitneyi*	7	several 100	P	A	1975	N		
Guam Flycatcher *Myiagra freycineti*	7	450	A	A	1981	C4		extinct 1983
Algerian Nuthatch *Sitta ledanti*	1	82	P	A	1982	C2		
Forty-spotted Pardalote *Pardalotus quadragintus*	6	850	A	B	1981	D		
Amani Sunbird *Anthreptes pallidigaster*	4	2900–4700	P	B	(1978)	E		
White-breasted White-eye *Zosterops albogularis*	6	50	A	A	1962	E		extinct?
Seychelles White-eye *Z. modestus*	4b	c.100	A	A	mid 70s	E		
Mauritius Olive White-eye *Z. chloronothus*	4b	275	P	A	1974/75	D		
Stitchbird *Notiomystis cincta*	6	4000–5000	A	A	(1985)	E		
Kauai Oo *Moho braccatus*	7	2	A	A	1981	C4	A	
Gough Bunting *Rowettia goughensis*	4b	200	P	A	1974	C2		
Tristan Bunting *Nesospiza acunhae*	4b	7120–9240	A	A	1970s; 82/83	E; D		
Grosbeak Bunting *N. wilkinsi*	4b	230	P	A	1974; 82/83	C2		
Mangrove Finch *Camarhynchus heliobates*	3	100–200	A	A	1974	N		
Kirtland's Warbler *Dendroica kirtlandii*	2	167	C	A	1987	B3	A	

Appendix 29.2 cont.

	Zoogeo. region	Estimate	Popn. compnt.	Range	Year	Method	Prev. est.	Remarks refs.
Akepa *Loxops coccineus*	7	15930	A	A	1976–83	C4		
Akikiki *Oreomystis bairdi*	7	1650	A	A	1981	C4	B	
Nukupuu *Hemignathus lucidus*	7	28	A	A	1980	C4		
Akiapolaau *H. munroi*	7	1500	A	A	early 80s	C4		
Maui Parrotbill *Pseudonestor xanthophrys*	7	500	A	A	1980	C4		
Ou *Psittirostra psittacea*	7	403	A	A	1977–79; 81	C4		
Nohia Finch *Telespyza ultima*	7	1318	A	A	1973	C4	A	
Laysan Finch *T. cantans*	7	15100	A	A	1984; (85)	C4	A	
Palila *Loxioides bailleui*	7	2020	A	A	1984	C4	A	
Poo Uli *Melamprosops phaeosoma*	7	140	A	A	early 80s	C4		
Crested Honeycreeper *Palmeria dolei*	7	3800	A	A	early 80s	C4		
Yellow-shouldered Blackbird *Agelaius xanthomus*	3	940–1220	A	A	1982	B2		
Red Siskin *Carduelis cucullata*	3	600–800	A	B	1981	D		
Clarke's Weaver *Ploceus golandi*	4	1000–2000	P	A	(1984)	E		
Mauritius Fody *Foudia rubra*	4b	250	P	A	1974/75	C3		
Seychelles Fody *F. sechellarum*	4b	750–1100	A	A	1964–65	D		
Rodrigues Fody *F. flavicans*	4b	60	P	A	1983	C3	A	
Rarotonga Starling *Aplonis cinerascens*	7	c.100	A	A	1984	E		
Bali Starling *Leucopsar rothschildi*	5	55	A	A	1988	B2	A	
Grand Comoro Drongo *Dicrurus fuscipennis*	4b	c.100	A	A	1985	E		
Saddleback *Creadion carunculatus*	6	2300	A	A	(1981)	N		
Marianas Crow *Corvus kubaryi*	7	1650	A	A	1981; 82	D		
Hawaiian Crow *C. hawaiiensis*	7	10	A	A	1985	A?	A	

Zoogeographic Regions—These are as adopted by Mountfort (1988) and numbered:

1 Palaearctic Region   2 Nearctic Region   3 Neotropical Region Afrotropical Region   4 Mainland Africa   4a Madagascar

4b Africa's related islands   5 Oriental Region   6 Australasian Region   7 Pacific Ocean Islands

Estimate—Most recent published estimate of population size in the wild.

Population Component—A—total individuals in population   BA—breeding adults   BP—breeding pairs or nests   P—pairs   C—singing males

D—breeding units

Range—A—estimate applies to the whole of the species' range   B—estimate applies to a delimited portion (> 50 per cent) of the species' range.

Year—The year in which the most recent published estimate was made. Where the exact year of the survey is unclear the date of the publication is given in brackets. In cases where the surveys upon which the estimate was made took place over a series of years, the beginning and end years of the series are separated by a hyphen. Where the estimate was made in a season encompassing parts of 2 years, these are separated by a slash. Where discrete populations of the species were surveyed in different years, the years are separated by a semi-colon.

Method of Estimation—A—inventory of the whole population   B—direct estimation of population size from: B1—mark/recapture information

B2—counts of individuals or nests   B3—mapping methods, e.g. Common Bird Census   C—estimation of population size by extrapolating from a census in a defined area(s) where the total range of the species was also mapped. Census methods: C1—mark/recapture   C2—counts of individuals or nests

C3—mapping techniques   C4—line transects   D—estimation of population size by extrapolation from a survey where either the relationship between the number counted and the number present was not known even for the survey area and/or where the total range of the species was unsurveyed and therefore had to be estimated   E—population size 'estimated' or guessed at   N—no method given

Previous Estimate—Indicates the availability of comparable estimates made 2 or more years previously for the whole (A) or > 50 per cent (B) of the species' range.

References—As given by Collar and Andrew (1988).

Appendix 29.3 Threatened species for which demographic parameters have been published

	Demographic parameter									
	1	2	3	4	5	6	7	8	9	10
Sphenisciformes										
Yellow-eyed Penguin	A	A	Y	Y	—	—	Y	A	B	CMR
Jackass Penguin	?	B	Y	Y	—	Y	Y	B	B	MR
Procellariiformes										
Short-tailed Albatross	—	—	—	—	—	—	—	A	—	—
Cahow	—	—	—	—	—	—	—	A	—	—
Dark-rumped Petrel	—	—	Y	Y	—	—	Y	A	—	MR
Pycroft's Petrel	—	—	Y	Y	—	—	—	—	—	—
Black Petrel	—	B	Y	Y	—	—	—	—	—	—
Pelecaniformes										
Dalmatian Pelican	—	B	Y	Y	—	—	—	A	B	—
Abbott's Booby	—	—	Y	Y	—	—	—	A	—	—
Galapagos Flightless Cormorant	B	A	Y	Y	1E	—	Y	A	U	CMR
Ascension Frigatebird	—	—	—	—	—	—	—	A	—	—
Ciconiiformes										
Northern Bald Ibis	B	B	—	—	—	—	—	A	—	—
Southern Bald Ibis	—	B	Y	Y	1E	Y	—	A	—	—
Anseriiformes										
Hawaiian Goose	—	—	—	Y	—	—	—	—	—	—
Laysan Duck	B	—	Y	—	1E	—	—	—	—	—
Falconiformes										
California Condor	B	—	Y	/	?	Y	Y	A	—	MR
Red Kite	B	A	Y	/	—	—	—	A	U	MR
White-tailed Eagle	—	B	—	—	—	—	—	A	—	NMG
Cape Vulture	—	B	—	—	—	—	—	A	—	CMR
Spanish Imperial Eagle	—	—	—	—	—	—	—	A	—	—
Galapagos Hawk	B	B	Y	Y	—	Y	Y	C	U?	MR
Lesser Kestrel	—	B	Y	Y	—	—	—	A	—	—
Mauritius Kestrel	B	—	Y	/	—	—	—	A	—	—
Galliformes										
Malleefowl	—	—	Y	—	—	—	—	—	—	—
Gruiiformes										
Black-necked Crane	—	—	Y	—	—	—	—	—	—	—
Red-crowned Crane	—	—	—	—	—	—	—	C	—	—
Whooping Crane	B	A	Y	Y	?	—	Y?	A	B?	CMR/I
White-naped Crane	—	—	—	—	—	—	—	C	—	—
Siberian Crane	—	—	Y	—	—	—	—	C	—	—
Wattled Crane	—	B	Y	/	—	—	Y	A	—	—
Lord Howe Island Woodhen	—	—	—	—	—	—	—	C	B	MR
Inaccessible Rail	—	—	Y	—	—	—	—	—	—	—
Corncrake	—	—	Y	—	Y	—	—	—	—	—
Takahe	B	B	Y	Y	—	—	—	B	—	CMR/I
Houbara Bustard	—	—	Y	Y	Y	—	—	—	—	—
Charadriiformes										
Piping Plover	—	B	Y	Y	1E	Y	—	A	B	MR
Hooded Plover	—	—	—	—	—	—	—	A	—	—
New Zealand Shore Plover	B?	—	—	—	—	—	—	A	A?	—

	Demographic parameter									
	1	2	3	4	5	6	7	8	9	10
Black Stilt	B	—	Y	Y	Y	Y	—	B	—	I
Audouin's Gull	—	—	Y	/	—	—	—	A	—	—
Damara Tern	—	B?	Y	Y	—	—	—	B	—	—
Columbiformes										
Pink Pigeon	B	—	Y	/	—	—	—	A	—	—
Psittaciformes										
Mauritius Parakeet	B	—	—	—	—	—	—	A	—	—
Puerto Rican Amazon	B	B	Y	Y	1E?	Y?	—	A	—	I
Kakapo	B	—	Y	Y	—	—	—	A	—	—
Strigiformes										
Spotted Owl	—	—	—	—	—	—	—	A	U?	MR
Coraciiformes										
Red-cockaded Woodpecker	B	—	—	—	—	Y	—	A	B	—
Passeriformes										
Seychelles Magpie-robin	A	A	Y	Y	—	Y	Y	A	U	MR
Seychelles Warbler	—	—	—	—	—	—	—	—	—	MR
Chatham Island Black Robin	A?	A?	Y	/	?	Y	Y	A	U	I
Seychelles Paradise-flycatcher	—	—	Y	Y	—	—	—	—	—	—
Forty-spotted Pardalote	—	—	Y	/	—	—	—	—	—	—
Kirtland's Warbler	B	B	Y	Y	Y	Y	Y	B	B	MR
Palila	B	—	Y	Y	—	—	—	—	—	—
Yellow-shouldered Blackbird	—	B	Y	Y	—	—	—	B/C	—	CMR
Mauritius Fody	—	—	Y	—	—	—	—	—	—	—
Rodrigues Fody	—	—	Y	Y	—	—	—	—	—	—
Hawaiian Crow	B	—	Y	Y	—	—	—	—	—	—

Demographic parameters:

1 Proportion of population that attempts to breed (A, proportion known by age; B, overall proportion known)

2 Age at first breeding (A, proportion known by age; B, minimum age known in wild)

3 Hatching success (Y indicates estimate available)

4 Fledgling success (Y). / indicates studies where success at the egg and chick stage were not separated

5 Mayfield estimate (Y) of the proportion of eggs that gave rise to fledged young (1E indicates studies in which breeding success was calculated only from nests followed from the date the first egg was laid)

6 Frequency of renesting after breeding failure (Y)

7 Frequency of renesting within a season after successful breeding (Y)

8 Overall estimate of productivity by the following methods:
   A—inventory (young reared per breeding pair per season)
   B—product of nesting success × fledgling success × proportion of pairs attempting to breed
   C—the ratio of young to adults at the end of the breeding season

9 Mean prebreeding dispersal distance (U, unbiased estimate; B, biased estimated).

10 Annual survival rate (underlined if estimated for more than one age class) by the following methods:
   MR—minimum return rates of marked birds
   CMR—mark-recapture (i.e. where markedness has been used to measure the efficiency of resighting)
   I—inventory (birds not marked but counted by age and survival inferred from changes in numbers)
   NMG—no method given
   References: as given by Collar and Andrew (1988).

Synthesis

30 Concluding remarks

IAN NEWTON

Environmental limiting factors

Perhaps the main contribution of research to bird population management is to identify the factors that limit population sizes. Without knowledge of the limiting factors, the management of àny bird population is bound to be hit and miss. It is important to distinguish between the external (environmental) factors that influence populations and the internal (demographic) features that these factors affect. External factors include resources (notably food supplies) and natural enemies (predators, parasites). Internal (demographic) features include the rates of births and deaths, immigration, and emigration—the net effects of which mediate the influence of external factors to determine population trend.

Enormous effort has been devoted over the years to measuring the demographic parameters of bird populations, notably birth and death rates, to the considerable neglect of the external factors that ultimately determine population level. For many species of birds, demographic detail is the only information we have on the population. But while we study populations by demography alone, we will never understand what ultimately determines their average level, and why this level varies greatly between areas. Two populations may be identical in rates of births, deaths, and movements, and yet persist indefinitely at quite different densities. To understand the differences in population mean densities in a species requires a study of the external factors. Also it is the external factors that have to be changed by management before change in population level can be achieved. In other words, the study of environmental limiting factors is essential for effective population management.

The distribution of any species is restricted by the distribution of its habitat, and within that habitat the availability of food and other resources provides an ultimate ceiling on numbers. The key question is which species or populations are currently limited by their resources and which are held below this level by natural or human restraints. Another important question is where in the annual cycle the limiting bottleneck is imposed. For migrant birds limitation can occur either in the breeding or wintering areas, or at a staging post en route (see Evans; Cooke and Gooch, this volume). The season when limitation occurs may also change over time, as feeding conditions change (Bairlein, this volume). Thus the period of greatest food shortage for Woodpigeons *Columba palumbus* in Britain, which determines the population level, formerly occurred in January–February, when the birds fed on clover, but now falls in November–December, when the birds feed on oilseed rape. This change results from changes in the grown acreages of clover and oilseed rape (Inglis *et al.* quoted by Feare, this volume).

Where food is limiting, such limitation is likely only at certain times of year or only in certain years. Some species of high latitudes are cut back so severely by food becoming unavailable in hard winters that they can then experience several years of unrestrained increase until they next decline (for Heron *Ardea cinerea* see Lack 1966). It is not only the quantity and availability of food which is important, but also its quality. This is a particular problem for herbivores which at certain times of year have to eat food of low digestibility and nutritive content. The amount of food available for any one species may be limited because of competition from other species dependent on the same supplies, as in some mixed flocks of waders (Evans, this volume). In addition, birds can store large amounts of food as reserves within their bodies, so that (at least in the larger species) the food acquired at one time of year can affect performance (breeding or survival) at another. Food can be considered limiting whenever it prevents (or restricts) a population from increasing.

Evidence for resource limitation comes from those species in which spatial or time trends of densities parallel similar trends in food supplies (Newton 1980). Some of the earliest evidence for temporal changes was provided by Reinikainen (1937) in mid Finland. He travelled the same route by ski each Sunday in March for 11 years, counted the Crossbills *Loxia curvirostra* met on his journeys, and estimated the cone crops of spruce. The number of Crossbills seen each year was strongly correlated with the size of the spruce crop, the highest number of birds being twenty times the lowest, with an increase of this order occurring from one year to the next (Fig. 30.1). More recent evidence for spatial variation in breeding density, linked with food supplies, has been provided by Watson and Langslow (1988). They found that the density of Golden Eagles breeding in different parts of Scotland correlated with the amount of carrion available to the birds in winter (Fig. 30.1). Other examples from this conference include some other raptors (Newton), seabirds (Croxall and Rothery), and storks (Bairlein).

In some bird species, densities were related not to food supplies directly (which were not measured), but to some determinant or correlate of food supplies, such as rainfall (Boyd, this volume) or soil fertility. Thus songbirds generally breed at greater density in woods growing on rich soils than in similar woods on poor soils, and waterbirds are generally more

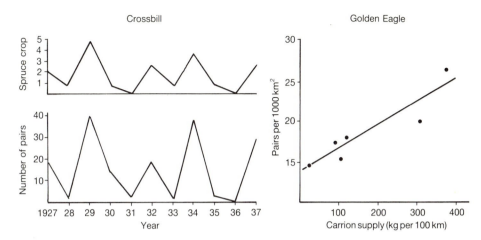

Fig. 30.1. Relationship between breeding density and food supply. Left: density of Crossbill pairs in different years in relation to size of spruce crop in one area of northern Finland (from Reinikainen 1937). Crossbills in number of pairs per 120 km; spruce crop classed in five categories. Right: Density of Golden Eagle pairs in six regions of Scotland in relation to winter carrion supply (from Watson and Langslow 1988).

abundant on eutrophic than on oligotrophic lakes (von Haartman 1971). For a few bird species, food supplies have been manipulated experimentally, and in some cases were followed by an appropriate change in density (various tits *Parus* spp., van Balen 1980; Jansson *et al.* 1981; Källander 1981). The correlative and experimental findings taken together provide a now substantial body of evidence that the densities of some bird species are influenced by food supplies.

That nest sites form another resource which can limit the breeding densities of certain species is also shown by both observational and experimental evidence. Thus in some species breeding pairs are scarce or absent in areas where nest sites are scarce or absent, but which seem suitable in other respects; and the provision of artificial nest sites is sometimes followed by an increase in the numbers of breeding pairs. The species concerned mostly use special sites (such as tree holes) and include Kestrel *Falco tinnunculus* (Cavé 1968; Village 1990), Wood Duck *Aix sponsa* (McLaughlin and Grice 1952) and Pied Flycatcher *Ficedula hypoleuca* (Sternberg 1972). In woods with nest boxes, this last species has reached densities equivalent to 2000 pairs per km^2, greater than all other species combined in those woods (von Haartman 1971). In one area of Holland, the Kestrel population increased from less than 20 to more than 100 pairs in one year, following the provision of nest boxes (Cavé 1968). However, as Blondel (this volume) has reminded us, limitation by resources may be much more frequent in man-modified landscapes, which are so often deficient in natural predators, than in primaeval habitats, where many bird species show low densities but predators are relatively more numerous.

For far fewer bird species is evidence available that numbers are held by natural enemies permanently below the level that resources would permit. The regulation of Partridge *Perdix perdix* densities at different equilibria, depending on predator pressure, has been described, modelled, and confirmed by experiment (Potts and Aebischer, this volume). Similarly, the limitation of Red Grouse *Lagopus l. scoticus* densities by one or other of two types of parasites has also been described, modelled, and confirmed by experiment (Hudson and Dobson, this volume). One of the parasites involved (trichostrongyle worms) acted mainly to reduce fecundity in the grouse but in high infection years also increased mortality.

Another important experiment showing the limitation of gamebird numbers by predators was described by Marcström *et al.* (1988). This work was done over 9 years on two islands off northern Sweden, which were connected to the mainland in winter by ice. On one island, mammalian predators were killed and on the other they were left, reversing the treatments after 5 years and each year monitoring the numbers of gamebirds present. Where predators were removed the numbers of adult and young Capercaillie *Tetrao urogallus* and Blackgrouse *Lyrurus tetrix* present at both seasons were increased over those on the control island, and the results were repeated after reversal of treatments. The conclusion was that predators reduced both the breeding populations and the breeding success of these two game species.

For many birds it is hard to tell whether numbers are limited by resources or natural enemies, without proper field experimentation in which one or other is manipulated against an appropriate control. Studying the causes of mortality will not necessarily help. Imagine that the density of a territorial species was limited by habitat quality/food supply, so that surplus individuals were forced into suboptimal habitat, where they were eaten by predators—the 'doomed surplus' model of Errington (1946) which was later applied to Red Grouse by Jenkins *et al.* (1964). From a study of mortality, one would conclude that the presence of predators limited numbers; however, the real limiting factor was habitat quality/food supply, which determined the density of territories in optimal habitat. To change density in the long term would entail a change of habitat, not of predators, and in each case surplus individuals would be removed by predators or any other mortality agent available locally. In other words, the factors that cause most mortality are not necessarily those that ultimately determine population level. The need for experiments to determine the true limiting factor is obvious. This is true not just for predators but also for parasites, which must be removed from populations before their full affects can be revealed (Hudson and Dobson, this volume).

Any assessment of the causes of mortality is itself not straightforward. A bird weakened by food shortage may succumb to disease, but just before death it may fall victim to a predator. In such cases, food shortage is the underlying (ultimate) cause of death while predation or disease are the immediate (proximate) causes. To cite one example, trained Goshawks *Accipiter gentilis* flown at Woodpigeon *Columba palumbus* flocks tended to select the weaker individuals (Kenward 1977). This was most apparent with pigeons that were chased, as weak ones tended to lag behind, but less so with pigeons caught by surprise without a chase. By taking weak individuals, whose survival chances were low, the impact of Goshawks on the pigeon population was reduced. In fact 28 per cent of victims examined after capture were found to be already starving, with body condition beyond the point of no return. For these birds, therefore, food shortage was the underlying cause of death, and predation merely the immediate cause.

So far, I have assumed that a population may be limited by one factor only: food shortage, predation, or parasitism, as the case may be. In reality, no one factor is likely to account wholly for a given population level. During a period of food shortage, for example, some individuals may starve, while other non-starving individuals may die from other causes. In such cases the limiting factor can be considered as that factor which, once removed, will permit the biggest difference in numbers. Again this factor can best be revealed by appropriate field experiments.

To sum up, a shift in emphasis in bird population work is needed, with more attention paid to environmental limiting factors, and with experiments forming a stronger component of such work. Specific projects could involve the manipulation of bird densities, of resource levels, or of predator numbers and parasite loads. A telling experiment cuts much more ice than any amount of descriptive work. The main problem with field experiments is in achieving a desirable level of replication.

Demographic parameters

From the foregoing discussion it is clear that a population could be managed perfectly well from a knowledge of the limiting factors alone, without any study of its demography. So detailed studies of reproductive and mortality rates, which have dominated bird population work for many years, are not essential to population management. They are helpful, however, and for declining populations can provide useful supplementary information. However, provided that the inputs (births + immigrants) to a population balance the losses (deaths + emigrants), population level can be sustained on a wide range of reproductive or survival rates (Perrins, this volume). Only when one or both of these rates falls below a critical threshold and when no further density-dependent compensation in other demographic parameters can occur, does the population decline.

Where fecundity and survival rates have been studied in the same species during both a period of increase (or stability) and of decline, such data have provided useful pointers to where the problem lies. Thus decline in a Puffin *Fratercula arctica* population was associated with decline in survival rate (Harris and Wanless, this volume), while decline in a Snow Goose *Anser caerulescens* population was associated with a change in both fecundity and first-year survival (Cooke and Gooch, this volume). Earlier comparative information of this type was obtained for Ospreys *Pandion haliaetus* in North America when populations were declining due to breeding failure caused by DDE (a residue of the insecticide DDT). The rate of population decline in different areas depended on the extent of reduction in breeding rate, and a production of at least 1.2–1.3 young per nesting pair was needed to prevent decline (Henny and Ogden 1970). Note, however, that in any bird species, changes in mean reproductive and death rates can also result from changes in the internal features of a population, such as age structure or genetic composition, and need not always imply some environmental influence.

I need hardly emphasise the need for further work on dispersal. Not only are movements between breeding sites the most neglected aspect of avian demography (Lebreton and Clobert, this volume), in nature they could often be crucial in the maintenance of local population densities. In any widespread species, we can expect that in some areas production of young is more than enough to offset mortality (net exports), whereas in other areas the reverse is true, so that densities can be maintained only by continual immigration (net imports) (Woolfenden and Fitzpatrick, this volume). As natural habitats become increasingly fragmented by human activities, and local extinctions become ever more likely, movements are likely to play an increasingly important role in population persistence. Yet we know little about dispersal patterns and their role in population regulation, even for common birds.

Moreover, failure to appreciate the importance of dispersal can easily lead to an exaggerated view of the role of birth and death rates in influencing population trends. Virtually all studies of population regulation in birds have been concerned with the dynamics of local populations. Yet almost all species have geographical ranges larger than an ecologist's study plot, and individual birds continually move in and out, on local or longer journeys. Such movements could cause much more rapid and pronounced increases in local density than could any rise in fecundity or survival. For example, the sudden appearance of large numbers of raptors in areas with rodent outbreaks, or of finches in areas with good seed crops.

Dispersal is central to the concept of 'source' and 'sink' areas, touched on above. A source area is where reproduction exceeds mortality, and a sink is where reproduction is insufficient to balance mortality, so that persistence depends on immigration. From this, two conditions follow (Pulliam 1988). First, if the surplus production in the source is large, and the per capita deficit in the sink is small, only a small fraction of a total population need occur in areas where reproduction exceeds local mortality. In this case, through movements, the bulk of a population could persist indefinitely in poor habitat. Second, the numbers in sink areas may depend more on conditions in nearby source areas than in the sink areas themselves. To explore these questions, we need more emphasis on spatial variation in demography, and particularly on dispersal, in population studies.

Density dependence

The old argument over whether animal populations are 'regulated' by density-dependent factors, or simply 'limited' by 'density-independent' ones (Andrewartha and Birch 1954; Lack 1966) can now be replaced by a more appropriate question: what is the relative importance of the two types of factors in causing the changes observed? In general, the more stable a population over time, the greater the density dependence in demographic parameters.

Sadly there are considerable problems in detecting density dependence in bird populations and in assessing its effects on population trends (Clobert and Lebreton, this volume). Existing methods of measuring density dependence rely not only on a long series of counts (which are not wholly independent of one another) but also on considerable variation in population size. Paradoxically, this means that the most stable populations, in which density dependence is most marked, are also those in which density dependence is most difficult to formally demonstrate. Secondly, as density dependence often involves competition, it may not operate over the full range of densities observed, but only at the highest levels, which seldom occur. Thirdly, any resource over which competition occurs may itself change in abundance from year to year, leading to variations in the population density at which competition arises. One way round this last problem is to express bird density as per unit of resource, rather than per unit of area, although the disadvantage of this approach is that measurement is needed of resources, as well as bird numbers.

A problem of a different kind, raised at this meeting, is what to count as population size where several species share the same resource or the same enemies (see Evans; Dobson and May, this volume). One species may decline in numbers not because its own density exceeds the critical level, but because its own total density and that of all its competitors do so. The effects of competition may fall equally or unequally on all the species present, and density dependence may be evident in individual species only while the total mixed population of all species is being considered. In other words, where interspecific competition occurs, the numbers of one species may decline in response to a rise in the numbers of another. Similar arguments hold where two or more species share the same parasites, as the infection rate of one species may depend partly on the numbers of another (Dobson and May, this volume). They also hold where two or more species share the same predator, as the predation rate on one species depends on the numbers of the others (see later). In this sense any group of species which share the same resources, parasites, or predators could be regarded as competitors, because a rise in the numbers of one species causes a reduction in the other.

Having detected density dependence, a second question is what effect this has on population trend. Not all density-dependent responses are strong enough to stabilize a population. Some may merely slow its growth and have little impact on subsequent numbers compared to other factors (e.g. clutch size variation in the Great Tit *Parus major*, Lack 1966; McCleery and Perrins, this volume). Other density-dependent responses may actually promote fluctuation in numbers, as when the response shows:

(1) 'over-compensation', where a change in fecundity/survival is so great that it more than offsets a change in density (e.g. juvenile survival in Sparrowhawk *Accipiter nisus*, Newton 1988);
(2) 'inverse density dependence' where a regulating factor acts more strongly at low densities than at high ones (e.g. the removal by predators of a constant number of prey individuals from an area, irrespective of prey density); or
(3) 'delayed density dependence', where the losses relate to some previous density, causing a lag in response (e.g. predator-prey cycles involving gamebirds and Goshawks *Accipiter gentilis*, Lindén and Wikman 1983).

Modelling provides an obvious means of assessing the effects of particular density-dependent responses on population trends.

Mathematical modelling

The modelling of population trends from knowledge of demographic parameters can be revealing, provided that all the relevant variables are known with sufficient accuracy. It is especially useful in the prediction of population trends which are known to be influenced by several factors at once (Potts and Aebischer, this volume), and in indicating how the same factor (such as predation) can produce population stability in one set of circumstances and fluctuation in another. The completeness of our understanding of a population may be checked by comparing a prediction from a model with the natural course of events observed in the field. In conservation and pest control, modelling also enables us to explore the likely consequences of different management options. Other advantages of modelling are that it forces us to make assumptions explicit (and therefore testable), define vague terms precisely, and validate demographic estimates (Lebreton and Clobert, this volume). In testing ideas, however, models are no substitute for field experiments.

Catastrophes and extinctions

Whatever the factors normally limiting population levels, catastrophic mortalities, causing sudden massive declines in populations, occur from time to time. A single outbreak of botulism on the Great Salt Lake in 1929 killed between one and three million waterfowl (Kalmbach 1935). Catastrophes can have a pronounced impact even on large populations (this volume, Croxall and Rothery; Smith *et al.*; Woolfenden and Fitzpatrick), and in small ones must sometimes cause extinctions (Green and Hirons, this volume). In models designed to analyse extinction risk, it is clearly important to take the possibility of periodic disasters into account (Woolfenden and Fitzpatrick). Extinctions are much more likely to result from ecological-demographic factors than from the genetic problems of inbreeding and limited gene pools (Lande 1988; Lande, this volume).

Estimates of minimal viable population sizes are often based on shaky data, and consider only genetic aspects. It would be a pity if such estimates were used by conservation planners to exclude some rare species from conservation attempts merely because, according to a model, the chances of success were small. Several bird species have been reduced to very few individuals and, after protection, have increased again. The present population of about 500 Laysan Ducks *Anas laysanensis* was supposedly derived from less than ten individuals (Moulton and Weller 1984), possibly a single brood (Lovejoy 1978), while the present world population of more than 100 Whooping Cranes *Grus americana* all descend from just 15 individuals (King 1981). Even in natural conditions populations must frequently derive from a few founder individuals which colonised a new area. An example is the Cattle Egret *Bubulcus ibis*, which was formerly confined to the Old World; then about 1930 a few turned up in Guyana, and subsequently multiplied to occupy much of South and North

America. So extreme low numbers, small gene pool and inbreeding do not inevitably spell extinction.

Social behaviour

Much confusion centres on the role of territorial and other social behaviour in population limitation (Smith *et al.*, this volume). Insofar as the number of territories in an area reflects the availability of resources, territorial behaviour is the means by which such resources are shared between certain members of a population, and by which surplus individuals are excluded, to die or leave. Other forms of dominance hierarchies serve the same result, in concentrating the effects of resource shortage on particular individuals, and resulting in the continual adjustment of numbers (or breeding) to resource levels. Thus stated, territorial and other dominance behaviour serves as the mechanism by which the effects of changes in resource levels are translated into changes in demography and population level. The whole sequence of events between changes in resources and population may thus be depicted as:

Change in resource level → change in proportion of birds excluded by dominance behaviour → change in demography (including dispersal) → change in population density.

Previous ideas on the possible 'self-regulation' of populations have added to the confusion. One such view is that the aggressive/spacing behaviour of a species is a strictly intrinsic feature, which can vary through time (leading to changes in numbers), but without reference to resources or natural enemies (Chitty 1967). Another view is that animals have specific behavioural mechanisms designed to keep their numbers below the resource limit (Wynne-Edwards 1962). Neither of these ideas finds much support in field evidence from birds, and the latter involves the difficult concept of group selection. A more acceptable view, which is consistent with the theory of evolution by natural (individual) selection, is that the spacing and dominance behaviour of animals has evolved as a means of maximizing individual fitness. As such behaviour in any population helps to separate the 'haves' from the 'have nots', and is responsive to changes in animal density and in resources, it inevitably functions incidentally in population limitation. In birds various forms of dominance behaviour, notably territorialism, often lead to 'contest competition', as opposed to 'scramble competition'. Further work in this field might usefully concentrate on finding exactly how the various forms of dominance behaviour in birds operate in adjusting the population numbers to the resources.

Some of the best work carried out to date on this question concerns wintering waders feeding on mud flats (Goss-Custard 1980). In these

birds, competition takes two forms, one resulting from progressive depletion of food supplies, and the other from aggression and other interference
between feeding birds. The effects of interference competition fall most
heavily on young subordinate birds, whose food intakes fall in consequence, which presumably accounts for their greater mortality compared
to that of adults. Whether dominance occurs with respect to food supplies,
nest sites, mating opportunities, or predator avoidance, it always resolves
to defence of the single resource—space.

Life history strategies

Birds span a wide range of life history types, from short-lived (r-selected)
mostly small-bodied species to long-lived (K-selected) mostly large-bodied
ones. For a given body size, tropical and subtropical species tend to show
more K-characteristics, with lower fecundity and higher survival, than
their temperate zone equivalents (Rowley and Russell, this volume). What
different problems do the r or K species present for conservation?

The small, short-lived species often live at high densities, but, because of
greater vulnerability to various mortality agents, are prone to greater
fluctuation, and hence to local extinction. Marked declines in reproduction are soon followed by declines in adult numbers. The large, long-lived
species are generally more stable in numbers, have a large non-breeding
component, and are more resilient to short-term reductions in reproductive rates. In general, however, they present greater problems in conservation: the low densities at which most such species live makes them
especially vulnerable, because of the large areas of habitat required to
sustain their populations. This is especially true of large raptors (Newton,
this volume), but also holds for the foraging areas of long-lived seabirds
and the breeding areas of waterfowl and waders (Evans, this volume).
Also, because they do not breed until they are several years old, long-lived
bird species inevitably contain a large proportion of 'immatures', which in
some species occupy different habitat from breeders. These immatures
have to be catered for in conservation, in addition to the breeding adults.
Perhaps most importantly, however, because of low reproductive rate and
long-deferred maturity, long-lived species take much longer to recover
their numbers after a decline than do short-lived species. Even in the best
conditions, the California Condor *Gymnogyps californianus* would take
several human lifetimes to reoccupy to the full the whole of its historic
range.

Role of natural enemies

From what we have heard at this meeting, there seems to be little difference
in the effects of predators and parasites on populations.

1. *Elimination of a population* has occurred where a vulnerable prey is suddenly confronted by a new predator (e.g. island seabirds exposed to introduced rats or cats; this volume, Croxall and Rothery; Jouventin and Weimerskirch), or where a vulnerable host is confronted by a new disease (Hawaian landbirds exposed to avian pox and malaria; this volume, Dobson and May).

2. *Limitation of a population* at a level lower than resources would permit, or even local elimination, has occurred where a vulnerable prey is exposed to generalist predators, supported at high density mainly by alternative prey (e.g. Partridge exposed to Fox and other predators; this volume, Potts and Aebischer). It has also occurred where a vulnerable host is exposed to a disease organism which is maintained at high level in the area by an alternative host (e.g. Red Grouse affected by the louping-ill virus also found in sheep; Duncan *et al.* 1978; Hudson and Dobson, this volume). In both these examples, the relationship is not a simple two-part prey-predator or parasite-host system, but involves a third component: alternative prey for the predators in the Partridge example and an alternative host for the parasite in the Red Grouse example. In situations such as these, the relationship between the two species of interest would usually be expected, on a year-to-year basis, to be density independent.

3. *Mortality, but no long-term reduction in breeding population* is probably the rule in the majority of predator-prey systems, and in many parasite-host systems. Examples include the effects of Sparrowhawks *Accipiter nisus* on Great Tits (Perrins and Geer 1980; McCleery and Perrins, this volume) and other song-birds (Newton 1986), and the effects of certain diseases on various birds (Lack 1954).

4. *Cycling of a population* has occurred where a prey species is exposed to a specialist predator, which has little or no alternative food (e.g. forest grouse exposed to predation by Goshawk *Accipiter gentilis* or Great Horned Owl *Bubo virginianus*, Lack 1954; Keith 1963; Linden and Wikman 1983); or where a host species is exposed to a specialist parasite which has no alternative host in the same area (e.g. Red Grouse exposed to trichostrongyle worms, Potts *et al.* 1984; Hudson and Dobson, this volume). Relationships such as these, on a year-to-year basis, show delayed density dependence.

So perhaps predators and parasites, despite their different life-styles, have the same basic range of effects on bird populations. However, parasites can normally multiply much more rapidly than their hosts, while predators normally reproduce more slowly. This may enable some balance in the parasite–host system (lesser virulence in parasite, greater resistance in host) to evolve more rapidly than the prey can evolve immunity to a predator. Thus, one consequence may be that species are less often eliminated by exposure to a new pathogen than by exposure to a new predator. There are several examples of populations becoming resistant to new killing diseases (e.g. myxomatosis in rabbits) but no known examples of a species acquiring a novel defence against some new predator. At best, individual birds may learn to avoid disturbance or predation risk, for example by moving elsewhere (Coulson, this volume).

Species that are hunted

The aim of the game manager is to harvest the maximum proportion of individuals from a population each year in a sustained way, without causing long-term decline. One controversial aspect of game management centres on predator control (practised for game birds in parts of Europe). Both conservationist and hunter are concerned with the long-term maintenance of breeding stock, which can usually be achieved without the control of predators. However, the hunter is also concerned with the production of as large a post-breeding surplus as possible. By control of predators that eat the eggs and chicks, he strives to increase the production of young and hence the potential harvest. For conservation purposes alone, in contrast, predation on eggs and young is irrelevant, providing that it does not reduce future breeding numbers. It is no paradox that predators might reduce the size of the seasonal peak population, but have no noticeable affect on breeding numbers at the seasonal low.

Another issue, especially relevant to hunted species, is the extent to which mortality from hunting is additive to natural mortality, rather than being compensated by reduced natural mortality (i.e. offset by improved survival in the non-shot birds). Results from different American duck species are conflicting (Nichols, this volume). For hunting loss to be compensated by reduced natural loss, two conditions must be fulfilled: first, the overall mortality has to be strongly density-dependent, and second the hunting loss must occur before or during, but not after, the main period of natural loss. Thus, if hunting is concentrated in the autumn, then there is ample time for compensation through reduced natural loss before the next spring. But if hunting is concentrated in late winter, when much natural mortality has already occurred, then it may become 'additive', resulting in population decline. The extent to which hunting loss is additive to natural loss in any given species may therefore depend partly on the relative timing of the two types of loss. The same considerations determine the extent to which any density-dependent and density-independent form of loss can oppose and compensate one another.

Rare species of conservation concern

The main question that arises with rare species is how to increase numbers. Many bird species have become rare because their habitats and associated food supplies have been destroyed by human activities. Habitat destruction involves not only complete removal, as when forest is felled or marsh is drained, but also degradation, as when a habitat is modified in such a way that its carrying capacity for particular bird species is reduced. Examples of habitat degradation include human overfishing which might reduce the food supply for certain seabirds, causing their numbers to fall; and modern forestry procedures, which remove dead trees, thus reducing the nest sites and limiting the numbers of hole-nesting birds.

In species limited by habitat (or resources within it), numbers can be raised either by increasing the extent of habitat available or by upgrading existing habitat so as to increase its carrying capacity. For some species both options may be practically impossible, because of conflict with other human interests, but for other species little more might be entailed than halting vegetation succession. For species limited by habitat, it is pointless to try and improve reproduction or survival, because the habitat will support only a certain number and any surplus must inevitably die or leave.

Alternatively, other bird species may be reduced below the level that the habitat would support, by some kind of overkill—from persecution or toxic chemicals, or by the introduction of a new predator or pathogen. In such populations, the only way to increase numbers is to engineer an increase in breeding or survival rates. For this, it is necessary to study the causes of breeding failure and mortality. Some of the causal factors might then be removed, enabling numbers to rise. In some species, improvements in breeding rates have been achieved by the provision of safer nest sites, nest-guarding, or the withdrawal of the pesticide DDT; in other species, improvements in survival have been achieved by protective legislation, by reductions in hunting pressure, or by the withdrawal of the pesticides aldrin and dieldrin. Such measures, which involve changing the factors that affect breeding and survival, are effective only for populations not at the time limited by resources.

In conserving any rare species, it is important at the outset to distinguish between these two causes of decline (resource restriction and overkill), because reversing the decline requires different management in each case. In the past, a lot of money spent on conservation may have been wasted by pursuing the wrong management option.

Pest species

The main problem is how to reduce the damage and this is usually seen as requiring a reduction in the pest's population (Feare, this volume). The same principles apply as to conservation, but in reverse. Habitat removal provides a long-term solution, but may be impractical, while killing requires a sustained commitment. Both methods may cause the loss of desirable species at the same time.

Many of the most renowned pest species, such as Starling *Sturnus vulgaris* and Quelea *Quelea quelea*, are short-lived, fast-breeding types, which can only be controlled (if at all) by a massive sustained effort, continued year after year without respite. Such programmes can easily become more expensive than the damage done. They are often continued, not because they are effective in reducing crop damage, but in response to political pressure from farmers. In contrast, some long-lived, slow-breeding species can more readily be reduced effectively, because of the longer period they take to recover. Some predators of game and fish provide examples, as they have been eliminated by human persecution from large parts of their range (Newton 1979). Whatever the natural rate of increase of a species, lasting decline can be achieved more readily if control is done at the start of a breeding season, when numbers are at their seasonal low, rather than after breeding, when numbers are at their peak. At that time huge numbers would have to be killed if control was to do any more than harvest the post-breeding surplus.

Alternative means of reducing damage can sometimes be found. Thus scaring birds from vulnerable crops can be effective, provided that alternative feeding areas are available nearby. The main need is for more species-specific control measures, which minimize the damage to habitats and to non-target species.

Island populations

Emphasis has been given to populations isolated on islands, either real islands or habitat fragments in otherwise unsuitable terrain. Most avian extinctions relate to island forms, and as many island species are endemic, conservation efforts must usually be centred on the islands themselves. The problem of habitat fragmentation in mainland situations means that many former widespread species are becoming increasingly split into small isolated units (Blondel, this volume). Even with dispersal, many such units are unlikely to persist, and will die out for demographic, genetic, or ecological reasons (Lande, this volume).

The proportion of habitat patches occupied by a given species at any one time will depend on the balance between the rate of dying out and the rate of recolonization in individual patches (Lande 1987). Any change in the size, number, or spacing of patches, which causes the rate of local die-outs to exceed the rate of local recolonizations could cause the extinction of a species despite the continuing presence of suitable habitat. In future, therefore, the conservation of certain species may be sustainable only with continued human help, involving frequent translocations of individuals between habitat fragments and repeated reintroductions to vacant sites not reached by colonists. Hence, more work is needed on the demography of small populations, on the methodology of reintroductions, and on the factors that determine the success or failure of such projects.

Scant attention has been paid at this meeting to the importance of maintaining distribution patterns. Yet the long-term security of any species must depend partly on the extent of its distribution. The organochlorine pesticide era was proof of this, as populations of certain species were eliminated within a few years from large parts of Europe and North America (Newton 1979; Cade *et al.* 1988). Such species survived only because small areas occurred within their extensive ranges where birds were less affected. The question of distribution is especially relevant to the Flamingo *Phoenicopterus ruber* in Western Europe. As a result of imagination and effort, the Camargue population has increased dramatically in recent decades (Johnson, Green, and Hirons, this volume), but with only two other (irregular) breeding sites, the continuance of the species in Western Europe is clearly not secure. It would be a pity if this beautiful species, which has added so much enjoyment to our stay in the Camargue, were to disappear from such a large sector of its range.

References

Andrewartha, H. G. and Birch, L. C. (1954). *The distribution and abundance of animals*. Chicago University Press.

Balen, J. H. van (1980). Population fluctuations of the Great Tit and feeding conditions in winter. *Ardea*, **68**, 143–64.

Cade, T. J., Enderson, J. H., Thelander, C. G., and White, C. M. (1988). Peregrine Falcon Populations: their management and recovery. The Peregrine Fund, Boise.

Cavé, A. J. (1968). The breeding of the Kestrel, *Falco tinnunculus* L., in the reclaimed area Oostelijk Flevoland. *Netherlands Journal of Zoology*, **18**, 313–407.

Chitty, D. (1967). The natural selection of self-regulatory behaviour in animal populations. *Proceedings of the Ecology Society of Australia*, **2**, 51–78.

Dempster, J. P. (1975). *Animal population ecology*. Academic Press, London.

Duncan, J. S., Reid, H. W., Mors, R., Phillips, J. D. P., and Watson, A. (1978). Ticks, louping ill and Red Grouse on moors in Speyside, Scotland. *Journal of Wildlife Management*, **42**, 500–5.

Errington, P. J. (1946). Predation and vertebrate populations. *Quarterly Review of Biology*, **21**, 144–77, 221–45.

Goss-Custard, J. D. (1980). Competition for food and interference among waders. *Ardea*, **68**, 31–52.

Haartman, L. von (1971). Population dynamics. In *Avian biology*, Vol. 1, (eds D. S. Farner and J. R. King), pp. 391–459. Academic Press, London.

Henny, C. J. and Ogden, J. C. 1970. Estimated status of Osprey populations in the United States. *Journal of Wildlife Management*, **34**, 214–17.

Jansson, C., Ekman, J., and Brömssen, A. von. (1981). Winter mortality and food supply in tits *Parus* spp. *Oikos*, **37**, 313–22.

Jenkins, D., Watson, A., and Miller, G. R. (1964). Predation and Red Grouse populations. *Journal of Applied Ecology*, **1**, 183–95.

Källander, H. (1981). The effects of provision of food in winter on a population of the Great Tit *Parus major* and the Blue Tit *P. caeruleus*. *Ornis Scandinavica*, **12**, 244–8.

Kalmbach, E. R. (1935). Will botulism become a world-wide hazard to wildfowl? *Journal of American Veterinary Medical Association*, **87**, 183–7.

Keith, L. B. (1963). *Wildlife's ten-year cycle*. University of Wisconsin Press, Madison.

Kenward, R. E. 1977. Predation on released Pheasants (*Phasianus colchicus*) by Goshawks in central Sweden. *Viltrevy*, **10**, 79–112.

King, W. B. (1981). *Endangered birds of the world*. Smithsonian Institution Press, Washington.

Lack, D. (1954). *The natural regulation of animal numbers*. Oxford University Press.

Lack, D. (1966). *Population studies of birds*. Oxford University Press.

Lande, R. (1987). Extinction thresholds in demographic models of territorial populations. *American Naturalist*, **130**, 624–35.

Lande, R. (1988). Genetics and demography in biological conservation. *Science*, **241**, 1455–60.

Linden, H. and Wikman, M. (1983). Goshawk predation on tetraonids: availability of prey and diet of the predator in the breeding season. *Journal of Animal Ecology*, **52**, 953–68.

Lovejoy, T. E. (1978). Genetic aspects of dwindling populations. A review. In *Endangered birds: Management techniques for preserving threatened species*, (ed. S. A. Temple), pp. 275–9. University of Wisconsin Press, Madison.

McLaughlin, C. L. and Grice, D. (1952). The effectiveness of large-scale erection of Wood Duck boxes as a management procedure. *Transactions of North American Wildlife Conference*, **17**, 242–59.

Marcström, V., Kenward, R. E., and Engren, E. (1988). The impact of predation on boreal tetraonids during vole cycles: an experimental study. *Journal of Animal Ecology*, **57**, 859–72.

Moulton, D. W. and Weller, M. W. (1984). Biology and conservation of the Laysan Duck (*Anas laysanensis*). *Condor*, **86**, 105–17.

Newton, I. (1979). *Population ecology of raptors*. Poyser, Berkhamsted.

Newton, I. (1980). The role of food in limiting bird numbers. *Ardea*, **68**, 11–30.

Newton, I. (1986). *The Sparrowhawk*. Poyser, Calton.

Newton, I. (1988). A key factor analysis of a Sparrowhawk population. *Oecologia*, **76**, 588–96.

Perrins, C. M. and Geer, T. A. (1980). The effect of Sparrowhawks on tit populations. *Ardea*, **68**, 133–42.

Potts, G. R., Tapper, S. C., and Hudson, P. J. (1984). Population fluctuations in Red Grouse: analysis of bag records and a simulation model. *Journal of Animal Ecology*, **53**, 21–36.

Pulliam, R. (1988). Sources, sinks and population regulation. *American Naturalist*, **132**, 652–61.

Reinikainen, A. (1937). The irregular migrations of the Crossbill, *Loxia c. curvirostra*, and their relation to the cone-crop of the conifers. *Ornis Scandinavica*, **14**, 55–64.

Sternberg, H. (1972). The origin and age composition of newly formed populations of Pied Flycatchers (*Ficedula hypoleuca*). *Proceedings of International Ornithology Congress*, **15**, 690–1.

Village, A. (1990). *The Kestrel*. Poyser, London.

Watson, J. and Langslow, D. R. (1988). Can food supply explain variation in nesting density and breeding success among Golden Eagles *Aquila chrysaetos*? In *Raptors in the modern world*, (ed. R. D. Chancellor), pp. 181–6. World Working Group on birds of prey and owls, London.

Wynne Edwards, V. C. (1962). *Animal dispersion in relation to social behaviour*. Oliver and Boyd, Edinburgh and London.

Index